Rough Fuzzy Hybridization
A New Trend In Decision-Making

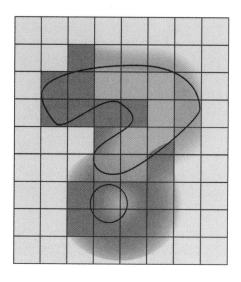

Springer
*Singapore
Berlin
Heidelberg
New York
Barcelona
Budapest
Hong Kong
London
Milan
Paris
Tokyo*

Rough Fuzzy Hybridization
A New Trend In Decision-Making

S. K. Pal
A. Skowron

editors

Professor S.K. Pal
Machine Intelligence Unit
Indian Statistical Institute
203 B.T. Road
Calcutta 700
India

Professor A. Skowron
Institute of Mathematics
University of Warsaw
Banacha 2
02-097 Warsaw
Poland

Library of Congress Cataloging-in-Publication Data

Rough-fuzzy hybridization: a new trend in decision-making / edited by
S. Pal, A. Skowron.
   p.   cm.
Includes bibliographical references.
ISBN 9814021008 (softcover)
   1. Artificial intelligence.   2. Fuzzy sets.   3. Rough sets.
4. Decision-making – Mathematical models.   5. Uncertainty.
I. Pal, Sankar K.   II. Skowron, Andrzej.
Q335.R73 1999
006.3—dc21
                                        98-33664
                                        CIP

ISBN 981-4021-00-8

This work is subject to copyright. All rights are reserved, whether the whole or part of the material is concerned, specifically the rights of translation, reprinting, reuse of illustrations, recitation, broadcasting, reproduction on micro-films or in any other way, and storage in databanks or in any system now known or to be invented. Permission for use must always be obtained from the publisher in writing.

© Springer-Verlag Singapore Pte. Ltd. 1999
Printed in Singapore

The publisher makes no representation, express or implied, with regard to the accuracy of the information contained in this book and cannot accept any legal responsibility or liability for any errors or omissions that may be made.

Typesetting: Camera-ready by editors
SPIN 10688185   5 4 3 2 1 0

To

Anshu, Arghya and Amita

S. K. Pal

To

Anna and Małgorzata

A. Skowron

Preface

The theory of fuzzy sets provides an effective means of describing the behavior of systems which are too complex or too ill-defined to admit precise mathematical analysis by classical methods and tools. It has shown enormous promise in handling uncertainties to a reasonable extent, particularly in decision-making models under different kinds of risks, subjective judgment, vagueness and ambiguity. Extensive applications of this theory to various fields, e.g., expert systems, control systems, pattern recognition and image processing, have already been well established.

More recently, the theory of rough sets has emerged as another major mathematical approach for managing uncertainty that arises from inexact, noisy, or incomplete information. It is turning out to be methodologically significant to the domains of artificial intelligence and cognitive sciences, especially in the representation of and reasoning with vague and/or imprecise knowledge, data classification, data analysis, machine learning, and knowledge discovery. The theory is also proving to be of substantial importance in many areas of applications.

It may be noted that fuzzy set theory hinges on the notion of a membership function on the domain of discourse, assigning to each object a grade of belongingness in order to represent an imprecise concept. The focus of rough set theory is on the ambiguity caused by limited discernibility of objects in the domain of discourse. The idea is to approximate any concept (a crisp subset of the domain) by a pair of exact sets, called the lower and upper approximations. But concepts, in such a granular universe, may well be imprecise in the sense that these may not be representable by crisp subsets. This led to a direction, among others, in which the notions of rough sets and fuzzy sets can be integrated, the aim being to develop a model of uncertainty stronger than either.

Research articles combining fuzzy set theory and rough set theory for developing efficient methodologies and algorithms for various real life decision making applications, have started to come out. Integrations of these theories with artificial neural networks, a biologically motivated powerful tool, are also being attempted with an aim of building more efficient and intelligent systems in soft computing paradigm.

The present volume provides a collection of nineteen articles (in the form of chapters) containing new material and describing, in a unified way, the basic concept and characterising features of these theories and integrations, with recent developments and significant applications. These articles are written by leading experts from all over the world. One may note that the significance of fuzzy set theory to handle uncertainties in different decision making problems is adequately established since late sixties. On the other hand, the theory of rough sets (a tool for handling uncertainties arising from granularity in the domain of discourse i.e., from the indiscernibility between objects in a set) is seen recently to draw the attention of researchers for its various applications. Keeping in mind that the theory of rough sets is relatively new, we have provided more emphasis on this by including articles either on rough sets or on their integration with fuzzy sets both in developing theory and showing applications.

Chapter 1 provides a tutorial article by J. Komorowski, Z. Pawlak, L. Polkowski and A. Skowron which describes basic notions of rough sets, characterizes the present state of rough set theory and applications as well as discusses a perspective of rough sets. This article is included for the convenience of understanding the rough set theory and hence, the remaining chapters of the book.

The second chapter by Z. Pawlak, the founder of rough set theory, explains the necessity of extending the theory to rough relations and rough functions to deal with real life ambiguous problems more efficiently, particularly when the rough sets alone are not adequate to handle them. Accordingly, the notions of rough (approximate) continuity, rough derivatives, rough integrals and rough differential equations are defined and investigated for discrete functions, i.e., functions defined and valued on the set of integers. The effect of discretization of the real life on the basic properties of real functions, such as continuity and differentiability has been studied. These results have significance to image processing and fuzzy control problems.

The next four chapters describe some of the theoretical developments of rough set theory along with its few variants, concerning certain decision making tasks along with indications of possible real life applications. In a decision support system it is desirable to have decomposition of information systems and decision tables when the data base becomes large i.e., when the number of objects as well as the average number of conditional attributes in rough set based decision rules becomes too high to classify new cases. The article of D. Ślęzak focuses on this issue and provides a new decomposition approach which enables to combine non-deterministic decision rules derived from subsets of conditional attributes. The foundation of the decomposition is based on the generalized decision function rules. Along with the conditional independence model of decomposition, the article provides a Bayesian-like decomposition network. In the article of A. Nakamura, the author has suggested a conflict logic with degrees (CLD) defined on the conflict space by considering the graded information (knowledge) and using monadic predicate symbols. Conflict is a dual concept of indiscernibility and is considered as a kind of dissimilarity. Though only the monadic predicate symbols are used to establish the syntax and semantics, yet the author has claimed that it can easily be extended to general predicate logic. Finally some decidability and undecidability results of CLD are presented with a remark on its practical use. L. Polkowski establishes, in his article, a relation between rough set theoretic approximation, viz, lower and upper approximation, and morphological approximation, viz., opening and closing. Using this relation a topological foundation of mathematical morphological operations has been developed from the topological theory of rough sets. An approximate collage theorem, well known for data compression, based on the topology of almost rough sets has been introduced as an application of the present theory. Some algorithms for construction of decision trees in the framework of local approach have been presented in the chapter by M. Moshkov. In particular, deterministic decision tree construction is discussed for an arbitrary information system with finite set of attributes from the problem description. The accuracy bounds and the complexity bounds of the algorithms are also discussed.

Chapters 7-9 are related to certain important issues combining fuzzy set and rough sets. A new concept of shadowed sets, which is induced by fuzzy sets and reveals interesting conceptual and algorithmic relationships between rough sets and fuzzy sets, is introduced in the article of W. Pedrycz. Unlike fuzzy sets, the shadowed sets do not have numerical membership values but rely on the basic concept of truth values in terms of "yes", "no" and "unknown". Since shadowed sets are obtained from fuzzy sets and by construction have good resemblances with rough sets, they can play a role of connector between fuzzy sets and rough sets. Some operations and applications of shadowed sets are also mentioned. In the chapter of Beaubouef et al., a relation between rough set theory and the concept of uncertainty in information theory has been established. A rough set metric of uncertainty, and rough entropy, have then been defined to measure the uncertainty in information theory. As an application of fuzzy-rough data base, some new fuzzy-rough information measures have also been discussed. The article of R. E. Kent describes soft concept analysis which is synonymous with enriched concept analysis. It provides, among others, a natural foundation for soft computation by unifying and explaining notions from soft computing terms of suitable generalized notions from formal concept analysis, rough set theory and fuzzy set theory.

Chapters 10-15 are addressed towards various interesting applications of rough set theory including those from pattern recognition area. The article of T. Mollestad and J. Komorowski deals with the problems of data mining. The performance of data mining or rule generation is highly dependent on the correctness or consistency of the training data set. The article is addressed to the problem of generating rules from incomplete, inconsistent information system. In particular, a framework for default rule extraction system is described using rough set theoretic approach. The corresponding algorithm has been tested on real life data set and is found to perform well in comparison with data mining systems currently in use. An application of rough classification to power system security analysis is described in the article of G. L. Torres et al. Here a systematic approach to present the compact knowledge of the system by reducing the set of rules is proposed. Information from a power system control has been considered as an example. In the paper of R. W. Swiniarski, two methods are explained to perform the tasks of feature extraction, selection and reduction. The first one uses principal component analysis and the other is based on rough set theory. Finally a process of combining these two is discussed for feature extraction and selection. The methods are illustrated by furnishing the results of texture recognition. An application to medical domain is demonstrated in the article of Słowiński and Stefanowski who, first of all, have discussed some difficulties associated with using rough set theory for medical data analysis. Then two heuristic approaches which lead to a better understanding of the medical information systems are described. During the generation of rules for decision making, emphasis has been given on testing their ability for classification and looking for correct clinical interpretation. The performance of these methods has been illustrated on a medical data. The article by S. Tsumoto discusses a rough set based method that can extract from medical databases rules, which plausibly represent medical experts' decision processes. The proposed method is evaluated on medical databases, the experimental results of which show that induced rules correctly represent experts' decision processes. In order to investigate the customer behavior and to predict the future programme of the company concerned, the article of Eiben et al. analyses a large date base. In particular, three models viz. logistic regression, rough data model and genetic algorithms (an efficient, robust, parallel searching technique based on the mechanism of natural evolution and genetics) have been used separately on the data set to perform the targeted job. The individual performance of these models along with some comparison is illustrated.

The remaining five chapters demonstrate different applications of rough-fuzzy hybridization along with integration of neural networks to result in a stronger soft computing paradigm. The major contribution in the article of Peters and Ramanna is modeling of software quality decision rules with rough and fuzzy petri nets, both individually and in a combined manner. The decision rules evaluate the consistency and correctness of assessment of software characteristics. These rules not only assess the software quality but also correctly indicate the degree or extent to which the software under consideration is to be changed. An interesting application in the area of acoustics is explained by Kostek in Chapter 17 where fuzzy logic and rough set theory are used to provide a knowledge base expert system to design hall acoustics. The hall acoustics is controlled both by objective and subjective parameters. The fuzzy logic is used for assessing the performance of acoustics under several parameters and on the other hand rough set theory is used to make a decision system from the subjectively quantized parameters. A similar application to signal processing is shown in the next paper by Czyzewski and Krolikowski who provides two approaches for automatic reduction of noise and distortion from audio signals. For the reduction of stationary and non stationary noises rule based systems are designed using fuzzy logic and rough set theory. Chapters 19 and 20 are, particularly, concerned with integration of fuzzy sets, rough sets and artificial neural networks for pattern recognition. For solving pattern classification problems a modular neural network model consisting of several subnetworks is employed in Chapter 19 by Sarkar and Yegnanarayana. In the process of obtaining final result the information collected from the subnetworks are quantized and these information are fused by fuzzy integral. To make a decision regarding the importance of the information of the subnetworks a fuzzy-rough set theoretic approach is described. In the last article, Mitra, Pal and Banerjee, have designed a rough-fuzzy-multilayer perceptron for developing a knowledge based network. Here fuzzy logic helps in handling uncertainties in the input description and output decision. The role of rough sets is in efficient encoding of crude domain knowledge in the form of rules. The investigation also provides a method of generating appropriate network architecture and improving the classification performance. Comparison with Baye's and k-NN classifiers is provided for speech data.

This comprehensive collection provides a cross sectional view of the research work that is being carried out applying the theory of rough sets and its fuzzy hybridization for developing various decision making algorithms; and makes the book unique of its kind. The book may be used either in a graduate level as a part in the subject of artificial intelligence and pattern recognition, or as a reference book for the research workers in the areas of rough sets, fuzzy sets, soft computing and their applications to various machine learning and decision making problems.

We take this opportunity to thank all the contributors for agreeing to write for the book. We owe a vote of thanks to Mr. Ian Shelley of Springer Verlag for his initiative and encouragement. The assistance provided by Mr. Suman K. Mitra and Mr. Indranil Dutta during the preparation of the book is also gratefully acknowledged. This work was initiated when Prof. S. K. Pal visited the Institute of Mathematics, University of Warsaw, Poland, under a INDO-POLISH collaboration project "Reasoning under Uncertainty about Complex Objects : Rough Set Theory and Fuzzy Set Theory", co-ordinated by the Dept. of Science and Technology (DST), India, and the Polish State Committee for Scientific Research (KBN), Poland.

January 1999

Sankar K. Pal
Andrzej Skowron

Table of Contents

Preface	vii

Rough Set Theory

Rough Sets: A Tutorial J. Komorowski, Z. Pawlak, L. Polkowski and A. Skowron	3
Rough Sets, Rough Function and Rough Calculus Z. Pawlak	99
Decomposition and Synthesis of Decision Tables with Respect to Generalized Decision Functions D. Ślęzak	110
Conflict Logic with Degrees A. Nakamura	136
Approximation Mathematical Morphology. Rough Set Approach L. Polkowski	151
Local Approach to Construction of Decision Trees M. Moshkov	163

Rough-Fuzzy Theory

Shadowed Sets: Bridging Fuzzy and Rough Sets W. Pedrycz	179
Information Measures for Rough and Fuzzy Sets and Application to Uncertainty in Relational Databases T. Beaubouef, F.E. Petry and G. Arora	200
Soft Concept Analysis R.E. Kent	215

Rough Set Application

A Rough Set Framework for Mining Propositional Default Rules T. Mollestad and J. Komorowski	233

Power System Security Analysis Based on Rough Classification 263
G. Lambert-Torres, R. Rossi, J.A. Jardini, A.P. Alves da Silva, and
V.H. Quintana

Rough Sets and Principal Component Analysis and Their Applications
in Data Model Building and Classification 275
R.W. Swiniarski

Medical Information Systems — Problems with Analysis and
Way of Solutions 301
K. Słowiński and J. Stefanowski

Induction of Expert Decision Rules Using Rough Sets and
Set-Inclusion 316
S. Tsumoto

Modelling Customer Retention with Statistical Techniques,
Rough Data Models, and Genetic Programming 330
A.E. Eiben, T.J. Euverman, W. Kowalczyk and F. Slisser

Fough-Fuzzy Application

A Rough Sets Approach to Assessing Software Quality:
Concepts and Rough Petri Net Models 349
J.F. Peters III and S. Ramanna

Assessment of Concert Hall Acoustics Using Rough Set and
Fuzzy Set Approach 381
B. Kostek

Application of Fuzzy Logic and Rough Sets to Audio Signal
Enhancement 397
A. Czyzewski and R. Krolikowski

Application of Fuzzy-Rough Sets in Modular Neural Networks 410
M. Sarkar and B. Yegnanarayana

Rough Fuzzy Knowledge-based Network — A Soft Computing
Approach 427
S. Mitra and S.K. Pal

Rough Set Theory

Rough Sets: A Tutorial

Jan Komorowski[1,5], *Zdzisław Pawlak*[4], *Lech Polkowski*[3,5], *Andrzej Skowron*[2,5]

[1] Department of Computer and Information Science
 Norwegian University of Science and Technology (NTNU)
 7034 Trondheim, Norway
 janko@idi.ntnu.no
[2] Institute of Mathematics, Warsaw University
 Banacha 2, 02-097 Warszawa, Poland
 skowron@mimuw.edu.pl
[3] Institute of Mathematics, Warsaw University of Technology
 Pl. Politechniki 1, 00-665 Warszawa, Poland
 polk@mimuw.edu.pl
[4] Institute of Theoretical and Applied Informatics, Polish Academy of Sciences
 ul. Bałtycka 5, 44-000 Gliwice, Poland
 zip@mimuw.edu.pl
[5] Polish-Japanese Institute of Information Technology
 Koszykowa 86, 02-008 Warszawa, Poland

Abstract

A rapid growth of interest in rough set theory [290] and its applications can be lately seen in the number of international workshops, conferences and seminars that are either directly dedicated to rough sets, include the subject in their programs, or simply accept papers that use this approach to solve problems at hand. A large number of high quality papers on various aspects of rough sets and their applications have been published in recent years as a result of this attention. The theory has been followed by the development of several software systems that implement rough set operations. In Sect. 12 we present a list of software systems based on rough sets. Some of the toolkits, provide advanced graphical environments that support the process of developing and validating rough set classifiers. Rough sets are applied in many domains, such as, for instance, medicine, finance, telecommunication, vibration analysis, conflict resolution, intelligent agents, image analysis, pattern recognition, control theory, process industry, marketing, etc.

Several applications have revealed the need to extend the traditional rough set approach. A special place among various extensions is taken by the approach that replaces indiscernibility relation based on equivalence with a tolerance relation.

In view of many generalizations, variants and extensions of rough sets a uniform presentation of the theory and methodology is in place. This tutorial paper is intended to fulfill these needs. It introduces basic notions and illustrates them with simple examples. It discusses methodologies for analyzing data and surveys applications. It also presents an introduction to logical, algebraic and topological aspects and major extensions to standard rough sets. It finally glances at future research.

Keywords: approximate reasoning, soft computing, indiscernibility, lower and upper approximations, rough sets, boundary region, positive region, rough membership function, decision rules, dependencies to a degree, patterns, feature extraction and selection, rough mereology.

Introduction

Rough set theory was developed by Zdzisław Pawlak [285, 290, 203] in the early 1980's. It deals with the classificatory analysis of data tables. The data can be acquired from measurements or from human experts. The main goal of the rough set analysis is to synthesize approximation of concepts from the acquired data. We show that first in the traditional approach and later how it evolves towards "information granules" under tolerance relation.

The purpose of developing such definitions may be twofold. In some instances, the aim may be to gain insight into the problem at hand by analyzing the constructed model, i.e. the structure of the model is itself of interest. In other applications, the transparency and explainability features of the model may be of secondary importance and the main objective is to construct a classifier that classifies unseen objects well. A logical calculus on approximate notions is equally important. It is based on the concept of "being a part to a degree" and is known as rough mereology (see e.g. [328, 330, 331, 332, 333, 340, 398, 335, 336]).

The overall modeling process typically consists of a sequence of several sub-steps that all require various degrees of tuning and fine-adjustments. In order to perform these functions, an environment to interactively manage and process data is required. An important feature of rough sets is that the theory is followed by practical implementations of toolkits that support interactive model development. Several software systems based on rough sets exist. For a list of these systems see Sect. 12.

The article consists of two parts. In Part I we discuss:

- classical rough set theory (Sections 1 to 5),
- the modeling process using rough sets which includes feature selection, feature extraction (by discretization, symbolic attribute value grouping, searching for relevant hyperplanes), rule synthesis and validation, (Sect. 7),
- some extensions to classical rough set approach (Sect. 8),
- some introductory information on algebraic and logical aspects of rough sets (Sect. 9),
- some relationships with other approaches (Sect. 10),
- a list of applications of rough sets (Sect. 11)
- a list of software systems that implement rough set methods (Sect. 12),
- and, finally, some conclusions including also considerations on future research.

In Part II we overview rough mereology developed as a tool for synthesis of objects satisfying a given specification to a satisfactory degree. The main goal of this approach is to develop methodology for construction of calculus on approximate concepts.

The tutorial attempts to address the needs of a broad readership. By combining informal introductions of each topic with simple examples, it should be accessible to all readers with interest in data analysis: from undergraduate students in computer science, to engineers, medical informatics scientists, to financial analysts, to social science researchers, etc. Since every informal exposition is followed by precise definitions, the tutorial is also an authoritative source for graduate students and researchers in the subject.

PartI
Rough sets

1 Information Systems

A data set is represented as a table, where each row represents a case, an event, a patient, or simply an object. Every column represents an attribute (a variable, an observation, a property, etc.) that can be measured for each object; the attribute may be also supplied by a human expert or user. This table is called an *information system*. More formally, it is a pair $\mathcal{A} = (U, A)$, where U is a non-empty finite set of *objects* called the *universe* and A is a non-empty finite set of *attributes* such that $a : U \to V_a$ for every $a \in A$. The set V_a is called the *value set* of a.

Example 1.1 A very simple information system is shown in Tab. 1. There are seven cases or objects, and two attributes (*Age* and Lower Extremity Motor Score *LEMS*).

	Age	LEMS
x_1	16-30	50
x_2	16-30	0
x_3	31-45	1-25
x_4	31-45	1-25
x_5	46-60	26-49
x_6	16-30	26-49
x_7	46-60	26-49

Table 1. An example information system.

The reader will easily notice that cases x_3 and x_4 as well as x_5 and x_7 have exactly the same values of conditions. The cases are (pairwise) *indiscernible* using the available attributes. □

In many applications there is an outcome of classification that is known. This *a posteriori* knowledge is expressed by one distinguished attribute called decision attribute; the process is known as supervised learning. Information systems of this kind are called *decision systems*. A decision system is any information system of the form $\mathcal{A} = (U, A \cup \{d\})$, where $d \notin A$ is the *decision attribute*. The elements of A are called *conditional attributes* or simply *conditions*. The decision attribute may take several values though binary outcomes are rather frequent.

Example 1.2 A small example decision table can be found in Tab. 2. The table has the same seven cases as in the previous example, but one decision attribute (*Walk*) with two possible outcomes has been added.

	Age	LEMS	Walk
x_1	16-30	50	Yes
x_2	16-30	0	No
x_3	31-45	1-25	No
x_4	31-45	1-25	Yes
x_5	46-60	26-49	No
x_6	16-30	26-49	Yes
x_7	46-60	26-49	No

Table 2. *Walk*: An example decision table

The reader may again notice that cases x_3 and x_4 as well as x_5 and x_7 still have exactly the same values of conditions, but the first pair has a different outcome (different value of the decision attribute) while the second pair also has the same outcome. □

The definitions to be synthesized from decision tables will be of the rule form "if *Age* is 16-30 and *LEMS* is 50 then *Walk* is Yes". Among the possible properties of the constructed rule sets, minimality (of the left hand side lengths of the rules) is one of the important issues. This is studied in the next section.

2 Indiscernibility

A decision system (i.e. a decision table) expresses all the knowledge about the model. This table may be unnecessarily large in part because it is redundant in at least two ways. The same or indiscernible objects may be represented several times, or some of the attributes may be superfluous. We shall look into these issues now.

The notion of equivalence is recalled first. A binary relation $R \subseteq X \times X$ which is reflexive (i.e. an object is in relation with itself xRx), symmetric (if xRy then yRx) and transitive (if xRy and yRz then xRz) is called an equivalence relation.

The *equivalence class* of an element $x \in X$ consists of all objects $y \in X$ such that xRy.

Let $\mathcal{A} = (U, A)$ be an information system, then with any $B \subseteq A$ there is associated an equivalence relation $IND_\mathcal{A}(B)$:

$$IND_\mathcal{A}(B) = \{(x, x') \in U^2 \mid \forall a \in B\ a(x) = a(x')\}$$

$IND_\mathcal{A}(B)$ is called the *B-indiscernibility relation*. If $(x, x') \in IND_\mathcal{A}(B)$, then objects x and x' are indiscernible from each other by attributes from B. The equivalence classes of the B-indiscernibility relation are denoted $[x]_B$. The subscript $\mathcal{A}$ in the indiscernibility relation is usually omitted if it is clear which information system is meant.

Some extensions of standard rough sets do not require transitivity to hold. See, for instance, [399]. Such a relation is called tolerance or similarity. This case will be discussed in Sect. 8.

Example 2.1 Let us illustrate how a decision table such as Tab. 2 defines an indiscernibility relation. The non-empty subsets of the conditional attributes are $\{Age\}$, $\{LEMS\}$ and $\{Age, LEMS\}$.

If we consider, for instance, $\{LEMS\}$, objects x_3 and x_4 belong to the same equivalence class and are indiscernible. (By the same token, x_5, x_6 and x_7 belong to another indiscernibility class.) The relation IND defines three partitions of the universe.

$$IND(\{Age\}) = \{\{x_1, x_2, x_6\}, \{x_3, x_4\}, \{x_5, x_7\}\}$$
$$IND(\{LEMS\}) = \{\{x_1\}, \{x_2\}, \{x_3, x_4\}, \{x_5, x_6, x_7\}\}$$
$$IND(\{Age, LEMS\}) = \{\{x_1\}, \{x_2\}, \{x_3, x_4\}, \{x_5, x_7\}, \{x_6\}\}$$

□

3 Set Approximation

An equivalence relation induces a partitioning of the universe (the set of cases in our example). These partitions can be used to build new subsets of the universe. Subsets that are most often of interest have the same value of the outcome attribute. It may happen, however, that a concept such as "*Walk*" cannot be defined in a crisp manner. For instance, the set of patients with a positive outcome cannot be defined crisply using the attributes available in Tab. 2. The "problematic" patients are objects x_3 and x_4. In other words, it is not possible to induce a crisp (precise) description of such patients from the table. It is here that the notion of rough set emerges. Although we cannot define those patients crisply, it is possible to delineate the patients that certainly have a positive outcome, the patients that certainly do not have a positive outcome and, finally, the patients that belong to a boundary between the certain cases. If this boundary is non-empty, the set is rough. These notions are formally expressed as follows.

Let $\mathcal{A} = (U, A)$ be an information system and let $B \subseteq A$ and $X \subseteq U$. We can approximate X using only the information contained in B by constructing the

B-lower and B-upper approximations of X, denoted $\underline{B}X$ and $\overline{B}X$ respectively, where $\underline{B}X = \{x \mid [x]_B \subseteq X\}$ and $\overline{B}X = \{x \mid [x]_B \cap X \neq \emptyset\}$.

The objects in $\underline{B}X$ can be with certainty classified as members of X on the basis of knowledge in B, while the objects in $\overline{B}X$ can be only classified as possible members of X on the basis of knowledge in B. The set $BN_B(X) = \overline{B}X - \underline{B}X$ is called the B-boundary region of X, and thus consists of those objects that we cannot decisively classify into X on the basis of knowledge in B. The set $U - \overline{B}X$ is called the B-outside region of X and consists of those objects which can be with certainty classified as do not belonging to X (on the basis of knowledge in B). A set is said to be *rough* (respectively *crisp*) if the boundary region is non-empty (respectively empty)[6].

Example 3.1 The most common case is to synthesize definitions of the outcome (or decision classes) in terms of the conditional attributes. Let $W = \{x \mid Walk(x) = \text{Yes}\}$, as given by Tab. 2. We then obtain the approximation regions $\underline{A}W = \{x_1, x_6\}$, $\overline{A}W = \{x_1, x_3, x_4, x_6\}$, $BN_A(W) = \{x_3, x_4\}$ and $U - \overline{A}W = \{x_2, x_5, x_7\}$. It follows that the outcome *Walk* is rough since the boundary region is not empty. This is shown in Fig. 1. □

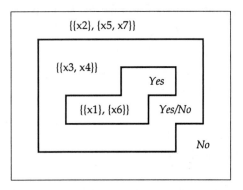

Fig. 1. Approximating the set of walking patients using the two conditional attributes *Age* and *LEMS*. Equivalence classes contained in the corresponding regions are shown.

One can easily show the following properties of approximations:

(1) $\underline{B}(X) \subseteq X \subseteq \overline{B}(X)$
(2) $\underline{B}(\emptyset) = \overline{B}(\emptyset) = \emptyset, \underline{B}(U) = \overline{B}(U) = U$
(3) $\overline{B}(X \cup Y) = \overline{B}(X) \cup \overline{B}(Y)$

---

[6] The letter B refers to the subset B of the attributes A. If another subset were chosen, e.g. $F \subseteq A$, the corresponding names of the relations would have been *F-boundary region, F-lower- and F-upper approximations*.

(4) $\underline{B}(X \cap Y) = \underline{B}(X) \cap \underline{B}(Y)$
(5) $X \subseteq Y$ implies $\underline{B}(X) \subseteq \underline{B}(Y)$ and $\overline{B}(X) \subseteq \overline{B}(Y)$
(6) $\underline{B}(X \cup Y) \supseteq \underline{B}(X) \cup \underline{B}(Y)$
(7) $\overline{B}(X \cap Y) \subseteq \overline{B}(X) \cap \overline{B}(Y)$
(8) $\underline{B}(-X) = -\overline{B}(X)$
(9) $\overline{B}(-X) = -\underline{B}(X)$
(10) $\underline{B}(\underline{B}(X)) = \overline{B}(\underline{B}(X)) = \underline{B}(X)$
(11) $\overline{B}(\overline{B}(X)) = \underline{B}(\overline{B}(X)) = \overline{B}(X)$

where $-X$ denotes $U - X$.

It is easily seen that the lower and the upper approximations of a set, are respectively, the interior and the closure of this set in the topology generated by the indiscernibility relation.

One can define the following four basic classes of rough sets, i.e., four categories of vagueness:

a) X is *roughly B-definable*, iff $\underline{B}(X) \neq \emptyset$ and $\overline{B}(X) \neq U$,
b) X is *internally B-undefinable*, iff $\underline{B}(X) = \emptyset$ and $\overline{B}(X) \neq U$,
c) X is *externally B-undefinable*, iff $\underline{B}(X) \neq \emptyset$ and $\overline{B}(X) = U$,
d) X is *totally B-undefinable*, iff $\underline{B}(X) = \emptyset$ and $\overline{B}(X) = U$.

The intuitive meaning of this classification is the following.

X is roughly *B-definable* means that with the help of B we are able to decide for some elements of U that they belong to X and for some elements of U that they belong to $-X$.

X is internally B-undefinable means that using B we are able to decide for some elements of U that they belong to $-X$, but we are unable to decide for any element of U whether it belongs to X.

X is externally B-undefinable means that using B we are able to decide for some elements of U that they belong to X, but we are unable to decide for any element of U whether it belongs to $-X$.

X is totally B-undefinable means that using B we are unable to decide for any element of U whether it belongs to X or $-X$.

Rough set can be also characterized numerically by the following coefficient

$$\alpha_B(X) = \frac{|\underline{B}(X)|}{|\overline{B}(X)|},$$

called the *accuracy of approximation*, where $|X|$ denotes the cardinality of $X \neq \emptyset$. Obviously $0 \leq \alpha_B(X) \leq 1$. If $\alpha_B(X) = 1$, X is *crisp* with respect to B (X is *precise* with respect to B), and otherwise, if $\alpha_B(X) < 1$, X is *rough* with respect to B (X is *vague* with respect to B).

4 Reducts

In the previous section we investigated one natural dimension of reducing data which is to identify equivalence classes, i.e. objects that are indiscernible using the available attributes. Savings are to be made since only one element of the equivalence class is needed to represent the entire class. The other dimension in reduction is to keep only those attributes that preserve the indiscernibility relation and, consequently, set approximation. The rejected attributes are redundant since their removal cannot worsen the classification. There is usually several such subsets of attributes and those which are minimal are called reducts. Computing equivalence classes is straightforward. Finding a minimal reduct (i.e. reduct with a minimal cardinality of attributes among all reducts) is NP-hard [400]. One can also show that the number of reducts of an information system with m attributes may be equal to

$$\binom{m}{\lfloor m/2 \rfloor}$$

This means that computing reducts it is a non-trivial task that cannot be solved by a simple minded increase of computational resources. It is, in fact, one of the bottlenecks of the rough set methodology. Fortunately, there exist good heuristics (e.g. [521, 522]) based on genetic algorithms that compute sufficiently many reducts in often acceptable time, unless the number of attributes is very high.

Example 4.1 Consider the following decision system (defined in Tab. 3): $\mathcal{A}' = (U, \{Diploma, Experience, French, Reference\} \cup \{Decision\})$. Let us consider only the conditional attributes i.e. an information system

$$\mathcal{A} = (U, \{Diploma, Experience, French, Reference\}).$$

For simplicity, each equivalence class contains one element. It appears that there is a minimal set of attributes $\{Experience, Reference\}$ which discerns objects in the same way as the full set of considered objects. The reader may check that the indiscernibility relation using the full set of attributes and the set $\{Experience, Reference\}$ is the same. The actual construction of minimal sets of attributes with such property will be soon revealed. □

	Diploma	Experience	French	Reference	Decision
x_1	MBA	Medium	Yes	Excellent	Accept
x_2	MBA	Low	Yes	Neutral	Reject
x_3	MCE	Low	Yes	Good	Reject
x_4	MSc	High	Yes	Neutral	Accept
x_5	MSc	Medium	Yes	Neutral	Reject
x_6	MSc	High	Yes	Excellent	Accept
x_7	MBA	High	No	Good	Accept
x_8	MCE	Low	No	Excellent	Reject

Table 3. *Hiring*: An example of an unreduced decision table.

Given an information system $\mathcal{A} = (U, A)$ the definitions of these notions are as follows. A *reduct* of $\mathcal{A}$ is a minimal set of attributes $B \subseteq A$ such that $IND_A(B) = IND_A(A)$. In other words, a reduct is a minimal set of attributes from A that preserves the partitioning of the universe and hence the ability to perform classifications as the whole attribute set A does.

Let $\mathcal{A}$ be an information system with n objects. The *discernibility matrix* of $\mathcal{A}$ is a symmetric $n \times n$ matrix with entries c_{ij} as given below.

$$c_{ij} = \{a \in A \mid a(x_i) \neq a(x_j)\} \text{ for } i, j = 1, ..., n$$

Each entry thus consists of the set of attributes upon which objects x_i and x_j differ.

A *discernibility function* $f_{\mathcal{A}}$ for an information system $\mathcal{A}$ is a Boolean function of m Boolean variables $a_1^*, ..., a_m^*$ (corresponding to the attribute s $a_1, ..., a_m$) defined as follows.

$$f_{\mathcal{A}}(a_1^*, ..., a_m^*) = \bigwedge \left\{ \bigvee c_{ij}^* \mid 1 \leq j \leq i \leq n, c_{ij} \neq \emptyset \right\}$$

where $c_{ij}^* = \{a^* \mid a \in c_{ij}\}$. The set of all prime implicants[7] of $f_{\mathcal{A}}$ determines the set of all reducts of $\mathcal{A}$.

Example 4.2 The discernibility function for the information system $\mathcal{A}$ defined in Tab. 3 is:

$$\begin{aligned}
f_{\mathcal{A}}(d, e, f, r) = &(e \vee r)(d \vee e \vee r)(d \vee e \vee r)(d \vee r)(d \vee e)(e \vee f \vee r)(d \vee e \vee f) \\
&(d \vee r)(d \vee e)(d \vee e)(d \vee e \vee r)(e \vee f \vee r)(d \vee f \vee r) \\
&(d \vee e \vee r)(d \vee e \vee r)(d \vee e \vee r)(d \vee e \vee f)(f \vee r) \\
&(e)(r)(d \vee f \vee r)(d \vee e \vee f \vee r) \\
&(e \vee r)(d \vee e \vee f \vee r)(d \vee e \vee f \vee r) \\
&(d \vee f \vee r)(d \vee e \vee f) \\
&(d \vee e \vee r)
\end{aligned}$$

where each parenthesized tuple is a conjunction in the Boolean expression, and where the one-letter Boolean variables correspond to the attribute names in an obvious way. After simplification, the function is $f_{\mathcal{A}}(d, e, f, r) = er$. (The notation er is a shorthand for $e \wedge r$.)

Let us also notice that each row in the above discernibility function corresponds to one column in the discernibility matrix. This matrix is symmetrical with the empty diagonal. So, for instance, the last but one row says that the sixth object (more precisely, the sixth equivalence class) may be discerned from the seventh one by any of the attributes *Diploma*, *French* or *Reference* and by any of *Diploma*, *Experience* or *French* from the eight one. □

---
[7] An implicant of a Boolean function f is any conjunction of literals (variables or their negations) such that if the values of these literals are true under an arbitrary valuation v of variables then the value of the function f under v is also true. A prime implicant is a minimal implicant. Here we are interested in implicants of monotone Boolean functions only i.e. functions constructed without negation.

If we instead construct a Boolean function by restricting the conjunction to only run over column k in the discernibility matrix (instead of over all columns), we obtain the so-called *k-relative discernibility function*. The set of all prime implicants of this function determines the set of all *k-relative reducts* of $\mathcal{A}$. These reducts reveal the minimum amount of information needed to discern $x_k \in U$ (or, more precisely, $[x_k] \subseteq U$) from all other objects.

Using the notions introduced above, the problem of supervised learning, (i.e., the problem where the outcome of the classification is known), is to find the value of the decision d that should be assigned to a new object which is described with the help of the conditional attributes. We often require the set of attributes used to define the object to be minimal. For the example Tab. 3 it appears that {*Experience, Reference*} and {*Diploma, Experience*} are two minimal sets of attributes that uniquely define to which decision class an object belongs. The corresponding discernibility function is relative to the decision. The notions are now formalized.

Let $\mathcal{A} = (U, A \cup \{d\})$ be given. The cardinality of the image $d(U) = \{k \mid d(x) = k, x \in U\}$ is called the *rank of d* and is denoted by $r(d)$. Let us further assume that the set V_d of values of decision d is equal to $\{v_d^1, \ldots, v_d^{r(d)}\}$.

Example 4.3 Quite often the rank is two, e.g., {Yes, No} or {Accept, Reject}. It can be an arbitrary number, however. For instance in the *Hiring* example, we could have rank three if the decision had values in the set {Accept, Hold, Reject}. □

The decision d determines a partition $CLASS_\mathcal{A}(d) = \{X_\mathcal{A}^1, \ldots, X_\mathcal{A}^{r(d)}\}$ of the universe U, where $X_\mathcal{A}^k = \{x \in U \mid d(x) = v_d^k\}$ for $1 \leq k \leq r(d)$. $CLASS_\mathcal{A}(d)$ is called the *classification of objects in $\mathcal{A}$ determined by the decision d*. The set $X_\mathcal{A}^i$ is called the *i-th decision class of $\mathcal{A}$*. By $X_\mathcal{A}(u)$ we denote the decision class $\{x \in U \mid d(x) = d(u)\}$, for any $u \in U$.

Example 4.4 There are two decision classes in each of the running example decision systems, i.e., {Yes, No} and {Accept, Reject}, respectively. The partitioning of the universe for the *Walk* table is $U = X^{Yes} \cup X^{No}$, where $X^{Yes} = \{x_1, x_4, x_6\}$ and $X^{No} = \{x_2, x_3, x_5, x_7\}$. For the *Hiring* table we have $U = X^{Accept} \cup X^{Reject}$, where $X^{Accept} = \{x_1, x_4, x_6, x_7\}$ and $X^{Reject} = \{x_2, x_3, x_5, x_8\}$. The notation X^{Yes} and X^{No} is a shorthand for X^1 and X^2, respectively. □

If $X_\mathcal{A}^1, \ldots, X_\mathcal{A}^{r(d)}$ are the decision classes of $\mathcal{A}$, then the set $\underline{B}X_1 \cup \ldots \cup \underline{B}X_{r(d)}$ is called the *B-positive region of $\mathcal{A}$* and is denoted by $POS_B(d)$.

Example 4.5 A quick check, left to the reader, reveals that $\underline{A}X^{Yes} \cup \underline{A}X^{No} \neq U$ while $\underline{A}X^{Accept} \cup \underline{A}X^{Reject} = U$. This is related to the fact that for the decision system in Tab. 2 a unique decision cannot be made for objects x_3 and x_4 while in case of the other table all decisions are unique. □

This important property of decision systems is formalized as follows. Let $\mathcal{A} = (U, A \cup \{d\})$ be a decision system. The *generalized decision in $\mathcal{A}$* is the function

$\partial_A : U \longrightarrow \mathcal{P}(V_d)$ defined by $\partial_A(x) = \{i \mid \exists x' \in U \; x' \; IND(A) \; x \text{ and } d(x) = i\}$. A decision table $\mathcal{A}$ is called *consistent (deterministic)* if $|\partial_A(x)| = 1$ for any $x \in U$, otherwise $\mathcal{A}$ is *inconsistent (non-deterministic)*.

It is easy to see that a decision table $\mathcal{A}$ is consistent if, and only if, $POS_A(d) = U$. Moreover, if $\partial_B = \partial_{B'}$, then $POS_B(d) = POS_{B'}(d)$ for any pair of non-empty sets $B, B' \subseteq A$.

Example 4.6 The $\mathcal{A}$-positive region of $\mathcal{A}$ in the *Walk* decision system is a proper subset of U, while in the *Hiring* decision system the corresponding set is equal to the universe U. The first system is non-deterministic, the second one - deterministic. □

We have introduced above the notion of *k-relative* discernibility function. Since the decision attribute is so significant, it is useful to introduce a special definition for its case. Let $\mathcal{A} = (U, A \cup \{d\})$ be a consistent decision table and let $M(\mathcal{A}) = (c_{ij})$ be its discernibility matrix. We construct a new matrix $M^d(\mathcal{A}) = (c^d_{ij})$ assuming $c^d_{ij} = \emptyset$ if $d(x_i) = d(x_j)$ and $c^d_{ij} = c_{ij} - \{d\}$, otherwise. Matrix $M^d(\mathcal{A})$ is called *the decision-relative discernibility matrix of* $\mathcal{A}$. Construction of *the decision-relative discernibility function* from this matrix follows the construction of the discernibility function from the discernibility matrix. It has been shown [400] that the set of *prime implicants* of $f^d_M(\mathcal{A})$ defines the set of all *decision-relative reducts* of $\mathcal{A}$.

Example 4.7 The *Hiring* decision table in Tab. 4 is now used to illustrate the construction of the corresponding decision-relative discernibility matrix and function. The rows are reordered for convenience putting the accepted objects in the top rows. The corresponding discernibility matrix in Tab. 5 is symmetrical

	Diploma	Experience	French	Reference	Decision
x_1	MBA	Medium	Yes	Excellent	Accept
x_4	MSc	High	Yes	Neutral	Accept
x_6	MSc	High	Yes	Excellent	Accept
x_7	MBA	High	No	Good	Accept
x_2	MBA	Low	Yes	Neutral	Reject
x_3	MCE	Low	Yes	Good	Reject
x_5	MSc	Medium	Yes	Neutral	Reject
x_8	MCE	Low	No	Excellent	Reject

Table 4. *Hiring*: The reordered decision table.

and the diagonal is empty, and so are all the entries for which the decisions are equal.

The resulting simplified decision-relative discernibility function is $f^d_M(\mathcal{A}) = ed \vee er$. From the definition of the decision-relative matrix it follows that selecting one column of the indiscernibility matrix, e.g., corresponding to $[x_1]$,

	$[x_1]$	$[x_4]$	$[x_6]$	$[x_7]$	$[x_2]$	$[x_3]$	$[x_5]$	$[x_8]$
$[x_1]$	∅							
$[x_4]$	∅	∅						
$[x_6]$	∅	∅	∅					
$[x_7]$	∅	∅	∅	∅				
$[x_2]$	e,r	d,e	d,e,r	e,f,r	∅			
$[x_3]$	d,e,r	d,e,r	d,e,r	d,e,f	∅	∅		
$[x_5]$	d,r	e	e,r	d,e,f,r	∅	∅	∅	
$[x_8]$	d,e,f	d,e,f,r	d,e,f	d,e,r	∅	∅	∅	∅

Table 5. *Hiring*: The decision-relative discernibility matrix.

and simplifying it gives a minimal function that discerns $[x_1]$ from objects belonging to the corresponding decision class from objects belonging to the other decision classes. For example, the first column gives a Boolean function $(e \vee r)(d \vee e \vee r)(d \vee r)(d \vee e \vee f)$ which after simplification becomes $ed \vee rd \vee re \vee rf$. The reader can check that, for instance, "if *Reference* is Excellent and *French* is Yes then *Decision* is Accept" is indeed the case for x_1. It is rather illuminating to notice that if there is any other object for which "*Reference* is Excellent" and "*French* is Yes" hold, then the decision will also be "Accept". Indeed, this is the case for x_6. □

If a Boolean function such as in the case of k-relative discernibility function is constructed by restricting the conjunction to run only over these entries of the column that corresponds to objects with a decision different from the decision on x_k then the (k,d)-*relative discernibility function* is obtained. Decision rules with minimal descriptions of their left hand sides may be constructed from prime implicants of these functions (see Sect. 7.3).

Example 4.8

Figures 2 to 5 display these four types of indiscernibility. It is possible to consider other kinds of reducts, e.g. reducts that preserve the positive region and then use the same Boolean reasoning method to compute these reducts.

□

5 Rough Membership

In classical set theory, either an element belongs to a set or it does not. The corresponding membership function is the characteristic function for the set, i.e. the function takes values 1 and 0, respectively. In the case of rough sets, the notion of membership is different. The *rough membership function* quantifies the degree of relative overlap between the set X and the equivalence $[x]_B$ class to which x belongs. It is defined as follows:

$$\mu_X^B : U \longrightarrow [0,1] \text{ and } \mu_X^B(x) = \frac{|[x]_B \cap X|}{|[x]_B|}$$

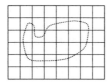

Fig. 2. Not relative to a particular case (or object) and not relative to the decision attribute. The full indiscernibility relation is preserved. Reducts of this type are minimal attribute subsets that enable us to discern all cases from each other, up to the same degree as the full set of attributes does.

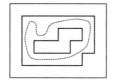

Fig. 3. Not relative to a particular case (or object) but relative to the decision attribute. All regions with the same value of the generalized decision ∂_A are preserved. Reducts of this type are minimal conditional attribute subsets $B \subseteq A$ that for all cases enable us to make the same classifications as the full set of attributes does, i.e. $\partial_A = \partial_B$.

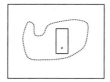

Fig. 4. Relative to case (or object) x but not relative to the decision attribute. Reducts of this type are minimal conditional attribute subsets that enable us to discern case x from all other cases up to the same degree as the full set of conditional attributes does.

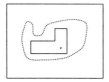

Fig. 5. Relative to case (or object) x and relative to the decision attribute. Our ability to discern case x from cases with different generalized decision than x is preserved. Reducts B of this type are minimal conditional attribute subsets that enable us to determine the outcome of case x, up to the same degree as the full set of attributes does, i.e. $\partial_A(x) = \partial_B(x)$.

The rough membership function can be interpreted as a frequency-based estimate of $\Pr(x \in X \mid u)$, the conditional probability that object x belongs to set X, given knowledge u of the information signature of x with respect to attributes B, i.e. $u = Inf_B(x)$ (see e.g. [519], [305], [303], [529]).

The formulae for the lower and upper set approximations can be generalized to some arbitrary level of precision $\pi \in (\frac{1}{2}, 1]$ by means of the rough membership function [536], as shown below.

$$\underline{B}_\pi X = \{x \mid \mu_X^B(x) \geq \pi\}$$

$$\overline{B}_\pi X = \{x \mid \mu_X^B(x) > 1 - \pi\}$$

Note that the lower and upper approximations as originally formulated are obtained as a special case with $\pi = 1.0$.

Approximations of concepts are constructed on the basis of background knowledge. Obviously, concepts are also related to unseen so far objects. Hence it is very useful to define parameterized approximations with parameters tuned in the searching process for approximations of concepts. This idea is crucial for construction of concept approximations using rough set methods.

Rough sets can thus approximately describe sets of patients, events, outcomes, etc. that may be otherwise difficult to circumscribe.

6 Dependency of Attributes

Another important issue in data analysis is discovering dependencies between attributes. Intuitively, a set of attributes D depends totally on a set of attributes C, denoted $C \Rightarrow D$, if all values of attributes from D are uniquely determined by values of attributes from C. In other words, D depends totally on C, if there exists a functional dependency between values of D and C.

Formally dependency can be defined in the following way. Let D and C be subsets of A.

We will say that D depends on C in a degree k $(0 \leq k \leq 1)$, denoted $C \Rightarrow_k D$, if

$$k = \gamma(C, D) = \frac{|POS_C(D)|}{|U|},$$

where

$$POS_C(D) = \bigcup_{X \in U/D} \underline{C}(X),$$

called a *positive region* of the partition U/D with respect to C, is the set of all elements of U that can be uniquely classified to blocks of the partition U/D, by means of C.

Obviously

$$\gamma(C, D) = \sum_{X \in U/D} \frac{|\underline{C}(X)|}{|U|}.$$

If $k = 1$ we say that D depends totally on C, and if $k < 1$, we say that D depends partially (in a degree k) on C.

The coefficient k expresses the ratio of all elements of the universe, which can be properly classified to blocks of the partition U/D, employing attributes C and will be called the *degree of the dependency*.

It can be easily seen that if D depends totally on C then $IND(C) \subseteq IND(D)$. This means that the partition generated by C is finer than the partition generated by D. Let us notice that the concept of dependency discussed above corresponds to that considered in relational databases.

Summing up: D is *totally (partially)* dependent on C, if employing C all *(possibly some)* elements of the universe U may be uniquely classified to blocks of the partition U/D.

7 Concept Approximation Construction: The Modeling Process

One of the main goals of machine learning, pattern recognition, knowledge discovery and data mining as well as of fuzzy sets and rough sets is to synthesize approximations of target concepts (e.g. decision classes) from the background knowledge (represented e.g. in the form of decision tables). It is usually only possible to search for approximate descriptions of target concepts due to incomplete knowledge about them (e.g. positive and negative examples of concept objects are given).

Approximate descriptions of concepts are constructed from some primitive concepts. It is furthermore well known that target concept descriptions defined directly by Boolean combinations of descriptors of the form $a = v$ (when a is and attribute and $a \in V_a$ are often not of good approximation quality. Feature selection and feature extraction problems are often approached by searching for relevant primitive concepts and are well known approaches in machine learning, KDD and other areas as (see, e.g. [145, 207, 91]).

In the case of feature selection relevant features are sought among the given features e.g. among descriptors $a = v$ where a is a relevant attribute. In Sect. 7.1 rough set-based methods for feature selection are briefly discussed.

The feature extraction problem is implemented as a search for some new features that are more relevant for classification and are defined (in some language) by means of the existing features.

These new features can be e.g. of the form $a \in [0.5, 1)$ or $2a + 3b > 0.75$. Their values on a given object are computed from given values of conditional attributes on the object. The new features are often binary taking value 1 on a given object iff the specified condition is true on this object. In the case of symbolic value attributes we look for new features like $a \in \{\text{French, English, Polish}\}$ with value 1 iff a person speaks any of these languages. The important issues in feature extraction are problems of discretization of real value attributes, grouping of symbolic (nominal) value attributes, searching for new features defined by hyperplanes or more complex surfaces defined over existing attributes. In Section 7.2 discretization based on rough set and Boolean reasoning approach is discussed. Some other approaches to feature extraction that are based on Boolean reasoning are also discussed. All cases of feature extraction problem mentioned above may be described in terms of searching for relevant features in a particular language of features. Boolean reasoning plays the crucial role of an inference engine for feature selection problems.

Feature extraction and feature selection are usually implemented in a pre-processing stage of the whole modeling process. There are some other aspects related to this stage of modeling such as, for instance, elimination of noise from the data or treatment of missing values. More information related to these problems can be found in [337, 338] and in the bibliography included in these books.

In the next stage of the synthesis of target concept approximations descriptions of the target concepts are constructed from the extracted relevant features (relevant primitive concepts) by applying some operations. In the simplest case when Boolean connectives ∨ and ∧ are chosen these descriptions form the so-called decision rules. In Sect. 7.3 we give a short introduction to methods for decision rule synthesis that are based on rough set methods and Boolean reasoning. Two main cases of decision rules are discussed: exact (deterministic) and approximate (non-deterministic) rules. More information on decision rule synthesis and using rough set approach the reader may find in [337, 338] and in the bibliography included in these books.

Finally, it is necessary to estimate the quality of constructed approximations of target concepts. Let us observe that the "building blocks" from which different approximations of target concepts are constructed may be inconsistent on new, so far unseen objects (i.e. some objects from the same class may be classified to disjoint concepts). This creates a necessity to develop methods for resolving these inconsistencies. The quality of target concept approximations can be considered acceptable if the inconsistencies may be resolved by using these methods. In Sect. 7.4 some introductory comments on this problem are presented and references to rough set methods that resolve conflicts among different decision rules by voting for the final decision are given.

7.1 Significance of Attributes and Approximate Reducts

One of the first ideas [290] was to consider as relevant features those in the *core* of an information system, i.e. features that belong to the intersection of all reducts of the information system. It can be easily checked that several definitions of relevant features that are used by machine learning community [3] can be interpreted by choosing a relevant decision system corresponding to the information system.

Another approach is related to dynamic reducts (see e.g. [18]) i.e. conditional attribute sets appearing "sufficiently often" as reducts of samples of the original decision table. The attributes belonging to the "majority" of dynamic reducts are defined as relevant. The value thresholds for "sufficiently often" and "majority" needs to be tuned for the given data. Several of the reported experiments show that the set of decision rules based on such attributes is much smaller than the set of all decision rules and the quality of classification of new objects is increasing or at least not significantly decreasing if only rules constructed over such relevant features are considered.

It is also possible to consider as relevant features those from some approximate reducts of sufficiently high quality. As it follows from the considerations concerning reduction of attributes, they can be not equally important and some of them can be eliminated from an information table without loosing information contained in the table. The idea of attribute reduction can be generalized by an introduction of the concept of *significance of attributes*, which enables an evaluation of attributes not only by a two-valued scale, *dispensable – indispensable*, but by associating with an attribute a real number from the [0,1] closed interval; this number expresses the importance of the attribute in the information table.

Significance of an attribute a in a decision table $\mathcal{A} = (\mathcal{U}, \mathcal{C} \cup \mathcal{D})$ (with the decision set D) can be evaluated by measuring the effect of removing of an attribute $a \in C$ from the attribute set C on the positive region defined by the table $\mathcal{A}$. As shown previously, the number $\gamma(C, D)$ expresses the degree of dependency between attributes C and D, or accuracy of approximation of U/D by C. We can ask how the coefficient $\gamma(C, D)$ changes when an attribute a is removed, i.e., what is the difference between $\gamma(C, D)$ and $\gamma((C - \{a\}, D)$. We can normalize the difference and define the significance of an attribute a as

$$\sigma_{(C,D)}(a) = \frac{(\gamma(C,D) - \gamma(C - \{a\}, D))}{\gamma(C,D)} = 1 - \frac{\gamma(C - \{a\}, D)}{\gamma(C,D)},$$

Thus the coefficient $\sigma(a)$ can be understood as the error of classification which occurs when attribute a is dropped. The significance coefficient can be extended to the set of attributes as follows:

$$\sigma_{(C,D)}(B) = \frac{(\gamma(C,D) - \gamma(C - B, D))}{\gamma(C,D)} = 1 - \frac{\gamma(C - B, D)}{\gamma(C,D)},$$

denoted by $\sigma(B)$, if C and D are understood, where B is a subset of C.

If B is a reduct of C, then $\sigma(C-B) = 0$, i.e., removing any reduct complement from the set of conditional attributes enables to make decisions with certainty, whatsoever.

Any subset B of C can be treated as an *approximate reduct* of C, and the number

$$\varepsilon_{(C,D)}(B) = \frac{(\gamma(C,D) - \gamma(B, D))}{\gamma(C,D)} = 1 - \frac{\gamma(B,D)}{\gamma(C,D)},$$

denoted simply as $\varepsilon(B)$, will be called an *error of reduct approximation*. It expresses how exactly the set of attributes B approximates the set of condition attributes C (relatively to D).

The concept of approximate reduct (with respect to the positive region) is a generalization of the reduct concept. A minimal subset B of condition attributes C, such that $\gamma(C, D) = \gamma(B, D)$, or $\varepsilon_{(C,D)}(B) = 0$ is a reduct (preserving the positive region). The idea of an approximate reduct can be useful in those cases when a smaller number of condition attributes is preferred over the accuracy of classification on training data. This can allow to increase the classification accuracy on testing data. The error level of reduct approximation should be tuned for a given data set to achieve this effect.

Section 7.3 introduces several other methods of reduct approximation that are based on other measures than positive region. Experiments show that by tuning the approximation level one can, in most cases, increase the classification quality of new objects. It is important to note once again that Boolean reasoning may be used to compute these different types of reducts and to extract relevant approximations from them (see e.g. [411]).

7.2 Discretization and Some Other Feature Extraction Methods

The discretization step determines how coarsely we want to view the world. For instance, temperature, which is usually measured in real numbers, can be discretized into two, three or more, but finitely many, intervals. Another example could be heart-beat rate at rest. Although the parameter is already expressed as discrete value (i.e. a natural number), medical doctors will usually not distinguish among, say 68 or 72 beats per minute, and classify it as normal. On the other hand, 48 to 56 beats per second is considered low, (but normal for a trained long-distance runner) while 120 to 140 beats will be very fast and abnormal unless it is the rate for a fetus in a certain digestional stage. One can easily see that the selection of appropriate intervals and partitioning of attribute value sets is a complex problem and its complexity can grow exponentially in the number of attributes to be discretized. Discretization is a step that is not specific to the rough set approach but that most rule or tree induction algorithms currently require for them to perform well.

A number of successful approaches to the problem of finding effective heuristics for real value attributes quantization (discretization) has been proposed by machine learning, pattern recognition and KDD researchers see, e.g. [45, 67, 90, 228, 280, 347].

The rough set community has been also committed to constructing efficient algorithms for new feature extraction, in particular the efforts have been focused on discretization and symbolic attribute value grouping (see e.g. [177, 178, 45, 240, 180, 241, 243, 233]).

Applications of rough set methods combined with Boolean reasoning [27] have been developed for extraction of new features from data tables under an assumption that these features belong to a predefined set.

The most successful among these methods are:

- discretization techniques (see e.g. [240, 241, 233, 242, 234, 243]),
- methods of partitioning (grouping) of nominal (symbolic) attribute value sets (see e.g. [233, 243, 237, 238, 239]) and
- combinations of the above methods (see e.g. [237, 238, 239]).

Searching for new features expressed by multi-modal formulae (see e.g. [14, 15, 16]) needs also to be mentioned here as a successful method for feature extraction.

The results reported in the above cited papers show that the discretization problems and symbolic value partition problems are of high computational complexity (i.e. NP-complete or NP-hard). This clearly justifies the need to design efficient heuristics.

We will concentrate on the basic discretization methods based on the rough set and Boolean reasoning approaches. In the discretization of a decision table $\mathcal{A} = (U, A \cup \{d\})$, where $V_a = [v_a, w_a)$ is an interval of reals, we search for a partition P_a of V_a for any $a \in A$. Any partition of V_a is defined by a sequence of the so-called cuts $v_1 < v_2 < ... < v_k$ from V_a. Hence, any family of partitions $\{P_a\}_{a \in A}$ can be identified with a set of cuts. In the discretization process we search for a set of cuts satisfying some natural conditions.

Example 7.1 Let us consider a (consistent) decision system (Tab. 6 (a)) with two conditional attributes a and b and seven objects $u_1, ..., u_7$. The values of attributes on these objects and the values of decision d are presented in Tab. 6. Geometrical interpretation of objects and decision classes are shown in Fig.6.

$\mathcal{A}$	a	b	d
u_1	0.8	2	1
u_2	1	0.5	0
u_3	1.3	3	0
u_4	1.4	1	1
u_5	1.4	2	0
u_6	1.6	3	1
u_7	1.3	1	1

(a)

$\Longrightarrow$

$\mathcal{A}^P$	a^P	b^P	d
u_1	0	2	1
u_2	1	0	0
u_3	1	2	0
u_4	1	1	1
u_5	1	2	0
u_6	2	2	1
u_7	1	1	1

(b)

Table 6. Discretization process. (a) The original decision system $\mathcal{A}$. (b) $\mathbf{P}$-discretization of $\mathcal{A}$, where $\mathbf{P} = \{(a, 0.9), (a, 1.5), (b, 0.75), (b, 1.5)\}$

The sets of possible values of a and b are defined by:

$$V_a = [0, 2) \, ; V_b = [0, 4) \, .$$

The sets of values of a and b on objects from U are given by

$$a(U) = \{0.8, 1, 1.3, 1.4, 1.6\};$$
$$b(U) = \{0.5, 1, 2, 3\},$$

respectively.

We will describe a discretization process that returns a partition of the value sets of conditional attributes into intervals. The partition is done in such a way that if the name of the interval containing an arbitrary object is substituted for any object instead of its original value in $\mathcal{A}$ a consistent decision system is also obtained. In this way the size of the value attribute sets in a decision system is reduced.

In our example the following intervals for condition attributes are obtained:
$[0.8.1)$; $[1, 1.3)$; $[1.3, 1.4)$; $[1.4, 1.6)$ for a;
$[0.5, 1)$; $[1, 2)$; $[2, 3)$ for b
defined by objects in decision system. The reader may notice that we do not consider intervals $[0, 0.5)$, $[1.6, 2)$ for a and $[0, 0.5)$, $[3, 4)$ for b. The reason for that will be clear later.

The idea of cuts is used now. Cuts are pairs (a, c) where $c \in V_a$. We will restrict our considerations for cuts defined by the middle points of the intervals defined above. The following cuts are obtained (see Fig. 7):

$(a, 0.9)$; $(a, 1.15)$; $(a, 1.35)$; $(a, 1.5)$;
$(b, 0.75)$; $(b, 1.5)$; $(b, 2.5)$.

Any cut defines a new conditional attribute with binary values. For example, the attribute corresponding to the cut $(a, 1.2)$ is equal to 0 if $a(x) < 1.2$, otherwise is equal to 1. Hence, objects positioned on different sides of the straight line $a = 1.2$ are discerned by this cut. The reader may now see why some of the above mentioned intervals have been eliminated from our considerations: cuts positioned in these intervals will not discern any pair of objects in the table.

Any set P of cuts defines a new conditional attribute a_P for any a. One should consider a partition of the value set of a by cuts from P and put the unique names for the elements of these partition. Lets take the following set of cuts: $P = \{(a, 0.9), (a, 1.5), (b, 0.75), (b, 1.5)\}$. This set assigns all values of a less then 0.9 to the interval named 0, all values in the all values in the interval $[0.9, 1.5)$ to the interval 1 and all values from $[1.5, 4)$ to the interval 2. An analogous construction is done for b. The values of the new attributes a_P and b_P are shown in Tab. 6 (b).

The next natural question is: How to construct a set of cuts with a minimal number of elements discerning all pairs of objects to be discerned? We will show that this can be done using Boolean reasoning.

Let us introduce a Boolean variable corresponding to any attribute a and any interval determined by a. In our example the set of Boolean variables defined by $\mathcal{A}$ is equal to

$$VB(\mathcal{A}) = \{p_1^a, p_2^a, p_3^a, p_4^a, p_1^b, p_2^b, p_3^b\};$$

where $p_1^a \sim [0.8; 1)$ of a (i.e. p_1^a corresponds to the interval $[0.8; 1)$ of attribute a); $p_2^a \sim [1; 1.3)$ of a; $p_3^a \sim [1.3; 1.4)$ of a; $p_4^a \sim [1.4; 1.6)$ of a; $p_1^b \sim [0.5; 1)$ of b; $p_2^b \sim [1; 2)$ of b; $p_3^b \sim [2; 3)$ of b (see Fig. 7).

Let us recall that a valuation of propositional variables is any function from the set of propositional variables into $\{0, 1\}$. Now one can easily observe that there is a one-to-one correspondence between the set of valuations of propositional variables defined above for a given $\mathcal{A}$ and the set of cuts in $\mathcal{A}$. The rule is as follows: (i) for any cut choose the interval containing it and next the propositional variable corresponding to it; (ii) for any propositional variable choose a cut in the interval corresponding to the variable. For example, the set of cuts $\mathbf{P} = \{(a, 0.9), (a, 1.5), (b, 0.75), (b, 1.5)\}$ corresponds to the valuation assigning 1

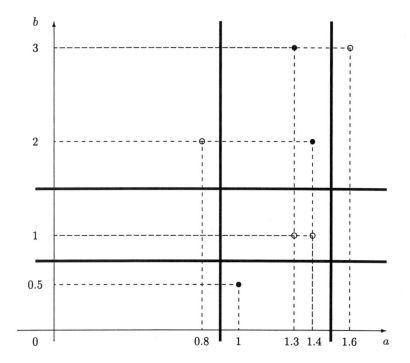

Fig. 6. A geometrical representation of data and cuts.

to the propositional variables: p_1^a, p_4^a, p_1^b, p_2^b only. Having this correspondence, we will say that the Boolean formula built from the propositional variables is satisfied by a given set of cuts iff it is satisfied by the valuation corresponding to that set (i.e. taking value 1 only on variables corresponding to cuts from this set).

Now, using our example, we will show how to built a Boolean formula $\Phi^{\mathcal{A}}$, called the discernibility formula for a given $\mathcal{A}$ and with the following property: *the set of prime implicants of $\Phi^{\mathcal{A}}$ defines uniquely the family of all minimal set of cuts discerning objects in $\mathcal{A}$.* Moreover, any valuation satisfying this formula determines the set of cuts discerning all object pairs to be discerned.

Having in mind the discernibility matrix for $\mathcal{A}$, one can see that we should choose at least one cut on one of the attributes appearing in the entry (x_i, x_j) of the discernibility matrix of $\mathcal{A}$ for any objects x_i and x_j discernible by conditional attributes which have different decisions.

The discernibility formulae $\psi(i, j)$ for different pairs (u_i, u_j) of discernible

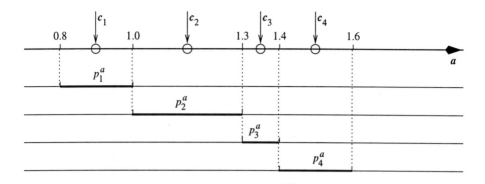

Fig. 7. The set of cuts $(a, c_1), (a, c_2), (a, c_3), (a, c_4)$ on a, the set of propositional variables $p_1^a, p_2^a, p_3^a, p_4^a$ and the set of intervals corresponding to these variables in $\mathcal{A}$ (see Example 7.1)

objects from $U \times U$ with different decisions have the following form:

$$\psi(2,1) = p_1^a \vee p_1^b \vee p_2^b; \qquad \psi(2,4) = p_2^a \vee p_3^a \vee p_1^b;$$
$$\psi(2,6) = p_2^a \vee p_3^a \vee p_4^a \vee p_1^b \vee p_2^b \vee p_3^b; \quad \psi(2,7) = p_2^a \vee p_1^b;$$
$$\psi(3,1) = p_1^a \vee p_2^a \vee p_3^b; \qquad \psi(3,4) = p_2^a \vee p_2^b \vee p_3^b;$$
$$\psi(3,6) = p_3^a \vee p_4^a; \qquad \psi(3,7) = p_2^b \vee p_3^b;$$
$$\psi(5,1) = p_1^a \vee p_2^a \vee p_3^a; \qquad \psi(5,4) = p_2^b;$$
$$\psi(5,6) = p_4^a \vee p_3^b; \qquad \psi(5,7) = p_3^a \vee p_2^b;$$

For example, formula $\psi(5,6)$ is true on the set of cuts if there exists a cut $p_1 = (a, c)$ on V_a in this set such that $c \in [1.4, 1.6)$ or a cut $p_2 = (b, c)$ on V_b that $c \in [2, 3)$.

The discernibility formula $\Phi^{\mathcal{A}}$ in CNF form is given by taking all the above conditions:

$$\Phi^{\mathcal{A}} = \left(p_1^a \vee p_1^b \vee p_2^b\right) \wedge \left(p_1^a \vee p_2^a \vee p_3^b\right) \wedge \left(p_1^a \vee p_2^a \vee p_3^a\right) \wedge \left(p_2^a \vee p_3^a \vee p_1^b\right)$$
$$\wedge p_2^b \wedge \left(p_2^a \vee p_2^b \vee p_3^b\right) \wedge \left(p_2^a \vee p_3^a \vee p_4^a \vee p_1^b \vee p_2^b \vee p_3^b\right) \wedge \left(p_3^a \vee p_4^a\right)$$
$$\wedge \left(p_4^a \vee p_3^b\right) \wedge \left(p_2^a \vee p_1^b\right) \wedge \left(p_2^b \vee p_3^b\right) \wedge \left(p_3^a \vee p_2^b\right).$$

Transforming formula $\Phi^{\mathcal{A}}$ to the DNF form we obtain four prime implicants:

$$\Phi^{\mathcal{A}} = \left(p_2^a \wedge p_4^a \wedge p_2^b\right) \vee \left(p_2^a \wedge p_3^a \wedge p_2^b \wedge p_3^b\right)$$
$$\vee \left(p_3^a \wedge p_1^b \wedge p_2^b \wedge p_3^b\right) \vee \left(p_1^a \wedge p_4^a \wedge p_1^b \wedge p_2^b\right).$$

If we decide to take e.g. the last prime implicant $S = \{p_1^a, p_4^a, p_1^b, p_2^b\}$, we obtain the following set of cuts

$$\mathbf{P}(S) = \{(a, 0.9), (a, 1.5), (b, 0.75), (b, 1.5)\}.$$

The new decision system $\mathcal{A}^{\mathbf{P}(S)}$ is represented in Table 6 (b). □

A more formal description of the discretization problem is now presented.

Let $\mathcal{A} = (\mathcal{U}, A \cup \{d\})$ be a decision system where $U = \{x_1, x_2, \ldots, x_n\}$; $A = \{a_1, \ldots, a_k\}$ and $d : U \to \{1, \ldots, r\}$. We assume $V_a = [l_a, r_a) \subset \Re$ to be a real interval for any $a \in A$ and $\mathcal{A}$ to be a consistent decision system. Any pair (a, c) where $a \in A$ and $c \in \Re$ will be called a *cut on* V_a. Let $\mathbf{P}_a$ be a partition on V_a (for $a \in A$) into subintervals i.e. $\mathbf{P}_a = \{[c_0^a, c_1^a), [c_1^a, c_2^a), \ldots, [c_{k_a}^a, c_{k_a+1}^a)\}$ for some integer k_a, where $l_a = c_0^a < c_1^a < c_2^a < \ldots < c_{k_a}^a < c_{k_a+1}^a = r_a$ and $V_a = [c_0^a, c_1^a) \cup [c_1^a, c_2^a) \cup \ldots \cup [c_{k_a}^a, c_{k_a+1}^a)$. Hence any partition $\mathbf{P}_a$ is uniquely defined by the set of cuts: $\{(a, c_1^a), (a, c_2^a), \ldots, (a, c_{k_a}^a)\} \subset A \times \Re$ and often identified with it.

Any set of cuts $\mathbf{P} = \bigcup_{a \in A} \mathbf{P}_a$ defines from $\mathcal{A} = (U, A \cup \{d\})$ a new decision system $\mathcal{A}^{\mathbf{P}} = (\mathcal{U}, A^{\mathbf{P}} \cup \{d\})$ called $\mathbf{P}$-*discretization of* $\mathcal{A}$, where $A^{\mathbf{P}} = \{a^{\mathbf{P}} : a \in A\}$ and $a^{\mathbf{P}}(x) = i \Leftrightarrow a(x) \in [c_i^a, c_{i+1}^a)$ for $x \in U$ and $i \in \{0, \ldots, k_a\}$.

Two sets of cuts $\mathbf{P}', \mathbf{P}$ are equivalent, i.e. $\mathbf{P}' \equiv_\mathcal{A} \mathbf{P}$, iff $A^{\mathbf{P}} = A^{\mathbf{P}'}$. The equivalence relation $\equiv_\mathcal{A}$ has a finite number of equivalence classes. In the sequel we will not discern between equivalent families of partitions.

We say that the set of cuts $\mathbf{P}$ is $\mathcal{A}$-*consistent* if $\partial_\mathcal{A} = \partial_{\mathcal{A}^{\mathbf{P}}}$, where $\partial_\mathcal{A}$ and $\partial_{\mathcal{A}^{\mathbf{P}}}$ are generalized decisions of $\mathcal{A}$ and $\mathcal{A}^{\mathbf{P}}$, respectively. The $\mathcal{A}$-consistent set of cuts $\mathbf{P}^{irr}$ is $\mathcal{A}$-*irreducible* if $\mathbf{P}$ is not $\mathcal{A}$-consistent for any $\mathbf{P} \subset \mathbf{P}^{irr}$. The $\mathcal{A}$-consistent set of cuts $\mathbf{P}^{opt}$ is $\mathcal{A}$-*optimal* if $card(\mathbf{P}^{opt}) \leq card(\mathbf{P})$ for any $\mathcal{A}$-consistent set of cuts $\mathbf{P}$.

One can show [240] that the decision problem of checking if for a given decision system $\mathcal{A}$ and an integer k there exists an irreducible set of cuts $\mathbf{P}$ in $\mathcal{A}$ such that $card(\mathbf{P}) < k$ is NP-complete. The problem of searching for an optimal set of cuts $\mathbf{P}$ in a given decision system $\mathcal{A}$ is NP-hard.

However, one can construct efficient heuristics returning semi-minimal sets of cuts [240, 241, 244, 233, 237, 238, 239, 319]. Here we discuss the simplest one based on the Johnson strategy. Using this strategy one can look for a cut discerning the maximal number of object pairs (with different decisions), next one can eliminate all already discerned object pairs and repeat the procedure until all object pairs to be discerned are discerned. It is intersecting to note that this can be realized by computing the minimal relative reduct of the corresponding decision system.

Again we will explain this idea using our example.

From a given decision system one can construct a new decision system $\mathcal{A}^*$ having as objects all pairs of objects from $\mathcal{A}$ with different decision values, so all object pairs to be discerned. We are adding one more object *new* on which all constructed new conditional attributes have value 0 and on which the decision value is also 0. The new decision is equal to 1 on all other objects in the new decision system. The set of condition attributes in the new decision system $\mathcal{A}^*$ is equal to the set of all attributes defined by all cuts (or all propositional variables considered above). These attributes are binary. The value of the new attribute corresponding to a cut (a, c) on the pair (u_i, u_j) is equal to 1 iff this cut is discerning objects (u_i, u_j) (i.e. $min(a(u_i), a(u_j)) < c < max(a(u_i), a(u_j))$) and 0 otherwise. One can formulate this condition in another way. The value of the new attribute corresponding to the propositional variable p_s^a on the pair (u_i, u_j)

is equal to 1 iff the interval corresponding to p_s^a is included in $[min(a(u_i), a(u_j)), max(a(u_i), a(u_j))]$ and 0 otherwise.

The resulting new decision system $\mathcal{A}^*$ is shown in Tab. 7.

Objects in $\mathcal{A}^*$ are all pairs (x_i, x_j) discernible by the decision d. One more object is included, namely *new* with all values of attributes equal to 0. This allows formally to keep the condition: *"at least one occurrence of 1 (for conditional attributes) appears in any row for any subset of columns corresponding to any prime implicant"*.

The relative reducts of this table correspond exactly to the prime implicants of the function $\Phi^{\mathcal{A}}$ (for the proof see e.g. [233]).

Our *"MD heuristic"* is based on searching for a cut with maximal number of object pairs discerned by this cut [240], [237]. The idea is analogous to the Johnson approximation algorithm and can be formulated as follows:

$\mathcal{A}^*$	p_1^a	p_2^a	p_3^a	p_4^a	p_1^b	p_2^b	p_3^b	d^*
(u_1, u_2)	1	0	0	0	1	1	0	1
(u_1, u_3)	1	1	0	0	0	0	1	1
(u_1, u_5)	1	1	1	0	0	0	0	1
(u_4, u_2)	0	1	1	0	1	0	0	1
(u_4, u_3)	0	0	1	0	0	1	1	1
(u_4, u_5)	0	0	0	0	0	1	0	1
(u_6, u_2)	0	1	1	1	1	1	1	1
(u_6, u_3)	0	0	1	1	0	0	0	1
(u_6, u_5)	0	0	0	1	0	0	1	1
(u_7, u_2)	0	1	0	0	1	0	0	1
(u_7, u_3)	0	0	0	0	0	1	1	1
(u_7, u_5)	0	0	1	0	0	1	0	1
new	0	0	0	0	0	0	0	0

Table 7. Decision system $\mathcal{A}^*$ constructed from $\mathcal{A}$

ALGORITHM MD-heuristic (Semi-optimal family of partitions)

Step 1. *Construct the table $\mathcal{A}^*$ from $\mathcal{A}$ and erase the last row (i.e. a "new" element) from $\mathcal{A}^*$; set $\mathcal{B} := \mathcal{A}^*$;*

Step 2. *Choose a column from $\mathcal{B}$ with the maximal number of occurrences of 1's;*

Step 3. *Delete from $\mathcal{B}$ the column chosen in Step 2 and all rows marked in this column by 1;*

Step 4. ***If** $\mathcal{B}$ is non-empty then go to Step 2 **else** Stop.*

In our example the algorithm is choosing first p_2^b next p_2^a and finally p_4^a. Hence $S = \{p_2^a, p_4^a, p_2^b\}$ and the resulting set of cuts $P = \{(a, 1.15), (a, 1.5), (b, 1.5)\}$. Fig 8 is showing the constructed set of cuts (marked by bold lines).

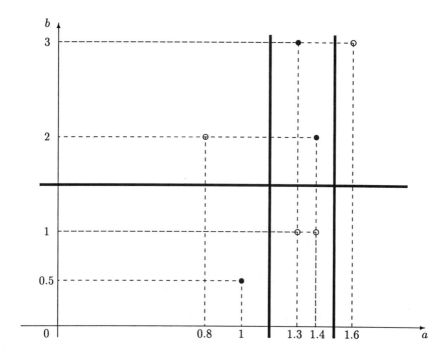

Fig. 8. The minimal set of cuts of $\mathcal{A}$

The algorithm based on Johnson's strategy described above is searching for a cut which discerns the largest number of pairs of objects (MD-heuristic). Then we move the cut c from $\mathcal{A}^*$ to the resulting set of cuts $\mathbf{P}$ and remove from U^* all pairs of objects discerned by c. Our algorithm is continued until $U^* = \{new\}$. Let n be the number of objects and let k be the number of attributes of decision system $\mathcal{A}$. Then $card\,(\mathcal{A}^*) \leq (n-1)\,k$ and $card\,(U^*) \leq \frac{n(n-1)}{2}$. It is easy to observe that for any cut $c \in \mathcal{A}^*$ we need $O\,(n^2)$ steps to find the number of all pairs of objects discerned by c. Hence the straightforward realization of this algorithm requires $O\,(kn^2)$ of memory space and $O(kn^3)$ steps to determine one *cut*, so it is not feasible in practice. The MD-heuristic presented in [236] determines the best cut in $O\,(kn)$ steps using $O\,(kn)$ space only. This heuristic is very efficient with respect to the time necessary for decision rules generation as well as with respect to the quality of unseen object classification. (see e.g [233, 241, 237]).

Let us observe that the new features in the considered case of discretization are of the form $a \in V$, where $V \subseteq V_a$ and V_a is the set of values of attribute a.

One can extend the presented approach (see e.g. [243], [237], [238], [238],) to the case of symbolic (nominal, qualitative) attributes as well as to the case when in a given decision system nominal and numeric attribute appear. The received heuristics are of very good quality.

Experiments for classification methods (see [238]) have been carried over decision systems using two techniques called *"train-and-test"* and *"n-fold-cross-validation"*. In Table 8 some results of experiments obtained by testing the proposed methods MD and MD-G for classification quality on well known data tables from the "UC Irvine repository" are shown. The results reported in [95] are summarized in columns labeled by S-ID3 and C4.5 in Table 8. It is interesting to compare those results with regard to the classification quality. Let us note that the heuristics MD and MD-G are also very efficient with respect to the time complexity.

Names of Tables	Classification accuracies			
	S-ID3	C4.5	MD	MD-G
Australian	78.26	85.36	83.69	84.49
Breast (L)	62.07	71.00	69.95	69.95
Diabetes	66.23	70.84	71.09	76.17
Glass	62.79	65.89	66.41	69.79
Heart	77.78	77.04	77.04	81.11
Iris	96.67	94.67	95.33	96.67
Lympho	73.33	77.01	71.93	82.02
Monk-1	81.25	75.70	100	93.05
Monk-2	69.91	65.00	99.07	99.07
Monk-3	90.28	97.20	93.51	94.00
Soybean	100	95.56	100	100
TicTacToe	84.38	84.02	97.7	97.70
Average	78.58	79.94	85.48	87.00

Table 8. The quality comparison between decision tree methods. MD: MD-heuristics; MD-G: MD-heuristics with symbolic value partition

In case of real value attributes one can search for features in the feature set containing the characteristic functions of half-spaced determined by hyperplanes or parts of spaces defined by more complex surfaces in multidimensional spaces. In [241], [233], [239] genetic algorithms have been applied in searching for semi-optimal hyperplanes or second order surfaces. The reported results are showing substantial increase in the quality of classification of unseen objects but we pay for that spending more time in searching for the semi-optimal hyperplanes.

In all of these cases one can use a general "board game" determined by the corresponding discernibility matrix in searching for optimal, in a sense, features and apply the following general scheme. For each entry of the discernibility matrix for discernible objects x and y one should consider the set of all formulas (from a considered language of features) discerning these objects. From the discernibility matrix the Boolean function(s) is (are) constructed, in a standard way [400], with the following property: the prime implicants of these functions determine the problem solutions. Using this general scheme one can invent much easier efficient heuristics searching for semi-prime implicants, and hence semi-optimal solutions, because they can be extracted by manipulation on Boolean formulas with a simple structure. The experimental results are supporting this claim (see e.g. [237], [238]). One of the possible strategy in searching for semi-optimal solutions is to search for short prime implicants because using *the minimum description length principle*, one can expect that from them the decision algorithms with high quality of unseen object classification can be built.

Boolean reasoning can also be used as a tool to measure the complexity of approximate solution of a given problem. As a complexity measure of a given problem one can consider the complexity of the corresponding to that problem Boolean function (represented by the number of variables, number of clauses, etc.).

7.3 Decision Rule Synthesis

The reader has certainly realized that the reducts (of all the various types) can be used to synthesize *minimal* decision rules. Once the reducts have been computed, the rules are easily constructed by overlaying the reducts over the originating decision table and reading off the values.

Example 7.2 Given the reduct $\{Diploma, Experience\}$ in the Tab. 4, the rule read off the first object is "if *Diploma* is MBA and *Experience* is Medium then *Decision* is Accept". □

We shall make these notions precise.

Let $\mathcal{A} = (U, A \cup \{d\})$ be a decision system and let $V = \bigcup \{V_a \mid a \in A\} \cup V_d$. Atomic formulae over $B \subseteq A \cup \{d\}$ and V are expressions of the form $a = v$; they are called *descriptors* over B and V, where $a \in B$ and $v \in V_a$. The set $\mathcal{F}(B, V)$ of formulae over B and V is the least set containing all atomic formulae over B and V and closed with respect to the propositional connectives $\wedge$ (conjunction), $\vee$ (disjunction) and $\neg$ (negation).

Let $\varphi \in \mathcal{F}(B, V)$. $\| \varphi_{\mathcal{A}} \|$ denotes the meaning of φ in the decision table $\mathcal{A}$ which is the set of all objects in U with the property φ. These sets are defined as follows:

1. if φ is of the form $a = v$ then $\| \varphi_{\mathcal{A}} \| = \{x \in U \mid a(x) = v\}$
2. $\| \varphi \wedge \varphi'_{\mathcal{A}} \| = \| \varphi_{\mathcal{A}} \| \cap \| \varphi'_{\mathcal{A}} \|$; $\| \varphi \vee \varphi'_{\mathcal{A}} \| = \| \varphi_{\mathcal{A}} \| \cup \| \varphi'_{\mathcal{A}} \|$; $\| \neg \varphi_{\mathcal{A}} \| = U - \| \varphi_{\mathcal{A}} \|$

The set $\mathcal{F}(B, V)$ is called the set of *conditional formulae of* $\mathcal{A}$ and is denoted $\mathcal{C}(B, V)$.

A *decision rule* for $\mathcal{A}$ is any expression of the form $\varphi \Rightarrow d = v$, where $\varphi \in \mathcal{C}(B, V)$, $v \in V_d$ and $\| \varphi_{\mathcal{A}} \| \neq \emptyset$. Formulae φ and $d = v$ are referred to as the *predecessor* and the *successor* of decision rule $\varphi \Rightarrow d = v$.

Decision rule $\varphi \Rightarrow d = v$ is *true* in $\mathcal{A}$ if, and only if, $\| \varphi_{\mathcal{A}} \| \subseteq \| d = v_{\mathcal{A}} \|$; $\| \varphi_{\mathcal{A}} \|$ is the set of objects *matching* the decision rule; $\| \varphi_{\mathcal{A}} \| \cap \| d = v_{\mathcal{A}} \|$ is the set of objects *supporting* the rule.

Example 7.3 Looking again at Tab. 4, some of the rules are, for example:

$Diploma =$ MBA $\land$ $Experience =$ Medium $\Rightarrow$ $Decision =$ Accept
$Experience =$ Low $\land$ $Reference =$ Good $\Rightarrow$ $Decision =$ Reject
$Diploma =$ MSc $\land$ $Experience =$ Medium $\Rightarrow$ $Decision =$ Accept

The first two rules are true in Tab. 4 while the third one is not true in that table.
□

Let us assume that our decision table is consistent. One can observe that by computing (k, d)–relative reducts for $x_k \in U$ it is possible to obtain the decision rules with minimal number of descriptors on their left hand sides among rules true in $\mathcal{A}$. It is enough for any such prime implicant to create the left hand side of the rule as follows: construct a conjunction of all descriptors $a = v$ where a is in prime implicant and v is the value of a on x_k.

Several numerical factors can be associated with a synthesized rule. For example, the support of a decision rule is the number of objects that match the predecessor of the rule. Various frequency-related numerical quantities may be computed from such counts like the *accuracy coefficient* equal to

$$\frac{||| \varphi_{\mathcal{A}} || \cap || d = v_{\mathcal{A}} |||}{||| \varphi_{\mathcal{A}} |||}$$

(see e.g. [108], [207], [13]).

The main challenge in inducing rules from decision tables lies in determining which attributes should be included in the conditional part of the rule. Although we can compute minimal decision rules, this approach results in rules that may contain noise or other peculiarities of the data set. Such detailed rules will be over-fit and will poorly classify unseen cases. More general, i.e. shorter rules should be rather synthesized which are not perfect on known cases (influenced by noise) but can be of high quality on new cases. Several strategies implementing this idea have been implemented. They are based on different measures like boundary region thinning (see e.g. [536], [394]), preserving up to a given threshold the positive region (see e.g. [394]), entropy (see [411], [408]). One can also use reduct approximations, i.e. attribute subsets that in a sense "almost" preserve e.g. the indiscernibility relation. One way of computing approximations is first to compute reducts for some random subsets of the universe of a given decision system and next to select the most stable reducts, i.e. reducts that occur in most of the subsystems. These reducts, called *dynamic reducts*, are usually inconsistent for the original table, but the rules synthesized from them are more tolerant to noise and other abnormalities; they perform better on unseen cases since they cover more general patterns in the data [13], [18]. Another approach is related to searching for patterns almost included in the decision classes combined with decomposition of decision tables into regular domains (see e.g. [235, 238, 244, 246, 247, 522]). One can also search for default rules. For a presentation of generating default rules see [215, 213, 214] and [135] who investigate synthesis of default rules or normalcy rules and some implementations of heuristics that search for such reducts.

One particularly successful method based on the re-sampling approach is called dynamic reducts. It is implemented in the ROSETTA system [276].

For a systematic overview of rule synthesis see e.g. [435], [13], [108], [394].

7.4 Rule Application

When a set of rules have been induced from a decision table containing a set of training examples, they can be inspected to see if they reveal any novel relationships between attributes that are worth pursuing for further research. Furthermore, the rules can be applied to a set of unseen cases in order to estimate their classificatory power.

Several application schemes can be envisioned. Let us consider one of the simplest which has shown to be useful in practice.

1. When a rough set classifier is presented with a new case, the rule set is scanned to find applicable rules, i.e. rules whose predecessors match the case.
2. If no rule is found (i.e. no rule "fires"), the most frequent outcome in the training data is chosen.
3. If more than one rule fires, these may in turn indicate more than one possible outcome.
4. A voting process is then performed among the rules that fire in order to resolve conflicts and to rank the predicted outcomes. A rule casts as many votes in favor of its outcome as its associated support count. The votes from all the rules are then accumulated and divided by the total number of votes cast in order to arrive at a numerical measure of certainty for each outcome. This measure of certainty is not really a probability, but may be interpreted as an approximation to such, if the model is well calibrated.

For a systematic overview of rule application methods see e.g. [421], [13], [108, 109].

8 Rough Sets and Tolerance Relations

We discuss in this section extensions of rough sets based on tolerance relations but we would like to mention that many other generalizations have been studied like abstract approximation spaces [40], [205], [403], (see also Sect. 9); nondeterministic information systems (see e.g. [198], [284], [272], [343], [271]); recently developed extensions of rough set approach to deal with preferential ordering on attributes (criteria) in multi-criteria decision making [105], [103], [104]; an extension based on reflexive relations (as models for object closeness, only) [428]; extensions of rough set methods for incomplete information systems [174], [175]; formal languages approximations [144], [282], [283]; neighborhood systems [182], [183], [184]; extensions of rough sets for distributed systems and multi-agent systems (see e.g. [349], [350], [349], [336]). For discussion of other possible extensions see [336].

Tolerance relations provide an attractive and general tool for studying indiscernibility phenomena. The importance of those phenomena had been noticed by Poincare and Carnap. Studies have led to the emergence of such approaches to indiscernibility in rough set community.

We present only some examples of problems related to an extension of rough sets by using tolerance relations instead of equivalence relations as a model for indiscernibility. More details the reader can find e.g. in [29, 39, 40, 98, 123, 139, 149, 262, 263, 268, 269, 270, 164, 165, 166, 182, 189, 201, 247, 246, 251, 341, 337, 338, 342, 295, 343, 394, 399, 397, 403, 427, 428, 438, 439, 466, 493, 494, 523, 525, 527, 526, 529, 542, 543]. Let us also note that there are many interesting results on relationships between similarity and fuzzy sets (see e.g. [530, 71, 72, 74, 75, 76, 86, 87, 122, 377, 386, 387, 492]). Problems of similarity relations are also related to problems of clustering (see e.g. [62, 512]).

We call a relation $\tau \subseteq X \times U$ a *tolerance relation* on U if (i) τ is *reflexive*: $x\tau x$ for any $x \in U$ (ii) τ is *symmetric*: $x\tau y$ implies $y\tau x$ for any pair x, y of elements of U.

The pair (U, τ) is called a *tolerance space*. It leads to a metric space with the distance function

$$d_\tau(x,y) = \min\{k : \exists_{x_0, x_1, \ldots, x_k} x_0 = x \wedge x_k = y \wedge (x_i \tau x_{i+1} \text{ for } i = 0, 1, \ldots, k-1)\}$$

Sets of the form $\tau(x) = \{y \in U : x\tau y\}$ are called *tolerance sets*.

One can easily generalize the definitions of the lower and upper approximations of sets by substituting tolerance classes for equivalence classes of the indiscernibility relation. We obtain the following formulae for the τ- approximations of a given subset X of the universe U:

$\underline{\tau}X = \{x \in U : \tau(x) \subseteq X\}$ and $\overline{\tau}X = \{x \in U : \tau(x) \cap X \neq \emptyset\}$.

However, one can observe that when we are dealing with tolerances we have a larger class of definable sets than in case of equivalence relations. For example one could take as primitive definable sets the tolerance classes of some iterations of tolerance relations or the equivalence classes of the relation defined from the tolerance relation τ by: $xIND_\tau y$ iff $dom_\tau(x) = dom_\tau(y)$ where $dom_\tau(x) = \cap\{\tau(z) : x \in \tau(z)\}$. Moreover, the presented above definition of the set approximations is not unique. For example in [40] approximations of sets have been defined which are more close in a sense to X than the classical ones. They can be defined as follows:

$$\tau_* X = \{x \in U : \exists y(x\tau y \& \tau(y) \subseteq X)\}$$

and

$$\tau^* X = \{x \in U : \forall y(x\tau y \Rightarrow \tau(y) \cap X \neq \emptyset)\}.$$

One can check that $\underline{\tau}X \subseteq \tau_* X \subseteq X \subseteq \tau^* X \subseteq \overline{\tau}X$

This approximations are closely related to the Brouwerian ortho–complementation (see [40]). One can take for any set X as its ortho-complementation the set $X^\# = \{x \in U : \forall h \in X(\neg(x\tau h))\} \subseteq X^c$ where $X^c = U - X$ and to find formulas (see [40]) expressing the new approximations using this kind of complementation. Let us observe that the condition $\neg(x\tau h)$ inside of the above formula can be interpreted as the discernibility condition for x, h.

Hence in the process of learning of the concept approximations we have more possibilities for approximation definition, approximation tuning and primitive definable sets choosing when we deal with tolerance relations than in case of equivalence relations. However, we pay for this because it is harder from computational point of view to search for relevant approximations in this larger space.

There has been made a great effort to study properties of logical systems based on similarity relations (see e.g. [530, 72, 74, 76, 87, 251, 262, 263, 149, 269, 270, 343, 386, 387, 439, 493, 494, 496, 526]).

There is a great need for algorithmic tools suitable for relevant tolerance relation discovery from data, to tune the parameters of these relations or set approximations to obtain approximations of analyzed concepts of satisfactory quality. Recently, results in this direction have been reported (see e.g. [98, 247, 246, 427, 428, 165]) with promising experimental results for extracting patterns from data. Tolerance relations can be interpreted as graphs and several problems of searching for relevant patterns in data are strongly related to graph problems (see e.g. [247, 246]. These problems are NP-complete or NP-hard however several efficient heuristics have been developed to extract relevant patterns from data. Practitioners will look very much for logical systems helping to infer relevant tolerance relations and this is a challenge for logicians.

Let us recall some previous observations related to concept approximations.

The lower and upper approximations are only examples of the possible approximations. In terminology of machine learning they are approximations of subsets of objects known from training sample. However, when one would like to deal with approximations of subsets of all objects (including also new i.e. unseen so far objects) some techniques have been proposed to construct set approximations suitable for such applications. The best known among them is the technique called the boundary region thinning related to the variable precision rough set approach [536]; another technique is used in tuning of decision rules. For instance, achieving better quality on new objects classification by introducing some degree of inconsistency on training objects. This technique is analogous to the well known techniques for decision tree pruning. The discussed approaches can be characterized in the following way: parameterized approximations of sets are defined and by tuning these parameters better approximations of sets or decision rules are obtained. Some of the above methods can be extended to tune concept approximations defined by tolerance relations. Further research in this direction will certainly lead to new interesting results.

One extension of rough set approach is based on recently developed rough mereology ([328, 330, 331, 332, 333, 330, 334, 335, 336]). The relations to be a part to a degree (discovered from data) are defining tolerance relations (defining so called rough inclusions) used to measure the closeness of approximated concepts. Tolerance relations play an important role in the process of schemes construction defining approximations of target concepts by some primitive ones. Contrary to classical approaches these schemes are "derived" from data by applying some algorithmic methods. The reader can look for more details in the section of the paper on rough mereological approach.

Tolerance relations can be defined from information systems or decision tables. Hence the reduction problems of information necessary to define tolerances relations arise (see e.g. [402, 403, 438, 439, 397]). We will briefly present an idea of this approach. By a *tolerance information system* [403] we understand a triple $\mathcal{A}' = (U, A, \tau)$ where $\mathcal{A}' = (U, A)$ is an information system and τ is a tolerance relation on *information vectors* $\text{Inf}_B(x) = \{(a, a(x)) : a \in B\}$ where $x \in U$, $B \subseteq A$. In particular, a tolerance information system can be realized as a pair $(\mathcal{A}, D)$ where $\mathcal{A} = (U, A)$ is an information system, while $D = (D_B)_{B \subseteq A}$ and $D_B \subseteq INF(B) \times INF(B)$ is a relation, called *the discernibility relation*, satisfying the following conditions:

(i) $INF(B) \times INF(B) - D_B$ is a tolerance relation;
(ii) $((u - v) \cup (v - u)) \subseteq (u_0 - v_0) \cup (v_0 - u_0))$ & $uD_Bv \to u_0D_Bv_0$ for any $u, v, u_0, v_0 \in INF(B)$ i.e. D_B is monotonic with respect to the discernibility property;
(iii) $non(uD_Cv)$ implies $non(u|B\ D_B\ v|B)$ for any $B \subseteq C$ and $u, v \in INF(C)$

where $INF(B) = \{\text{Inf}_B(x) : x \in U\}$ and if $u \in INF(C)$ and $B \subseteq C \subseteq A$ then $u|B = \{(a, w) \in u : a \in B\}$ i.e. $u|B$ is the restriction of u to B. A (B, D_B)-tolerance τ_B is defined by

$$y\tau_B x \quad \text{iff} \quad non(\text{Inf}_B(x) D_B \text{Inf}_B(y)).$$

A (B, D_B)-*tolerance function* $I[B, D_B] : U \longrightarrow \mathcal{P}(U)$ is defined by $I[B, D_B](x) = \tau_B(x)$ for any $x \in U$.

The set $I[B, D_B](x)$ is called *the tolerance set of* x. The relation $INF(B) \times INF(B) - D_B$ expresses similarity of objects in terms of accessible information about them. The set $RED(\mathcal{A}, D)$ is defined by

$$\{B \subseteq A : I[A, D_A] = I[B, D_B] \quad \text{and} \quad I[A, D_A] \neq I[C, D_C] \quad \text{for any} \quad C \subset B\}$$

Elements of $RED(\mathcal{A}, D)$ are called *tolerance reducts of* $(\mathcal{A}, D)$ (or, *tolerance reducts*, in short). It follows from the definition that the tolerance reducts are minimal attribute sets preserving (A, D_A) - tolerance function. The tolerance reducts of $(\mathcal{A}, D)$ can be constructed in an analogous way as reducts of information systems. The problem of minimal tolerance reduct computing is NP-hard [403]. However again some efficient heuristics for computing semi-minimal reducts can be constructed. The method can be extended for computing so called relative tolerance reducts and other objects [439]. It is possible to apply Boolean reasoning to the object set reduction in tolerance information systems. This is based on the notion of an absorbent [466]. A subset $Y \subseteq X$ is an *absorbent* for a tolerance relation τ (τ-*absorbent*, in short) if and only if for each $x \in X$ there exists $y \in Y$ such that $x\tau y$. The problem of minimal absorbent construction for a given tolerance information system can be easily transformed to the problem of minimal prime implicant finding for a Boolean function corresponding to this system. Hence, again the problem of minimal absorbent construction is NP-hard so efficient heuristics have been constructed to find sub-minimal absorbents for tolerance information systems.

The presented methods of information reduction in tolerance information systems create some step towards practical applications. However, more research in this direction should still be done.

Further progress in investigations on tolerance information systems will have impact on applications of rough sets in many areas like granular computing, case based reasoning, process control, scaling continuous decisions etc.

We have discussed in this section some problems related to rough set approach based on tolerance approach. We have pointed out some interesting problems to be investigated.

9 Algebraic and Logical Aspects of Rough Sets

One of the basic algebraic problem related to rough sets can be characterized as follows.

Let Σ be a class of information systems, Γ – a class of algebraic structures and e – a mapping form Σ into Γ. We say that Σ is e-*dense* in Γ if for any algebra Alg from Γ there exists and information system $\mathcal{A}$ in Σ such that Alg is isomorphic to $e(\mathcal{A})$ (or a sub-algebra of $e(\mathcal{A})$). If Σ is e-*dense* in Γ then we say that the representation theorem for Σ (relatively to e and Γ) holds.

From this definition it follows that to formulate the representation theorem first one should choose the mapping e and the class Γ. They should be chosen as "natural" for the considered class of information systems. The mapping e endows the information systems with a natural algebraic structure. We will show some examples of natural algebraic structures for information systems to give the reader some flavor of the research going on. The reader interested in study of algebraic characterizations of rough sets should refer to [270, 269, 278, 79] and papers cited in these articles.

Let us recall that a definable set in an information system $\mathcal{A}$ is any union of discernibility classes of $IND(A)$. The first observation is that the set $DE(\mathcal{A})$ of all definable sets in $\mathcal{A}$ endowed with set theoretical operations: union, intersection and complementation forms a Boolean algebra with the empty set as 0 and the universe U as 1. The equivalence classes of the indiscernibility relation are the only atoms of this Boolean algebra. Let us note that definability of sets in incomplete information systems (i.e. attributes are partial functions on objects) has also been investigated [33].

For any information system $\mathcal{A} = (U, A)$ one can define the family $RS(\mathcal{A})$ of *rough sets* i.e. pairs $(\underline{A}X, \overline{A}X)$ where $X \subseteq U$. Hence two questions arise. How to characterize the set of all rough sets in a given information system? What are the "natural" algebraic operations on rough sets?

To answer the first question let us assign to any rough set $(\underline{A}X, \overline{A}X)$ the pair $(\underline{A}X, BN_A X)$. One can easily see that the boundary region $BN_A X = \overline{A}X - \underline{A}X$ does not contain any singleton discernibility class (i.e. a class with one object only). Let us consider the set $\mathbf{Z}$ of all pairs (Y, Z) from $\mathcal{P}(U) \times \mathcal{P}(U)$ such that for some $X \subseteq U$ we have $Y = \underline{A}X$ and $Z = BN_A X$. One can observe that the set **Z** can characterized as the set of all pairs of definable sets in $\mathcal{A}$ which are disjoint and the second element of any pair does not contain any singleton indiscernibility class of $IND(A)$.

From the point of view of algebraic operations one can choose another representation of rough sets. Let us recall that the lower approximation of a given set X is the set of all objects which can be with certainty classified as belonging to X on the basis of knowledge encoded in A and the set theoretical complement of the upper approximation of X is the set of all objects in U which can be with certainty rejected as belonging to X on the basis of knowledge encoded in A. Hence to any rough set $(\underline{A}X, \overline{A}X)$ in $\mathcal{A}$ one can assign a pair $(\underline{A}X, U - \overline{A}X)$. It happens that one can define some "natural" operations on such pairs of sets. First one easily will guess that $(\emptyset, U)$ corresponds to the smallest rough set and $(U, \emptyset)$ corresponds to the largest rough set. To define operations on such representations of rough sets let us imagine that we have two experts able to deliver answers about objects (observed through "glasses" of A) if they belong to some concepts i.e. subsets of U. Can we now define an approximate fusion of this concepts? There are several possibilities. We can treat as the lower approximation of the concept (representing concepts of two agents) the intersection of the lower approximations of two concepts using a rule: if both experts classify with certainty the observed object to their concepts we will treat this object as belonging with certainty to a concept being a fusion of those two. We will reject the observed object as belonging to the upper approximation of the concept being the fusion of two concepts if at least one of the experts will reject it with certainty as belonging to the corresponding concept. Hence we obtain the following definition of the algebraic operation on considered representations of rough sets: $(X_1, X_2) \wedge (Y_1, Y_2) = (X_1 \cap Y_1, X_2 \cup Y_2)$. The reader can immediately find interpretation for another operation: $(X_1, X_2) \vee (Y_1, Y_2) = (X_1 \cup Y_1, X_2 \cap Y_2)$. Let us consider one more example. How we can built a model for the complementation of a concept observed by an expert on the basis of his judgments? We again have several possibilities. The first model is the following: if the expert is classifying with certainty an observed object as belonging to a concept then we are rejecting it with certainty as belonging to the concept but if the expert is rejecting with certainty an observed object as belonging to a concept we are classifying it with certainty to the concept. Hence we have the following definition of one argument negation operation $\sim$: $\sim (X_1, X_2) = (X_2, X_1)$. However, now the reader will observe that there are some other possibilities to build a model for the complement of the concept to which the expert is referring e.g. by assuming $\neg(X_1, X_2) = (U - X_1, X_1)$ or $\div(X_1, X_2) = (X_2, U - X_2)$. The defined operations are not random operations. We are now very close (still the operation corresponding to implication should be defined properly!) to examples of known algebras, like Nelson or Heyting algebras, intensively studied in connection with different logical systems. The reader can find formal analysis of relationships of rough sets with Nelson, Heyting, Lukasiewicz, Post or double Stone algebras e.g. in [278] and in particular, the representation theorems for rough sets in different classes of algebras. Let us also note that the properties of defined negation operations are showing that they correspond to well known negations studied in logic: strong (constructive) negation or weak (intuitionistic) negation.

Algebraic structures relevant for construction of generalized approximation spaces have been also investigated e.g. in [40]. In [40] it is shown that the general structure of po-sets augmented with two sub-posets consisting of "inner definable" elements and "outer definable" elements is sufficient to define inner and outer approximation maps producing the best approximation from the bottom (lower approximation) and from the top (upper approximation) of any element with respect to the poset. By imposing De Morgan law it is received a duality between inner approximation space and outer approximation space. This class of De Morgan structures includes degenerate and quasi Brouwer-Zadeh posets, which are generalizations of topological spaces and preclusivity spaces, respectively. In the former case the approximable concepts are described as points of posets whereas in the later case the approximable concepts are described by subsets of a given universe. The classical, Pawlak approach coincides with the class of all clopen topologies or with the class of all preclusivity spaces induced from equivalence relations.

There is another research direction based on information systems [270]. The aim is to study information algebras and information logics corresponding to information systems. First, so called information frames are defined. They are relational structures consisting parameterized families of binary relations over the universe of objects. These relations are e.g indiscernibility relations corresponding to different subsets of attributes. Many other interesting frames can be found e.g. in [270]. If a frame is extended by adding e.g. set theoretical operations new algebraic structure called (concrete) information algebra is received. The information algebras in the abstract form are Boolean algebras augmented with some parameterized families of operations reflecting relevant properties of frames and in consequence of information systems. The main problems studied are related to the representation theorems for information algebras as well as to construction and properties of logical systems with semantics defined by information algebras [270].

An attempt to define rough algebras derived from rough equality is presented e.g. in [10].

For more readings on algebraic aspects of (generalized) approximation spaces the reader is referred to [106], [127], [357], [362], [363], [361], [128], [47], [495], [509], [39], [507].

There is a number of results on logics reflecting rough set aspects (for the bibliography see [338], [270]). Among these logics there are propositional as well as predicate logics. They have some new connectives (usually modal ones) reflecting different aspects of approximations. On semantical level they are allowing to express e.g. how the indiscernibility classes (or tolerance classes) interact with interpretations of formulas in a given model M. For example, in case of necessity connective the meaning $(\Box \alpha)_M$ of the formula α in the model M is the lower approximation of α_M, in case of possibility connective $(\langle\rangle \alpha)_M$ it is the upper approximation of α_M, i.e. the interpretation of α in M. Many other connectives have been introduced and logical systems with these connectives have been characterized. For example in predicate logic one can consider also rough quantifiers [168]. The results related to the completeness of axiomatization, decidability as well as expressibility of these logical systems are typical results. More information on rough logic the reader can find in [270], in particular in [11] a review of predicate rough logic is presented. Many results on information logics, in particular characterization theorems, can be found e.g. in [496].

Some relationships of rough algebras with many-valued logics have been shown e.g. in [10], [261]. For example in [10] soundness and completeness of 3-valued Łukasiewicz logic with respect to rough semantics has been proven. The rough semantics has been defined by rough algebras [10] (i.e. a special kind of topological quasi-Boolean algebra) [508]. Relationships of rough sets with 4-valued logic are presented in [261] and with quantum logic in [40].

We would like to mention several attempts to use rough set logics for reasoning about knowledge (see e.g. [364], [360], [365], [371], [372], [373], [297]).

Properties of dependencies in information systems have been studied by many researchers see e.g. [256], [257], [32], [82], [129], [258], [369], [370], [259], [505], [367], [368].

Finally we would like to mention a research direction related to so called rough mereological approach for approximate synthesis of objects satisfying a given specification to a satisfactory degree. We will discuss some aspects of this approach in Part II of this tutorial. Let us note here that one of the perspective for applied logic is to look for algorithmic methods of extracting logical structures from data e.g. relational structures corresponding to relevant feature extraction [404], default rules (approximate decision rules see e.g. [215]), connectives for uncertainty coefficients propagation and schemes of approximate reasoning. This is very much related to rough mereological approach and is crucial for many applications, in particular in knowledge discovery and data mining [91], calculi on information granules and computing with words [531], [532].

For more readings on logical aspects of rough sets the reader is referred to [270], [65], [11], [232], [79], [8], [278], [336], [496], [78], [148], [340], [304], [64], [193], [231], [332], [495],[230], [401], [493], [494], [359], [266], [267], [265], [357], [89], [356], [366], [88], [262], [264], [272].

10 Relationships with Other Approaches

Some interesting results on relationships of rough sets with other approaches to reasoning under uncertainty have been reported. In this section we point out on applications of rough sets in decision analysis, data mining and knowledge discovery, we present a comparison of some experimental results received by applying some machine learning techniques and rough set methods, we discuss some relationships of rough sets and fuzzy sets, we present some consequences of relationships of rough set approach with the Dempster-Shafer theory of evidence and finally we overview some hybrid methods and systems.

There have been studied also relationships of rough sets with other approaches e.g. with mathematical morphology (see e.g. [334, 323, 324, 325, 327]), statistical and probabilistic methods (see e.g. [286], [519], [440], [167], [315],[321], [528], [80], [81], [83]), concept analysis (see e.g. [100], [141], [277], [279]).

10.1 Decision analysis

Decision analysis is a discipline providing various tools for modeling decision situation in view of explaining them or prescribing actions increasing the coherence between the possibilities offered by the situation, and goals and value systems of the agents involved. Mathematical decision analysis consists in building a functional or a relational model. The functional model has been extensively used within the framework of multi–attribute utility theory. The relational is known e.g. in the form of an out ranking relation or a fuzzy relation (see [302], [378], [379], [380], [381], [9]).

Both modeling and explanation/prescription stages are also crucial operations et elaboration of a systematic and rational approach to modeling and solving complex decision problems [2], [302].

Rough set approach proved to be a useful tool for solving problems in decision analysis in particular in the analysis of multi–criteria decision problems related to:

(i) multi–criteria sorting problems;
(ii) multi–criteria, multi–sorting problems;
(iii) multi–criteria description of objects.

The case (i) can be described as decision problems related to the decision table with one decision. One can expect the following results from the rough set analysis of decision table: (i) evaluation of importance of particular attributes; (ii) construction of minimal subsets of independent attributes which can not be eliminated without disturbing the ability of approximating the sorting decisions; (iii) computing the relevant attributes i.e. core of the attribute set; (iv) elimination of redundant attributes from the decision table; (v) generation of sorting rules from the reduced decision table; they involve the relevant attributes only and explain a decision policy of the agent (decision maker or expert) in particular how to solve conflicts between decision rules voting for different decision when new objects are matched by these rules (see e.g. [421], see also 7.4). The multi–criteria sorting problems represents the largest class of decision problems to which the rough set approach has been successfully used. The applications concern many domains (see Sect. 11).

In the case (ii) we deal with decision tables with more than one decision (received from different agents). Using rough set methods one can measure the degree of consistency of agents, detect and explain discordant and concordant parts of agent's decision policies, evaluate the degree of conflict among the agents, and construct the preference models (sorting rules) expressed in common terms (conditional attributes) in order to facilitate a mutual understanding of the agents [302].

In the case (iii) the primary objective is to describe a decision situation. The rough set approach to the decision situation description is especially well suited when minimal descriptions in terms of attributes is of primary concern. Another important problem analyzed by rough set methods is conflict analysis [295]. If agents are not explicitly represented in the information system one can look for discovery of dependencies among conditional attributes interpreted as consequences of decisions represented by objects. Again, rough set methodology can be used to solve this type of problems [302].

For more readings on rough set approach to decision analysis see e.g. [421], [424], [425], [423], [293], [296], [301], [302].

Let us also note that some extensions of rough set approach have been proposed for dealing with preferential ordering of attributes (criteria) (see e.g. [103]).

10.2 Rough Sets and Data Mining

Rough set theory has proved to be useful in Data Mining and Knowledge Discovery. It constitutes a sound basis for data mining applications. The theory offers mathematical tools to discover hidden patterns in data. It identifies partial or total dependencies (i.e. cause–effect relations) in data bases, eliminates redundant data, gives approach to null values, missing data, dynamic data and others. The methods of data mining in very large data bases using rough sets have been developed.

There are some important steps in the synthesis of approximations of concepts related to the construction of (i) relevant primitive concepts from which approximations of more complex concepts will be constructed; (ii) (closeness) similarity measures between concepts; (iii) operations for construction of more complex concepts from primitive ones.

These problems can be solved by combining the classical rough set approach and recent extensions of rough set theory. Methods for solving problems arising in the realization of these steps are crucial for knowledge discovery and data mining (KDD) [91] as well.

There have been done in last years a substantial progress in developing rough set methods for data mining and knowledge discovery (see the cited in Sect. 11 cases and e.g. [238], [237], [246], [247], [235], [539], [46], [160], [161], [318], [320],[214], [337], [338], [239], [162], [522], [410], [435], [276], [189], [125], [66], [190], [124], [209], [147], [163], [243], [248], [249], [385], [241], [242], [188]), [481], [245], [52], [159], [158], [244], [399], [409], [392], [406], [240], [479], [480], [537], [394], [538], [534]).

In particular new methods for extracting patterns from data (see e.g. [165], [247], [214], [235]), decomposition of decision tables (see e.g. [245], [409], [249], [247], [410]) as well as a new methodology for data mining in distributed and multi-agent systems (see e.g. [336] and Part II of this article) have been developed.

In Sect. 11 there are reported many successful case studies of data mining and knowledge discovery based on rough set methods and the reader can find there more references to papers on data mining.

10.3 Comparison with Some Results in Machine Learning

Recently several comparison studies have been reported showing that the results received by using software systems based on rough set methods are fully comparable with those obtained by using other systems (see e.g. [108], [115], [109], [13], [238], [435], [161]) for object classifying. Let us recall one of a method recently reported in [238]. Table 8 presents the results of experiments obtained by using the methods based on rough sets and Boolean reasoning and the methods reported in [95]. One can compare those results with regard to the classification qualities. MD and MD-G heuristics are developed using rough set methods and Boolean reasoning.

Several papers are comparing the results received by applying statistical methods and comparing them with the results received by rough set methods (see e.g. [233], [34], [498]). In the future more research should be done to recognize proper areas for application of hybrid methods based on rough sets and statistical methods.

10.4 Rough Sets and Fuzzy Sets

Rough set theory and fuzzy set theory are complementary. It is natural to combine the two models of uncertainty (vagueness for fuzzy sets and coarseness for rough sets) in order to get a more accurate account of imperfect information [71]. The results concerning relationships between rough sets and fuzzy sets are presented e.g. in [71], [76], [22], [303], [36], [37], [38], [41], [42], [43], [185], [187], [206], [288], [292], [294], [506], [307], [344], [345], [388], [189], [524].

Rough set methods provide approximate descriptions of concepts and they can be used to construct approximate description of fuzzy concepts as well. This is very important for more compressed representation of concepts, rules, patterns in KDD because using fuzzy concepts one can describe these items in a more compact way. Moreover, these descriptions can be more suitable for communication with human being.

In rough set theory approximations of sets are defined relatively to a given background knowledge represented by data tables (information systems, decision tables) with the set of attributes A.

The rough membership function μ_X^B where $X \subseteq U$ and $B \subseteq A$ can be used to define approximations and the boundary region of a set, as shown below:

$$\underline{B}(X) = \{x \in U : \mu_X^B(x) = 1\},$$
$$\overline{B}(X) = \{x \in U : \mu_X^B(x) > 0\},$$
$$BN_B(X) = \{x \in U : 0 < \mu_X^B(x) < 1\}.$$

The rough membership function has the following properties [303]:

a) $\mu_X^B(x) = 1$ iff $x \in \underline{B}(X)$,
b) $\mu_X^B(x) = 0$ iff $x \in -\overline{B}(X)$,
c) $0 < \mu_X^B(x) < 1$ iff $x \in BN_B(X)$,

d) If $IND(B) = \{(x,x) : x \in U\}$, then $\mu_X^B(x)$ is the characteristic function of X,
e) If $xIND(B)y$, then $\mu_X^B(x) = \mu_X^B(y)$,
f) $\mu_{U-X}^B(x) = 1 - \mu_X^B(x)$ for any $x \in U$,
g) $\mu_{X \cup Y}^B(x) \geq \max(\mu_X^B(x), \mu_Y^B(x))$ for any $x \in U$,
h) $\mu_{X \cap Y}^B(x) \leq \min(\mu_X^B(x), \mu_Y^B(x))$ for any $x \in U$,
i) If **X** is a family of pairwise disjoint sets of U, then $\mu_{\cup \mathbf{X}}^B(x) = \sum_{X \in \mathbf{X}} \mu_X^B(x)$ for any $x \in U$,

The above properties show clearly the difference between fuzzy and rough memberships. In particular properties g) and h) show that the rough membership can be regarded formally as a generalization of fuzzy membership. Let us observe that the "rough membership", in contrast to the "fuzzy membership", has probabilistic flavor.

It has been shown [303] that the formulae received from inequalities g) and h) by changing them into equalities are not true in general. This important observation is a simple consequence of the properties of set approximation: the calculus of set approximations in rough set theory is intensional. Namely, it is impossible to find a function independent from a given background knowledge that will allow to compute the values of the rough membership function for the intersection of sets (or union of sets) having only the values of the membership function for the argument sets. This property is showing some more differences between rough membership functions and fuzzy membership functions.

Rough set and fuzzy set approaches create many possibilities for hybridization. The number of reported results in this direction is continuously increasing. Let us mention some of them.

Combining rough sets and fuzzy sets allows to obtain rough approximations of fuzzy sets as well as approximations of sets by means of fuzzy similarity relations [DP4]. Let us consider the second case. Fuzzy rough sets (see e.g. [76]) are defined by membership function on the universe of objects U by

$$\mu_{\overline{S}(X)}(x) = sup_{\omega \in X} \mu_S(x, \omega)$$

$$\mu_{\underline{S}(X)}(x) = inf_{\omega \notin X}(1 - \mu_S(x, \omega))$$

where S is a fuzzy indistinguishability relation (fuzzy similarity relation) [76] and $x \in U$.

In this case we consider the fuzzy similarity relations instead of the (crisp) indiscernibility relations. In case of crisp indiscernibility (i.e. equivalence relation) relation we obtain

$$\mu_{\overline{S}(X)}(x) = 1 \text{ iff } x \in \overline{S}(X);$$

$$\mu_{\underline{S}(X)}(x) = 1 \text{ iff } x \in \underline{S}(X);$$

where $\overline{S}(X)$ and $\underline{S}(X)$ denote the upper and the lower approximations of X with respect to the indiscernibility relation S.

There are other interesting relationships of rough set with fuzzy sets (see e.g. [69], [70], [76]). For example, Ruspini's entailment can be expalined in terms of rough deduction. In the rough set approach the indistinguishability notion is basic, while in Ruspini's fuzzy logic, it is the idea of closeness. It has been shown [71] that by introducing lower and upper approximations of fuzzy sets we come close to Caianiello's C-calculus [71]. Fuzzy rough sets are allowing to put together fuzzy sets and modal logic (see e.g. graded extensions of S5 system by Nakamura [76]). Rough aspects of fuzzy sets are also discussed in [12].

Rough–fuzzy hybridization methods give a tool for solving problems in KDD. In the sequel we will describe examples some tools and some research problems related to this topic.

The classical rough set approach is based on crisp sets. A generalization of rough set approach for handling different types of uncertainty has been proposed e.g. in [422]. It has been observed that the synthesized (extracted) features e.g. cuts [248], hyperplanes [233] can be tuned into more relevant features for classification when they are substituted by fuzzy cuts and fuzzy hyperplanes, respectively. This is related to the following property: points which are close to a cut or to a hyperplane can be hardly classified to half spaces corresponding to this cut or hyperplane because of e.g. noise influencing the position of points. The same idea can be extended to decision rules or pattern description [247]. Further investigations of techniques transforming crisp concepts (features) into fuzzy ones will certainly show more interesting results.

Let us mention another source for rough–fuzzy hybridization. These approaches can be characterized in the following way: parameterized approximations of sets are defined and by tuning these parameters approximations of fuzzy sets are received. Recently proposed shadowed sets for fuzzy sets [306] use this technique. Fuzzy membership function is represented by a family of parameterized functions with the same domain but only with three possible values. These functions correspond to the parameterized lower, upper and boundary region by a threshold determining the size of shadowed region. The size of this region can be tuned up in the process of learning.

One of the main problems in soft computing is to find methods allowing to measure the closeness of concept extensions. Rough set methods can also be used to measure the closeness of (fuzzy) concepts.

In classical rough set approach sets are represented by definable sets, i.e. unions of indiscernibility classes. Extension of this approach have been proposed by several researchers (for references see e.g. [164], [247]). Instead of taking an equivalence relations as the indiscernibility relations the tolerance relation (or even more arbitrary binary relation [105]) is considered. This leads to a richer family of definable sets but it is harder (from computational complexity point of view) to construct of concept approximations of high quality. Searching problems for optimal tolerance relations are NP-complete or NP-hard [247]. However, it has been possible to develop efficient heuristics searching for relevant tolerance relation(s) that allow to extract interesting patterns in data (see e.g. [164], [247]). The reported results are promising. A successful realization of this approach is possible because in the rough set approach relevant tolerance relations determining patterns can be extracted from the background knowledge represented in the form of data tables. The extracted patterns can be further fuzzyfied and applied in construction of concept approximation [235].

Rough set methods can be used to define fuzzy concepts approximately. In this case one should look for relevant α-cuts of the fuzzy set and treating these cuts as decision classes to find their approximations with respect to known conditional features. Once again problem of choosing relevant cuts is analogous to the problem of relevant feature extraction. From computational complexity point of view it is a hard problem and can be solved approximately by discovery of learning strategies. One can observe that the relevant cuts should be "well" approximated (i.e. new objects with high chance should be properly classified to them) as well as they should give together "good" approximation of the target fuzzy set.

The most general case is related to methods for approximate description of fuzzy concepts by fuzzy concepts. One can look at this issue as the searching problem for an approximate calculus on approximate concepts (information granules) [531], [532]. This calculus should allow to construct approximate descriptions of fuzzy concepts from approximate descriptions of known ones. One possible approach to solve this problem is to use fuzzy set methods based on t-norms and co-norms to define closeness of fuzzy concepts and to perform fusion of fuzzy concepts [68]. In practical applications there is a need to develop methods returning approximations of target concepts satisfying to a satisfactory degree given specifications (constraints) from approximations of some primitive (known) concepts. An approach to solve this problem has been recently proposed as rough mereology (see e.g. [332], [398], [336]). In this approach rules for propagation of uncertainty coefficients have to be learned form the available background knowledge represented by data tables. Another interesting property of this approach is that the construction of schemes for approximate description of the target concepts should be stable. This means that "small" changes in approximation quality of primitive concepts should give sufficiently "small" changes in approximation quality of constructed approximation of the target concepts. In [247] there are mentioned possible applications of this approach e.g. to the decomposition of large data tables.

Rough set approach combined with rough mereology can be treated as an inference engine for computing with words and granular computing [531], [532]. For example, the construction of satisfactory target fuzzy concept approximations from approximations of the input (primitive) fuzzy concepts can be realized in the following stages:

- first the fuzzy primitive (input) and the target (output) concept are represented by relevant families of cuts;
- next by using rough set methods the appropriate approximations of cuts are constructed in terms of available (conditions) measurable features (attributes) related to concepts;
- the approximations of input cuts obtained in stage 2 are used to construct schemes defining to a satisfactory degree the approximations of output cuts

from approximated input cuts (and other sources of background knowledge) [332], [398], [336];
- the constructed family of schemes represents satisfactory approximation of the target concept by the input concepts; (in this step more compact descriptions of the constructed family of schemes can be created, if needed).

Progress in this direction seems to be crucial for further developments in soft computing and KDD.

10.5 Rough Sets and the Dempster-Shafer Theory of Evidence

We only present one example of applications for the decision rule synthesis implied by the relationships between rough set methods and Dempster-Shafer's theory of evidence [391]. More details on the relationships between rough sets and Dempster-Shafer's theory the reader can find in [395]. In particular an interpretation of the Dempster-Shafer rule of combination by a simple operation on decision tables can be found in [395]. Some other aspects of relationships of rough sets and evidence theory are discussed in [518], [517], [525], [192], [191], [143].

In [395] it has been shown that one can compute a basic probability assignment (bpa) $m_\mathcal{A}$ for any decision table $\mathcal{A}$ assuming

$$m_\mathcal{A}(\emptyset) = 0 \quad \text{and} \quad m_\mathcal{A}(\theta) = \frac{|\{x \in U : \partial_\mathcal{A}(x) = \theta\}|}{|U|}$$

where $\emptyset \neq \theta \subseteq \Theta_\mathcal{A} = \{i : d(x) = i \text{ for some } x \in U\}$.
Hence some relationships between belief functions $Bel_\mathcal{A}$ and $Pl_\mathcal{A}$ related to the decision table $\mathcal{A}$ can be shown [395]:

$$Bel_\mathcal{A}(\theta) = \frac{|\underline{\mathcal{A}} \bigcup_{i \in \theta} X_i|}{|U|} \quad \text{and} \quad Pl_\mathcal{A}(\theta) = \frac{|\overline{\mathcal{A}} \bigcup_{i \in \theta} X_i|}{|U|}$$

for any $\theta \subseteq \Theta_\mathcal{A}$.

The belief functions related to decision tables can be applied to generate strong approximate decision rules. One of the possible approach is to search for solutions of the following problem:

APPROXIMATION PROBLEM (AP)
INPUT: A decision table $\mathcal{A} = (U, A \cup \{d\})$, k – positive integer and rational numbers $\varepsilon, tr \in (0, 1]$.
OUTPUT: Minimal (with respect to the inclusion) sets $B \subseteq A$ and $\theta \subseteq \Theta_\mathcal{A}$ of cardinality at most k satisfying the following conditions:
  (i) $|Pl_{\mathcal{A}|B}(\theta) - Bel_{\mathcal{A}|B}(\theta)| < \varepsilon$
  (ii) $Bel_{\mathcal{A}|B}(\theta) > tr$.
where $\mathcal{A}|B = (U, B \cup \{d\})$.

The above conditions (i) and (ii) are equivalent to

$$|\overline{B}\bigcup_{i\in\theta} X_i - \underline{B}\bigcup_{i\in\theta} X_i| < \varepsilon |U| \quad \text{and} \quad |\underline{B}\bigcup_{i\in\theta} X_i| > tr|U|$$

respectively. Hence we are looking for "small" sets B and θ such that the boundary region (with respect to B) corresponding to $\cup_{i\in\theta} X_i$ is "small" (less than $\varepsilon|U|$) and the lower approximation of $\cup_{i\in\theta} X_i$ is "sufficiently large" (greater than $tr|U|$) so one can expect that the rules for this region will be strong. The solution for the above problem can be obtained by developing efficient heuristics.

10.6 Hybrid Methods and Systems

It is an experience of soft computing community that hybrid systems combining different soft computing techniques into one system can often improve the quality of the constructed system. This has also been claimed in case of rough set methods that combined with neural networks, genetic algorithms, statistical inference tools or Petri nets may give better solutions. A number of papers on hybrid systems showing the results which bear out this claim have been published. To be specific: adding statistical tools can improve the quality of decision rules induced by rough set methods (see e.g. [28]). Rough set based data reduction can be very useful in preprocessing of data input to neural networks. Several other methods for hybridization of rough sets and neural networks have been developed (see e.g. in [132], [133], [134], [453], [210], [456], [457], [462]). Decision algorithms synthesized by rough set methods can be used in designing neural networks (see e.g. [248], [250], [412], [460], [461], [462]). Rough set ideas can lead to new models of neurons (see e.g. [195], [196], [197], [462]). Optimization heuristics based on evolutionary programming can efficiently generate rough set constructs like reducts, patterns in data, decision rules (see e.g. [521], [249], [522]). Rough set methods can be useful in specifying concurrent systems from which corresponding Petri nets can be automatically generated (see e.g. [405], [406], [444]). Rough sets combined with fuzzy sets and Petri nets give an efficient method for designing clock information systems (see e.g. [311], [312]). Rough set approach to mathematical morphology leads to a generalization called analytical morphology ([329], [396], [334]), mathematical morphology of rough sets ([323], [324], [324], [327]) as well as to an idea of approximate compression of data ([326], [327]).

Moreover hybridization of rough set methods with classical methods like principal component analysis, Bayesian methods, 2D FFT (see e.g. [449], [448], [451], [452]) or wavelets (see e.g. [513]) leads to classifiers of better quality .

11 Applications and Case Studies

There have been developed different software systems based on rough set methods (see e.g. [338], [473] and Sect. 12). There are numerous areas of successful applications of rough set software systems. Many interesting case studies are reported. Let us mention some of them:

- **MEDICINE:**
  - Treatment of duodenal ulcer by HSV ([300], [93], [94], [426], [420]);
  - Analysis of data from peritoneal lavage in acute pancreatitis ([413], [414]);
  - Supporting of therapeutic decisions ([436]);
  - Knowledge acquisition in nursing ([30], [514], [109]);
  - Diagnosis of pneumonia patients ([313]);
  - Medical databases (e.g. headache, meningitis, CVD) analysis ([476], [478], [479], [491], [481], [482], [483], [484], [485], [489], [486], [487], [467], [468], [469], [470], [471], [472]);
  - Image analysis for medical applications ([219], [132], [134];)
  - Surgical wound infection ([138]);
  - Classification of histological pictures ([132]);
  - Preterm birth prediction ([515], [116], [516], [110], [111], [109]);
  - Medical decision-making on board space station Freedom (NASA Johnson Space Center) ([109]);
  - Verification of indications for treatment of urinary stones by extra-corporeal shock wave lithotripsy (ESWL) ([417]);
  - Analysis of factors affecting the occurrence of breast cancer among women treated in US military facilities (reported by W. Ziarko);
  - Analysis of factors affecting the differential diagnosis between viral and bacterial meningitis ([490], [539]);
  - Developing an emergency room for diagnostic check list – A case study of appendicitis([383]);
  - Analysis of medical experience with urolithiasis patients treated by extracorporeal shock wave lithotripsy ([415]);
  - Diagnosing in progressive encephalopathy ([281], [502], [501]);
  - Automatic detection of speech disorders ([61]);
  - Rough set-based filtration of sound applicable to hearing prostheses ([57];
  - Classification of tooth surfaces ([159], the EUFIT'96 competition);
  - Discovery of attribute dependencies in experience with multiple injured patients ([418]);
  - Modeling cardiac patient set residuals ([273]);
  - Multistage analysis of therapeutic experience with acute pancreatitis ([416]);
  - Breast cancer detection using electro-potentials ([452]);
  - Analysis of medical data of patients with suspected acute appendicitis ([34]);
  - Attribute reduction in a database for hepatic diseases ([464]);
  - EEG signal analysis ([513]);
- **ECONOMICS, FINANCE AND BUSINESS:**
  - Evaluation of bankruptcy risk ([430], [429], [104]);
  - Company evaluation ([222]);
  - Bank credit policy ([222]);
  - Prediction of behavior of credit card holders ([510]);

- Drafting and advertising budget of a company ([222]);
- Customer behavior patterns ([320], [539]);
- Response modeling in database marketing ([498]);
- Analysis of factors affecting customer's income level ([539] also reported by Tu Bao Ho);
- Analysis of factors affecting stock price fluctuation ([102]);
- Discovery of strong predictive rules for stock market ([540], [17]);
- Purchase prediction in database marketing ([497]);
- Modeling customer retention ([163], [84]);
- Temporal patterns ([158]);
- Analysis of business databases ([316], [162], [318]);
- Rupture prediction in a highly automated production system ([456]);

– **ENVIRONMENTAL CASES:**
- Analysis of a large multi-species toxicity database ([140]);
- Drawing premonitory factors for earthquakes by emphasizing gas geochemistry ([465]);
- Control conditions on a polder ([376]);
- Environmental protection ([107], [109]);
- Global warming: influence of different variables on the earth global temperature ([112]);
- Global temperature stability ([117], [109]);
- Programming water supply systems ([5], [382], [105]);
- Predicting water demands in Regina ([6], [5]);
- Prediction of oceanic upwelling off the Mauretanian cost using sea surface temperature images, and real and model meteorological data (reported by I. Duentsch);
- Prediction of slope-failure danger level from cases ([99]);

– **ENGINEERING:**
- **Control:** The design and implementation of rough and rough–fuzzy controllers ([217], [541], [535], [218], [463], [220], [223], [322], [225], [226], [299], [227]), [499], [310], [221]);
- **Signal and image analysis:**
    * Noise and distortion reduction in digital audio signal ([52], [53], [54], [57]) [55]);
    * Filtration and coding of audio ([59]);
    * Audio signal enhancement ([60]);
    * Recognition of musical sounds ([153]);
    * Detection and interpolation of impulsive distortions in old audio recordings ([50], [58]);
    * Subjective assessment of sound quality ([151], [152]);
    * Assessment of concert hall acoustics ([154]);
    * Classification of musical timbres and phrases ([150], [155], [156];
    * Mining musical databases ([157]);
    * Image analysis ([219], [132], [134]);

            * Converting a continuous tone image into a half-tone image using error diffusion and rough set methods ([459]);
            * Texture analysis ([250], [447], [449], [458], [451], [454]);
            * Voice recognition ([19], [56], [51], [61], [55]);
            * Classification of animal voices ([159], EUFIT'96 competition);
            * Handwritten digit recognition ([445], [15], [121], [16], [450]);
            * Optical character recognition ([49]);
            * Circuits synthesis and fault detection ([211], [212]);
            * Vibro-acoustic technical diagnostics ([252], [253], [254], [255], [199]);
            * Intelligent scheduling ([137]);
    - **Others:**
        * Preliminary wind-bracing in steel skeleton structure ([7]);
        * Technical diagnostics of mechanical objects ([437]);
        * Decision supporting for highly automated production system ([457]);
        * Estimation of important highway parameters ([195], [196]);
        * Time series analysis of highway traffic volumes ([195], [196]);
        * Real–time decision making ([407], [309]);
        * Material analysis ([130], [317]);
        * Power system security analysis ([176]);
- **INFORMATION SCIENCE:**
    - **Software engineering:**
        * Qualitative analysis of software engineering data ([384]);
        * Assessing software quality ([311]);
        * Software deployability ([308]);
        * Knowledge discovery form software engineering data([385]);
    - Information retrieval ([431], [432], [433], [98]);
    - Data mining from musical databases ([157]);
    - Analysis and synthesis of concurrent systems ([405], [406], [441], [442], [444]);
    - Integration RDMS and data mining tools using rough sets ([97],[249], [244]);
    - Rough set model of relational databases ([374], [20], [21], [188], [190]);
    - Cooperative knowledge base systems ([349], [350], [351], [352]);
    - Natural language processing ([114], [216], [115], [113], [109]);
    - Cooperative information system re–engineering ([441], [443]);
- **DECISION ANALYSIS:** (see cases and applications in this section);
- **SOCIAL SCIENCES, OTHERS:**
    - Conflict analysis ([287], [291], [295], [63]);
    - Social choice functions ([96], [260]);
    - Rough sets in librarian science ([390]);
    - Rough sets–based study of voter preference ([118]);
    - Analysis of test profile performance ([353]);
    - On–line prediction of volleyball game progress ([446], [455]);

- **MOLECULAR BIOLOGY:** Discovery of functional components of proteins from amino–acid sequences ([480], [488]);
- **CHEMISTRY: PHARMACY** Analysis of relationship between structure and activity of substances ([169], [170], [171], [172], [173]);

12 Software Systems

We enclose the list of software systems based on rough sets. The reader can find more details in [338]. It was possible to identify the following rough set systems for data analysis:

- Datalogic/R, http://ourworld.compuserve.com/homepages/reduct
- Grobian (Roughian), e-mail: I.Duentsch@ulst.ac.uk, ggediga@luce.psycho.Uni-Osnabrueck.DE
- KDD-R: Rough Sets-Based Data Mining System, e-mail: ziarko@cs.uregina.ca
- LERS—A Knowledge Discovery System , e-mail: jerzy@eecs.ukans.edu
- PRIMEROSE, e-mail: tsumoto@computer.org
- ProbRough — A System for Probabilistic Rough Classifiers Generation, e-mail: {zpiasta,lenarcik}@sabat.tu.kielce.pl
- Rosetta Software System, http://www.idi.ntnu.no/~aleks/rosetta/
- Rough Family - Software Implementation of the Rough Set Theory, e-mail: Roman.Slowinski@cs.put.poznan.pl, Jerzy.Stefanowski@cs.put.poznan.pl
- RSDM: Rough Sets Data Miner, e-mail: {cfbaizan, emenasalvas}@.fi.upm.es
- RoughFuzzyLab - a System for Data Mining and Rough and Fuzzy Sets Based Classification, e-mail: rswiniar@saturn.sdsu.edu
- RSL - The Rough Set Library, ftp://ftp.ii.pw.edu.pl/pub/Rough/
- TAS: Tools for Analysis and Synthesis of Concurrent Processes using Rough Set Methods, e-mail: zsuraj@univ.rzeszow.pl
- Trance: a Tool for Rough Data Analysis, Classification, and Clustering, e-mail:wojtek@cs.vu.nl

PartII
Rough Mereology: Approximate Synthesis of Objects

We would like to give here a brief account of the rough mereological approach to approximate reasoning [328, 330, 331, 332, 333, 340, 398, 335, 336]. We propose this formalization as a tool for solving multi–agent or distributed applications related to approximate reasoning and to calculi on information granules [532], [531].

13 Rough Mereology

Rough mereology offers the general formalism for the treatment of partial containment. Rough mereology can be regarded as a far - reaching generalization of mereology of Leśniewski [181]: it does replace the relation of being a (proper) part with a hierarchy of relations of being a part in a degree. The basic notion is the notion of a rough inclusion.

A real function $\mu(X, Y)$ on a universe of objects U with values in the interval $[0, 1]$ is called *a rough inclusion* when it satisfies the following conditions:

(A) $\mu(x, x) = 1$ for any x (meaning normalization);
(B) $\mu(x, y) = 1$ implies that $\mu(z, y) \geq \mu(z, x)$ for any triple x, y, z (meaning monotonicity);
(C) $\mu(x, y) = 1$ and $\mu(y, x) = 1$ imply $\mu(x, z) \geq \mu(y, z)$ for any triple x, y, z (meaning monotonicity);
(D) there is n such that $\mu(n, x) = 1$ for any x. An object n satisfying (D) is a μ-null object.

We let $x =_\mu y$ iff $\mu(x, y) = 1 = \mu(y, x)$ and $x \neq_\mu y$ iff $non(x =_\mu y)$.

We introduce other conditions for rough inclusion:

(E) **if** objects x, y have the property :
  if $z \neq_\mu n$ and $\mu(z, x) = 1$
  then there is $t \neq_\mu n$ with $\mu(t, z) = 1 = \mu(t, y)$
  then it follows that: $\mu(x, y) = 1$.
  (E) is an inference rule: it is applied to infer the relation of being a part from the relation of being a subpart.
(F) For any collection Γ of objects there is an object x with the properties:
  (i) if $z \neq_\mu n$ and $\mu(z, x) = 1$ then there are $t \neq_\mu n$, $w \in \Gamma$ such that
  $$\mu(t, z) = \mu(t, w) = \mu(w, x) = 1;$$
  (ii) if $w \in \Gamma$ then $\mu(w, x) = 1$;
  (iii) if y satisfies the above two conditions then $\mu(x, y) = 1$.

Any x satisfying F(i) is called a *set of objects* in Γ; if, in addition, x satisfies F(ii,iii), then x is called the *class of objects* in Γ. These notions allow for representations of collections of objects as objects.

We interpret the formula: $\mu(x, y) = r$ as the statement: *x is a part of y in degree at least r*.

The formula $x = class(\mu_U)\{x_1, x_2, ..., x_k\}$ is interpreted as the statement that the object x is composed (designed, synthesized) from parts $x_1, x_2, ..., x_k$. In mereology of Leśniewski the notions of a part, an element, and a subset are all equivalent: one can thus interpret the formula $\mu(x, y) = r$ as the statement: x is an element (a subset) of y in degree r; if $y = class(\mu)\Gamma$, then $\mu(x, y) = r$ means that x is a member of the collection Γ to a degree r.

A standard choice of an appropriate measure can be based on the frequency count; the formal rendering of this idea is the *standard rough inclusion* function defined for two sets $X, Y \subseteq U$ by the formula

$$\mu(X,Y) = \frac{card(X \cap Y)}{card(X)}$$

when X is non-empty, 1 otherwise. This function satisfies all of the above axioms for rough inclusion.

Relations to fuzzy containment:

Fuzzy containment may be defined in a fuzzy universe U endowed with fuzzy membership functions μ_X, μ_Y by the formula :

$$\sigma(X \subseteq Y) = \inf_Z \{I(\mu_X(Z), \mu_Y(Z))\}$$

for a many - valued implication I. We quote a result in [332] which shows that rough inclusions generate a class of fuzzy containments stable under residual implications of the form $\vec{T}$ where T is a continuous t-norm [73] viz.: for any rough inclusion μ on U, the function

$$\sigma(X,Y) = \inf_Z \{\vec{T}(\mu(Z,X), \mu(Z,Y))\}.$$

is also a rough inclusion. The impact of this is that in models of rough mereology which implement σ as the model rough inclusion, we have the composition rule of the form:

if $\sigma(x,y,r)$ and $\sigma(y,z,s)$ then $\sigma(x,z,T(r,s))$.

Hence we can develop an associative calculus of partial containment.

13.1 Rough inclusions from information systems

Rough inclusions can be generated from a given information system $\mathcal{A}$; for instance, for a given partition $P = \{A_1, \ldots, A_k\}$ of the set A of attributes into non-empty sets $A_1, \ldots, A_k$, and a given set $W = \{w_1, \ldots, w_k\}$ of weights, $w_i \in [0,1]$ for $i = 1, 2, \ldots, k$ and $\sum_{i=1}^{k} w_i = 1$ we let

$$\mu_{o,P,W}(x,y) = \sum_{i=1}^{k} w_i \cdot \frac{\|IND_i(x,y)\|}{\|A_i\|}$$

where $IND_i(x,y) = \{a \in A_i : a(x) = a(y)\}$. We call $\mu_{o,P,W}$ a *pre-rough inclusion*.

The function $\mu_{o,P,W}$ is rough-invariant i.e. if $a(x) = a(x')$ and $a(y) = a(y')$ for each $a \in A$ then $\mu_{o,P,W}(x,y) = \mu_{o,P,W}(x',y')$. $\mu_{o,P,W}$ can be extended to a rough inclusion on the set 2^U [332] e.g. via the formula: $\mu(X,Y) = T\{\bot \{\mu_{o,P,W}(x,y) : y \in Y\} : x \in X\}$ where T is a t-norm and $\bot$ is a t-conorm.
An advantage of having rough inclusions in this form is that we can optimize weights w_i in the learning stage.

	hat	ker	pig
x_1	1	0	0
x_2	0	0	1
x_3	0	1	0
x_4	1	1	0

Table 9. The information system H

Example 13.1 Consider an information system H
The table below shows values of the initial rough inclusion $\mu_{o,P,W}(x,y) = \frac{card(IND(x,y))}{3}$ i.e. we consider the simplest case when $k = 1$, $w_1 = 1$.

	x_1	x_2	x_3	x_4
x_1	1	0.33	0.33	0.66
x_2	0.33	1	0.33	0.00
x_3	0.33	0.33	1	0.66
x_4	0.66	0.00	0.66	1

Table 10. Initial rough inclusion for H

□

Example 13.2 In addition to the information system (agent) H from Example 1 we consider agents B and HB. Together with H they form the string $\mathbf{ag} = (H)(B)(HB)$ i.e. HB takes objects: x sent by H and y sent by B and assembles a complex object xy.

	pis	cut	kni	cr
y_1	1	1	1	1
y_2	1	0	0	0
y_3	0	0	1	0
y_4	1	1	1	0

Table 11. The information system B

	har	lar	off	tar
x_1y_1	1	0	1	0
x_1y_3	0	1	0	0
x_2y_1	1	0	0	0
x_2y_3	0	1	0	1

Table 12. The information system HB

The values of the initial rough inclusion $\mu = \mu_{o,P,W}$ are calculated for (B) and (BH) by the same procedure as in Example 13.1. □

13.2 Approximate mereological connectives

An important ingredient in our scheme is related to the problem of rough mereological connectives: given information systems $\mathbf{A}, \mathbf{B}, \mathbf{C}, ...$ we will say that $\mathbf{A}$ *results from* $\mathbf{B}, \mathbf{C}, ...$ if there exists a (partial) mapping (an *operation*) $o_\mathbf{A}$: $U_\mathbf{B} \times U_\mathbf{C} \times ... \to U_\mathbf{A}$ with $rng\ o_A = U_\mathbf{A}$ i.e. any $x \in U_\mathbf{A}$ is of the form $o_\mathbf{A}(y, z, ...)$ where $y \in U_\mathbf{B}$ and $z \in U_\mathbf{C}, ...$. In the case when pre - rough inclusions $\mu_A, \mu_B, \mu_C, ...$ are defined in respective universes $U_\mathbf{A}, U_\mathbf{B}, U_\mathbf{C}, ...$ there arises the problem of *uncertainty propagation* i.e. we have to decide in what way is the measure μ_A related to measures $\mu_B, \mu_C, ...$. Formally, we have to find an (approximation) to a function f satisfying the following property:

---

for any $\epsilon_1, \epsilon_2, \epsilon_3, ... \in [0, 1]$:
for any $x, x_1 \in U_\mathbf{B}, y, y_1 \in U_\mathbf{C}, ...$:
  if $\mu_\mathbf{B}(x, x_1) \geq \epsilon_1$, $\mu_\mathbf{C}(y, y_1) \geq \epsilon_2$ and ...
  then $\mu_\mathbf{A}(o_A(x, y, ...), o_A(x_1, y_1, ...)) \geq f(\epsilon_1, \epsilon_2, ...)$

In practice, it is unrealistic to expect the existence of a function f satisfying the above condition globally; therefore, we localize this notion. To this end, let us select a subset $S_A \subseteq U_A$ of objects which we will call *standard objects*; we will use the symbol $xyz...$ to denote the object $o_A(x, y, ...) \in U_A$. Given a standard $s = xyz...$, we will call a function f an (approximation to) *rough mereological connective relative to* o_A and s in case it satisfies the condition:

---

for any $\epsilon_1, \epsilon_2, \epsilon_3, ... \in [0, 1]$:
for any $x_1 \in U_\mathbf{B}, y_1 \in U_\mathbf{C}, ...$:
  if $\mu_\mathbf{B}(x_1, x) \geq \epsilon_1$, $\mu_\mathbf{C}(y_1, y) \geq \epsilon_2$ and ...
  then $\mu_\mathbf{A}(o_A(x_1, y_1, ...), s) \geq f(\epsilon_1, \epsilon_2, ...)$

We outline an algorithm which may be used to extract from information systems approximations to uncertainty functions (rough mereological connectives).

Example 13.3 We will determine an approximation to the mereological connective at the standard $x_1 y_1$ in Table 12 of Example 13.2 i.e. a function f such that (for simplicity of notation we omit subscripts of μ):

$$\text{if } \mu(x, x_1) \geq \epsilon_1 \text{ and } \mu(y, y_1) \geq \epsilon_2$$
$$\text{then } \mu(xy, x_1 y_1) \geq f(\epsilon_1, \epsilon_2), \text{ for any pair } \epsilon_1, \epsilon_2.$$

The following tables show conditions which f is to fulfill.

□

x	$\mu(x,x_1)$	y	$\mu(y,y_1)$	$\mu(xy, x_1y_1)$
x_1	1	y_1	1	1
x_1	1	y_2	0.25	0.5
x_1	1	y_3	0.25	0.25
x_1	1	y_4	0.75	1
x_2	0.33	y_1	1	0.75
x_2	0.33	y_2	0.25	0.25
x_2	0.33	y_3	0.25	0.00
x_2	0.33	y_4	0.75	0.5

Table 13. The conditions for f (first part)

x	$\mu(x,x_1)$	y	$\mu(y,y_1)$	$\mu(xy, x_1y_1)$
x_3	0.33	y_1	1	0.75
x_3	0.33	y_2	0.25	0.25
x_3	0.33	y_3	0.25	0.25
x_3	0.33	y_4	0.75	0.75
x_4	0.66	y_1	1	1
x_4	0.66	y_2	0.25	0.5
x_4	0.66	y_3	0.25	0.25
x_4	0.66	y_4	0.75	1

Table 14. The conditions for f (second part)

This full set T_0 of conditions can be reduced: we can find a minimal set T of vectors of the form $(\varepsilon_1', \varepsilon_2', \varepsilon)$ such that if f satisfies the condition $f(\varepsilon_1', \varepsilon_2') = \varepsilon$ for each $(\varepsilon_1', \varepsilon_2', \varepsilon) \in T$ then f extends by the formula (1), below.

The following algorithm produces a minimal set T of conditions.

Algorithm
Input: table T_0 of vectors

$$(\mu(x,x_1), \mu(y,y_1), \mu(xy, x_1y_1));$$

Step 1. For each pair $(\mu(x,x_1) = \varepsilon_1, \mu(y,y_1) = \varepsilon_2)$, find $\varepsilon(\varepsilon_1, \varepsilon_2) = \min\{\varepsilon : \varepsilon_1' \geq \varepsilon_1, \varepsilon_2' \geq \varepsilon_2, (\varepsilon_1', \varepsilon_2', \varepsilon) \in T_0\}$. Let T_1 be the table of vectors $(\varepsilon_1, \varepsilon_2, \varepsilon(\varepsilon_1, \varepsilon_2))$.
Step 2. For each ε^* such that $(\varepsilon_1, \varepsilon_2, \varepsilon(\varepsilon_1, \varepsilon_2) = \varepsilon^*) \in T_1$, find: $row(\varepsilon^*) = (\varepsilon_1^*, \varepsilon_2^*, \varepsilon^*)$ where $(\varepsilon_1^*, \varepsilon_2^*, \varepsilon^*) \in T_1$ and if $(\varepsilon_1', \varepsilon_2', \varepsilon^*) \in T_1$ then $\varepsilon_1' \geq \varepsilon_1^*, \varepsilon_2' \geq \varepsilon_2^*$.
Output: table T of vectors of the form $row(\varepsilon)$.

One can check that Table 15 shows a minimal set T of vectors for the case of Tables 13, 14.

One can extract from the algorithm the synthesis formula of f from conditions

ε_1	ε_2	ε
0.66	0.75	1
0.33	0.75	0.5
0.66	0.25	0.25
0.33	0.25	0.00

Table 15. A minimal set T of vectors

T_0 :

$$f(\varepsilon_1, \varepsilon_2) = \min \{\varepsilon' : (\varepsilon_1', \varepsilon_2', \varepsilon') \in T_0 \wedge (\varepsilon_1' \geq \varepsilon_1) \wedge (\varepsilon_2' \geq \varepsilon_2)\} \quad (1)$$

14 Reasoning in Multi–Agent Systems

We outline here basic ideas of reasoning under uncertainty by intelligent units (agents) in multi - agent systems. Schemes based on these ideas may be - in our opinion - applied in the following areas of application, important for development of automated techniques [4]:

- computer-aided manufacturing or computer-aided design [4], [48],[126] where a complex object=a final artifact (assembly) is produced (designed) from inventory (elementary) parts by a dedicated team of agents.
- logistics [77] where complex structures are organized from existing elementary structures (units) to perform a task according to a given specification.
- adaptive control of complex systems [200], [398] where the task consists in maintaining a given constraint (specification) by adaptive adjustment of behavior of some parts (organs, physiological processes etc.).
- business re-engineering [77], [434] where the task is to adaptively modify a complex object (structure, organization, resources, etc.) according to the current economic situation (specification).
- cooperative/distributed problem solving including planning, dynamic task assignment etc. [77], [85], [434] where the task is to organize a system of agents into a scheme of local teams for solving a problem (specification).
- automated fabrication [31] where the task is to build complex objects (e.g. mechanisms) by layer-after -layer synthesis.
- preliminary stage of design process [348] where the approximate reasoning about objects and processes is crucial as it is carried out in an informal, often natural, language.

The general scheme for approximate reasoning can be represented by the following tuple

$$Appr\_Reas = (Ag, Link, U, St, Dec\_Sch, O, Inv, Unc\_mes, Unc\_prop)$$

where

(i) The symbol Ag denotes the set of agent names.

(ii) The symbol $Link$ denotes a set of non-empty strings over the alphabet Ag; for $v(ag) = ag_1 ag_2 ... ag_k ag \in Link$, we say that $v(ag)$ defines an *elementary synthesis scheme* with the *root* ag and the *leaf agents* $ag_1, ag_2, ..., ag_k$. The intended meaning of $v(ag)$ is that the agents $ag_1, ag_2,..,ag_k$ are the children of the agent ag which can send to ag some simpler constructs for assembling a more complex artifact. The relation $\leq$ defined via $ag \leq ag'$ iff ag is a leaf agent in $v(ag')$ for some $v(ag')$, is usually assumed to be at least an ordering of Ag into a type of an acyclic graph; we assume for simplicity that $(Ag, \leq)$ is a tree with the root $root(Ag)$ and leaf agents in the set $Leaf(Ag)$.

(iii) The symbol U denotes the set $\{U(ag) : ag \in Ag\}$ of *universes* of agents.

(iv) The symbol St denotes the set $\{St(ag) : ag \in Ag\}$ where $St(ag) \subset U(ag)$ is the set of *standard objects* at the agent ag.

(v) The symbol O denotes the set $\{O(ag) : ag \in Ag\}$ of *operations* where $O(ag) = \{o_i(ag)\}$ is the set of *operations at* ag.

(vi) The symbol Dec_Sch denotes the set of *decomposition schemes*; a particular decomposition scheme dec_sch_j is a tuple

$$(\{st(ag)_j : ag \in Ag\}, \{o_j(ag) : ag \in Ag\})$$

which satisfies the property that if $v(ag) = ag_1 ag_2 ... ag_k ag \in Link$ then

$$o_j(ag)(st(ag_1)_j, st(ag_2)_j, .., st(ag_k)_j) = st(ag)_j$$

for each j.

The intended meaning of dec_sch_j is that when any child ag_i of ag submits the standard construct $st(ag_i)_j$ then the agent ag assembles from $st(ag_1)_j$, $st(ag_2)_j$, ..., $st(ag_k)_j$ the standard construct $st(ag)_j$ by means of the operation $o_j(ag)$.

The rule dec_sch_j establishes therefore a decomposition scheme of any standard construct at the agent $root(Ag)$ into a set of consecutively simpler standards at all other agents. The standard constructs of leaf agents are *primitive (inventory) standards*. We can regard the set of decomposition schemes as a skeleton about which the approximate reasoning is organized. Any rule dec_sch_j conveys a certain knowledge that standard constructs are synthesized from specified simpler standard constructs by means of specified operations. This ideal knowledge is a reference point for real synthesis processes in which we deal as a rule with constructs which are not standard: in adaptive tasks, for instance, we process new, unseen yet, constructs (objects, signals).

(vii) The symbol Inv denotes the *inventory set* of primitive constructs. We have $Inv = \cup\{U(ag) : ag \in Leaf(Ag)\}$.

(viii) The symbol Unc_mes denotes the set $\{Unc\_mes(ag) : ag \in Ag\}$ of *uncertainty measures of agents*, where $Unc\_mes(ag) = \{\mu_j(ag)\}$ and $\mu_j(ag) \subseteq U(ag) \times U(ag) \times V(ag)$ is a relation (possibly function) which determines a distance between constructs in $U(ag)$ valued in a set $V(ag)$; usually, $V(ag) = [0, 1]$, the unit interval.

(ix) The symbol Unc_prop denotes the set of *uncertainty propagation rules* $\{Unc\_prop(v(ag)) : v(ag) \in Link\}$; for $v(ag) = ag_1 ag_2 ... ag_k ag \in Link$, the set $Unc\_prop(v(ag))$ consists of functions $f_j : V(ag_1) \times V(ag_2) \times ... \times V(ag_k) \longrightarrow V(ag)$ such that

$$\text{if } \mu_j(ag_i)(x_i, st(ag_i)_j) = \varepsilon_i \text{ for } i = 1, 2, .., k$$
$$\text{then } \mu_j(ag)(o_j(x_1, x_2, .., x_k), st(ag)_j) = \varepsilon \geq f_j(\varepsilon_1, \varepsilon_2, .., \varepsilon_k).$$

The functions f_j relate values of uncertainty measures at the children of ag and at ag.

This general scheme may be adapted to the particular cases.

As an example, we will interpret this scheme in the case of a fuzzy controller. In its version due to Mamdani [200] in its simplest form, we have two agents: *input, output*, and standards of agents are expressed in terms of linguistic labels like *positively small, negative, zero* etc. Operations of the agent *output* express the control rules of the controller e.g. the symbol $o(positively\ small, negative) = zero$ is equivalent to the control rule of the form if $st(input)_i$ is *positively small* and $st(input)_j$ is *negative* then $st(output)_k$ is *zero*. Uncertainty measures of agents are introduced as fuzzy membership functions corresponding to fuzzy sets representing standards i.e. linguistic labels. An input construct (signal) $x(input)$ is fuzzyfied i.e. its distances from input standards are calculated and then the fuzzy logic rules are applied. By means of these rules uncertainty propagating functions are defined which allow for calculating the distances of the output construct $x(output)$ from the output standards. On the basis of these distances the construct $x(output)$ is evaluated by the defuzzyfication procedure.

The process of synthesis by a scheme of agents of a complex object x which is an approximate solution to a requirement Φ consists in our approach of the two communication stages viz. the top - down communication/negotiation process and the bottom - up synthesis process. We outline the two stages here.

In the process of top - down communication, a requirement Φ received by the scheme from an external source is decomposed into approximate specifications of the form

$$(\Phi(ag), \varepsilon(ag))$$

for any agent ag of the scheme. The intended meaning of the approximate specification $(\Phi(ag), \varepsilon(ag))$ is that a construct $z \in U(ag)$ satisfies $(\Phi(ag), \varepsilon(ag))$ iff there exists a standard $st(ag)$ with the properties that $st(ag)$ satisfies the predicate $\Phi(ag)$ and

$$\mu(ag)(z, st(ag)) \geq \varepsilon(ag).$$

The uncertainty bounds of the form $\varepsilon(ag)$ are defined by the agents viz. the root agent $root(Ag)$ chooses $\varepsilon(root(Ag))$ and $\Phi(root(Ag))$ as such that according to it any construct x satisfying $(\Phi(root(Ag)), \varepsilon(root(Ag)))$ should satisfy the external requirement Φ in an acceptable degree. The choice of $(\Phi(root(Ag)), \varepsilon(root(Ag)))$ can be based on the previous learning process; the other agents choose their approximate specifications in negotiations within each elementary scheme $v(ag)$

$\in$ *Link*. The result of the negotiations is successful when there exists a decomposition scheme dec_sch_j such that for any $v(ag) \in Link$, where $v(ag) = ag_1 ag_2...ag_k ag$, from the conditions $\mu(ag_i)(x_i, st(ag_i)_j) \geq \varepsilon(ag_i)$ and $st(ag_i)_j$ satisfies $\Phi(ag_i)$ for $i = 1, 2, .., k$, it follows that $\mu(ag)(o_j(x_1 x_2, .., x_k), st(ag)_j) \geq \varepsilon(ag)$ and $st(ag)_j$ satisfies $\Phi(ag)$.

The uncertainty bounds $\varepsilon(ag)$ are evaluated on the basis of uncertainty propagating functions whose approximations are extracted from information systems of agents.

The synthesis of a complex object x is initiated at the leaf agents: they select primitive constructs (objects) and calculate their distances from their respective standards; then, the selected constructs are sent to the parent nodes of leaf agents along with vectors of distance values. The parent nodes synthesize complex constructs from the sent primitives and calculate the new vectors of distances from their respective standards. Finally, the root agent $root(Ag)$ receives from its children the constructs from which it assembles the final construct and calculates the distances of this construct from the root standards. On the basis of the found values, the root agent classifies the final construct x with respect to the root standards as eventually satisfying $(\Phi(root(Ag)), \varepsilon(root(Ag)))$.

Our approach is analytic: all logical components (uncertainty measures, uncertainty functions etc.) necessary for the synthesis process are extracted from the empirical knowledge of agents represented in their information systems; it is also intensional in the sense that rules for propagating uncertainty are local as they depend on a particular elementary synthesis scheme and on a particular local standard.

We will now give a more detailed account of the process of synthesis.

15 Synthesis Schemes

Synthesis agents.
We start with the set Ag of synthesis agents and the set Inv of inventory objects. Any synthesis agent ag has assigned a *label* $lab(ag) = \{U(ag), A(ag), St(ag), L(ag), \mu_o(ag), F(ag)\}$ where: $U(ag)$ is the universe of objects at ag, $A(ag) = \{U(ag), A(ag), V(ag)\}$ is the information system of ag, $St(ag) \subseteq U(ag)$ is the set of *standard objects* (standards) at ag, $L(ag)$ is a set of *unary predicates* at ag (specifying properties of objects in $U(ag)$). Predicates of $L(ag)$ are constructed as formulas in $C(A(ag), V)$ (i.e. Boolean combinations of descriptors over $A(ag)$ and V); $\mu_o(ag) \subseteq U(ag) \times U(ag) \times [0, 1]$ is a pre-rough inclusion at ag generated from $A(ag)$; $F(ag)$ is a set of functions at ag called *mereological connectives* (cf. below, the notion of a *(C,Φ,ε)*− scheme). Synthesis agents reason about objects by means of the approximate logic of synthesis.

Approximate logic of synthesis.
Consider a synthesis agent ag. The symbol b_{ag} will denote the variable which

runs over objects in U_{ag}. A *valuation* v_X where X is a set of synthesis agents is a function which assigns to any b_{ag} for $ag \in X$ an element $v_X(b_{ag}) \in U_{ag}$. The symbol v_{ag}^x denotes $v_{\{ag\}}$ with $v_{\{ag\}}(b_{ag}) = x$.

We now define *approximate specifications* at ag as formulas of the form $\langle st(ag), \Phi(ag), \varepsilon(ag)\rangle$ where $st(ag) \in St(ag), \Phi(ag) \in L(ag)$ and $\varepsilon(ag) \in [0,1]$. We say that $v = v_{\{ag\}}$ *satisfies a formula* $\alpha = \langle st(ag), \Phi(ag), \varepsilon(ag)\rangle$, symbolically $v \models \alpha$, in case $\mu(ag)(v(b_{ag}), st(ag)) \geq \varepsilon$ and $st(ag) \models \Phi(ag)$. We write $x \models \langle st(ag), \Phi(ag), \varepsilon(ag)\rangle$ iff $v_{ag}^x \models \langle st(ag), \Phi(ag), \varepsilon(ag)\rangle$. The meaning of $\alpha = \langle st(ag), \Phi(ag), \varepsilon(ag)\rangle$ is thus the set $[\alpha]_{ag}$ of objects x satisfactorily (as determined by $\varepsilon(ag)$) close to a standard (viz. $st(ag)$) satisfying $\Phi(ag)$. How the agents cooperate is determined by a chosen scheme; selection of a scheme is itself an adaptive process of design [398].

The synthesis language
Link. The synthesis agents are organized into a hierarchy (which may be an empty relation in case of autonomous agents system). We describe this hierarchy in a language *Link* over the alphabet Ag. The agents $ag_1, ag_2, .., ag_k, ag_0$ in Ag form the string $\mathbf{ag} = ag_1 ag_2 ... ag_k ag_0 \in Link$ if and only if there exist a mapping $\rho(\mathbf{ag}) : U(ag_1) \times ... \times U(ag_k) \longmapsto U(ag_0)$ (meaning: the agent ag_0 can assemble by means of $\rho(\mathbf{ag})$ the object $\rho(\mathbf{ag})(x_1, ..., x_k) \in U(ag_0)$ from any tuple $(x_1 \in U(ag_1), ..., x_k \in U(ag_k))$.

Elementary constructions.
If $\mathbf{ag} = ag_1 ag_2 ... ag_k ag_0 \in Link$, then the pair

$$c = (\mathbf{ag}, \{\langle st(ag_i), \Phi(ag_i), \varepsilon(ag_i)\rangle : i = 0, 1, .., k\})$$

will be called an *elementary construction*. We write: $Ag(c) = \{ag_0, ag_1, ..., ag_k\}$, $Root(c) = ag_0$, $Leaf(c) = Ag(c) - \{ag_0\}$.

Constructions.
For elementary constructions c, c' with $Ag(c) \cap Ag(c') = \{ag\}$ where $ag = Root(c) \in Leaf(c')$, we define *the ag-composition* $c \star_{ag} c'$ of c and c' with $Root(c \star_{ag} c') = Root(c')$, $Leaf(c \star_{ag} c') = (Leaf(c) - \{ag\}) \cup (Leaf(c'), Ag(c \star_{ag} c') = Ag(c) \cup Ag(c')$. A *construction* is any expression C obtained from a set of elementary constructions by applying the composition operation a finite number of times.

(C, Φ, ε)-schemes.
For an elementary construction $c = c(\mathbf{ag})$ as above, we define a $(c, \Phi, \varepsilon)-scheme$ as

$$(\mathbf{ag}, \{\langle st(ag_i), \Phi(ag_i), \varepsilon(ag_i)\rangle : i = 0, 1, .., k\}$$

where $f(ag_0) \in F(ag_0)$ satisfies the condition:

if $\mu(ag_i)(x_i, st(ag_i)) \geq \varepsilon(ag_i)$ for $i = 1, 2, .., k$
then
$$\mu_o(ag)(x, st(ag_0)) \geq (\varepsilon(ag_1), \varepsilon(ag_2), .., \varepsilon(ag_k))$$
$$\geq \varepsilon(ag_0).$$

A construction C composed of elementary constructions $c_1, .., c_m$, c_o with $Root(C) = Root(c_o) = ag_o$ is the *support of a* (C, Φ, ε)-*scheme* when each c_i is the support of a $(c_i, \Phi_i, \varepsilon_i)$-*scheme*, where $\Phi_i = \Phi(Root(c_i))$, $\varepsilon_i = \varepsilon(Root(c_i))$, $\Phi = \Phi(ag_0)$ and $\varepsilon = \varepsilon(ag_0)$. The (C, Φ, ε)-scheme **c** defines a function $F_{\mathbf{c}}$ called *the output function* of **c** given by $F_{\mathbf{c}}(v_{Leaf(C)}) = x$ where $x \in U(ag_0)$ is the unique object produced by C from $v_{Leaf(C)}$. The following statement expresses the sufficiency criterion of synthesis by a scheme of an object satisfying the approximate requirement $\langle st(ag_o), \Phi(ag_o), \varepsilon(ag_o) \rangle$.

Theorem 15.1 (*the sufficiency criterion of synthesis*). For any valuation v_X on the set X of leaf agents $ag(1), .. ., ag(m)$ of the (C, Φ, ε)-scheme with $ag_o = Root(C)$ such that

$$v(b_{ag(i)}) \models \langle st(ag(i)), \Phi(ag(i)), \varepsilon(ag(i)) \rangle$$

for $i = 1, 2, ..., m$, we have

$$F_{\mathbf{c}}(v_X) \models \langle st(ag_o), \Phi(ag_o), \varepsilon(ag_o) \rangle$$

∎

Let us emphasize the fact that the functions $f(ag)$, called *mereological connectives* above, are expected to be extracted from experiments with samples of objects (see Example 3, above). The above property allows for an easy to justify correctness criterion of a given (C, Φ, ε)-scheme provided that all parameters in this scheme have been chosen properly. The searching process for these parameters and synthesis of an uncertainty propagation scheme satisfying the formulated conditions constitutes the main and not easy part of synthesis (and design as well).

16 Mereological Controllers

The approximate specification (Φ, ε) can be regarded as an invariant to be kept over the universe of global states (complex objects) of the distributed system. A mereological controller generalizes the notion of a fuzzy controller. The control problems can be divided into several classes depending on the model of controlled object. In this work we deal with the simplest case. In this case, the model of a controlled object is the (C, Φ, ε)-scheme **c** which can be treated as a model of the unperturbed by noise controlled object whose states are satisfying the approximate specification (Φ, ε).

We assume the leaf agents of the (C, Φ, ε)-scheme **c** are partitioned into two disjoint sets, namely the set $Un\_control(\mathbf{c})$ of *uncontrollable (noise) agents* and the set $Control(\mathbf{c})$ of controllable *agents*.

We present now two examples of a control problem for a given (C, Φ, ε)-scheme.

(OCP) OPTIMAL CONTROL PROBLEM:

Input: (C, Φ, ε)-scheme **c**; information about actual valuation v of leaf agents i.e. the values $v(b_{ag})$ for any $ag \in Control(\mathbf{c})$ and a value ε' such that $F_{\mathbf{c}}(v) \models \langle st(ag_{\mathbf{c}}), \Phi, \varepsilon' \rangle$.

Output: A new valuation v' such that $v'(b_{ag}) = v(b_{ag})$ for $ag \in Un\_control(\mathbf{c})$ and $F_{\mathbf{c}}(v') \models \langle st(ag_{\mathbf{c}}), \Phi, \varepsilon_0 \rangle$ where $\varepsilon_0 = \sup\{\delta : F_{\mathbf{c}}(w) \models \langle st(ag_{\mathbf{c}}), \Phi, \delta \rangle$ for some w such that $w(b_{ag}) = v(b_{ag})$ for $ag \in Un\_control(\mathbf{c})\}$.

These requirements can hardly be satisfied directly. A relaxation of (OCP) is

---

(CP) ∇-CONTROL PROBLEM

Input: (C, Φ, ε)-scheme **c**; information about actual valuation v of leaf agents (i.e. the values $v(b_{ag})$ for any $ag \in Control(\mathbf{c})$) and a value ε' such that $F_{\mathbf{c}}(v) \models \langle st(ag_{\mathbf{c}}), \Phi, \varepsilon' \rangle$.

Output: A new valuation v' such that $v'(b_{ag}) = v(b_{ag})$ for $ag \in Un\_control(\mathbf{c})$ and $F_{\mathbf{c}}(v') \models \langle st(ag_{\mathbf{c}}), \Phi, \varepsilon_0 \rangle$ where $\varepsilon_0 \rangle \varepsilon' + \nabla$ for some given threshold ∇.

We will now describe the basic idea on which our controllers of complex dynamic objects represented by distributed systems of intelligent agents are built. The main component of the controller are Δ - *incremental rules*.

Δ-*rules* have the form:

$$(\Delta(\mathbf{ag}))\ (\Delta\varepsilon(ag_{i_1}), ..., \Delta\varepsilon(ag_{i_r}))$$
$$= h(\varepsilon(ag), -\Delta\varepsilon(ag), \varepsilon(ag_1), ..., \varepsilon(ag_k))$$

where $ag_{i_1}, ..., ag_{i_r}$ are all controllable children of ag (i.e. children of ag having descendents in $Control(\mathbf{c})$), $h : R^{k+2} \to R^r$ and R is the set of reals.

Approximations to the function h are extracted from experimental data.

The meaning of $\Delta(\mathbf{ag})$ is : if $x' \models \langle st(ag), \Phi(ag), \varepsilon'(ag) \rangle$ for $x' \in U(ag)$ where $\varepsilon'(ag) = \varepsilon(ag) + \Delta\varepsilon(ag)$ then if the controllable children $ag_{i_1}, ..., ag_{i_r}$ of ag will issue objects $y_{i_1}, ..., y_{i_r}$ with $y_{i_j} \models \langle st(ag_{i_j}), \Phi(ag_{i_j}), \varepsilon(ag_{i_j}) + \Delta\varepsilon(ag_{i_j}) \rangle$ for $j = 1, ..., r$ where $(\Delta\varepsilon(ag_{i_1}), ..., \Delta\varepsilon(ag_{i_r})) = h(\varepsilon(ag), -\Delta\varepsilon(ag), \varepsilon(ag_1), ..., \varepsilon(ag_k))$ then the agent ag will construct an object y such that $y \models \langle st(ag), \Phi(ag), \varepsilon \rangle$ where $\varepsilon \geq \varepsilon(ag)$.

In the above formula, we assume $\Delta\varepsilon(ag) \leq 0$ and $\Delta\varepsilon(ag_{i_1}) \geq 0, ..., \Delta\varepsilon(ag_{i_r}) \geq 0$. The above semantics covers the case when Δ - rules allow to compensate in one step the influence of noise.

$\Delta-$ rules can be composed in an obvious sense. $\Delta(\mathbf{ag}\star\mathbf{ag'})$ denotes the composition of $\Delta(\mathbf{ag})$ and $\Delta(\mathbf{ag}')$ over $\mathbf{ag} \star \mathbf{ag}'$. The variable $\Delta(\mathbf{c})$ will run over compositions of $\Delta-$ rules over $\mathbf{c}$. We can sum up the above discussion in a counterpart of Theorem 15.1 which formulates a goodness - of - controller criterion.

Theorem 16.1 (*the sufficiency criterion of correctness of the controller*). Let $F_\mathbf{c}(v) \models \langle st(ag_\mathbf{c}), \Phi, \varepsilon \rangle$ where v is the valuation of leaf agents of the (C, Φ, ε)-scheme $\mathbf{c}$ and let $F_\mathbf{c}(v') \models \langle st(ag_\mathbf{c}), \Phi, \varepsilon' \rangle$ where v' is a valuation of leaf agents of $\mathbf{c}$ such that $v'(b_{ag}) = v(b_{ag})$ for $ag \in Control(\mathbf{c})$, $\varepsilon' < \varepsilon$. If $\{\varepsilon_{new}(ag)\}$ is a new assignment to agents defined by a composition $\Delta(\mathbf{c})$ of some Δ-rules such that $\varepsilon_{new}(ag) = \varepsilon(ag)$ for $ag \in Un\_control(\mathbf{c})$, $\varepsilon_{new}(ag_\mathbf{c}) = \varepsilon$ and $\{x_{ag} : ag \in Control(\mathbf{c})\}$ is the set of control parameters (inventory objects) satisfying $x_{ag} \models \langle st(ag), \Phi(ag), \varepsilon_{new}(ag)\rangle$ for $ag \in Control(\mathbf{c})$ then for the object $x_{new} = F_\mathbf{c}(v_1)$ constructed over the valuation v_1 of leaf agents in $\mathbf{c}$ such that $v_1(b_{ag}) = v'(b_{ag})$ for $ag \in Un\_control(\mathbf{c})$ and $v_1(b_{ag}) = x_{ag}$ for $ag \in Control(\mathbf{c})$ it holds that $x_{new} \models \langle st(ag_\mathbf{c}), \Phi, \varepsilon\rangle$. ∎

The meaning of Theorem 16.1 is that the controllable agents are able to compensate of noise which perturbs the state $\langle st(ag_\mathbf{c}), \Phi, \varepsilon \rangle$ to $\langle st(ag_\mathbf{c}), \Phi, \varepsilon' \rangle$ in case the search in the space of Δ - rules results in finding a composition of them which defines new, better uncertainty coefficients $\varepsilon_{new}(ag)$ for $ag \in Control(\mathbf{c})$; the new valuation v_1 defined by $\{\varepsilon_{new}(ag) : ag \in Control(\mathbf{c})\}$ satisfies the state $\langle st(ag_\mathbf{c}), \Phi, \varepsilon\rangle$.

The above approach can be treated as a first step towards modeling complex distributed dynamical systems. We expect that it can be extended to solve control problem for complex dynamical systems i.e. dynamical systems which are distributed, highly nonlinear, with vague concepts involved in their description. One can hardly expect that classical methods of control theory can be successfully applied to such complex systems.

17 Adaptive Calculus of Granules

Within paradigm of rough mereology one may formalize adaptive calculus of granules [335]. The metaphor of a granule, already present in rough set theory, has been recently advocated as a central notion of soft computing [532], [531].

17.1 Information Granules

We would like to present a general view on the problem of information granule construction and information granule calculus. Our main claims are :

- granules can be identified with finite relational structures (finite models);
- composition operations of granules (knowledge fusion) can be represented by operations on finite models;

- granules are fused, transformed and converted into decision by intelligent computing units (agents) or their schemes;
- schemes of agents are seen as decision classifiers and may be regarded as terms over granules and operations on them whose values are decisions;
- structure of granules and composition operations as well as the agent scheme and conversion operations on granules should be adaptively learned from data (the accessible information about objects).

We propose to realize this program on the basis of rough mereology. Let us mention that fuzzy set approach can be treated as a particular case of this methodology. One can consider also this approach as a basis for feature extraction in pattern recognition and machine learning. In the feature extraction process for instance we would like to develop learning strategies extracting from initial accessible information $Inf_A(x_i)$ about object x_i and decision $d(x_i)$ on x_i (for $i = 1, ..., n$ where n is the number of objects in data table) appropriate granules (finite structures) in the form of e.g. a finite model M and valuations v_i as well as formulas α_i for $i = 1, ..., k$ expressible in a language of the signature the same as M (and such that total size of these formulas is as small as possible) for which the following conditions hold

$d(x_i) \neq d(x_j)$ implies $\{\alpha_p : M, v_i \models \alpha_p\} \neq \{\alpha_p : M, v_j \models \alpha_p\}$
for any $i \neq j; i, j = 1.,,,, k$.

We start our presentation assuming the simple form of information granules defined by information systems. We have a variety of indiscernibility spaces $\{U/IND_B : B \subseteq A\}$; the Boolean algebra generated over the set of atoms U/IND_B by means of set - theoretical operations of union, intersection and complement is said to be the *B-algebra* $CG(B)$ *of B - pre-granules*. Any member of $CG(B)$ is called a *B - pre-granule*.

We have an alternative logical language in which we can formalize the notion of an *information pre-granule*; for a set of attributes $B \subseteq A$, we recall the definition of the B-logic L_B: elementary formulas of L_B are of the form (a, v) where $a \in B$ and $v \in V_a$. Formulas of L_B are built from elementary formulas by means of logical connectives $\vee, \wedge$; thus, each formula in DNF is represented as $\vee_{j \in J} \wedge_{i \in I_j} (a_i, v_i)$. The formulas of L_B, called *information pre-granules*, are interpreted in the set of objects U: the denotation $[(a, v)]$ of an elementary formula (a, v) is the set of objects satisfying the equation $a(x) = v$ i.e. $[(a, v)] = \{u \in U : a(u) = v\}$ and this is extended by structural induction viz. $[\alpha \vee \beta] = [\alpha] \cup [\beta]$, $[\alpha \wedge \beta] = [\alpha] \cap [\beta]$ for $\alpha, \beta \in L_B$.

Clearly, given a B -pre-granule $G \in CG(B)$, there exists an information pre-granule α_G of L_B such that $[\alpha_G] = G$.

An atom of the Boolean algebra $CG(B)$ will be called an *elementary B-pre-granule*; clearly, for any atom G of $CG(B)$ there exists an *elementary information pre-granule* α_G of the form $\wedge_{a \in B}(a, v_a)$ such that $[\alpha_G] = G$.

For given non-empty sets $B, C \subseteq A$, a pair (G_B, G_C) where $G_B \in CG(B)$ and $G_C \in CG(C)$ is called a (B,C) - *granule of knowledge*. There exists therefore an *information granule* $(\alpha_{G_B}, \alpha_{G_C})$ such that $\alpha_{G_B} \in L_B, \alpha_{G_C} \in L_C, [\alpha_{G_B}] = G_B$ and $[\alpha_{G_C}] = G_C$. If G_B, G_C are atoms then the pair (G_B, G_C) is called an *elementary*

(B,C) - granule.

One can associate with any granule (G',G) where $G' \in CG(B'), G \in CG(B)$ a rule $\alpha_G \Longrightarrow \alpha_{G'}$ [335].

The notion of a granule corresponds to the logical content (*logic of knowledge*) of the information system; however, there is more to the notion of a granule of knowledge: we have to take into account the restrictions which on the choice of *good granules* are imposed by the structure and demands of two interfaces: *input_interface* which controls the input objects (signals) and *output_interface* which controls the output objects (actions, signals).

Consider a granule (G,G'); let $G = [\alpha_G], G' = [\alpha_{G'}]$. There are two characteristics of the granule (G,G') important in applications to adaptive synthesis of complex objects viz. the characteristic whose values measure what part of $[G']$ is in $[G]$ (the *strength of the covering* of the rule $\alpha_G \Longrightarrow \alpha_{G'}$ and the characteristic whose values measure what part of $[G]$ is in $[G']$ (the *strength of the support* for the rule $\alpha_G \Longrightarrow \alpha_{G'}$).

To select sufficiently strong rules, we would set thresholds : $tr, tr' \in [0,1]$. We define then, by analogy with machine learning techniques, two characteristics:

(ρ) $\rho(\alpha, \alpha') = \mu([\alpha], [\alpha'])$;
(η) $\eta(\alpha, \alpha') = \mu([(\alpha')], [\alpha])$ for any (B,C)-information granule (α, α')

and we call a $(\mu, B, C, tr, tr')-$ *information granule of knowledge* any (B,C)-information granule (α, α') such that

(i) $\rho(\alpha, \alpha') \geq tr$ and
(ii) $\eta(\alpha, \alpha') \geq tr'$.

The set of all $(\mu, B, C, tr, tr')-$ granules corresponding to $(\mu, B, C, tr, tr')-$ information granules generates a Boolean algebra of (μ, B, C, tr, tr')-*granules of knowledge*. Let us observe that given sets $B, C \subseteq A, \alpha \in L_B$ and $\alpha' \in L_C$, we may define the value $Gr(B, C, \mu, tr, tr', \alpha, \alpha')$ to be $TRUE$ in the case when the pair (α, α') is an $(\mu, B, C, tr, tr')-$ granule of knowledge. In this way we define the relation Gr which we call the *granulation relation* induced by the triple *(input_interface, logic of knowledge, output_interface)* related to the information system **A**.

The functions η, ρ and thresholds tr, tr' introduced above have been used to present an example of the interface between two sources of information e.g. between input information source and inner agent knowledge. This is necessary because very often the exact interpretation of information from one source into another one is impossible. Different methods for constructing and tuning these interfaces up, crucial for granular computing, can be developed using rough set and fuzzy set approaches.

Rough inclusion μ_0 may enter our discussion of a granule and of the relation Gr in each of the following ways:

- concerning the definitions $(\eta), (\rho)$ of functions η and ρ, we may replace in them the rough membership function μ with μ_0 possibly better fitted to a context.

- the function μ_0 can be extended to a rough inclusion μ^* by means of a formula above, relative to a t-conorm $\perp$ and a t-norm $\top$ suitably chosen. This rough inclusion μ^* can now be used to measure the information granule closeness.

17.2 Information Granules of the Form (Φ, ε)

In applications it is often convenient to use another language for information granules description allowing for more compressed description of information granules. Let us assume that a pre-rough inclusion is generated over a set B of attributes. Let μ^* be the rough inclusion generated from μ_0. Now, a new information granule language $\mathbf{L}_B$ for $B \subseteq A$ consists of pairs (Φ, ε) where Φ is an elementary B-information pre-granule and $\varepsilon \in [0, 1]$. By $[\Phi, \varepsilon]_B$ we denote the B-granule $[\alpha_{\Phi,\varepsilon}]$ where $\alpha_{\Phi,\varepsilon}$ is a B-information granule being a disjunction of all $\wedge u$ where u is of the form $\{(a, v) : a \in B, v \in V_a\}$ (where V_a is the value set of a) and $\mu^*([\wedge u], [\Phi]) \geq \varepsilon$.

We say that
$$x \models_{ag} (\Phi, \varepsilon) \text{ iff } \mu^*([\wedge u(x)], [\Phi]) \geq \varepsilon$$
where $u(x)$ is $\{(a, a(x)) : a \in B\}$.

One can also consider more general information granules taking instead of formulas (Φ, ε) sets $(\Phi, \varepsilon_1), ..., (\Phi, \varepsilon_n)$ where $\varepsilon_1 \leq ... \leq \varepsilon_n$ are interpreted as closeness degrees to Φ.

17.3 Synthesis in Terms of Granules

We adopt here our scheme for approximate synthesis and we refer to the notation therein. We will show how the synthesis process over this scheme can be driven by granule exchange among agents. We include the following changes in comparison to this scheme.

1. We introduce an additional agent Cs (the *customer*), where $Cs \notin Ag$, whose actions consist in issuing approximate specifications Ψ describing the desired object (signal, action etc.) to be synthesized by the scheme (Inv, Ag) where Inv is the set of inventory objects of agents from Ag.
2. We assume that the *customer* − *root of Ag interface* produces approximate formulas of the form

$$(\Phi(root(Ag)), \varepsilon(root(Ag)))$$

for the root agent $root(ag)$ of the agent scheme corresponding to approximate specifications Ψ in the sense that any object satisfying the formula

$$(\Phi(root(Ag)), \varepsilon(root(Ag)))$$

is regarded as satisfying the approximate specification Ψ to a satisfactory degree.

3. Let $[\Phi(ag), \varepsilon(ag)]_{ag} =$
$$\{x \in U(ag) : x \models_{ag} (\Phi(ag), \varepsilon(ag))\}.$$

4. For a given specification $(\Phi(ag), \varepsilon(ag))$ at ag, k-ary operation $o(ag)$ at ag and a mereological connective $f(ag)$ (see above) the decomposition process returns a sequence
$$(\Phi_1, \varepsilon_1), ..., (\Phi_k, \varepsilon_k)$$
of specifications (in the language of ag) for agents $ag_1,...,ag_k$ satisfying
$$\emptyset \neq o(ag)([\Phi_1, \varepsilon_1]_{ag} \times ... \times [\Phi_k, \varepsilon_k]_{ag})$$
$$\subseteq [\Phi(ag), f(ag)(\varepsilon_1, ..., \varepsilon_k)]_{ag}]$$
where $f(ag)(\varepsilon_1, ..., \varepsilon_k) \geq \varepsilon(ag)$.

5. Next, information granules $\alpha_1, ..., \alpha_k$ at $ag_1, ..., ag_k$, respectively, are chosen in such a way that granules
$$[\alpha_1]_{ag_1}, ..., [\alpha_k]_{ag_k}$$
and respectively
$$[\Phi_1, \varepsilon_1]_{ag}, ..., [\Phi_k, \varepsilon_k]_{ag}$$
are sufficiently close (in the tuning process parameters ρ, η, tr, tr' are fixed). The closeness of the granules should guarantee that the following inclusion is true
$o(ag)([\alpha_1]_{ag_1} \times ... \times [\alpha_k]_{ag_k})$
$$\subseteq [\Phi(ag), f(ag)(\varepsilon_1, ..., \varepsilon_k)]_{ag}.$$
Formulae $\alpha_1,...,\alpha_k$ at $ag_1,...,ag_k$ are described in the form
$$(\Phi(ag_1), \varepsilon(ag_1)), ..., (\Phi(ag_k), \varepsilon(ag_k))$$
for agents $ag_1,...,ag_k$, respectively.

The meaning of an expression $x \models_S (\Phi, \varepsilon)$ is that an agent scheme S is yielding at the $root(S)$ the object x. A goodness - of - granule synthesis scheme criterion can be formulated in the following:

Theorem 17.1 ([335]). Let S be an agent scheme satisfying the following conditions:

(i) the root of S denoted by $root(S)$ has attached a specification
$$(\Phi(root(S)), \varepsilon(root(S)));$$

(ii) any non-leaf and non-root agent ag of S satisfies conditions stated in (4)-(5);
(iii) for any leaf agent ag of S the attached specification $(\Phi_{ag}, \varepsilon_{ag})$ is satisfied by some object from the inventory object set INV.

Then the scheme S is yielding at $root(S)$ an object x satisfying
$$x \models_S (\Phi(root(S)), \varepsilon(root(S))).$$

∎

17.4 Adaptivity of calculus of granules

The adaptivity of our scheme is due to the several factors. Among them are

- The possibility of changing the parameters

$$\mu, tr, tr', B, C$$

in the granulation predicate

$$Gr(ag)(B, C, \mu, tr, tr', \alpha, \alpha')$$

for any agent $ag \in Ag$.
- The possibility of new granule formation at any agent $ag \in Ag$ in the dynamic process of synthesis.
- The possibility of forming new rough inclusion at any agent $ag \in Ag$ in the dynamic process of synthesis e.g. by choosing $\top, \bot$ in the definition of μ.

In conclusions we discuss some other potential applications of rough mereological approach.

18 Conclusions

The rough set approach to data analysis has many important advantages. Some of them are listed below.

- Synthesis of efficient algorithms for finding hidden patterns in data.
- Identification of relationships that would not be found using statistical methods.
- Representation and processing of both qualitative and quantitative parameters and mixing of user-defined and measured data.
- Reduction of data to a minimal representation.
- Evaluation of the significance of data.
- Synthesis of classification or decision rules from data.
- Legibility and straightforward interpretation of synthesized models.

Most algorithms based on the rough set theory are particularly suited for parallel processing, but in order to exploit this feature fully, a new computer organization based on rough set theory is necessary.

Although rough set theory has many achievements to its credit, nevertheless several theoretical and practical problems require further attention.

Especially important is widely accessible efficient software development for rough set based data analysis, particularly for large collections of data.

Despite of many valuable methods of efficient, optimal decision rule generation methods from data, developed in recent years based on rough set theory - more research here is needed, particularly, when quantitative and quantitative attributes are involved. Also an extensive study of a new approach to missing data is very important. Comparison to other similar methods still requires due attention, although important results have been obtained in this area. Particularly interesting seems to be a study of the relationship between neural network and rough set approach to feature extraction from data.

Last but not least, rough set computer is badly needed for more advanced applications. Some research in this area is already in progress.

We would like to stress some areas of research related to the rough mereological approach. They are in particular important for further development of rough set theory and soft computing. They can be characterized as new algorithmic methods for inducing structures of information granules and information granule calculus from data (also in distributed and multi-agent environments). Among them are adaptive algorithmic methods for:

- extracting logical (algebraic) structures of information granules from data: this belongs to the process of searching for a model couched in logical (algebraic) terms;
- constructing interfaces among various knowledge structures: this group of problems is relevant in granular computing; as observed above, granules of knowledge are the result, among other factors, of uncertainty immanent to interfaces among various sources of knowledge;
- extracting distance functions for similarity measures from data: here we would like to have clustering methods based on closeness measures to construct aggregation - based models;
- inducing exact dependencies: this group of problems belong to the second step i.e. searching for elements of the model and dependencies among them; exact dependencies constitute the skeleton along which we organize schemes for approximate reasoning;
- inducing approximate dependencies: here we search for approximate i.e. close to exact dependencies and possible ways of expressing them like described in this collection and literature default rules, templates, rough classifiers, rough mereological connectives etc.;
- inducing networks of dependencies: emulation of schemes of approximate reasoning including also algorithmic methods for inducing concurrent data models from data.

We propose rough mereology as a general framework for investigations in these directions. Taking this point of view the research should be concentrated around two main groups of problems, namely methods of adaptive learning from data of components of schemes of approximate reasoning (like standards, rough inclusions, mereological connectives, decomposition schemes etc. (see Sect. 2.4)) and adaptive learning of schemes of approximate reasoning (synthesis schemes of agents).

The research results in the above mentioned areas will also have important impact on development of new methods for KDD, in particular for development of algorithmic methods for pattern extraction from data (also in multi-agent environment) and extracting from data calculus for approximate reasoning on sets of extracted patterns (e.g. algorithmic methods for large data table decomposition, synthesis of global laws from local findings).

The advantage of employing various sources of knowledge and various structures of knowledge in data mining and knowledge discovery implies that new algorithmic methods are desirable for hybrid systems in which rough set methods will be applied along with methods based on (one or more of) the following: fuzzy sets; neural nets; evolutionary strategies; statistical reasoning; belief nets; evidence theory.

Problems of knowledge discovery should be studied also from the point of view of complexity. The following topics seem to be important:

- analysis of complexity of knowledge discovery processes: complexity of extracting classes of problems solvers;
- analysis of complexity of problems of approximate reasoning: complexity of feature and model extraction/selection, complexity of data mining processes, complexity of knowledge discovery processes;
- quality of heuristics for hard problems of approximate reasoning: quality - complexity trade - off;
- refinements of complexity theory for problems of approximate reasoning: classification issues, relations to minimal length description.

Let us observe that this analysis would require in many cases transgressing the classical complexity theory.

In addition to areas of application discussed in the paper we would like to point to some areas where perspectives of applications of rough sets are promising as borne out by the current research experience. Among them are applications in: data mining and knowledge discovery; process control; case based reasoning; conflict analysis and negotiations; natural language processing; software engineering.

Finally, the progress in the discussed above rough merology, being an extension of rough set theory, should bring forth:

- new computation model based on information granulation and granule calculus;
- new software systems for mentioned above important applications;
- hardware developments: a rough mereological processor and a rough mereological computer.

We are convinced that progress in the above areas is of the utmost importance for creating new methods, algorithms, software as well as hardware systems which prove the applicability of rough set techniques to challenging problems of Data Mining, Knowledge Discovery and other important areas of applications.

Acknowledgments

The results presented in this tutorial are due to a long term collaborative effort of several international groups. The authors wish to thank their colleagues who contributed to this tutorial both directly and indirectly. Several results from joint work were re-used here and citations provided as detailed as it was judged feasible.

This research has been supported in part by the European Union 4th Framework Telematics project CARDIASSIST, by the ESPRIT-CRIT 2 project #20288, by the Norwegian Research Council (NFR) grants #74467/410, #110177/730 and by the Polish National Research Committee (KBN) grant #8T11C01011.

References

1. A. Aamodt, J. Komorowski (Eds.) (1995), *Proc. Fifth Scandinavian Conference on Artificial Intelligence.* Trondheim, Norway, May 29-31, Frontiers. In: Artifical Intelligence and Applications **28**, IOS Press.
2. R.L. Ackoff, M.W. Sasieni (1968), *Fundamentals of operational research.* Wiley, New York.
3. H. Almuallim, T. G. Dietterich (1994), *Learning Boolean concepts in the presence of many irrelevant features.* Artificial Intelligence **69(1-2)**, pp. 279-305.
4. S. Amarel et al. (1991), *PANEL on AI and design.* In: Twelfth International Joint Conference on Artificial Intelligence (IJCAI-91), Morgan Kaufmann, San Mateo, pp. 563-565.
5. A. An, C. Chan, N. Shan, N. Cercone, W. Ziarko (1997), *Applying knowledge discovery to predict water-supply consumption.* IEEE Expert **12/4**, pp. 72-78.
6. A. An, N. Shan, C. Chan, N. Cercone, W. Ziarko (1995), *Discovering rules from data for water demand prediction.* In: Proceedings of the Workshop on Machine Learning in Engineering (IJCAI'95), Montreal, pp. 187-202; see also, Journal of Intelligent Real-Time Automation, Engineering Applications of Artificial Intelligence **9/6**, pp. 645-654.
7. T. Arciszewski, W. Ziarko (1987), *Adaptive expert system for preliminary design of wind-bracings in steel skeleton structures.* In: Second Century of Skyscraper, Van Norstrand, pp. 847-855.
8. P. Balbiani (1998), *Axiomatization of logics based on Kripke models with relative accessibility relations.* In: Orłowska [270], pp. 553-578.
9. C.A. Bana e Costa (1990), *Readings in multiple-criteria decision aid.* Springer-Verlag, Berlin.
10. M. Banerjee (1997), *Rough sets and 3-valued Lukasiewicz logic.* Fundamenta Informaticae **32/1**, pp. 213-220.
11. M. Banerjee, M.K. Chakraborty (1998), *Rough logics: A survey with further directions.pedrycz@ee.ualberta.ca* In: Orłowska [270] pp. 579-600.
12. M. Banerjee, S.K. Pal (1996), *Roughness of a fuzzy set.* Information Sciences Informatics and Comp. Sc. **93/3-4**. pp. 235-246.
13. J. Bazan (1998), *A comparison of dynamic non-dynamic rough set methods for extracting laws from decision tables.* In: Polkowski and Skowron [337], pp. 321-365.
14. J. Bazan, H.S. Nguyen, T.T. Nguyen, A. Skowron, J. Stepaniuk (1994), *Some logic and rough set applications for classifying objects.* Institute of Computer Science, Warsaw University of Technology, ICS Research Report **38/94**.
15. J. Bazan, H.S. Nguyen, T.T. Nguyen, A. Skowron, J. Stepaniuk (1995), *Application of modal logics and rough sets for classifying objects.* In: M. De Glas, Z. Pawlak (Eds.), Proceedings of the Second World Conference on Fundamentals of Artificial Intelligence (WOCFAI'95), Paris, July 3-7, Angkor, Paris, pp. 15-26.

16. J.G. Bazan, H.S. Nguyen, T.T. Nguyen, A. Skowron, J. Stepaniuk (1998), *Synthesis of decision rules for object classification*. In: Orłowska [270], pp. 23–57.
17. J. Bazan, A. Skowron, P. Synak (1994), *Market data analysis*. Institute of Computer Science, Warsaw University of Technology, ICS Research Report **6/94**.
18. J. Bazan, A. Skowron P. Synak (1994), *Dynamic reducts as a tool for extracting laws from decision tables*. Proc. Symp. on Methodologies for Intelligent Systems, Charlotte, NC, USA, Oct. 16-19, LNAI, **869**, Springer-Verlag, pp. 346–355.
19. D. Brindle (1995), *Speaker-independent speech recognition by rough sets analysis*. In: Lin, Wildberger [194], pp. 101–106.
20. T. Beaubouef, F.E. Petry (1993), *A rough set model for relational databases*. In: Ziarko [537], pp. 100–107.
21. T. Beaubouef, F.E. Petry (1994), *Rough querying of crisp data in relational databases*. In: Ziarko [537], pp. 85–88.
22. S. Bodjanova (1997), *Approximation of fuzzy concepts in decision making*. Fuzzy Sets and Systems **85**, pp. 23–29.
23. P. Borne, G. Dauphin–Tanguy, C. Sueur, S. El Khattabi (Eds.), *Proceedings of IMACS Multiconference, Computational Engineering in Systems Applications (CESA'96)*. July 9–12, Lille, France, Gerf EC Lille – Cite Scientifique **3/4**.
24. B. Bouchon–Meunier, M. Delgado, J.L. Verdegay, M.A. Vila, R.R. Yager (1996), *Proceedings of the Sixth International Conference. Information Processing Management of Uncertainty in Knowledge–Based Systems (IPMU'96)*. July 1-5, Granada, Spain **1–3**, pp. 1–1546.
25. B. Bouchon–Meunier, R.R. Yager (1998), *Proc. Seventh Conference on Information Processing and Management of Uncertainty in Knowledge-Based Systems, (IPMU'98)*. July 6–10, Université de La Sorbonne, Paris, France, Edisions, E.D.K., Paris, pp. 1–1930.
26. Z. Bonikowski, E. Bryniarski, and U. Wybraniec–Skardowska (1998), *Extensions and intensions in the rough set theory*. Information Sciences **107**, 149–167.
27. F. M. Brown (1990), *Boolean Reasoning*. Kluwer Academic Publishers, Dordrecht.
28. C. Browne, I. Düntsch, G. Gediga (1998), *IRIS revisited, A comparison of discriminant enhanced rough set data analysis*. In: Polkowski and Skowron [338], pp. 347–370.
29. E. Bryniarski E., U. Wybraniec-Skardowska (1997), *Generalized rough sets in contextual spaces*. In: Lin and Cercone [189] pp. 339–354.
30. A. Budihardjo, J.W. Grzymała–Busse, L. Woolery (1991), *Program LERS–LB 2.5 as a tool for knowledge acquisition in nursing*. In: Proceedings of the Fourth International Conference on Industrial Engineering Applications of Artificial Intelligence Expert Systems, Koloa, Kauai, Hawaii, June 2–5, pp. 735–740.
31. M. Burns (1993), *Resources: Automated Fabrication. Improving Productivity in Manufacturing*. Prentice Hall, Englewood Cliffs, NJ.
32. W. Buszkowski, E. Orłowska (1998), *Indiscernibility-based formalisation of dependencies in information systems*. In: Orłowska [270], pp. 293–315.
33. W. Buszkowski (1998), *Approximation spaces and definability for incomplete information systems*. In: Polkowski and Skowron [339], pp. 115–122.
34. U. Carlin, J. Komorowski, A. Øhrn (1998), *Rough set analysis of patients with suspected acute appendicitis*. In: Bouchon–Meunier and Yager [25], pp. 1528–1533.
35. J. Catlett (1991), *On changing continuos attributes into ordered discrete attributes*. In: Y. Kodratoff, (Ed.), Machine Learning-EWSL-91, Proc. of the European Working Session on Learning, Porto, Portugal, March 1991, LNAI, pp. 164–178.

36. G. Cattaneo (1996), *A unified algebraic approach to fuzzy algebras and rough approximations*. In: R. Trappl (Ed.), Proceedings of the 13th European Meeting on Cybernetics and Systems Research (CSR'96), April 9-12, The University of Vienna **1**, pp. 352-357.
37. G. Cattaneo (1996), *Abstract rough approximation spaces (Bridging the gap between fuzziness and roughness)*. In: Petry and Kraft [314], pp. 1129-1134.
38. G. Cattaneo (1996), *Mathematical foundations of roughness and fuzziness*, In: Tsumoto, Kobayashi, Yokomori, Tanaka, and Nakamura [475]. pp. 241-247.
39. G. Cattaneo (1997), *Generalized rough sets. Preclusivity fuzzy-intuitionistic (BZ) lattices*. Studia Logica **58**, pp. 47-77.
40. G. Cattaneo (1998), *Abstract approximation spaces for rough theories*. In: Polkowski and Skowron [337], pp. 59-98.
41. G. Cattaneo, R. Giuntini, R. Pilla (1997), *MVBZ algebras and their substructures. The abstract approach to roughness and fuzziness*. In: Mares, Meisar, Novak, and Ramik [204], pp. 155-161.
42. G. Cattaneo, R. Giuntini, R. Pilla (1997), *$MVBZ^{dM}$ and Stonian algebras (Applications to fuzzy sets and rough approximations)*. In: Fuzzy Sets and Systems.
43. M.K. Chakraborty, M. Banerjee (1997), *In search of a common foundation for rough sets and fuzzy sets*. In: Zimmermann [545], pp. 218-220.
44. Y.-Y. Chen, K. Hirota, J.-Y. Yen (Eds.) (1996), *Proceedings of 1996 ASIAN FUZZY SYSTEMS SYMPOSIUM - Soft Computing in Intelligent Systems Information Processing*. December 11-14, Kenting, Taiwan, ROC.
45. M.R. Chmielewski, J.W. Grzymala-Busse (1994), In: Lin [186] pp. 294-301.
46. J. Cios, W. Pedrycz, R.W. Swiniarski (1998), *Data Mining in Knowledge Discovery*, Academic Publishers.
47. S. Comer (1991), *An algebraic approach to the approximation of information*. Fundamenta Informaticae **14**, pp. 492-502.
48. J.H. Connolly J.H., E.A. Edmunds E.A. (1994), *CSCW and Artificial Intelligence*. Springer - Verlag, Berlin.
49. A. Czajewski (1998), *Rough sets in optical character recognition*. In: Polkowski and Skowron [339], pp. 601-604.
50. A. Czyżewski (1995), *Some methods for detection interpolation of impulsive distortions in old audio recordings*. In: IEEE ASSP Workshop on Applications of Signal Processing to Audio Acoustics Proceedings, October 15-18, New York, USA.
51. A. Czyżewski (1996), *Speaker-independent recognition of digits - Experiments with neural networks, fuzzy logic rough sets*. Journal of the Intelligent Automation and Soft Computing **2/2**, pp. 133-146.
52. A. Czyżewski (1996), *Mining knowledge in noisy audio data*. In: Simoudis, Han, and Fayyad [393] pp. 220-225.
53. A. Czyżewski (1997), *Learning algorithms for audio signal enhancement - Part I, Neural network implementation for the removal of impulse distortions*. Journal of the Audio Engineering Society **45/10**, pp. 815-831.
54. A. Czyżewski (1997), *Learning algorithms for audio signal enhancement - Part II, Rough set method implementation for the removal of hiss*. Journal of the Audio Engineering Society **45/11**, pp. 931-943.
55. A. Czyżewski (1998), *Soft processing of audio signals*. In: Polkowski and Skowron [338], pp. 147-165.
56. A. Czyżewski, A. Kaczmarek (1995), *Speaker-independent recognition of isolated words using rough sets*. In: Wang [503], pp. 397-400.

57. A. Czyżewski, B. Kostek (1996), *Rough set-based filtration of sound applicable to hearing prostheses*. In: Tsumoto, Kobayashi, Yokomori, Tanaka, and Nakamura [475], pp. 168–175.
58. A. Czyżewski, B. Kostek (1996), Restoration of old records employing artificial intelligence methods. Proceedings of IASTED International Conference – Artificial Intelligence, Expert Systems Neural Networks, August 19–21, Honolulu, Hawaii, USA, pp. 372–375.
59. A. Czyżewski, R. Królikowski (1997), *New methods of intelligent filtration coding of audio*. In: 102nd Convention of the Audio Engineering Society, March 22–25, preprint **4482**, Munich, Germany .
60. A. Czyżewski, R. Królikowski (1998), Applications of fuzzy logic rough sets to audio signal enhancement. In this book.
61. A. Czyżewski, R. Królikowski, P. Skórka (1996), *Automatic detection of speech disorders*. In: Zimmermann [544], pp. 183–187.
62. B.V. Dasarathy (Ed.) (1991), *Nearest Neighbor Pattern Classification Techniques*. IEEE Computer Society Press.
63. R. Deja (1996), *Conflict analysis*. In: Tsumoto, Kobayashi, Yokomori, Tanaka, and Nakamura [475], pp. 118–124.
64. S. Demri (1996), *A class of information logics with a decidable validity problem*. In: Lecture Notes in Computer Science **1113**, Springer-Verlag pp. 291–302.
65. S. Demri, E. Orłowska (1998), *Logical analysis of indiscernibility*. In: Orłowska [270], pp. 347–380.
66. J.S. Deogun, V.V. Raghavan, A. Sarkar, H. Sever (1997), *Data mining: Trends in research and development*. In: Lin and Cercone [189]. pp. 9–45.
67. J. Dougherty, R. Kohavi, M. Sahami (1995), *Supervised Unsupervised Discretization of Continuous Features*. Proceedings of the Twelfth International Conference on Machine Learning, Morgan Kaufmann, San Francisco, CA, pp. 194–202.
68. D. Dubois, H. Prade (1980), *Fuzzy Sets and Systems: Theory and Applications*. Academic Press.
69. D. Dubois, H. Prade (1987), *Twofold fuzzy sets and rough sets - Some issues in knowledge representation*. Fuzzy Sets and Systems **23**, pp. 3–18.
70. D. Dubois, H. Prade (1990), *Rough fuzzy sets and fuzzy rough sets*. International J. General Systems **17**, pp. 191–209.
71. D. Dubois, H. Prade (1992), *Putting rough sets and fuzzy sets together*. In: Słowiński [419], pp. 203–232.
72. D. Dubois, H. Prade (1992), *Comparison of two fuzzy set-based logics: similarity logic and possibilistic logic*. In: Proceedings FUZZ-IEEE'95, Yokohama, Japan, pp. 1219–1236.
73. D. Dubois, H. Prade, R.R. Yager(Eds.) (1993), *Readings in Fuzzy Sets for Intelligent Systems*, Morgan Kaufmann, San Mateo.
74. D. Dubois, F. Esteva, P. Garcia, L. Godo, and H. Prade (1997), *Logical approach to interpolation based on similarity*. Journal of Approximate Reasoning (to appear) (available as IIIA Research Report 96-07).
75. D. Dubois, H. Prade (1998), *Similarity-based approximate reasoning*. In: J.M. Zurada, R.J. Marks II, and X.C.J. Robinson (Eds.), *Proceedings of the IEEE Symposium*, Orlando, FL, June 17–July 1st, IEEE Press, pp. 69–80.
76. D. Dubois, H. Prade (1998), *Similarity versus preference in fuzzy set-based logics*. In: Orłowska [270] pp. 440–460.
77. E.H. Durfee (1988), *Coordination of Distributed Problem Solvers*. Kluwer Academic Publishers, Boston.

78. I. Düntsch (1997), *A logic for rough sets*. Theoretical Computer Science **179/1-2** pp. 427–436.
79. I. Düntsch (1998), *Rough sets and algebras of relations*. In: Orłowska [270], pp. 95–108.
80. I. Düntsch, G. Gediga, G. (1997), *Statistical evaluation of rough set dependency analysis*. International Journal of Human-Computer Studies **46**, pp. 589–604.
81. I. Düntsch, G. Gediga (1997), *Non-invasive data analysis*. Proceedings of the Eighth Ireland Conference on Artificial Intelligence, Derry, pp. 24–31.
82. I. Düntsch, G. Gediga (1997), *Algebraic aspects of attribute dependencies in information systems*. Fundamenta Informaticae **29**, pp. 119–133.
83. I. Düntsch, G. Gediga (1998), *IRIS revisited: A comparison of discriminant and enhanced rough set data analysis*. In: Polkowski and Skowron [337], pp. 347–370.
84. A.E. Eiben, T.J. Euverman, W. Kowalczyk, F. Slisser (1998), *Modelling customer retention with statistical techniques, rough data models, genetic programming*. In this book.
85. E. Ephrati, J.S. Rosenschein (1994), *Divide and conquer in multi-agent planning*. In: Proceedings of the Twelfth National Conference on Artificial Intelligence (AAAI'94) AAAI Press/MIT Press, Menlo-Park, CA, pp. 375–380.
86. F. Esteva, P. Garcia, L. Godo (1993), *On the relatonsship between preference and similarity-based approaches to possibilistic reasoning*. In: Proceedings of the 2nd IEEE International Conference on Fuzzy Systems (FUZZ-IEEE'93), San Francisco, CA, March 28–April 1st, pp. 918–923.
87. F. Esteva, P. Garcia, L. Godo (1997), *Similarity reasoning: Logical grounds and applications*. In: Zimmermann [545], pp. 113–117.
88. L. Farinas del Cerro, E. Orłowska (1985), *DAL–A logic for data analysis*. Theoretical Computer Science **36**, pp. 251–264; see also, Corrigendum, ibidem **47** (1986), pp. 345.
89. L. Farinas del Cerro, H. Prade (1986), *Rough sets, twofold fuzzy sets and modal logic – Fuzziness in indiscernibility and partial information*. In: A. Di Nola, A.G.S. Ventre (Eds.), The Mathematics of Fuzzy Systems, Verlag TUV Rheinland, Köln, pp. 103–120.
90. U.M. Fayyad, K.B. Irani (1992), *The attribute election problem in decision tree generation*. Proc. of AAAI-92, July 1992, San Jose, CA., MIT Press, pp. 104-110.
91. U. Fayyad, G. Piatetsky-Shapiro, G. (Eds.) (1996), *Advances in Knowledge Discovery and Data Mining*. MIT/AAAI Press.
92. U.M. Fayyad, R. Uthurusamy (Eds.) (1995), *Proceedings of the First International Conference on Knowledge Discovery Data Mining (KDD'95)*. Montreal, August, AAAI Press, Menlo Park, CA, pp. 1–354.
93. J. Fibak, Z. Pawlak, K. Słowiński, R. Słowiński (1986), *Rough sets based decision algorithm for treatment of duodenal ulcer by HSV*. Bull. Polish Acad. Sci. Biological Sci. **34/10-12**, pp. 227–246.
94. J. Fibak, K. Słowiński, R. Słowiński (1986), *The application of rough sets theory to the verification of treatment of duodenal ulcer by HSV*. In: Proceedings of the Sixth International Workshop on Expert Systems their Applications, Agence de l'Informatique, Paris, pp. 587–599.
95. J. Friedman, R. Kohavi, Y. Yun (1996), *Lazy decision trees*. Proc. of AAAI-96, pp. 717–724.
96. M. Fedrizzi, J. Kacprzyk, H. Nurmi (1996), *How different are social choice functions, A rough sets approach*. Quality & Quantity **30**, pp. 87–99.

97. M.C. Fernandes–Baizan, E. Menasalvas Ruiz, M. Castano (1996), *Integrating RDMS data mining capabilities using rough sets*. In: Bouchon–Meunier, Delgado, Verdegay, Vila, and Yager [24] **2**, pp. 1439–1445.
98. K. Funakoshi, T.B. Ho (1996), *Information retrieval by rough tolerance relation*. In: Tsumoto, Kobayashi, Yokomori, Tanaka, and Nakamura [475], pp. 31–35.
99. H. Furuta, M. Hirokane, Y. Mikumo (1998), *Extraction method based on rough set theory of rule-type knowledge from diagnostic cases of slope-failure danger levels*. In: Polkowski and Skowron [338], pp. 178–192.
100. B. Ganter, R. Wille (1996), *Formale Begriffsanalyse: Mathematische Grundlagen*. Springer–Verlag, Berlin, pp. 1–286.
101. M. L. Geleijnse, A. Elhendy, R. van Domburg et al. (1996), *Prognostic Value of Dobutamine-Atropine Stress Technetium-99m Sestamibi Perfusion Scintigraphy in Patients with Chest Pain*. Journal of the American College of Cardiologists, **28/2**, August, pp. 447–454.
102. R. Golan, W. Ziarko (1995), *A methodology for stock market analysis utilizing rough set theory*. In: Proceedings of IEEE/IAFE Conference on Computational Intelligence for Financial Engineering, New York City, pp. 32–40.
103. S. Greco, B. Matarazzo, R. Słowiński (1997), *Rough set approach to multi-attribute choice ranking problems*. Institute of Computer Science, Warsaw University of Technology, ICS Research Report **38/95** (1995); see also, G. Fandel, T. Gal (Eds.), Multiple Criteria Decision Making: Proceedings of 12th International Conference in Hagen, Springer–Verlag, Berlin, pp. 318–329.
104. S. Greco, B. Matarazzo, R. Słowiński (1998), *A new rough set approach to evaluation of bankruptcy risk*. In: C. Zopounidis (Ed.), New Operational Tools in the Management of Financial Risks, Kluwer Academic Publishers, Dordrecht pp. 121–136.
105. S. Greco, B. Matarazzo, R. Słowinski (1998), *Rough approximation of a preference relation in a pairwise comparison table*. In: Polkowski and Skowron [338] pp. 13–36.
106. J.W. Grzymała-Busse (1986), *Algebraic properties of knowledge representation systems*. In: Z.W. Ras, M. Zemankova (Eds.), Proceedings of the ACM SIGART International Symposium on Methodologies for Intelligent Systems (ISMIS'86), Knoxville, Tennessee, October 22–24, ACM Special Interest Group on Artificial Intelligence, pp. 432–440.
107. J.W. Grzymała–Busse (1993), *ESEP, an expert system for enviromental protection*. In: Ziarko [537], pp. 499–508.
108. J.W. Grzymała-Busse (1997), *A new version of the rule induction system LERS*. Fundamenta Informaticae **31**, pp. 27–39.
109. J.W. Grzymała–Busse (1998), *Applications of the rule induction system LERS*. In: Polkowski and Skowron [337], pp. 366–375.
110. J.W. Grzymała-Busse L.K. Goodwin (1997), *Predicting preterm birth risk using machine learning from data with missing values*. In: S. Tsumoto (Ed.), Bulletin of International Rough Set Society **1/2**, pp. 17–21.
111. J.W. Grzymała–Busse, L.K. Goodwin (1996), *A comparison of less specific versus more specific rules for preterm birth prediction*. In: Proceedings of the First Online Workshop on Soft Computing WSC1 on the Internet, served by Nagoya University, Japan, August 19–30, pp. 129–133.
112. J.W. Grzymała–Busse, J.D. Gunn (1995), *Global temperature analysis based on the rule induction system LERS*. In: Proceedings of the Fourth International Workshop on Intelligent Information Systems, Augustów, Poland, June 5–9, 1995, Institute od Computer Science, Polish Academy of Sciences, Warsaw, pp. 148–158.

113. J.W. Grzymała-Busse, L.J. Old (1997), *A machine learning experiment to determine part of speach from word-endings*. In: Ras and Skowron [355], pp. 497–506.
114. J.W. Grzymała-Busse, S. Than (1993), *Data compression in machine learning applied to natural language*. Behavior Research Methods, Instruments & Computers **25**, pp. 318–321.
115. J.W. Grzymała-Busse, C.P.B. Wang (1996), *Classification rule induction based on rough sets*. In: [314], pp. 744–747.
116. J.W. Grzymała-Busse, L. Woolerly (1994), *Improving prediction of preterm birth using a new classification scheme rule induction*. In: Proceedings of the 18th Annual Symposium on Computer Applications in Medical Care (SCAMC), November 5-9, Washington D. C. pp. 730–734.
117. J.D. Gunn, J.W. Grzymała-Busse (1994), *Global temperature stability by rule induction. An interdisciplinary bridge*. Human Ecology **22**, pp. 59–81.
118. M. Hadjimichael, A. Wasilewska (1993), *Rough sets-based study of voter preference in 1988 USA presidential election*. In: Słowiński [419], pp. 137–152.
119. D. Heath, S. Kasif, S. Salzberg (1993), *Induction of oblique decision trees*. Proc. 13th International Joint Conf. on AI, Chambery, France, pp. 1002-1007.
120. R.C. Holt (1993), *Very simple classification rules perform well on most commonly used datasets*. Machine Learning **11**, pp. 63–90.
121. Y. Hou, W. Ziarko (1996), *A rough sets approach to handwriting classification*. In: Tsumoto, Kobayashi, Yokomori, Tanaka, and Nakamura [475], pp.372–382.
122. U. Höhle (1988), *Quotients with respect to similarity relations*. Fuzzy Sets and Systems **27**, pp. 31–44.
123. X. Hu, N. Cercone (1995), *Rough set similarity based learning from databases*. In: Fayyad and Uthurusamy [92], pp. 162–167.
124. X. Hu, N. Cercone (1997), *Learning maximal generalized decision rules via discretization, generalization, and rough set feature selection*. In: Proceedings of the Ninth IEEE International Conference on Tools with Artificial Intelligence, Newport Beach, CA, pp. 548–557.
125. X. Hu, N. Cercone, W. Ziarko (1997), *Generation of multiple knowledge from databases based on rough set theory*. In: Lin and Cercone [189], pp. 109–121.
126. T. Ishida (1994), *Parallel, Distributed and Multiagent Production Systems*. Lecture Notes in Computer Science **878**, Springer–Verlag, Berlin.
127. T. Iwiński (1987), *Algebraic approach to rough sets*. Bull. Polish Acad. Sci. Math. **35**, pp. 673–683.
128. T. Iwiński (1988), *Rough order and rough concepts*. Bull. Polish Acad. Sci. Math. pp. 187–192.
129. J. Järvinen (1997), *A Representation of dependence spaces and some basic algorithms*. Fundamenta Informaticae **29/4**, pp. 369–382.
130. A.G. Jackson, S.R. Leclair, M.C. Ohmer, W. Ziarko, H. Al-Kamhawi (1996), *Rough sets applied to material data*. Acta Metallurgica et Materialia, pp.44–75.
131. J. Jelonek, K. Krawiec, R. Słowiński, *Rough set reduction of attributes and their domains for neural networks*. In: [CI], pp.339–347
132. J. Jelonek, K. Krawiec, R. Słowiński, J. Stefanowski, J. Szymas, (1994), *Neural networks rough sets – Comparison combination for classification of histological pictures*. In: Ziarko [537], pp. 426–433.
133. J. Jelonek, K. Krawiec, R. Słowinski, J. Stefanowski, J. Szymas (1994), *Rough sets as an intelligent front-end for the neural network*. In: Proceedings of the First National Conference on Neural Networks their Applications **2**, Częstochowa, Poland, pp. 116–122.

134. J. Jelonek, K. Krawiec, R. Słowiński, J. Szymas (1995), *Rough set reduction of features for picture-based reasoning.* In: Lin and Wildberger [194], pp.89–92.
135. T.-K. Jensen, J. Komorowski A. Øhrn (1998), *Improving Mollestad's algorithm for default knowledge discovery.* In: Polkowski and Skowron [339], pp. 373–380.
136. G. John, R. Kohavi, K. Pfleger (1994), *Irrelevant features subset selection problem.* Proc. of the Twelfth International Conference on Machine Learning, Morgan Kaufmann, pp. 121–129.
137. J. Johnson (1998), *Rough mereology for industrial design.* In: Polkowski and Skowron [339], pp. 553–556.
138. M. Kandulski, J. Marciniec, K. Tukałło (1992), *Surgical wound infection – Conductive factors and their mutual dependencies.* In: Słowiński [419], pp. 95–110.
139. J.D. Katzberg, W. Ziarko W. (1996), *Variable precision extension of rough sets.* Fundamenta Informaticae **27**, pp. 155–168.
140. K. Keiser, A. Szladow, W. Ziarko (1992), *Rough sets theory applied to a large multispecies toxicity database.* In: Proceedings of the Fifth International Workshop on QSAR in Environmental Toxicology, Duluth, Minnesota.
141. R.E. Kent (1996), *Rough concept analysis: A synthesis of rough sets and formal concept analysis.* Fundamenta Informaticae **27/2–3**, pp. 169–181.
142. R. Kerber (1992), *Chimerge. Discretization of numeric attributes.* Proc. of the Tenth National Conference on Artificial Intelligence, MIT Press, pp. 123–128.
143. S. Kłopotek, S. Wierzchoń (1997), *Qualitative versus quantitative interpretation of the mathematical theory of evidence.* In: Ras and Skowron [355], pp. 391–400.
144. S. Kobayashi, T, Yokomori (1997), *Approximately learning regular languages with respect to reversible languages: A rough set based analysis.* In: Wang [504], pp. 91–94.
145. Y. Kodratoff, R. Michalski (1990), *Machine learning, An Artificial Intelligence approach* **3**, Morgan Kaufmann, 1990.
146. J. Komorowski, Z.W. Ras (Eds.) (1993), *Proceedings of the Seventh International Symposium on Methodologies for Intelligent Systems (ISMIS'93)*, Trondheim, Norway, June 15–18, Lecture Notes in Computer Science **689**, Springer–Verlag, Berlin.
147. J. Komorowski, J. Żytkow (Eds.) (1997), *The First European Symposium on Principles of Data Mining and Knowledge Discovery (PKDD'97).* June 25–27, Trondheim, Norway, Lecture Notes in Artificial Intelligence **1263**, Springer-Verlag, Berlin.
148. B. Konikowska (1997), *A logic for reasoning about relative similarity.* Studia Logica **58/1**, pp. 185–226.
149. B. Konikowska B. (1998), *A logic for reasoning about similarity.* In: Orłowska [270] pp. 461–490.
150. B. Kostek (1995) *Computer based recognition of musical phrases using the rough set approach.* In: Wang [503], pp. 401–404.
151. B. Kostek (1997) *Soft set approach to the subjective assessment of sound quality.* In: Proceedings of the Symposium on Computer Science and Engineering Cybernetics, August 18–23, Baden-Baden, Germany, The International Institute for Advanced Studies in Systems Research and Cybernetics (to appear).
152. B. Kostek (1997), *Sound quality assessment based on the rough set classifier.* In: Zimmermann [545], pp. 193–195.
153. B. Kostek (1998), *Soft computing-based recognition of musical sounds.* In: Polkowski and Skowron [338], pp. 193–213.
154. B. Kostek (1998), *Assesment of concert hall acoustics using rough set fuzzy approach.* In this book.

155. B. Kostek, A. Czyżewski (1996), *Automatic classification of musical timbres based on learning algorithms applicable to Cochlear implants*. In: Proceedings of IASTED International Conference – Artificial Intelligence, Expert Systems and Neural Networks, August 19-21, Honolulu, Hawaii, USA, pp. 98-101.
156. B. Kostek, M. Szczerba (1996), *Parametric representation of musical phrases*. In: 101st Convention of the Audio Engineering Society, November 8-11, Los Angeles, California, USA, preprint **4337**.
157. B. Kostek, M. Szczerba, A. Czyżewski (1997), *Rough set based analysis of computer musical storage*. In: Proceedings of the International Conference on Computing Intelligent Multimedia Applications (ICCIMA'97), February 10-13, Brisbane, Australia.
158. W. Kowalczyk (1996), *Analyzing temporal patterns with rough sets*. In: Zimmermann [544], pp. 139-143.
159. W. Kowalczyk (1996), *Analyzing signals with AI-techniques, two case studies*. A contribution to the International Competition for Signal Analysis Processing by Intelligent Techniques held during the Fourth European Congress on Intelligent Techniques and Soft Computing (EUFIT'96), September 2-5, 1996, Aachen, Germany; available by anonymous ftp from ftp.cs.vu.nl, the directory /pub/wojtek.
160. W. Kowalczyk (1998), *Rough data modelling, A new technique for analyzing data*. In: Polkowski and Skowron [337], pp. 400-421.
161. W. Kowalczyk (1998), *An empirical evaluation of the accuracy of rough data models*. In: Bouchon-Meunier and Yager [25], pp. 1534-1538.
162. W. Kowalczyk, Z. Piasta (1998), *Rough sets-inspired approach to knowledge discovery in business databases*. In: The Second Pacific-Asian Conference on Knowledge Discovery Data Mining, (PAKDD'98), Melbourne, Australia, April 15-17.
163. W. Kowalczyk, F. Slisser (1997), *Analyzing customer retention with rough data models*. In: Komorowski and Żytkow [147], pp. 4-13.
164. K. Krawiec, K., R. Słowiński, and D. Vanderpooten (1996), *Construction of rough classifiers based on application of a similarity relation*. In: Tsumoto, Kobayashi, Yokomori, Tanaka, and Nakamura [475] pp. 23-30.
165. K. Krawiec, R. Słowiński, and D. Vanderpooten (1998), *Learning decision rules from similarity based rough approximations*. In: Polkowski and Skowron [338], pp. 37-54.
166. M. Kretowski, J. Stepaniuk J. (1996), *Selection of objects and attributes, a tolerance rough set approach*. In: Proceedings of the Poster Session of Ninth International Symposium on Methodologies for Intelligent Systems, Zakopane Poland, June 10-13, pp. 169-180.
167. E. Krusińska, R. Słowiński, J. Stefanowski (1992), *Discriminant versus rough set approach to vague data analysis*, Applied Stochastic Models and Data Analysis **8**, pp. 43-56.
168. M. Krynicki, L. Szczerba (1998), On the logic with rough quantifier. In: Orłowska [270], pp. 601-613.
169. J. Krysiński (1990), *Rough set approach to the analysis of structure-activity relationship of quaternary imidazolium compounds*. Arzneimittel Forschung / Drug Research **40/11**, pp. 795-799.
170. J. Krysiński (1992), *Grob Mengen Theorie in der Analyse der Struktur Wirkungs Beziehungen von quartaren Pyridiniumverbindungen*. Pharmazie **46/12**, pp.878-881.
171. J. Krysiński (1992), *Analysis of structure-activity relationships of quaternary ammonium compounds*. In: Słowiński [419], pp. 119-136.

172. J. Krysiński (1995), *Application of the rough sets theory to the analysis of structure-activity relationships of antimicrobial pyridinium compounds.* Die Pharmazie **50**, pp. 593-597.
173. J. Krysiński (1995), *Rough sets in the analysis of the structure-activity relationships of antifungal imidazolium compounds.* Journal of Pharmaceutical Sciences **84/2**, pp. 243-247.
174. M. Kryszkiewicz (1995), *Rough set approach to incomplete information systems.* In: Wang [503], pp. 194-197.
175. M. Kryszkiewicz (1997), *Generation of rules from incomplete information systems.* In: Komorowski and Żytkow [147], pp. 156-166.
176. G. Lambert-Torres, R. Rossi, J. A. Jardini, A. P. Alves da Silva, V. H. Quintana (1998), *Power system security analysis based on rough classification.* In this book.
177. A. Lenarcik, Z. Piasta (1992), *Discretization of condition attributes space* In: Słowiński [419], pp. 373-389.
178. A. Lenarcik, Z. Piasta (1993), *Probabilistic approach to decision algorithm generation in the case of continuous condition attributes.* Foundations of Computing and Decision Sciences **18/3-4**, pp. 213-223.
179. A. Lenarcik, Z. Piasta (1994), *Deterministic rough classifiers.* In: Lin [186]. pp. 434-441.
180. A. Lenarcik, Z. Piasta (1997), *Probalilistic rough classifiers with mixture of discrete and continuous attributes,* In: Lin and Cercone [189] pp. 373-383.
181. S. Leśniewski (1992), *Foundations of the general theory of sets,* In: Surma, Srzednicki, Barnett, Rickey (Eds.), Stanislaw Leśniewski Collected Works, Kluwer Academic Publishers, Dordrecht, pp. 128-173.
182. T.Y. Lin (1989), *Neighborhood systems and approximation in database and knowledge base systems.* In: Proceedings of the 4th International Symposium on Methodologies for Intelligent Systems.
183. T.Y. Lin (1989), *Granular computing on binary relations I.* In: Polkowski and Skowron [337], pp. 107-121.
184. T.Y. Lin (1989), *Granular computing on binary relations II.* In: Polkowski and Skowron [337], pp. 122-140.
185. T.Y. Lin (1994), *Fuzzy reasoning and rough sets.* In: Ziarko [537], pp. 343-348.
186. T.Y. Lin (Ed.) (1994), *Proceedings of the Third International Workshop on Rough Sets Soft Computing (RSSC'94).* San Jose State University, San Jose, California, USA, November 10-12.
187. T.Y. Lin (1996), *Neighborhood systems - A qualitative theory for fuzzy and rough sets.* In: P.P. Wang (Ed.), Advances in Machine Intelligence and Soft Computing **4**, pp. 132-155.
188. T.Y. Lin (1996), *An overview of rough set theory from the point of view of relational databases.* In: S. Tsumoto (Ed.), Bulletin of International Rough Set Society **1/1**, pp. 23-30.
189. T. Y. Lin, N. Cercone (Eds.) (1997), *Rough sets and data mining. Analysis of imprecise data.* Kluwer Academic Publishers, Boston.
190. T.Y. Lin, R. Chen (1997), *Finding reducts in very large databases.* In: Wang [504], pp. 350-352.
191. T.Y. Lin, J.C. Liau (1997), *Probabilistics multivalued random variables - Belief functions and granular computing.* In: Zimmermann [545], pp. 221-225.
192. T.Y. Lin, J.C. Liau (1997), *Belief functions based on probabilistic multivalued random variables.* In: Wang [504], pp. 269-272.

193. T.Y. Lin, Q. Liu (1996), *First-order rough logic I, Approximate reasoning via rough sets*. Fundamenta Informaticae **27**, pp. 137–153.
194. T.Y. Lin, A.M. Wildberger (Eds.) (1995), *Soft Computing: Rough Sets, Fuzzy Logic, Neural Networks, Uncertainty Management, Knowledge Discovery*. Simulation Councils, Inc., San Diego, CA.
195. P. Lingras (1996), *Rough neural networks*. In: Bouchon–Meunier, Delgado, Verdegay, Vila, and Yager [24] **2**, pp. 1445–1450.
196. P. Lingras (1996), *Learning using rough Kohonen neural networks classifiers*. In: Borne, Dauphin–Tanguy, Sueur, and El Khattabi [23], pp. 753–757.
197. P. Lingras (1997), *Comparison of neofuzzy rough neural networks*. In: Wang [504], pp. 259–262.
198. W. Lipski (1981), *On databases with incomplete information*. Journal of the ACM **28**, pp. 41–70.
199. T. Løken (1998), *Vibration analysis using Rosetta, A practical application of rough sets*. (Project in Technical Report), Knowledge Systems Group, The Norwegian University of Science Technology, Trondheim, Norway.
200. E.H. Mamdani, S. Assilian (1975), *An experiment in linguistic synthesis with a fuzzy logic controller*. International Journal of Man - Machine Studies **7**, pp. 1–13.
201. S. Marcus (1994), *Tolerance rough sets, Cech topologies, learning processes*. Bull. Polish Acad. Sci. Tech. Sci. **42/3**, pp. 471–487.
202. W. Marek, Z. Pawlak (1976), *Information storage and retrieval: Mathematical foundations*. Theoretical Computer Science **1**, pp. 331-354.
203. W. Marek, Z. Pawlak (1984), *Rough sets and information systems*. Fundamenta Informaticae **17**, pp. 105-115.
204. M. Mares, R. Meisar, V. Novak, and J. Ramik (Eds.) (1997), *Proceedings of the Seventh International Fuzzy Systems Assotiation World Congress (IFSA '97)*. June 25–29, Academia, Prague **1**, pp. 1–529.
205. P. Maritz (1996), *Pawlak and topological rough sets in terms of multifunctions*. Glasnik Matematicki **31/51**, pp. 159–178.
206. S. Miyamoto (1996), *Fuzzy multisets and application to rough approximation of fuzzy sets*. In: Tsumoto, Kobayashi, Yokomori, Tanaka, and Nakamura [475], pp. 255–260.
207. T.M. Michell (1997), *Machine Learning*. Mc Graw-Hill, Portland.
208. D. Michie, D.J. Spiegelhalter, C.C. Taylor, (Eds.) (1994), *Machine learning, Neural and Statistical Classification*. Ellis Horwood, New York.
209. M. Millan, F. Machuca (1997), *Using the rough set theory to exploit the data mining potential in relational databases systems*. In: Wang [504], pp. 344–347.
210. S. Mitra, M. Banerjee (1996), *Knowledge-based neural net with rough sets*. In: T. Yamakawa et al. (Eds.), Methodologies for the Conception, Design, Application of Intelligent Systems, Proceedings of the Fourth International Conference on Soft Computing (IIZUKA'96), Iizuka, Japan 1996, World Scientific, pp. 213–216.
211. M. Modrzejewski (1993), *Feature selection using rough sets theory*. In: P.B. Brazdil (Ed.), Proceedings of the European Conference on Machine Learning, pp. 213–226.
212. T. Łuba, J. Rybnik (1992), *Rough sets some aspects of logical synthesis*. In: R. Słowiński [419], pp. 181–202.
213. T. Mollestad (1997), *A Rough set approach to data mining, extracting a logic of default rules from data*. Ph.D. thesis, Norwegian University of Science and Technology.
214. T. Mollestad J. Komorowski (1998), *A rough set framework for propositional default rules data mining*. In this book.

215. T. Mollestad A. Skowron (1996), *A rough set framework for data mining of propositional default rules*. In: Pawlak and Ras [354], pp. 448–457. Full version available at http://www.idi.ntnu.no.
216. H. Moradi, J.W. Grzymała-Busse, J. Roberts (1995), *Entropy of English text, Experiments with humans machine learning system based on rough sets*. In: Wang [503], pp. 87–88.
217. A. Mrózek (1985), *Information systems and control algorithms*. Bull. Polish Acad. Sci. Tech. Sci. **33**, pp. 195–212.
218. A. Mrózek (1992), *Rough sets in computer implementation of rule-based control of industrial processes*. In: Słowiński [419], pp. 19–31.
219. A. Mrózek, L. Płonka (1993), *Rough sets in image analysis*. Foundations of Computing Decision Sciences **18/3–4**, pp. 259–273.
220. A. Mrózek, L. Płonka (1994), *Knowledge representation in fuzzy rough controllers*. In: M. Dąbrowski, M. Michalewicz, Z.W. Ras (Eds.), Proceedings of the Third International Workshop on Intelligent Information Systems, Wigry, Poland, June 6–10. Institute of Computer Science, Polish Academy of Sciences, Warsaw, pp. 324–337.
221. A. Mrózek, L. Płonka (1998), *Rough sets in industrial applications*. In: Polkowski and Skowron [338], pp. 214–237.
222. A. Mrózek, K. Skabek (1998), *Rough sets in economic applications*. In: Polkowski and Skowron [338], pp. 238–271.
223. A. Mrózek, L. Płonka, R. Winiarczyk, J. Majtan (1994), *Rough sets for controller synthesis*. In: Lin [186], pp. 498–505.
224. A. Mrózek, L. Płonka, J. Kędziera (1996), *The methodology of rough controller synthesis*. In: In: Petry and Kraft [314], pp. 1135–1139.
225. T. Munakata (1995), *Commercial industrial AI a future perspective on rough sets*. In: Lin and Wildberger [194], pp. 219–222.
226. T. Munakata (1995), *Rough control, Basic ideas applications*. In: Wang [503], pp. 340–343.
227. T. Munakata (1997), *Rough control, A perspective*. In: Lin and Cercone [189], pp. 77–88.
228. S. Murthy, S. Kasif, S. Saltzberg, R. Beigel,(1993), *OC1. Randomized induction of oblique decision trees*. Proc. of the Eleventh National Conference on AI, pp. 322–327.
229. A. Nakamura (1993), *On a multi-modal logic based on the graded classifications*. In: [FCDS'93] pp. 275–292.
230. A. Nakamura (1994), *Fuzzy quantifiers and rough quantifiers*. In: P.P. Wang (Ed.), Advances in Fuzzy Theory and Technology **II** pp. 111–131.
231. A. Nakamura, *A rough logic based on incomplete information and its application*. International Journal of Approximate Reasoning **15/4**, pp. 367–378.
232. A. Nakamura, *Conflict logic with degrees*. In this book.
233. H. S. Nguyen (1997), *Discretization of Real Value Attributes, Boolean Reasoning Approach*. Ph.D. Dissertation, supervisor A. Skowron, Warsaw University (1997), pp. 1–90.
234. H.S. Nguyen(1998), *From optimal hyperplanes to optimal decision trees*. Fundamenta Informaticae **34**, 1998, pp. 145–174.
235. S.H. Nguyen (1998), *Data regularity analysis and applications in data mining*. Ph.D. Thesis (in preparation).

236. S.H. Nguyen, H.S. Nguyen (1996), *Some efficient algorithms for rough set methods*. In: Bouchon–Meunier, Delgado, Verdegay, Vila, and Yager [24], pp. 1451–1456.
237. H.S. Nguyen, S.H. Nguyen (1998), *Discretization methods in data mining*. In: Polkowski and Skowron [337], pp. 451–482.
238. H.S. Nguyen, S.H. Nguyen (1998), *Pattern extraction from data*. Fundamenta Informaticae **34**, 1998, pp. 129–144.
239. H.S. Nguyen, S.H. Nguyen (1998), *Pattern extraction from data*. In: Bouchon–Meunier and Yager [25], pp. 1346–1353.
240. H.S. Nguyen, A. Skowron (1995), *Quantization of real values attributes: Rough set and Boolean reasoning approaches*. In: Wang [503], pp. 34–37.
241. H.S. Nguyen, S.H. Nguyen, A. Skowron (1996), *Searching for Features defined by Hyperplanes*. In: Raś and Michalewicz [354], pp. 366–375.
242. H.S. Nguyen, A. Skowron (1996), *Quantization of real value attributes: Rough set and Boolean reasoning approach*. In: S. Tsumoto (Ed.), Bulletin of International Rough Set Society **1/1** (1996), pp. 5–16 also in Information Sciences (1998) (in print).
243. H.S. Nguyen, A. Skowron (1997), *Boolean reasoning for feature extraction problems*. In: Ras and Skowron [355], pp. 117–126.
244. S.H. Nguyen, T.T. Nguyen, A. Skowron, P. Synak (1996), *Knowledge discovery by rough set methods*. In: Nagib C. Callaos (Ed.), Proceedings of the International Conference on Information Systems Analysis and Synthesis (ISAS'96), July 22–26, Orlando, USA, pp. 26–33.
245. S.H. Nguyen, L. Polkowski, A. Skowron, P. Synak, J. Wróblewski (1996), *Searching for approximate description of decision classes*. In: Tsumoto, Kobayashi, Yokomori, Tanaka, and Nakamura [475], pp. 153–161.
246. S.H. Nguyen, A. Skowron (1997), *Searching for relational patterns in data*. In: Komorowski and Żytkow [147], pp. 265–276.
247. S.H. Nguyen, A. Skowron, P. Synak (1998), *Discovery of data patterns with applications to decomposition and classification problems*. In: Polkowski and Skowron [338], pp. 55–97.
248. H.S. Nguyen, M. Szczuka, D. Ślęzak (1997), *Neural network design, Rough set approach to real–valued data*. In: Komorowski and Żytkow [147], pp. 359–366.
249. S.H. Nguyen, A. Skowron, P. Synak, J. Wróblewski (1997), *Knowledge discovery in data bases, Rough set approach*. In: Mares, Meisar, Novak, and Ramik, [204], pp. 204–209.
250. T. Nguyen, R. Swiniarski, A. Skowron, J. Bazan, K. Thagarajan (1995), *Applications of rough sets, neural networks and maximum likelihood for texture classification based on singular value decomposition*. In: Lin and Wildberger [194], pp. 157–160.
251. J. Nieminen (1988), *Rough tolerance equality*. Fundamenta Informaticae **11/3**, pp. 289–296.
252. R. Nowicki, R. Słowiński, J. Stefanowski (1990), *Possibilities of applying the rough sets theory to technical diagnostics*. In: Proceedings of the Ninth National Symposium on Vibration Techniques Vibroacoustics, Kraków, December 12–14, AGH University Press, Kraków, pp. 149–152.
253. R. Nowicki, R. Słowiński, J. Stefanowski (1992), *Rough sets analysis of diagnostic capacity of vibroacoustic symptoms*. Journal of Computers Mathematics with Applications **24**, pp.109–123.

254. R. Nowicki, R. Słowiński, J. Stefanowski (1992), *Evaluation of vibroacoustic diagnostic symptoms by means of the rough sets theory.* Journal of Computers in Industry **20**, pp. 141–152.
255. R. Nowicki, R. Słowiński, J. Stefanowski (1992), *Analysis of diagnostic symptoms in vibroacoustic diagnostics by means of the rough set theory.* In: Słowiński [419], pp. 33–48.
256. M. Novotny (1998), *Dependence spaces of information systems.* In: Orłowska [270], pp. 193–246.
257. M. Novotny (1998), *Applications of dependence spaces.* In: Orłowska [270], pp. 247–289.
258. M. Novotny, Z. Pawlak (1991), *Algebraic theory of independence in information systems.* Fundamenta Informaticae **14**, pp. 454–476.
259. M. Novotny, Z. Pawlak (1992), *On problem concerning dependence space.* Fundamenta Informaticae **16/3–4**, pp. 275–287.
260. H. Nurmi, J. Kacprzyk, M. Fedrizzi (1996), *Theory methodology, Probabilistic, fuzzy rough concepts in social choice.* European Journal of Operational Research, pp. 264–277.
261. A. Obtułowicz (1988), *Rough sets and Heyting algebra valued sets.* Bull. Polish Acad. Sci. Math. **35/9–10**, pp. 667–671.
262. E. Orłowska (1984), *Modal logics in the theory of information systems.* Zeitschrift für Mathematische Logic und Grundlagen der Mathematik, **10/3**, pp.213–222.
263. E. Orłowska (1985), *A logic of indiscernibility relations.* In: A. Skowron (Ed.), Computation Theory, Lecture Notes in Computer Science **208**, pp. 177–186.
264. E. Orłowska (1985), *Logic of nondeterministic information.* Studia Logica **44**, pp. 93–102.
265. E. Orłowska (1988), *Kripke models with relative accessibility and their application to inferences from incomplete information.* In: G. Mirkowska, H. Rasiowa (Eds.), Mathematical Problems in Computation Theory, Banach Center Publications **21** pp. 329–339.
266. E. Orłowska (1990), *Kripke semantics for knowledge representation logics.* Studia Logica **49** pp. 255–272.
267. E. Orłowska (1990), *Verisimilitude based on concept analysis.* Studia Logica **49**, pp. 307–320.
268. E. Orłowska (1995), *Information algebras.* Lecture Notes in Computer Science **936**, Springer–Verlag, Berlin, pp. 55–65.
269. E. Orłowska (1998), *Introduction: What you always wanted to know about rough sets.* In: Orłowska [270] pp. 10–29.
270. E. Orłowska (Ed.) (1998), *Incomplete Information, Rough Set Analysis.* Physica–Verlag, Heidelberg.
271. E. Orłowska, M.W. Orłowski, *Maintenance of knowledge in dynamic information systems.* In: Słowiński [419], pp. 315–330.
272. E. Orłowska, Z. Pawlak (1984), *Representation of nondeterministic information.* Theoretical Computer Science **29** pp. 27–39.
273. A. Øhrn, S. Vinterbo, P. Szymański J. Komorowski (1997), *Modelling cardiac patient set residuals using rough sets.* Proc. AMIA Annual Fall Symposium (formerly SCAMC), Nashville, TN, USA, Oct. 25–29, pp. 203–207.
274. A. Øhrn J. Komorowski (1997), ROSETTA – *A Rough Set Toolkit for Analysis of Data.* Proc. Third International Joint Conference on Information Sciences, Durham, NC, USA, March 1–5, **3**, pp. 403–407.

275. A. Øhrn J. Komorowski (1998), *Analyzing The Prognostic Power of Cardiac Tests Using Rough Sets*. Proc. of the Invited Session on Intelligent Prognostic Methods in Medical Diagnosis and Treatment Planning at the Computational Engineering in Systems Applications conference – CESA'98, April Nabeul-Hammamet, Tunisia, 6 pages.
276. A. Øhrn, J. Komorowski, A. Skowron P. Synak (1998), *The Design and Implementation of a Knowledge Discovery Toolkit Based on Rough Sets - The* ROSETTA *system*. In: Polkowski and Skowron [337], pp. 376–399.
277. P. Pagliani (1993), *From concept lattices to approximation spaces, Algebraic structures of some spaces of partial objects*. Fundamenta Informaticae **18/1**, pp. 1–25.
278. P. Pagliani (1998), *Rough set theory and logic-algebraic structures*. In: Orłowska [270] pp. 109–190.
279. P. Pagliani (1998), *Modalizing relations by means of relations: A general framework for two basic approaches to knowledge discovery in databases*. In: Bouchon-Meunier and Yager [25], pp. 1175–1182.
280. Y.-H. Pao, I. Bozma (1986), *Quantization of numerical sensor data for inductive learning*. In: J. S. Kowalik (Ed.), Coupling Symbolic and Numerical Computing in Expert System. Elsevier Science Publ., Amsterdam, pp. 69–81.
281. P. Paszek, A. Wakulicz-Deja (1996), *Optimalization diagnose in progressive encephalopathy applying the rough set theory*. In: Zimmermann [544] **1**, pp. 192–196.
282. G. Paun, L. Polkowski, A. Skowron (1996), *Parallel communicating grammar systems with negotiations*. Fundamenta Informaticae **28/3-4**, pp. 315–330.
283. G. Paun, L. Polkowski, A. Skowron (1997), *Rough set approximations of languages*. Fundamenta Informaticae **32/2**, pp. 149–162.
284. Z. Pawlak, Z. (1981), *Information systems – theoretical foundations*. Information Systems **6**, pp. 205–218.
285. Z. Pawlak, Z. (1982), *Rough sets*. International Journal of Computer and Information Sciences **11**, pp. 341–356.
286. Z. Pawlak (1984), *Rough probability*. Bull. Polish Acad. Sci. Math. **132/9–10**, pp. 607–612.
287. Z. Pawlak (1984), *On conflicts*. Int. J. of Man-Machine Studies **21**, pp. 127–134.
288. Z. Pawlak (1985), *Rough sets and fuzzy sets*. J. of Fuzzy Sets and Systems **17**, pp. 99–102.
289. Z. Pawlak (1989), *Knowledge, reasoning and classification – A rough set perspective*. Bulletin of the European Association for Theoretical Computer Science (EATCS) **38**, pp. 199–210.
290. Z. Pawlak (1991), *Rough Sets – Theoretical Aspects of Reasoning about Data*. Kluwer Academic Publishers, Dordrecht.
291. Z. Pawlak (1993), *Anatomy of conficts*. Bulletin of the European Association for Theoretical Computer Science **50**, pp. 234-247.
292. Z. Pawlak (1994), *Hard and soft sets*. In: Ziarko [537], pp. 130–135.
293. Z. Pawlak (1994), *Decision analysis using rough sets*. International Trans. Opr. Res. **1/1**, pp. 107–114.
294. Z. Pawlak (1995), *Rough sets and fuzzy sets*. In: C. Jinshong (Ed.), Proceedings of ACM, Computer Science Conference, February 28 – March 2, Nashville, Tennessee, pp. 262–264.
295. Z. Pawlak (1997), *Conflict analysis*. In: Zimmermann [545], pp. 1589–1591.
296. Z. Pawlak (1997), *Rough set approach to knowledge-based decision support*. European Journal of Operational Research **2933**, pp. 1–10.

297. Z. Pawlak (1998), *Reasoning about data – A rough set perspective*. In: Polkowski and Skowron [339], pp. 25–34.
298. Z. Pawlak, J.W. Grzymała-Busse, R. Słowiński, and W. Ziarko (1995), *Rough sets*. Communications of the ACM **38(11)**, pp. 89–95.
299. Z. Pawlak, T. Munakata (1996), *Rough control, Application of rough set theory to control*. In: Zimmermann [544], pp. 209–218.
300. Z. Pawlak, K. Słowiński, R. Słowiński (1986), *Rough classification of patients after highly selected vagotomy for duodenal ulcer*. Journal of Man–Machine Studies **24**, pp. 413–433.
301. Z. Pawlak, R. Słowiński (1994), *Decision analysis using rough sets*. International Transactions in Operational Research. **1/1**, pp. 107–114.
302. Z. Pawlak, R. Słowiński (1994), *Rough set approach to multi-attribute decision analysis*. European Journal of Operational Research **72**, pp. 443–459 (Invited Review).
303. Z. Pawlak A. Skowron (1994), *Rough membership functions*. In: R. Yager, M. Fedrizzi J. Kacprzyk (Eds.), Advances in the Dempster-Shafer Theory of Evidence, Wiley, New York, pp. 251–271.
304. Z. Pawlak, A. Skowron,(Ed.)(1996), *Logic, algebra and computer science, Helena Rasiowa and Cecylia Rauszer in Memoriam*. Bulletin of the Section of Logic **25/3–4**, pp. 174–184.
305. Z. Pawlak, S.K.M. Wong, W. Ziarko (1988), *Rough sets: Probabilistic versus deterministic approach*. International Journal of Man–Machine Studies **29**, pp. 81–95.
306. W. Pedrycz (1998), *Computational Intelligence: An Introduction*. CRC Press, Boca Ratou.
307. W. Pedrycz (1998), *Shadowed sets: Bridging fuzzy and rough sets*. In this book.
308. J. F. Peters III, S. Ramanna (1998), *Software deployability decision system framework, A rough set approach*. In: Bouchon-Meunier and Yager [25], pp. 1539–1545.
309. J. F. Peters, A. Skowron, Z. Suraj, S. Ramanna, A. Paryzek (1998), *Modelling real – time decision – making systems with rough fuzzy Petri nets*. In Proceedings of EUFIT-98, Aachen (to appear).
310. J. F. Peters, K. Ziaei, S. Ramanna (1998), *Approximate time rough control, Concepts applications to satelite attitude control*. In: Polkowski and Skowron [339], pp. 491–498.
311. J. F. Peters III, S. Ramanna (1998), *A rough set approach to assessing software quality: Concepts and rough Petri net models*. In this book.
312. J. F. Peters III (1998), *Time clock information systems: Concepts and roughly fuzzy Petri net models*. In: Polkowski and Skowron [338], pp. 387–419.
313. G. I. Peterson (1994), *Rough classification of pneumonia patients using a clinical data-base*. In: Ziarko [537], pp. 412–419.
314. F.E. Petry, D.H. Kraft (1996), Proceedings of the Fifth IEEE International Conference on Fuzzy Systems (FUZZ-IEEE'96). September 8–11, New Orleans, Louisiana, pp. 1–2214.
315. Z. Piasta (1993), *Statistical and logical classifiers: A comparative study*. In: W. Ziarko (Ed.): Proceedings of the Second International Workshop on Rough Sets and Knowledge Discovery (RSKD'93). Banff, Alberta, Canada, October 12–15.
316. Z. Piasta (1996), *Rough classifiers in intelligent support of business decisions*. In: Proceedings of the First Polish Conference on Theory Applications of Artificial Intelligence (CAI'96), Łódź, Poland, pp. 103–111.
317. Z. Piasta (1998), *Transforming data into engineering knowledge with rough classifiers*. In: A.M. Brandt (Ed.), Optimization Methods for Material Design of

Cement–based Composites, Thomson Science & Professional, London (to appear).
318. Z. Piasta (1998), *Data mining knowledge discovery in marketing financial databases with rough classifiers*. Wydawnictwo Akademii Ekonomicznej we Wrocławiu, Wrocław (to appear, in Polish).
319. Z. Piasta, A. Lenarcik (1998), *Rule induction with probabilistic rough classifiers*. Machine Learning (to appear).
320. Z. Piasta, A. Lenarcik (1998), *Learning rough classifiers from large databases with missing values*. In: Polkowski and Skowron [338], pp. 483–499.
321. Z. Piasta, A. Lenarcik, S. Tsumoto (1997), *Machine discovery in databases with probabilistic rough classifiers*. In: S. Tsumoto (Ed.): Bulletin of International Rough Set Society **1/2**, pp. 51–57.
322. L. Płonka, A. Mrózek (1995), *Rule-based stabilization of the inverted pendulum*. Computational Intelligence: An International Journal **11/2**, pp.348–356.
323. L. Polkowski (1993), *Mathematical morphology of rough sets*. Bull. Polish Acad. Sci. Math. **41/3**, pp. 241–273.
324. L. Polkowski (1994), *Concerning mathematical morphology of almost rough sets*. Bull. Polish Acad. Sci. Tech. **42/1**, pp. 141–152.
325. L. Polkowski (1994), *Concerning mathematical morphology of rough sets*. Bull. Polish Acad. Sci. Tech. **42/1**, pp. 125–140.
326. L. Polkowski (1998), *Rough set approach to mathematical morphology, approximate compression of data*. In: Bouchon–Meunier and Yager [25], pp. 1183–1189.
327. L. Polkowski (1998), *Approximate mathematical morphology, rough set approach*. In this book.
328. L. Polkowski, A. Skowron (1994), *Rough mereology*. In: Proceedings of the Symposium on Methodologies for Intelligent Systems, Charlotte, NC, October 16–19, Lecture Notes in Artificial Intelligence **869**, Springer–Verlag, Berlin (1994) pp. 85–94.
329. L. Polkowski, A. Skowron (1995), *Rough mereology analytical mereology, New developments in rough set theory*. In: M. De Glas, Z. Pawlak (Eds.), Proceedings of the Second World Conference on Fundamentals of Artificial Intelligence (WOCFAI'95), July 13–17, Angkor, Paris, pp. 343–354.
330. L. Polkowski, A. Skowron (1996), *Rough mereological approach to knowledge – based distributed AI*. In: J.K. Lee, J. Liebowitz, and J.M. Chae (Eds.), Critical Technology, Proceedings of the Third World Congress on Expert Systems, February 5–9, Seoul, Korea, Cognizant Communication Corporation, New York, pp. 774–781.
331. L. Polkowski, A. Skowron (1996), *Implementing fuzzy containment via rough inclusions: Rough mereological approach to distributed problem solving*. In: Petry and Kraft [314], pp. 1147–1153.
332. L. Polkowski, A. Skowron (1996), *Rough mereology: A new paradigm for approximate reasoning*. International Journal of Approximate Reasoning **15/4**, pp. 333–365.
333. L. Polkowski, A. Skowron (1996), *Adaptive decision-making by systems of cooperative intelligent agents organized on rough mereological principles*. Journal of the Intelligent Automation and Soft Computing **2/2**, pp. 121-132.
334. L. Polkowski, A. Skowron (1998), *Rough mereology analytical morphology*. In: Orłowska [270], pp. 399–437.
335. L. Polkowski, A. Skowron (1998), *Towards adaptive calculus of granules*. In: Proceedings of the FUZZ-IEEE'98 International Conference, Anchorage, Alaska, USA, May 5–9, pp. 111–116.

336. L. Polkowski, A. Skowron (1998), *Rough sets: A perspective.* In: Polkowski and Skowron [337], pp. 31–58.
337. L. Polkowski, A. Skowron (Eds.) (1998), *Rough Sets in Knowledge Discovery 1: Methodology and Applications.* Physica-Verlag, Heidelberg.
338. L. Polkowski, A. Skowron (Eds.) (1998), *Rough Sets in Knowledge Discovery 2: Applications, Case Studies and Software Systems.* Physica-Verlag, Heidelberg.
339. L. Polkowski, A. Skowron (Eds.) (1998), *Proc. First International Conference on Rough Sets and Soft Computing (RSCTC'98.* Warszawa, Poland, June 22–27, Springer-Verlag, LNAI **1424**.
340. L. Polkowski, A. Skowron, J. Komorowski (1997), *Towards a rough mereology-based logic for approximate solution synthesis, Part 1.* Studia Logica **58/1**, pp. 143-184.
341. L. Polkowski, A. Skowron, and J. Żytkow (1995), *Tolerance based rough sets.* In: Lin and Wildberger [194], pp. 55–58.
342. J.A. Pomykała (1987), *Approximation operations in approximation space.* Bull. Polish Acad.Sci.Ser. Sci. Math. **35**, pp. 653–662.
343. J.A. Pomykała (1988), *On definability in the nondeterministic information system*, Bull. Polish Acad. Sci.Ser. Sci. Math., **36**, pp. 193–210.
344. M. Quafafou (1996), *Towards a transition from the crisp rough set theory to a fuzzy one.* In: Ras and Michalewicz [354], pp. 67–80.
345. M. Quafafou (1997), *α-RST: A generalization of rough set theory.* In: Wang [504], pp. 173–176.
346. E. Martienne, M. Quafafou (1998), *Vahueness and data reduction in concept learning.* In: E. Prade (Ed.), Proc. of the 13th European Conference on Arificial Intelligence (ECAI'98), August 23–28, Brighton, UK, Wiley, Chichester, pp. 351–355.
347. J.R. Quinlan (1986), *Induction of decision trees.* In: Machine Learning **1**, pp. 81–106.
348. R.B. Rao, S.C.-Y. Lu (1993), *Building models to support synthesis in early stage product design.* In: Proceedings of the Eleventh National Conference on Artificial Intelligence (AAAI'93), MIT Press, Cambridge, MA, pp. 277–282.
349. Z.W. Ras (1996), *Cooperative knowledge–based systems.* Journal of the Intelligent Automation and Soft Computing **2/2**, pp. 193–202.
350. Z.W. Ras (1997), *Collaboration control in distributed knowledge–based systems.* Information Sciences **96/3–4**, pp. 193–205.
351. Z.W. Ras (1998), *Answering non-standard queries in distributed knowledge–based systems.* In: Polkowski and Skowron [338], pp. 98–108.
352. Z.W. Ras, S. Joshi (1997), *Query answering system for an incomplete DKBS.* Fundamenta Informaticae **30/3–4**, pp. 313–324.
353. Z.W. Ras, A. Kudelska, N. Chilumula (1995), *Can we simplify international physical performance test profile using rough set approach?* In: Wang [503], pp. 393–396.
354. Z.W. Ras, M. Michalewicz, (Eds.), (1996), *Proceedings of the Ninth International Symposium on Methodologies for Intelligent Systems, Foundations of Intelligent Systems (ISMIS'96)*, October 15-18, Charlotte, NC, USA, Lecture Notes in Artificial Intelligence **1079**, Springer-Verlag, Berlin, pp. 1–664.
355. Z.W. Ras, A. Skowron (Eds.) (1997), *Proceedings of the Tenth International Symposium on Methodologies for Intelligent Systems, Foundations of Intelligent Systems (ISMIS'97)*, October 15-18, Charlotte, NC, USA, Lecture Notes in Artificial Intelligence **1325**, Springer-Verlag, Berlin, pp. 1–630.
356. H. Rasiowa (1986), *Rough concepts and multiple-valued logic.* In: Proceedings of the 16th ISMVL'86, Blacksburg, VA, IEEE Computer Society Press, pp. 282–288.

357. H. Rasiowa (1987), *Logic approximating sequences of sets.* In: Proceedings of Advanced Intern. School and Symp. on Math. Logic and its Applications, honorably dedicated to 80th anniversary of Kurt Gödel, Plenum Press, New York, pp. 167–186.
358. H. Rasiowa (1987), *An algebraic approach to some approximate reasonings.* In: Proceedings of the 17th ISMVL'87, Boston, MA, May 24–26, IEEE Computer Society Press, pp. 342–347.
359. H. Rasiowa (1990), *On approximation logics, A survey.* Jahrbuch 1990 Kurt Gödel Gessellschaft, Vienna, pp. 63–87.
360. H. Rasiowa (1991), *Mechanical proof systems for logic of reaching consensus by groups of intelligent agents.* Intern. Journ. of Approximate Reasoning **5/4** pp. 415–432.
361. H. Rasiowa, G. Epstein (1987), *Approximation reasoning and Scott's information systems.* In: Z. Ras, M. Zemankova (Eds.), Proceedings of the Second International Symposium on Methodologies for Intelligent Systems, Charlotte, N.C., October 14–17, North Holland, Amsterdam pp. 33–42.
362. H. Rasiowa, W. Marek (1987), *Gradual approximating sets by means of equivalence relations.* Bull. Polish Acad. Sci. Math. **35/3–4**, pp. 233–238.
363. H. Rasiowa, W. Marek (1987), *Approximating sets with equivalence relations.* Theoretical Computer Science **48**, pp. 145–152.
364. H. Rasiowa, W. Marek (1989), *On reaching consensus by groups of intelligent agents.* In: Z.W. Ras (Ed.), Proceedings of the Fourth International Symposium on Methodologies for Intelligent Systems (ISMIS'89), North Holland, Amsterdam, pp. 234–243.
365. H. Rasiowa, W. Marek (1992), *Mechanical proof systems for logic II: Consensus programs and their processing.* Journal of Intelligent Information Systems **2/2**, pp. 149–164.
366. H. Rasiowa (1985), A. Skowron, *Approximation logic.* In: Proceedings of Mathematical Methods of Specification and Synthesis of Software Systems Conference, Akademie Verlag **31**, Berlin pp. 123–139.
367. C. Rauszer (1985), *Remarks on logic for dependencies.* Bull. Polish Acad. Sci. Math. **33**, pp. 249–252.
368. C. Rauszer (1985), *Algebraic properties of functional dependencies.* Bull. Polish Acad. Sci. Math. **33**, pp. 561–569.
369. C. Rauszer (1993), *Dependencies in relational databases. Algebraic and logical approach.* Fundamenta Informaticae **19**, pp. 235–274.
370. C. Rauszer (1993), *An equivalence between theory of functional dependencies and a fragment of intuitionistic logic.* Bull. Polish Acad. Sci. Math. **33**, pp. 571–579.
371. C. Rauszer (1993), *Communication systems in distributed information systems.* In: Proceedings of the International Workshop on Intelligent Information Systems, June 7–11, 1993, Augustów, Poland, Institute of Computer Science, Polish Academy of Sciences, Warsaw, pp. 15–29.
372. C. Rauszer, (1993) *Approximation methods for knowledge representation systems.* In: J. Komorowski, Z.W. Ras (Eds.), Proceedings of the Seventh International Symposium on Methodologies for Intelligent Systems (ISMIS'93), Trondheim, Norway, June 15–18, 1993, Lecture Notes in Computer Science **689** (1993) pp. 326–337.
373. C. Rauszer (1994), *Knowledge representation systems for groups of agents.* In: J. Woleński (Ed.), Philosophical Logic in Poland, Kluwer Academic Publishers, Dordrecht, pp. 217–238.

374. C. Rauszer, W. Marek (1988), *Query optimization in the database distributed by means of product of equivalence relations*. Fundamenta Informaticae 11, pp. 241–286.
375. C. Rauszer, H. de Swart(1995), *Different approaches to knowledge, common knowledge and Aumann's theorem*. In: A. Laux, H. Wansing (Eds.), Knowledge and Belief in Philosophy and Artificial Intelligence, Akademie Verlag, Berlin, pp. 87–12.
376. A. Reinhard, B. Stawski, T. Weber, U. Wybraniec–Skardowska (1992), *An application of rough set theory in the control conditions on a polder*. In: Słowiński [419], pp. 331–362.
377. R.O. Rogriguez, P. Garcia, L. Godo (1995), *Similarity based models: Counterfactual and belif changes*. Research Report IIIA-95/5, IIIA-CSIS, University of Barcelona, Ballaterra, Spain.
378. B. Roy (1985), *Méthodologie Multicritère d'Aide à la Décision*. Economica, Paris.
379. B. Roy (1989), *Main sources of inaccurate determination, uncertainty and imprecision in decision models*. Mathematical and Computer Modelling 12, pp. 1245–1254.
380. B. Roy, D. Bouyssou (1993), *Aide Multicritère à la Décision: Méthods et Cas*. Economica, Paris.
381. B. Roy (1993), *Decision sicence or decision aid science*. European Journal of Operational Research 86/2, pp. 184–203.
382. B. Roy, R. Słowiński, W. Treichel (1992), *Multicriteria programming of water supply systems for rural areas*. Water Resources Bulletin 28/1, pp. 13–31.
383. S. Rubin, S. Vanderpooten, W. Michałowski, R. Słowiński (1996), *Developing an emergency room for diagnostic check list using rough sets – A case study of appendicitis*. In: J. Anderson, M. Katzper (Eds.), Simulation in the Medical Sciences, Proceedings of the Western Conference of the Society for Computer Simulation, Simulation Councils. Inc. San Diego, CA, pp. 19–24.
384. G. Ruhe (1996), *Qualitative analysis of software engineering data using rough sets*. In: Tsumoto, Kobayashi, Yokomori, Tanaka, and Nakamura [475], pp. 292–299.
385. G. Ruhe (1996), *Knowledge discovery from software engineering data: Rough set analysis its interaction with goal–oriented measurement*. In: Komorowski and Żytkow [147], pp. 167–177.
386. E.H. Ruspini (1991), *On truth, utility, and similarity*. Proceedings of the International Fuzzy Engineering Sympsoim (IFES'91), Yokohama, Japan, November 13–15, Fuzzy Engineering Towards Human Friendly Systems 1, pp. 42–50.
387. E.H. Ruspini (1990), *Similarity–based interpretations of fuzzy logic concepts*. Proceedings of the 2nd International Conference on Fuzzy Logic & Neural Networks (IIZUKA'90) Iizuka, Japan, July 20–24, 2, pp.735–780.
388. M. Sarkar, B. Yegnanarayana (1998), *Fuzzy–rough sets and fuzzy integrals in modular neural networks*. In this book.
389. J.A. Schreider (1975), *Equality, Resemblance and Order*. Mir Publishers, Moscow.
390. M. Semeniuk–Polkowska (1996), *Rough sets in librarian science (in Polish)*. Chair of Formal Linguistics, Warsaw University, pp. 1–109.
391. G. Shafer (1976), *A Mathematical Theory of Evidence*. Princeton University Press, Princeton.
392. N. Shan, W. Ziarko, H. Hamilton, N. Cercone (1995), *Using rough sets as tools for knowledge discovery*. In: Fayyad and Uthurusamy [92], pp. 263–268.
393. E. Simoudis, J. Han, U. Fayyad (Eds.) (1996), *Proceedings of the Second International Conference on Knowledge Discovery Data Mining (KDD'96)*. August 2–4, Portland, Oregon, USA, AAAI Press, Menlo Park, pp. 1–391.

394. A. Skowron (1995), *Synthesis of adaptive decision systems from experimental data*. In: Aamodt and Komorowski [1], pp. 220–238.
395. A. Skowron, J. Grzymała-Busse) (1994), *From rough set theory to evidence theory*. In: R.R. Yager, M. Fedrizzi, and J. Kacprzyk (Eds.), Advances in the Dempster-Shafer Theory of Evidence, John Wiley and Sons, New York, pp. 193-236.
396. A. Skowron, L. Polkowski (1996), *Analytical morphology: mathematical morphology of decision tables*. Fundamenta Informaticae **27/2–3**, pp.255-271.
397. A. Skowron, L. Polkowski (1997), *Synthesis of Decision Systems from Data Tables*. In: Lin and Cercone [189], pp. 259–300.
398. A. Skowron, L. Polkowski (1998), *Rough mereological foundations for design, analysis, synthesis, and control in distributive systems*. Information Sciences **104/1-2**, pp. 129–156.
399. A. Skowron, L. Polkowski, J. Komorowski (1996), *Learning Tolerance Relations by Boolean Descriptors, Automatic Feature Extraction from Data Tables*. In: Tsumoto, Kobayashi, Yokomori, Tanaka, and Nakamura [475], pp. 11–17.
400. A. Skowron, C. Rauszer (1992), *The Discernibility Matrices and Functions in Information Systems*. In: Słowiński [419], pp. 331–362.
401. A. Skowron, J. Stepaniuk (1991), *Towards an approximation theory of discrete problems*. Fundamenta Informaticae **15/2** pp. 187–208.
402. A. Skowron, J. Stepaniuk (1995), *Generalized approximation spaces*. In: Lin and Wildberger [194], pp. 18–21.
403. A. Skowron, J. Stepaniuk (1996), *Tolerance approximation spaces*. Fundamenta Informaticae **27**, pp. 245–253.
404. A. Skowron, J. Stepaniuk (1998), *Information granules and approximation spaces*. In: Bouchon–Meunier and Yager [25], pp. 1354–1361.
405. A. Skowron, Z. Suraj (1993), *Rough sets and concurrency*. Bull. Polish Acad. Sci. Tech. **41/3**, pp. 237–254
406. A. Skowron, Z. Suraj (1995), *Discovery of concurrent data models from experimental data tables, A rough set approach*. In: Fayyad and Uthurusamy [92], pp. 288–293.
407. A. Skowron, Z. Suraj (1996), *A parallel algorithm for real-time decision making, A rough set approach*. Journal of Intelligent Information Systems **7**, pp. 5–28.
408. D. Ślęzak (1996), *Approximate reducts in decision tables*. In: Bouchon–Meunier, Delgado, Verdegay, Vila, and Yager [24] **3**, pp. 1159–1164.
409. D. Ślęzak (1997), *Attribute set decomposition of decision tables*. In: Zimmermann [545] **1**, pp. 236–240.
410. D. Ślęzak (1998), *Decomposition and synthesis of decision tables with respect to generalized decision functions*. In this book.
411. D. Ślęzak (1998), *Searching for dynamic reducts in inconsistent ecision tables*. In: Bouchon–Meunier and Yager [25], pp. 1362–1369.
412. D. Ślęzak, M. Szczuka (1997), *Hyperplane-based neural networks for real-valued decision tables*. In: Wang [504], pp. 265–268.
413. K. Słowiński, E.S. Sharif (1993), *Rough sets approach to analysis of data of diagnostic peritoneal lavage applied for multiple injuries patients*. In: Ziarko [537], pp. 420–425.
414. K. Słowiński, R. Słowiński, J. Stefanowski (1988), *Rough sets approach to analysis of data from peritoneal lavage in acute pancreatitis*. Medical Informatics **13/3**, pp. 143–159.

415. K. Słowiński, J. Stefanowski (1996), *On limitations of using rough set approach to analyze non-trivial medical information systems*. In: Tsumoto, Kobayashi, Yokomori, Tanaka, and Nakamura [475], pp. 176–184.
416. K. Słowiński, J. Stefanowski (1998), *Multistage rough set analysis of therapeutic experience with acute pancreatitis*. In this book.
417. K. Słowiński, J. Stefanowski, A. Antczak, Z. Kwas (1995), *Rough sets approach to the verification of indications for treatment of urinary stones by extracorporeal shock wave lithotripsy (ESWL)*. In: Lin and Wildberger [194], pp. 93–96.
418. K. Słowiński, J. Stefanowski, W. Twardosz (1997), *Rough set theory rule induction techniques for discovery of attribute dependencies in experience with multiple injured patients*. Institute of Computer Science, Warsaw University of Technology, ICS Research Report 6/97.
419. R. Słowiński, (Ed.) (1992), *Intelligent Decision Support – Handbook of Applications and Advances of the Rough Sets Theory*. Kluwer Academic Publishers, Dordrecht.
420. K. Słowiński (1992), *Rough classification of HSV patients*. In: Słowiński [419], pp. 77–93.
421. R. Słowiński (1993), *Rough set learning of preferential attitude in multi-criteria decision making*. In: Komorowski and Ras [146], pp. 642–651.
422. R. Słowiński (1994), *Handling various types of uncertainty in the rough set approach*. In: Ziarko [537], pp. 366–376.
423. R. Słowiński (1994), *Rough set analysis of multi-attribute decision problems*. In: Ziarko [537], pp. 136–143.
424. R. Słowiński (1995), *Rough set approach to decision analysis*. AI Expert **10**, pp. 18–25.
425. R. Słowiński (1995), *Rough set theory and its applications to decision aid*. Belgian Journal of Operation Research, Francoro **35/3-4**, pp. 81–90.
426. R. Słowiński, K. Słowiński (1989), *An expert system for treatment of duodenal ulcer by highly selective vagotomy (in Polish)*. Pamiętnik 54. Jubil. Zjazdu Towarzystwa Chirurgów Polskich, Kraków I, pp. 223–228.
427. R. Słowiński, D. Vanderpooten (1995), *Similarity relation as a basis for rough approximations*. In: P. Wang (Ed.): Advances in Machine Intelligence & Soft Computing, Bookwrights, Raleigh NC (1997) pp. 17–33.
428. R. Słowiński, D. Vanderpooten (1999), *A generalized definition of rough approximations based on similarity*. IEEE Trans. on Data and Knowledge Engineering (to appear).
429. R. Słowiński, C. Zopounidis (1995), *Applications of the rough set approach to evaluation of bankruptcy risk*. International J. Intelligent Systems in Accounting, Finance & Management, **4/1**, pp. 27–41.
430. R. Słowiński, C. Zopounidis (1994), *Rough set sorting of firms according to bankruptcy risk*. In: M. Paruccini (Ed.), Applying Multiple Criteria Aid for Decision to Environmental Management, Kluwer Academic Publishers, Dordrecht, pp. 339–357.
431. P. Srinivasan (1989), *Approximations for information retrieval*. In: Z. W. Ras, (Ed.), Methodologies for Intelligent Systems 4, Elsevier, pp. 128–136.
432. P. Srinivasan (1989), *Intelligent information retrieval using rough set approximations*. Information Processing and Management **25**, pp. 347–361.
433. P. Srinivasan (1991), *The importance of rough approximations for information retrieval*. Journal of Man–Machine Studies **34**, pp. 657–671.

434. D. Sriram, R. Logcher, S. Fukuda (1991), *Computer - aided cooperative product development.* Lecture Notes in Computer Sience **492**, Springer–Verlag, Berlin.
435. J. Stefanowski (1998), *On rough set based approaches to induction of decision rules.* In: Polkowski and Skowron [337], pp. 500–529.
436. J. Stefanowski, K. Słowiński, R. Słowiński (1991), *Supporting of therapeutic decisions based on the rough sets theory.* In: Proceedings of the 16th Meeting of the EURO Working Group, Operational Research Applied to Health Services, the First Polish National Conference on Operational Research Applied to Health Systems, Książ 30.07–4.08, Wydawnictwo Politechniki Wrocławskiej, Wrocław, pp. 249–255.
437. J. Stefanowski, R. Słowiński, R. Nowicki (1992), *The rough sets approach to knowledge analysis for classification support in technical diagnostics of mechanical objects.* In: F. Belli, F.J. Radermacher (Eds.), Industrial & Engineering Applications of Artificial Intelligence Expert Systems, Lecture Notes in Economics Mathematical Systems **604**, Springer-Verlag, Berlin, pp. 324–334.
438. J. Stepaniuk (1996), *Similarity based rough sets and learning.* In: Tsumoto, Kobayashi, Yokomori, Tanaka, and Nakamura [475], pp. 18–22.
439. J. Stepaniuk (1998), *Approximation spaces, reducts and representatives.* In: Polkowski and Skowron [338], pp. 109–126.
440. J. Stepaniuk, J. Tyszkiewicz (1991), *Probabilistic properties of approximation problems.* Bull. Polish Acad. Sci. Tech. **39/3**, pp. 535–555.
441. Z. Suraj (1996), *An application of rough set methods to cooperative information systems re-engineering.* In: Tsumoto, Kobayashi, Yokomori, Tanaka, and Nakamura [475], pp. 364–371.
442. Z. Suraj (1996), *Discovery of concurrent data models from experimental tables, A rough set approach.* Fundamenta Informaticae **28/3–4**, pp. 353–376.
443. Z. Suraj (1997), *Reconstruction of cooperative information systems under cost constraints, A rough set approach.* In: Wang [504], pp. 399–402.
444. Z. Suraj (1998), *The synthesis problem of concurrent systems specified by dynamic information systems.* In: Polkowski and Skowron [338], pp. 420–450.
445. R. Swiniarski (1993), *Zernike moments: Their application for image recognition.* Internal report of San Diego State University, Department of Mathematical Computer Sciences, USA.
446. R. Swiniarski (1996), *Rough set expert system for on-line prediction of volleyball game progress for US olympic team.* In: B.D. Czejdo, I.I. Est, B. Shirazi, B. Trousse (Eds.), Proceedings of the Third Biennial European Joint Conference on Engineering Systems Design Analysis, July 1–4, Montpellier, France, pp. 15–20.
447. R. Swiniarski (1996), *Rough sets expert system for robust texture classification based on 2D fast Fourier transormation spectral features.* In: Tsumoto, Kobayashi, Yokomori, Tanaka, and Nakamura [475], pp. 419–425.
448. R. Swiniarski (1998), *Rough sets and principal component analysis and their applications in data model building and classification.* In this book.
449. R. Swiniarski (1997), *Design of nonlinear texture data model using localized principal components and rough sets. Application to texture classification.* In: Proceedings of International Symposium on Nonlinear Theory its Applications, Hawaii, USA, November 29 – December 3 (accepted).
450. R. Swiniarski (1998), *Intelligent feature extraction, Rough sets Zernike moments for data preprocessing feature extraction in handwritten digits recognition.* In: Proceedings of International Symposium on Engineering of Intelligent Systems, EIS98. University of Laguna, Tenerife, Spain, February 11–13.

451. R. Swiniarski (1998), *Texture recognition based on rough sets 2D FFT feature extraction*. In: Proceedings of World Automation Congress. Anchorage, Alaska, USA, May 9–14.
452. R. Swiniarski. (1998), *Rough sets Bayesian methods applied to cancer detection*. In: Polkowski and Skowron [339], pp. 609–616.
453. R. Swiniarski (1998), *Rough sets and neural networks application to handwritten character recognition by complex Zernike moments*. In: Polkowski and Skowron [339], pp. 617-624.
454. R. Swiniarski (1998), *Texture feature extraction, reduction and recognition based on rough sets*. In: Symposium on Object Recognition Scene Classification from Multispectral Multisensor Pixels. Columbus, Ohio, USA, July 6–10.
455. R. Swiniarski, A. Berzins (1996), *Rough sets for intelligent data mining, knowledge discovering and designing of an expert systems for on-line prediction of volleyball game progress*. In: Tsumoto, Kobayashi, Yokomori, Tanaka, and Nakamura [475], pp. 413–418.
456. R. Swiniarski, F. Hunt, D. Chalvet, D. Pearson (1997), *Feature selection using rough sets and hidden layer expansion for rupture prediction in a highly automated production system*. In: Proceedings of the 12th International Conference on Systems Science, September 12–15, Wroclaw, Poland; see also: Systems Science **23/1**.
457. R. Swiniarski, F. Hunt, D. Chalvet, D. Pearson (1995), *Intelligent data processing and dynamic process discovery using rough sets, statistical reasoning and neural networks in a highly automated production systems*. In: Proceedings of the First European Conference on Application of Neural Networks in Industry, August, Helsinki, Finland.
458. R. Swiniarski, J. Nguyen (1996), *Rough sets expert system for texture classification based on 2D spectral features*. In: B.D. Czejdo, I.I. Est, B. Shirazi, B. Trousse (Eds.), Proceedings of the Third Biennial European Joint Conference on Engineering Systems Design Analysis, July 1-4, Montpellier, France, pp. 3–8.
459. R. Swiniarski, H. Zeng (1998), *A new halftoning method based on error diffusion with rough set filtering*. In: Polkowski and Skowron [338], pp. 336–346.
460. M. Szczuka (1996), *Rough set methods for constructing neural network*. In: Proceedings of the Third Biennal Joint Conference On Engineering Systems Design Analysis, Session on Expert Systems, Montpellier, France, pp. 9–14.
461. M. Szczuka (1998), *Refining decision classes with neural networks*. In: Bouchon–Meunier and Yager [25], pp. 370–1375.
462. M. Szczuka (1998), *Rough sets and artificial neural networks*. In: Polkowski and Skowron [338], pp. 451–471.
463. A.J. Szladow, W. Ziarko (1992), *Knowledge-based process control using rough sets*. In: Słowiński [419], pp. 49–60.
464. H. Tanaka, Y. Maeda (1998), *Reduction methods for medical data*. In: Polkowski and Skowron [338], pp. 295–306.
465. J. Teghem, J.-M. Charlet (1992), *Use of 'rough sets' method to draw premonitory factors for earthquakes by emphasing gas geochemistry: The case of a low seismic activity context in Belgium*. In: Słowiński [419], pp. 165–180.
466. I. Tentush (1995), *On minimal absorbent sets for some types of tolerance relations*. Bull. Polish Acad. Sci. Tech. **43/1** pp. 79–88.
467. S. Tsumoto (1997), *Domain experts' interpretation of rules induced from clinical databases*. In: In: Zimmermann [545] **1**, pp. 1639–1642.

468. S. Tsumoto (1997), *Extraction of expert's decision process from clinical databases using rough set model.* In: Komorowski and Żytkow [147], pp. 58–67.
469. S. Tsumoto (1997), *Induction of positive negative deterministic rules based on rough set model.* In: Ras and Skowron [355], pp. 298–307.
470. S. Tsumoto (1997), *Empirical induction on medical expert system rules based on rough set model.* Ph.D. dissertation (in Japanese).
471. S. Tsumoto (1998), *Modelling diagnostic rules based on rough sets.* In: Polkowski and Skowron [339], pp. 475–482.
472. S. Tsumoto (1998), *Formalization induction of medical expert system rules based on rough set theory.* In: Polkowski and Skowron [338], pp. 307–323.
473. S. Tsumoto (Ed.) (1998), *Bulletin of International Rough Set Society* 2/1.
474. S. Tsumoto (1998), *Induction of expert decision rules using rough sets and set-inclusion.* In this book.
475. S. Tsumoto, S. Kobayashi, T. Yokomori, H. Tanaka, and A. Nakamura (Eds.) (1996), *Proceedings of the Fourth International Workshop on Rough Sets, Fuzzy Sets, and Machine Discovery (RSFD'96).* The University of Tokyo, November 6–8.
476. S. Tsumoto, H. Tanaka (1998), *PRIMEROSE, Probabilistic rule induction method based on rough set theory.* In: Ziarko [537], pp. 274–281 .
477. S. Tsumoto, H. Tanaka (1994), *Induction of medical expert system rules based on rough sets resampling methods.* In: Proceedings of the 18th Annual Symposium on Computer Applications in Medical Care, Journal of the AMIA **1** (supplement), pp. 1066–1070.
478. S. Tsumoto, H. Tanaka (1995), *PRIMEROSE, Probabilistic rule induction method based on rough set resampling methods.* In: Computational Intelligence: An International Journal **11/2**, pp. 389–405.
479. S. Tsumoto, H. Tanaka (1995), *Automated selection of rule induction methods based on recursive iteration of resampling.* In: Fayyad and Uthurusamy [92], pp. 312–317.
480. S. Tsumoto, H. Tanaka (1995), *Automated discovery of functional components of proteins from amino-acid sequences based on rough sets and change of representation.* In: Fayyad and Uthurusamy [92], pp. 318–324.
481. S. Tsumoto, H. Tanaka (1996), *Automated discovery of medical expert system rules from clinical databases based on rough sets.* In: Simoudis, Han, and Fayyad [393], pp. 63–69.
482. S. Tsumoto, H. Tanaka (1996), *Extraction of medical diagnostic knowledge based on rough set based model selection rule induction.* In: [RSFD'96] 426–436; see also: Chen, Hirota, and Yen [44], pp. 145–151.
483. S. Tsumoto, H. Tanaka (1996), *Automated induction of medical expert system rules from clinical databases based on rough sets.* In: Zimmermann [544], pp. 154–158.
484. S. Tsumoto, H. Tanaka (1996), *Domain knowledge from clinical databases based on rough set model.* In: In: Borne, Dauphin–Tanguy, Sueur, and El Khattabi [23], pp. 742–747.
485. S. Tsumoto, H. Tanaka (1996), *Incremental learning of probabilistic rules from clinical databases.* In: Bouchon–Meunier, Delgado, Verdegay, Vila, and Yager [24] **2**, pp. 1457–1462.
486. S. Tsumoto, H. Tanaka (1996), *Classification rule induction based on rough sets.* In: Petry and Kraft [314], pp. 748–754.
487. S. Tsumoto, H. Tanaka (1996), *Induction of expert system rules from databases based on rough set theory resampling methods.* In: Ras and Michalewicz [354], pp.

128–138.
488. S. Tsumoto, H. Tanaka (1996), *Machine discovery of functional components of proteins from amino-acid sequences based on rough sets change of representation.* In: Journal of the Intelligent Automation and Soft Computing **2/2**, pp. 169–180.
489. S. Tsumoto, H. Tanaka (1996), *PRIMEROSE3. Induction estimation of probabilistic rules from medical databases based on rough sets resampling methods.* In: M. Witten (Ed.), Computational Medicine, Public Health, Biotechnology, Building a Man in the Machine III, World Scientific, Singapore, pp. 1173–1189.
490. S. Tsumoto, W. Ziarko (1996), *The application of rough sets-based data mining technique to differential diagnosis of meningoencephalitis.* In: Ras and Michalewicz [354], pp. 438–447.
491. S. Tsumoto, W. Ziarko, N. Shan, H. Tanaka (1995), *Knowledge discovery in clinical databases based on variable precision rough sets model.* In: Proceedings of the 19th Annual Symposium on Computer Applications in Medical Care, New Orleans, Journal of American Medical Informatics Association Supplement, pp. 270–274.
492. A. Tversky (1997), *Features of similarity.* Psychological Review **84/4**, pp. 327–352.
493. D. Vakarelov (1991), *A modal logic for similarity relations in Pawlak knowledge representation systems.* Fundamenta Informaticae **15**, pp. 61–79.
494. D. Vakarelov (1991), *Logical analysis of positive and negative similarity relations in property systems.* In: M. De Glas, D. Gabbay (Eds.), *First World Conference on the Fundamentals of AI (WOCFAI'91)*, July 1–5, Paris, France, pp. 491–499.
495. D. Vakarelov (1995), *A duality between Pawlak's information systems and bi-consequence systems.* Studia Logica **55/1** pp. 205–228.
496. D. Vakarelov (1998), *Information systems, similarity relations and modal logic.* In: Orłowska [270] pp. 492–550.
497. D. Van den Poel, Z. Piasta (1998), *Purchase prediction in database marketing with the ProbRough system.* In: Polkowski and Skowron [339], pp. 593–600.
498. D. Van den Poel (1998), *Rough sets for database marketing.* In: Polkowski and Skowron [338], pp. 324–335.
499. I. Velasco, D. Teo, T.T. Lin (1997), *Design optimization of rough-fuzzy controllers using a genetic algorithm.* In: Wang [504], pp. 313–317.
500. S. Vinterbo, L. Ohno-Machado H. Fraser (1998), *Prediction of acute myocardial infarction using rough sets.* Submitted for publication.
501. A. Wakulicz-Deja, M. Boryczka, P. Paszek (1998), *Discretization of continuous attributes of decision system in mitochondrial encephalomyopathies.* In: Polkowski and Skowron [339], pp. 483–490.
502. A. Wakulicz-Deja, P. Paszek (1997), *Diagnose progressive encephalopathy applying the rough set theory.* International Journal of Medical Informatics **46**, pp. 119–127.
503. P.P. Wang (Ed.) (1995), *Proceedings of the International Workshop on Rough Sets Soft Computing at Second Annual Joint Conference on Information Sciences (JCIS'95).* Wrightsville Beach, North Carolina, 28 September – 1 October, pp. 1–679.
504. P.P. Wang (Ed.) (1997), *Proceedings of the Fifth International Workshop on Rough Sets Soft Computing (RSSC'97) at Third Annual Joint Conference on Information Sciences (JCIS'97).* Duke University, Durham, NC, USA, Rough Set & Computer Science **3**, March 1–5, pp. 1–449.

505. A. Wasilewska (1989), *Linguistically definable concepts and dependencies.* Journal of Symbolic Logic **54/2**, pp. 671–672.
506. A. Wasilewska (1996), *On rough and LT-fuzzy sets.* In: Chen, Hirota, and Yen [44], pp. 13–18.
507. A. Wasilewska (1997), *Topological rough algebras.* In: Lin and Cercone [189], pp. 411–425.
508. A. Wasilewska, L. Vigneron (1998), *Rough algebras and automated deduction.* In: Polkowski and Skowron [337], pp. 261–275.
509. A. Wasilewska, L. Vigneron (1995), *Rough equality algebras.* In: Wang [503], pp. 26–30.
510. R. Wilting (1997), *Predicting card credit. A research on predictable behaviour of clients applying for card credit by means of Rough Data Models.* (in Dutch). Master Thesis, Vrije Universiteit Amsterdam.
511. http://www.idi.ntnu.no/~aleks/rosetta/ – the ROSETTA WWW homepage.
512. D.A. Wilson, T.R. Martinez (1997), *Improved heterogeneous distance functions.* Journal of Artificial Intelligence Research **6**, pp. 1–34.
513. P. Wojdyłło (1998), *Wavelets, rough sets arificial neural networks in EEG analysis.* In: Polkowski and Skowron [339], pp. 444–449.
514. L. Woolery, J.W. Grzymała-Busse, S. Summers, A. Budihardjo (1991), *The use of machine learning program $LERS_LB$ 2.5 in knowledge acquisition for expert system development in nursing.* Computers nd Nursing **9**, pp. 227–234.
515. L. Woolery, M. Van Dyne, J.W. Grzymała-Busse, C. Tsatsoulis (1994), *Machine learning for development of an expert system to support nurses' assessment of preterm birth risk.* In: Nursing Informatics, An International Overview for Nursing in a Technological Era, Proceedings of the Fifth International Conference on Nursing Use of Computers Information Sci. June 17–22, San Antonio, TX, Elsevier, pp. 357–361.
516. L. Woolery, J.W. Grzymała-Busse (1994), *Machine learning for an expert system to predict preterm birth risk.* Journal of the American Medical Informatics Association 1, pp. 439–446.
517. S.K.M. Wong (1996), *Interval structure – A qualitative measure of uncertainty.* In: Lin and Wildberger [194], pp. 22–27.
518. S.K.M. Wong, P. Lingras (1989), *The compatibility view of Shafer-Dempster theory using the concept of rough set.* In: Z.W. Ras (Ed.), Proceedings of the Fourth International Symposium on Methodologies for Intelligent Systems (ISMIS'89), Charlotte, North Carolina, October 12–14, North Holland, pp. 33–42.
519. S.K.M. Wong, W. Ziarko (1986), *Comparison of the probabilistic approximate classification and the fuzzy set model.* Fuzzy Sets and Systems **21**, pp. 357–362.
520. S.K.M. Wong, W. Ziarko, L.W. Ye (1986), *Comparision of rough set and statistical methods in inductive learning.* Journal of Man–Machine Studies **24**, pp. 53–72.
521. J. Wróblewski (1995), *Finding minimal reducts using genetic algorithms.* In: Wang [503], pp. 186–189.
522. J. Wróblewski (1998), *Genetic algorithms in decomposition and classification problems.* In: Polkowski and Skowron [338], pp. 472–492.
523. U. Wybraniec-Skardowska (1989), *On a generalization of approximation space.* Bull. Polish Acad. Sci. Math. **37**, pp. 51–61.
524. Y.Y. Yao (1997), *Combination of rough and fuzzy sets based on alpha-level sets.* In: Lin and Cercone [189], pp. 301–321.
525. Y.Y. Yao (1998), *Generalized rough set models.* In: Polkowski and Skowron [337], pp. 286–318.

526. Y.Y. Yao, T.Y. Lin (1996), *Generalization of rough sets using modal logic.* Journal of the Intelligent Automation and Soft Computing **2**, pp. 103–120.
527. Y.Y. Yao, S.K.M. Wong (1995), *Generalization of rough sets using relationships between attribute values.* In: Wang [503], pp. 30–33.
528. Y.Y. Yao, S.K.M. Wong (1996), *Generalized probabilistic rough set models.* In: Chen, Hirota, and Yen [44], pp. 158–163.
529. Y.Y. Yao, S.K.M. Wong, T.Y. Lin (1997), *A review of rough set models.* In: T.Y. Lin, N. Cercone [189] pp. 47–75.
530. L.A. Zadeh (1971), *Similarity relations and fuzzy orderings.* Information Sciences **3**, pp. 177–200.
531. L.A. Zadeh (1996), *Fuzzy logic = computing with words.* IEEE Trans. on Fuzzy Systems **4**, pp. 103-111.
532. L.A. Zadeh (1997), *Toward a theory of fuzzy information granulation and its certainty in human reasoning and fuzzy logic.* Fuzzy Sets and Systems **90**, pp. 111-127.
533. E.C. Zeeman (1965), *The topology of brain and visual perception.* In: Fort, K.M. (Ed.): *Topology of 3-manifolds and related topics.* Prentice Hall, Englewood Cliffs N.J., pp. 240-256.
534. W. Ziarko (1991), *The discovery, analysis and representation of data dependencies in databases.* In: G. Piatetsky–Shapiro, W.J. Frawley (Eds.), Knowledge Discovery in Databases, AAAI Press/MIT Press, pp. 177–195.
535. W. Ziarko (1992), *Generation of control algorithms for computerized controllers by operator supervised training.* In: Proceedings of the 11th IASTED International Conference on Modelling, Identification Control. Innsbruck, Austria, pp. 510–513.
536. W. Ziarko (1993), *Variable precision rough set model.* J. of Computer and System Sciences, 46, pp. 39–59.
537. W. Ziarko (Ed.) (1994), *Rough Sets, Fuzzy Sets and Knowledge Discovery (RSKD'93).* Workshops in Computing, Springer–Verlag & British Computer Society, London, Berlin.
538. W. Ziarko (1994), *Rough sets and knowledge discovery: An overview.* In: Ziarko [537], pp. 11–15.
539. W. Ziarko (1998), *Rough sets as a methodology for data mining.* In: Polkowski and Skowron [337], pp. 554–576.
540. W. Ziarko, R. Golan, D. Edwards (1993), *An application of DATALOGIC/R knowledge discovery tool to identify strong predictive rules in stock market data.* In: Proceedings of AAAI Workshop on Knowledge Discovery in Databases, Washington, DC, pp. 89–101.
541. W. Ziarko, J. Katzberg (1989), *Control algorithms acquisition, analysis reduction, Machine learning approach.* In: Knowledge–Based Systems Diagnosis, Supervision Control, Plenum Press, Oxford, pp. 167–178.
542. W. Żakowski (1982), *On a concept of rough sets.* Demonstratio Mathematica **XV**, pp. 1129–1133.
543. W. Żakowski (1983), *Approximations in the space (U, Π).* Demonstratio Mathematica **XVI**, pp. 761–769.
544. H.-J. Zimmermann (1997), *Proceedings of the Fourth European Congress on Intelligent Techniques and Soft Computing (EUFIT'96).* September 2–5, Aachen, Germany, Verlag Mainz **1**, pp. 1–702.
545. H.-J. Zimmermann (1996), *Proceedings of the Fifth European Congress on Intelligent Techniques and Soft Computing (EUFIT'97).* Aachen, Germany, Verlag Mainz **1**, pp. 1–868.

Rough Sets, Rough Function and Rough Calculus

Zdzisław Pawlak

Institute of Theoretical and Applied Informatics Polish Academy of Sciences
ul. Baltycka 5, 44 000 Gliwice, Poland

1 Introduction

The concept of the rough set – a mathematical basis for reasoning about vagueness and uncertainty proved to be a natural instrument to inquire into many theoretical and practical problems related to data analysis. Although many serious real-life problems have been formulated and solved in the framework of rough set theory it seems that the extension of this theory to rough relations and rough functions is badly needed, for numerous applications can not be covered by the concepts of a rough set only.

The objective of this paper is to give some ideas concerning rough functions along the lines proposed by the author in [Pa1, Pa2, Pa4, Pa5, Pa6] and this paper is a modified version of [Pa6], where basic concepts of rough calculus have been proposed. Some similar concepts have been considered by Nakamura and Rosenfeld in [NR1].

It is interesting that ideas presented in this paper are not entirely new and their origin can be traced back to calculus of finite differences by George Boole (cf.[Bo1]).

Physical phenomena are usually described by differential equations. Solutions of these equations are real valued-functions, i.e., functions which are defined and valued on continuum of points. However, due to limited accuracy of measurements and computations, we are unable to observe (measure) or compute (simulate) exactly the abstract solutions. Consequently, we deal with approximate rather than exact solutions, i.e., we are using discrete and not continuous variables and functions.

Thus abstract mathematical models of physical systems are expressed in terms of real functions, whereas observed or computational models are described by data sets obtained as a result of measurements or computations - which use not real but rational numbers from a finite subset of ratinal numbers.

Hence an important question arises - what is the relationship between these two approaches, i.e., based on continuous or discrete mathematics philosophy?

Similar problems have been faced in image processing as perceived by Rosenfeld in [Ro3] and pursued by Nakamura and others in [NA1, NA2, NA3].

Another tool developed for discrete system analysis is the so called "cell-to-cell mapping theory" [Hs1], in which real numbers are replaced by intervals. Due to the lack of sound mathematical foundations, this method seems to be better suited to computer simulation than to prove theorems about discrete systems. It is worthwhile to mention that the idea of cell-to-cell mapping has found interesting application in the design and analysis of fuzzy controllers [PT1, Pa6, SC1].

Some aspects of the considered problems are also related to interval analysis first anticipated by Warmus in [Wa1, Wa2] and developed extensively by many authors recently.

Independently of practical problems caused by the "continuous versus discrete" antinomy, the philosophical question, of how to avoid the concept of infinity in mathematical analysis, has been tackled for a long time by logicians. Nonstandard analysis [Ro1], finistic analysis [My1] and infinitesimal analysis [CS1] provide various views on this topics.

In this paper we are going to investigate on the relationship between real and discrete functions based on the rough set philosophy. In particular we define rough (discrete) lower and upper representation of real functions and define and investigate some properties of these representations, such as rough continuity, rough derivatives, rough integral and rough differential equations - which can be viewed as discrete counterparts of real functions.

In particular we are interested how discretization of the real line effects basic properties of real functions, such as continuity, differentiability, etc. It turns out that some properties of real functions have counterparts in the case of discrete functions, but this is not always the case. The proposed approach is based on the rough set philosophy, in which the indiscernibility relation, defined in our case on the set of reals, is the starting point of our considerations.

The proposed approach differs essentially from numerical and approximation methods, even though we use, in some cases, similar terminology (e.g., approximation of function by another function) - for our attempt is based on functions defined and valued in the set of integers - however it has some overlaps with nonstandard, finistic and infinitesimal analysis, mentioned above.

Last but not least the proposed philosophy can be seen as a generalization of qualitative reasoning [Ku1, We1], where three-valued $(+, 0, -,$ i.e., increasing, not changing, decreasing) qualitative derivatives are replaced by more general concept of multi-valued qualitative derivatives, so that expressions such as "slowly increasing", "fast increasing", "very fast increasing" etc. can be used instead of only "increasing".

Ideas shown in this paper have been presented at the International Conference on Intelligent Systems, Augustow, June 5-10, 1995, Poland and Joint Conference on Information Sciences (JCIS'95), Wrightsville Beach, Sept 28 - Oct 1, 1995, North Carolina, USA.

2 Scale, Discretization and Indiscernibility

This section introduces the basic concept of our approach - the indiscernibility relation. As mentioned in the introduction, real-valued parameters of a physical system can be exactly measured or computed with some approximation only. Therefore, we will introduce the concept of a scale, which is a finite set of integers $\{0, 1, \ldots, n\}$ and is intended to be used as a set of measurement units, like kg, km, hr, etc. - and a mapping of the scale into the set of real numbers. Elements of the scale, i.e., measurement units, are understood as approximations of real numbers, inaccessible due to our lack of infinite precision of measurement or computation. Notice that the concept of the scale is similar to that of the landmark, used in the qualitative reasoning methods, but both concepts are used differently.

Every scale determines uniquely a partition of the real line, or, in other words, defines an equivalence relation on reals, called in what follows an iniscernibility relation. Elements of the same equivalence class of the indiscernibility relation are said to be indiscernible with respect to the scale, and can be expressed approximately only by units of the scale. Thus, due to the use of the assumed scale real-valued parameters are replaced by approximate, integer-valued parameters.

A more formal presentation of the above ideas is given below [Ob1].

Let $[n] = \{0, 1, \ldots, n\}$ be a set of natural numbers. A strictly monotonic function $d : [n] \to \mathbf{R}$, i.e., such that for all $i, j \in [n]$, $i < j$ implies $d(i) < d(j)$ will be called a *scale*.

Any scale $d : [n] \to \mathbf{R}$ is a finite increasing sequence of reals $x_0, x_1, \ldots, x_n$, such that $x_i = d(i)$, for very $i \in [n]$ - thus it can be seen as a *discretization* of the closed interval $R_n = \langle d(0), d(n) \rangle = \langle x_0, x_n \rangle$.

Given a scale $d : [n] \to \mathbf{R}$ then one can define two functions

$$d_*(x) = max\{i \in [n] : x_i \leq x\}$$
$$d^*(x) = min\{i \in [n] : x_i \geq x\}$$

for every $x \in R_n$.

On the interval $R_n = \langle x_0, x_n \rangle$ we define an equivalence relation I_d, called the *indiscernibility* relation, and defined thus

$$x I_d y \text{ iff } d_*(x) = d_*(y) \text{ and } d^*(x) = d^*(x).$$

The family of all equivalence classes of the relation I_d, or the partition of the interval R_n, is given below

$$\{x_0\}, (x_0, x_1), \{x_1\}, (x_1, x_2), \{x_2\}, \ldots, (x_{n-1}, x_n), \{x_n\}$$

where each equivalence classe $[x]_d$ is an interval such that $[x]_d = (x_i, x_{i+1})$ whenever $x_i < x < x_{i+1}$, and $[x_i]_d = \{x_i\}$ for all $i \in [n]$.

If $x_i < x < x_{i+1}$, then $I_{*d}(x) = d(d_*(x)) = x_i$ and $I_{*d}(x) = d(d^*(x)) = x_{i+1}$, i.e., $I_d^*(x)$ and $I_d^*(x)$ are the ends of the interval $\langle x_i, x_{i+1} \rangle$; if $x = x_i$, then $I_{*d}(x) = I_d^*(x) = x_i$.

The ends of the interval $\langle x_i, x_{i+1} \rangle$ are called the *lower* and the *upper d-approximation* of x, respectively.

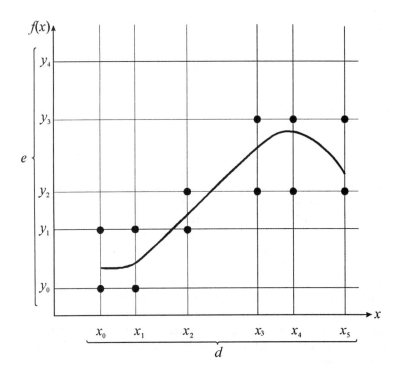

Fig. 1. The lower and upper approximation of a real function

The above discussed ideas are illustrated in Fig. 1.

Suppose we are given two scales $d : [n] \to \mathbf{R}$ and $e : [m] \to \mathbf{R}$, and let $f : R_n \to R_m$ be a function, where R_n, R_m denote the both side closed intervals $\langle x_0, x_n \rangle$, $\langle y_0, y_m \rangle$ respectively. We define its *lower rough representation* f_* with respect to d and e and its *upper rough representation* f^* with respect to d and e defined on $[n]$ and valued in $[m]$, as

$$f_*(i) = e_*(f(x))$$

$$f^*(i) = e^*(f(x))$$

where $d_*(x) = x_i$, for all $i \in [n]$ (see Fig. 1).

Thus with every real function one can associate two discrete functions; its lower and upper approximation. These approximations are uniquely determined by indiscernibility relations superimposed on the domain and range of the real function.

Let us observe that the just-defined approximations of real functions are different from those considered in approximation theory.

In what follows we are going to give some properties of discrete functions, defined and valued in the set of integers - mimicking some properties of real functions. It turns out that for this class of functions one can define concepts similar to that of real function, like continuity, derivatives, integrals, etc. These concepts display similar properties to those of real functions, and consequently discrete functions obtained as a result of measurements can be treated similarly to real functions.

We will start our consideration by defining rough (approximate) continuity for discrete functions.

3 Roughly Continuous Discrete Functions

The concept of continuity is strictly connected with real functions. Intuitively a function is continuous if a small change of its argument causes a small change of its value, or in other words - it cannot "vary too fast" [CJ1]. A similar idea can be employed also in the case of discrete functions, and we will say that a discrete function is roughly (approximately) continuous if a small change of its argument causes a small change of its value. In fact the concept of continuity of discrete functions has been used for a long time in qualitative reasoning [Ku1, We1] and others (cf. [Ch1, Pa1, Ro2]). Below the formal definition of roughly continuous function is given and some elementary properties of these functions are presented.

A discrete function $f : [n] \to [m]$ is *roughly continuous* iff for all $i, j \in [n]$, $|i - j| = 1$ implies $|f(i) - f(j)| \leq 1$.

The intermediate value property is valid for roughly continuous discrete functions as shown by the following proposition.

Proposition 1. *A discrete function $f : [n] \to [m]$ is roughly continuous iff for all $i, j \in [n], i \neq j$, and for every q between $f(i)$ and $f(j)$ there exist $p \in [n]$ between i and j for which $f(p) = q$.*

Thus the basic property of continuous real functions, the intermediate value theorem, after slight modifications is also valid for discrete functions. Hence it seems that the idea of continuity need not be necessarily attributed to real functions only, and can be extended to discrete functions.

4 Rough Derivatives and Rough Integrals of Discrete Functions

Now we are going to define two basic concepts in our approach to discrete functions, namely the rough derivative and the rough integral. It turns out that they display similar properties to "classical" derivatives and integrals. Let us observe that they are defined not on reals but on integers (representing finite set of data).

For a discrete function $f : [n] \to [m]$ we define the *rough derivative f'* as

$$f'(i) = \Delta f(i) = f(i+1) - f(i), \text{ for all } i \in [n-1].$$

We say that $f : [n] \to [m]$ has Darboux property if for every $i \in [n-1]$ we have that $f'(i) \in \{-1, 0, 1\}$. Thus for $f : [n] \to [m]$ having rough Darboux property and $i \in [n-1]$ the value $f'(i)$ is that $\alpha_i \in \{-1, 0, 1\}$ which makes $f(i+1) = f(i) + \alpha_i$.

Proposition 2. *A discrete function $f : [n] \to [m]$ is roughly continuous iff f has Darboux property.*

Directly from the definition of the rough derivative for discrete functions, we obtain the following counterpart of the well known theorem of differential calculus (cf. [Pa6]).

Proposition 3. *Let f and g be discrete function with domain $[n]$ and range $[m]$ respectively. Than for $f + g, fg$ and f/g we have*

a) $$(f+g)'(i) = f'(i) + g'(i),$$

b) $$(fg)'(i) = f'(i)g(i) + f(i)g'(i) + f'(i)g'(i),$$

c) $$(f/g)'(i) = \frac{f'(i)g(i) - f(i)g'(i)}{g^2(i) + g(i)g'(i)}.$$

¿From the definition of the rough derivative of discrete function and Proposition 3 we get the following proposition.

Proposition 4. *1) The rough derivative of a constant discrete function is equal to zero.*
2) If $f(i) = i + k$, where k is an integer constant, then $f'(i) = 1$.
3) If $f(i) = ki$, then $f'(i) = k$.
4) If $f(i) = k^i$, then $f'(i) = (k-1)k^i$; for $k = 2$ we have $f'(i) = 2^i$.
5) If $f(i) = i^k$, then $f'(i) = \sum_{j=0}^{k} \binom{k}{j} i^{k-j} - i^k$.
In particular, if $k = 2$ we get $f'(i) = 2i+1$; for $k = 3$ we have $f'(i) = 3i^2 + 3i + 1$, etc.

Higher order derivatives can be also defined in the same manner. In general, k-th rough derivative $f^{(k)}$ of a discrete function f is defined by the following well known formula in the difference calculus

$$f^{(k)}(i) = \sum_{j=0}^{k} \binom{k}{j}(-1)^j f(i+k-j).$$

The following egxample illustrates application of the above formula.

i	0	1	2	3	4	5
$f(i)$	1	1	3	4	2	1
$f^{(1)}(i)$	0	2	1	-2	-1	
$f^{(2)}(i)$	2	-1	-3	1		
$f^{(3)}(i)$	-3	-2	4			
$f^{(4)}(i)$	1	6				
$f^{(5)}(i)$	5					

Notice that f is a discrete function $f : [n] \to [m]$ defined on $n+1$ points, i.e., on the set $\{0, 1, \ldots, n\}$, and $f^{(k)} : [n-k] \to [m]$ is defined on $n-k+1$ points. Thus each discrete function $f : [n] \to [m]$ has at most derivatives up to the n-th order.

Consequently each discrete function $f : [n] \to [m]$ is uniquely defined by the set of the following initial conditions $f^{(n)}(0), f^{(n-1)}(0), \ldots, f^{(1)}(0), f^{(0)}(0)$, where $f^{(0)}(0) = f(0)$.

Some important properties of real functions are not valid for discrete functions, as shown by the following two propositions.

Proposition 5. *Assume that a discrete function $f : [n] \to [m]$ has a maximum (minimum) at $i \in (n)$, where $(n) = \{1, 2, \ldots, n-1\}$. Then not necessarily $f'(i) = 0$.*

Rolle's theorem does not hold for discrete functions, as shown by the proposition below.

Proposition 6. *Let $f : [n] \to [m]$ be a discrete, function, such that $f(0) = f(n) = 0$. Then not necessarily there exists $i \in (n)$ such that $f'(i) = 0$.*

We say that a discrete function f is *roughly smooth* if its first rough derivative is roughly continuous. It can be easily seen that for roughly smooth functions the above two propositions are valid, provided that they are slightly modified. Detailed discussion of this problem is left to the reader.

Next we define integration of discrete functions.

Let $f : [n] \to [m]$ be a discrete function. By a *rough integral* of f we mean the function

$$\int_{j=0}^{i} f(j)\Delta(j) = \sum_{j=0}^{i} f(j)\Delta(j)$$

where $\Delta(j) = (j+1) - j = 1$.

The following important property holds.

Proposition 7.

$$\int_{j=0}^{i} f'(j)\Delta(j) = f(i) + k$$

where k is an integer constant.

In other words

$$f(i) = f(0) + \sum_{j=0}^{i-1} f'(j)$$

or in recursive form

$$f(i+1) = f(i) + f'(i)$$

with the initial condition

$$f(0) = k.$$

This proposition can be used for solving rough differential equations, and will be discused in the next section.

The reader is advised to compare the concept of the rough derivative and the rough integral with corresponding concepts considered in [Bo1].

5 Rough Differential Equations

Starting from the notion of a rough derivative for discrete functions one can define a concept of differential equation for discrete functions, called in what follows a rough differential equation [Ob2] (see also [Bo1]). Rough differential equation, together with initial condition can be solved inductively by employing Proposition 7, which gives the relationship between initial condition, rough derivative and the solution.

Ordinary 1-st order differential equation is shown below

$$f'(x) = \Phi(x, f(x))$$

where Φ is a real valued function on the Cartesian product of reals.

Similarly one can define a *rough differential equation*, for discrete functions as

(∗) $$f'(i) = \Phi(i, f(i))$$

where Φ is an integer valued function defined on the Cartesian product $[n] \times [m]$.

Because $f'(i) = f(i+1) - f(i)$, the rough differential equation can be presented as

$$f(i+1) = \Phi(i, f(i)) + f(i)$$

which together with an initial condition

$$f(0) = j_0, \ j_0 \in [m]$$

defines uniquely the solution of the rough differential equation (∗).

Example 1. Consider a very simple rough differential equation given by the formula

(∗∗) $$f'(i) = 4i + 1$$

with the initial condition $f(0) = 2$.

By employing Proposition 3 one can easily show that the solution of this equation has the form

$$f(i) = f(0) + 2i^2 - i$$

We can also solve this equation by using Proposition 7. Suppose we are given the rough differential equation (∗∗) in tabular form, and we do not know its analytical presentation. In this case, by Proposition 7 we have

$$f(i+1) = f(i) + f'(i)$$

with $f(0) = 2$, which yields

$$f(0) = 2$$

$$f(1) = f(0) + f'(0) = 3$$

$$f(2) = f(1) + f'(1) = 8$$

$$f(3) = f(2) + f'(2) = 17$$

$$f(4) = f(3) + f'(3) = 30$$

$$f(5) = f(4) + f'(4) = 47$$

etc.

Thus we have two ways of solving rough differential equations. The first one is similar to that used in analysis, and it boils down to symbolic manipulation on formulas, whereas the second is suitable to functions presented in tabular form.

Ideas presented in this section can be easily extended for two-dimensional case (cf. [Gr1, Wa1]).

6 Conclusion

In this paper we have defined and investigated notions of rough (approximate) continuity, rough derivatives, rough integrals and rough differential equations for discrete functions, i.e., functions defined and valued on the set of integers. We have shown that the introduced concepts mirror some basic properties of calculus, and that discrete functions display properties similar to those of real functions, however this is not always the case.

However, it should be noted that the porposed approach essentialy differs from numerical methods because: firstly, our domains are finite hence we do not consider method convergence typical to numerical methods; secondly, rough differential equations should be derived from finite date sets in contrast to numerical methods obtained from given differential equations.

Many problems connected with the proposed approach still remain open. We did not cover much of material needed a serious consideration in connection with "rough (approximate) calculus". Nevertheless we hope that some fundamental notions have been clarified and sound foundations for further research and applications have been laid down.

7 Acknowledgments

The author wishes to express his thanks to anonymous referee for helpful sugestions and critical remarks.

References

[Bo1] Boole, G.: (1860). Treatise on the Calculus of Finite Differences. J. F. Moulton (ed.), Third Edition, G. E. Stechert and Co., New York, Printed in 1946 by the Murray Printing Company, Cambridge, Massachusets

[Bo2] Boxer, L.: Digitally Continuous Functions. Pattern Recognition Letters **15** (1994) 833–839

[Ch1] Chen, Li: The Necessary and Sufficient Condition and the Efficient Algorithms for Gradually Varied Fill. Chinese Science Bulletin **35/10** (1990) 870–873

[CJ1] Courant, R., John, F.: Introduction to Calculus and Analysis, 1, Interscience Publishers, A Division of John Wiley and Sons, Inc. New York, London, Sydney (1965)

[CS1] Chuaqui, R. Suppes, P.: Free-variable Axiomatic Foundations of Infinitesemal Analysis: a Fragment with finitary Consistency Proof. Journal of Symbolic Logic **60/1** (1995) 122–159

[Gr1] Grodzki, Z.: Rough Two-Dimentional Calculus. (manustript) (1997)

[Hs1] Hsu, C. S.: Cell-to-cell Mapping - A Metheod of Global Analysis for Non-linear Systems. Springer-Verlag, New York, Berlin, Heidelberg, London, Paris, Tokyo (1987)

[Ku1] Kuipers, B.: Qualitative Simulation. Artificial Intelligence **29** (1986) 289–338

[My1] Mycielski, J.: Analysis without Actual Infinity. Journal of Symbolic Logic **46** (1981) 625–633

[NA1] Nakamura, A., Aizawa, K.: Digital Circles, Computer Vision. Garphics and Image Processing **29** (1984) 242–255

[NA2] Nakamura, A., Aizawa, K.: Digital Images and Geometric Pictures, Digital Circles. Computer Vision, Garphics and Image Processing **30** (1985) 107–120

[NA3] Nakamura, A., Aizawa, K.: On the Recognition of Digital Pictures. IEEE Transaction on Pattern Analysis and Machine Intelligence Vol. PAMI-7 **6** (1985) 708–713

[NR1] Nakamura, A., Rosenfeld, A.: Digital Calculus. Information Sciences **98** (1997) 63–98

[Ob1] Obtułowicz, A.: Some Remarks on Rough Real Functions. ICS WUT Report 9/95 (1995)

[Ob2] Obtułowicz, A.: Differential Equations for Discrete Functions. ICS WUT Report 28/95 (1995)

[PT1] Papa, M., Tai, H. M., Shenoi, S.: (1994) Cell Mappings and Fuzzy Controller Design. Proceedings JCIS'94 (1994) 361

[Pa1] Pawlak, Z.: Rough Functions. Bull. PAS, Tech. Ser. **35/5-6** (1987) 249–251

[Pa2] Pawlak, Z.: Rough Sets, Rough Real Functions. ICS WUT Report 50/94 (1994)

[Pa3] Pawlak, Z.: Rough Real Functions and Rough Controllers. ICS WUT Report 1/95 (1994)

[Pa4] Pawlak, Z.: On Some Issues Connected with Roughly Continuous Functions. ICS WUT Report 21/95 (1995)

[Pa5] Pawlak, Z.: On Rough Dervatives, Rough Integrals and Rough Differential Equations. ICS WUT Report 41/95 (1995)

[Pa6] Pawlak, Z.: Rough Calculus. ICS WUT Raport, 58/95, also in: Proc. 2nd Annu. Joint Conf. on Information Sci. (1995) 34–35 (abstract)

[RS1] Rizk, Ch. G., Smith, S. M.: Automatic Design and Analysis of Fuzzy Logic Controllers Using Cell Mapping and Least Mean Square Algorithm. Proceedings JCIS'95 (1995) 328–431

[Ro1] Robinson, A.: Non-Standard Analysis, North-Holland Publishing Company (1970)

[Ro2] Rosenfeld, A.: Continous Functions on Digital Pictures. Pattern Rseognition Letters **4** (1986) 177–184

[Ro3] Rosenfeld, A.: Picture Languages. New York, Academic Press (1979)

[SC1] Smith, S. M., Comer, D. J., Nokleby, B.: A Computational Approach to Fuzzy Logic Controllers Design and Analysis Using Cell State Space Methods. In: A. Kandel, G. Langoltz (eds.), Fuzzy Control Systems, CRC Press (1994) 397–428

[Wa1] Warmus, M.: Calculus of Approximations. Bull. Acad, Pol. Sci., Cl. III **4(5)** (1956) 253–259

[Wa2] Warmus, M.: Approximations and Inequilities in the Calculus of Approximations: Classifications of Approximate Numbers. Bull. Acad, Pol. Sci., Cl. III **9(4)** (1961) 241–245

[We1] Werthner, H.: Qualitative Reasoning – Modeling and the Generation of Behavior. Springer-Verlag, Wien, New York (1994)

Decomposition and Synthesis of Decision Tables with respect to Generalized Decision Functions

Dominik Ślęzak

Warsaw University
Banacha 2, 02-097 Warsaw
Phone: +48 (22) 658-34-49, Fax: +48 (22) 658-34-48
email: slezak@alfa.mimuw.edu.pl

Abstract: An approach to the attribute set decomposition of decision tables is proposed. It enables to combine non-deterministic decision rules based on the generalized decision functions for different subsets of conditions. Optimal decomposition onto conditions subsets is proposed to be searched from Bayesian-like networks. Computational complexity of searching for such networks is discussed.

1 Introduction

In recent years rough set approach, originated by [8], turned out to be very effective as applicable to data mining and decision support systems. However, real-life problems require reconsidering rough set tools in view of large data bases, where the number of objects as well as the average number of conditional attributes in rough set based decision rules becomes too high to classify new cases. For these and also other purposes a great effort has been spent on initial decomposition of information systems and decision tables with large number of objects and attributes (see e.g. [6], [7], [18]). The aim of such a decomposition is to store the knowledge within smaller subtables, with their local rules of reasoning. They can be applied to any object being tested, after examining its relevance to particular subsystems. Then the final decision is computed by synthesis of answers from local sources.

We propose a new approach to decomposition, which enables to combine non-deterministic decision rules derived from different subsets of conditional attributes. According to introduced methods, the process of searching for such a decomposition can be reduced to the problem of finding optimal decomposition networks (with properties similar to Bayesian networks (see e.g. [9])) - more technically, minimal Markov-like boundaries under appropriate ordering over conditional attributes (compare e.g. with [12]). The tools for finding optimal decomposition are discussed with respect to their computational complexity. They are based on the minimum description length principle, what turns out to improve both the exactness of synthesis of information (compare e.g. with [10]) and the effectiveness of reasoning process (compare e.g. with [2], [15]).

The paper is organized as follows. In Section 2 we introduce criteria for attribute oriented decomposition of decision tables, based on the generalized decision functions. Argumentation for our way of stating the problem of finding optimal decomposition is given with respect to applications, as well as from computational and rough set theory points of view.

Section 3 contains the construction of conditional independence model for decomposition, its properties and potential applications.

In Section 4 we consider Bayesian-like network approach, adopted from probability theory (see e.g. [9]), which enables to represent knowledge about possible decomposition solutions in terms of directed acyclic graphs, named as decomposition networks.

Section 5 is devoted to the analysis of complexity of searching for minimal Markov-like boundaries under fixed ordering over conditional attributes (compare e.g. with [12]). The characterization is given in terms of well known rough set optimization tasks, to take the advantage of software techniques developed so far (see e.g. [20]).

In Section 6 we introduce conditional valuations for decomposition networks (compare with [11]). The correspondence to probability theory is redrawn in view of information propagation in causal nets (see e.g. [19]).

In Section 7 we finish the description of optimal decomposition network by proposing some heuristics for finding potentially best ordering over conditional attributes.

Section 8 closes the paper with some remarks and directions for further research.

2 Attribute decomposition

The purpose of decomposition of large information systems is to state a distributed environment of cooperative agents able to combine their knowledge, if necessary. In a special case of reasoning under uncertainty, by an agent we may understand some local source of information which can be used for approximate classification of objects under consideration. Each agent is equipped with its own decision rules which can be applied to objects. The main property of a distributed system should be the possibility of combining knowledge coming from particular agents - what, in practice, means that if several agents managed to classify some object, local answers remain not contradictive to each other and their synthesis reduces the uncertainty corresponding to reasoning in a degree appropriate for given problem.

While reasoning about objects from the domain specified by our needs, we are usually forced to base just on information gathered by the analysis of some sample of that domain. Such an approach is the main paradigm of rough sets theory ([8]), where any sample of objects, stored within an information system, is assumed to be the only kind of knowledge able to use for cases outside the sample. Stated what we would like to reason about, an information system can be described as a tuple $\mathbf{A} = (U, A)$, where the universe U is the sample of objects from some larger domain and each attribute $a \in A$, corresponding to some feature which is important with respect to objects classification, is identified with function $a : U \to V_a$ onto the set of its possible values.

Talking about decomposition within rough set framework, one should realize that the source of information is not connected just with given universe of objects U but, what is much more important actually, with their values on attributes. Thus, we are more likely to define an agent as able to referring to some subset of A than to attach to him some subuniverse of objects. Such a choice is additionally justified by the fact that while reasoning about objects outside U (called from now as "new objects") we refer not to particular elements of the universe but to equivalence classes of indiscernibility relation, defined, for arbitrary subset $B \subseteq A$, as

$$IND(B) = \{(u_1, u_2) \in U \times U : Inf_B(u_1) = Inf_B(u_2)\} \tag{1}$$

where information function Inf_B from U to Cartesian product $\times_{a_{i_j} \in B} V_{a_{i_j}}$, denoted from now as V_B, such that

$$Inf_B(u) = \left(a_{i_1}(u), .., a_{i_{|B|}}(u)\right) \tag{2}$$

is consistent with linear suborder $B = \left(a_{i_1}, .., a_{i_{|B|}}\right)$ induced by fixed order $A = (a_1, .., a_{|A|})$.[1]

Agent knowledge, whose synthesis is stated as opposite to decomposition, should be referred to the task of given information system. Such a task is usually connected with distinguished decision attribute to predict by using decomposed decision rules. By a decision table we understand triple $\mathbf{A} = (U, A, d)$, where, for arbitrary $B \subseteq A$, conditional information about decision attribute $d \notin A$, with a discrete value set V_d, is expressed by the generalized decision function $\partial_B^d : V_B \to 2^{V_d}$ such that

$$\partial_B^d(w_B) = \{d(u) : Inf_B(u) = w_B\} \tag{3}$$

where for any w_B with $Inf_B^{-1}(w_B) = \emptyset$ we put $\partial_B^d(w_B) = \emptyset$.

Remark. One can argue that putting $\partial_B^d(w_B) = \emptyset$ for value combinations w_B not occurring in data is too strong with respect to inductive inference applications. In fact, one can propose many alternative solutions appropriate for different inference approaches. Still, we would like to keep the above one because it is, first, consistent with definition (3) itself. Moreover, generalized decision function generates decision rules, which can be treated as fundamental for reasoning about new objects by analogy to those from the universe. For each $w_B \in V_B$, we obtain boolean implication

$$\bigwedge_{j=1,..,|B|} \left(a_{i_j} = w_B^{\downarrow \{a_{i_j}\}}\right) \Rightarrow \bigvee_{k=1,..,|\partial_B^d(w_B)|} (d = v_k) \tag{4}$$

---

[1] By $|B|$ we denote cardinality of given subset $B \subseteq A$, so, in particular, $|A|$ is the cardinality of A.

where $w_B^{\downarrow\{a_{i_j}\}} \in V_{a_{i_j}}$ is the projection of w_B onto $V_{a_{i_j}}$, given by

$$(v_{i_1},..,v_{i_j},..,v_{i_{|B|}}) \mapsto v_{i_j} \tag{5}$$

One can describe such a rule as that for each object, if it has vector value w_B on B, then its decision must be one of those in $\partial_B^d(w_B)$. One can also observe that given implication would not be satisfied any more in case we reject any element of the disjunction in its head. Thus, one can interpret putting $\partial_B^d(w_B) = \emptyset$ as that for a new object which is not indiscernible from any element of the universe there should be no decision rule classifying it in any sense. In other words, for any conditional vector value remaining inconsistent with our decision table, we can put nothing but the empty set of possibilities at the head of corresponding decision rule.

Now, let us consider more than one decision table $\mathbf{B}_i = (U, B_i, d)$, with their decision rules encoded by generalized decision functions $\partial_{B_i}^d$, $i \in I$. For any object with vector value $w_A \in V_A$, information delivered by each of $\mathbf{B}_i$ is now projected onto $w_A^{\downarrow B_i} \in V_{B_i}$,[2] and local decision rule becomes to correspond to $\partial_{B_i}^d \left(w_A^{\downarrow B_i}\right)$. The most natural way of coming back to decision table $\mathbf{A} = (U, A, d)$ is to combine local knowledge due to synthesis rule generated by intersection of local generalized decision function set values (from now, abbreviated as GDF-sets):

$$\bigwedge_{i=1,...,|A|} \left(a_i = w_A^{\downarrow\{a_i\}}\right) \Rightarrow \bigwedge_{i=1,2} \left(\bigvee_{k=1,...,\left|\partial_{B_i}^d\left(w_A^{\downarrow B_i}\right)\right|} (d = v_k)\right) \tag{6}$$

where we start from two-set decomposition onto decision tables $\mathbf{B}_1$, $\mathbf{B}_2$, such that $B_1 \cup B_2 = A$. In such a case, the whole problem becomes to verify whether we did not lose some information by decomposing and then combining decision tables $\mathbf{B}_i$, $i = 1, 2$, back to $\mathbf{A}$. In other words, we have to check whether for each vector value $w_A \in V_A$ there is equality

$$\partial_A^d(w_A) = \partial_{B_1}^d\left(w_A^{\downarrow B_1}\right) \cap \partial_{B_2}^d\left(w_A^{\downarrow B_2}\right) \tag{7}$$

Remark. For a consistent decision table $\mathbf{A}$ it means that the intersection in (7) can have at most one element and it remains non-empty iff corresponding vector value w_A occurs in $\mathbf{A}$.

Remark. One can easily see that inclusion "$\subseteq$" is always satisfied. However, this is not what we are interested in - we would like to find such a decomposition onto decision tables $\mathbf{B}_i$ that, for arbitrary object, there is no danger of giving as a possible decision a value that would have been denied by information generated by decision table $\mathbf{A}$. In other words, usually, it may happen that intersection of GDF-sets for foregoing objects contain more possible decision values than it is actually. Thus, keeping the decision knowledge on the initial level requires actually inclusion

$$\partial_A^d \supseteq \partial_{B_1}^d \cap \partial_{B_2}^d \tag{8}$$

where we omit universal quantifier over elements of V_A, for simplicity of notation.

---
[2] Such a projection is defined analogously as by (5).

Remark. In [18] we were considering a weaker form of decomposition, where criterion (7) had to be satisfied only for vector values w_A occurring in **A** (i.e. with non empty $Inf_A^{-1}(w_A)$). Such an approach enabled us to deal with new objects which did not fit to any indiscernibility class of $IND(A)$ but could be classified locally, with respect to both $IND(B_1)$ and $IND(B_2)$. Now, by extending (7) onto the whole V_A, we require that if $Inf_A^{-1}(w_A) = \emptyset$, then

$$\partial_{B_1}^d \left(w_A^{\downarrow B_1}\right) \cap \partial_{B_2}^d \left(w_A^{\downarrow B_2}\right) = \emptyset \qquad (9)$$

Although stronger, we show in following sections that such a decomposition criterion is still valid frequently enough for real-life data. Moreover, (9) preserves us from basing on information which is actually not contained in a decision table - if there are no cases corresponding to w_A, then there will not be any suggestions as about possible decision values.

Example 1. Let us consider decision table

$$
\begin{array}{c|ccccc|c}
A & a_1 & a_2 & a_3 & a_4 & a_5 & d \\
u_1 & No & No & No & No & No & 1 \\
u_2 & No & No & Yes & No & Yes & 1 \\
u_3 & No & No & Yes & No & No & 2 \\
u_4 & No & Yes & No & Yes & No & 2 \\
u_5 & No & Yes & No & No & No & 3 \\
u_6 & Yes & No & Yes & No & Yes & 3
\end{array} \qquad (10)
$$

For $B_1 = \{a_1, a_2, a_3, a_5\}$ and $B_2 = \{a_1, a_3, a_4, a_5\}$ condition (7) is satisfied, what can be seen from the following

$$
\begin{array}{cccccc}
\{a_1,a_3,a_5\} & \{a_2\} & \{a_4\} & \partial_{B_1}^d & \partial_{B_2}^d & \partial_{B_1 \cup B_2}^d \\
No,No,No & No & No & \{1\} & \{1,3\} & \{1\} \\
No,No,No & No & Yes & \{1\} & \{2\} & \emptyset \\
No,No,No & Yes & No & \{2,3\} & \{1,3\} & \{3\} \\
No,No,No & Yes & Yes & \{2,3\} & \{2\} & \{2\} \\
No,Yes,No & No & No & \{2\} & \{2\} & \{2\} \\
No,Yes,Yes & No & No & \{1\} & \{1\} & \{1\} \\
Yes,Yes,Yes & No & No & \{3\} & \{3\} & \{3\}
\end{array} \qquad (11)
$$

where we check only pairs of $w_1 \in V_{B_1}$ and $w_2 \in V_{B_2}$ which occur in **A** (otherwise we have always equality $\emptyset = \emptyset$, since GDF-set is put as empty for any non-occurring vector value), such that they agree on $B_1 \cap B_2$. Projection onto intersection is shown in the first column and then completed to all needed combinations.

3 Decomposition based on conditional independence

Resuming considerations from previous section, we proposed, for given decision table $\mathbf{A} = (U, A, d)$, decomposition criterion (7) (or, equivalently, (8)) for subsets $B_1, B_2 \subseteq A$, $B_1 \cup B_2 = A$, to generate corresponding subtables without losing global decision information. Now, we are going to draw some directions for decomposition optimization, i.e. to state which attribute subsets satisfying (7) are the most interesting in view of rough set based reasoning.

It turns out that to reason about any new object we have to find indiscernibility class of universe elements labeled with its vector value on conditions. Thus, as one could see in table (10), there is quite a high risk that a new object does not fit to any of conditional values patterns - in fact, this is one of the main reasons for developing algorithms for finding minimal decision rules or minimal subsets of conditions, called relative or decision reducts, which enable to reason about decision (see e.g. [17]). In our approach, the average decrease of non-classification risk is expressed by the following decomposition quality measure

$$Q(B_1, B_2) = \frac{1}{2} \cdot \left(\frac{|V_{B_1}^*|}{|V_{B_1}|} + \frac{|V_{B_2}^*|}{|V_{B_2}|} \right) \tag{12}$$

where by $V_{B_i}^* \subseteq V_{B_1}$ we denote the set of vector values on B_i which occur in $\mathbf{A}$. Thus, we would like to search for such B_1, B_2 that the set of vector values occurring over them is possibly close to the set of all possible vector values.

One should realize that optimal decomposition problem stated as maximization of function (12) under condition (7) is far from real applications yet. First of all, we do not know how to search for such subsets, the more so as one may like to decompose initial decision table onto more than two subtables. Thus, for stating more firm background for decomposition framework, we should examine what properties of criterion (7) may be helpful. The only thing we would like to mention before is that for optimization of formula (12) one should try to keep subsets B_1, B_2 as small as possible. In fact, such a tendency can be compared with the minimal description length princinple. Thus, we are going to perform optimization process with respect to cardinalities of decomposition leaves B_1, B_2, using quantity of $Q(B_1, B_2)$ just for numerical illustration.

We decide to adopt the notion of conditional independence, known e.g. from probability theory (see e.g. [9]), as the most suitable in view of expressive power of the following results.

Definition 1. Given decision table $\mathbf{A} = (U, A, d)$, for subsets $X, Y, Z \subseteq A$, we say that X and Y are conditionally independent under Z with respect to intersection (denoted by $I_{\mathbf{A}}^{\cap}(X, Y/Z)$) iff $X \cap Y \subseteq Z$ and

$$\partial_{XUZ}^d \left(w_A^{\downarrow X \cup Z} \right) \cap \partial_{YUZ}^d \left(w_A^{\downarrow Y \cup Z} \right) = \partial_{XUYUZ}^d \left(w_A^{\downarrow X \cup Y \cup Z} \right) \tag{13}$$

for each combination of $w_A \in V_A$. In particular, Z may be empty. Then we say that X and Y are independent, what means that for each combination of $w_X \in V_X$ and $w_Y \in V_Y$ there is

$$\partial_X^d \left(w_A^{\downarrow X} \right) \cap \partial_Y^d \left(w_A^{\downarrow Y} \right) = \partial_{XUY}^d \left(w_A^{\downarrow X \cup Y} \right) \tag{14}$$

Remark. By putting $A = X \cup Y \cup Z$, $B_1 = X \cup Z$ and $B_2 = Y \cup Z$, we obtain that criterions (7) and (13) are equivalent. Analogously as before, we write

$$\partial^d_{X \cup Y \cup Z} = (\supseteq) \ \partial^d_{X \cup Z} \cap \partial^d_{Y \cup Z} \qquad (15)$$

when it does not lead to misunderstandings.

Any decision table can be understood as a conditional independence model (called from now as a decomposition model) with respect to the notion introduced above. It means that each $\mathbf{A} = (U, A, d)$ induces the set of triples of subsets of A for which condition (13) is satisfied, called from now as conditional independence statements or, for short, CI-statements.

Example 2. Since it would be quite a long list to write down all CI-statements valid for decision table (10) let us just note that besides CI-statement shown by (11) in Example 1, we have also such statements like e.g.

$$I^\cap_\mathbf{A}(\{a_1\}, \{a_2, a_4, a_5\} / \{a_3\}), \ I^\cap_\mathbf{A}(\{a_1, a_2, a_4\}, \{a_5\} / \{a_3\}) \qquad (16)$$

In case of $I^\cap_\mathbf{A}(\{a_2\}, \{a_4\} / \{a_1, a_3, a_5\})$ we could decompose initial table onto subtables

$$\begin{array}{ccccc}
a_1 & a_2 & a_3 & a_5 & \partial^d_{B_1} \\
No & No & No & No & \{1\} \\
No & No & Yes & Yes & \{1\} \\
No & No & Yes & No & \{2\} \\
No & Yes & No & No & \{2,3\} \\
Yes & No & Yes & Yes & \{3\}
\end{array}
\quad
\begin{array}{ccccc}
a_1 & a_3 & a_4 & a_5 & \partial^d_{B_2} \\
No & No & No & No & \{1,3\} \\
No & Yes & No & Yes & \{1\} \\
No & Yes & No & No & \{2\} \\
No & No & Yes & No & \{2\} \\
Yes & Yes & No & Yes & \{3\}
\end{array}
\qquad (17)$$

where the quality was equal to

$$Q(\{a_1, a_2, a_3, a_5\}, \{a_1, a_3, a_4, a_5\}) = 0.31$$

For comparison, quality of decomposition based on both CI-statements from (16) is equal to 0.53.

Proposition 2. *For any $X, Y, Z, W \subseteq A$,*

$$I^\cap_\mathbf{A}(X, Y \cup Z/W) \Rightarrow I^\cap_\mathbf{A}(X, Y/W) \qquad (18)$$
$$I^\cap_\mathbf{A}(X, Y \cup Z/W) \Rightarrow I^\cap_\mathbf{A}(X, Y/Z \cup W) \qquad (19)$$
$$I^\cap_\mathbf{A}(X, Y/Z \cup W) \wedge I^\cap_\mathbf{A}(X, Z/W) \Rightarrow I^\cap_\mathbf{A}(X, Y \cup Z/W) \qquad (20)$$

Proof Without loss of generality, let us assume that subsets $X, Y, Z, W \subseteq A$ are pairwise disjoint. To simplify notation, we are going to operate with combinations of vector values w_X, w_Y, w_Z, w_W instead of projections from V_A onto corresponding attributes subsets. In first two cases, given that

$$\partial^d_{X \cup Y \cup Z \cup W} = \partial^d_{X \cup W} \cap \partial^d_{Y \cup Z \cup W} \qquad (21)$$

implication (18) is derived by rewriting $\partial^d_{X \cup Y \cup W}$, for arbitrary combination of w_X, w_Y and w_W, as

$$\bigcup_{w_Z \in V_Z} \partial^d_{X \cup Y \cup Z \cup W}(w_X, w_Y, w_Z, w_W) =$$
$$\bigcup_{w_Z \in V_Z} \left(\partial^d_{X \cup Y}(w_X, w_W) \cap \partial^d_{Y \cup Z \cup W}(w_Y, w_Z, w_W) \right) =$$
$$\partial^d_{X \cup Y}(w_X, w_W) \cap \bigcup_{w_Z \in V_Z} \partial^d_{Y \cup Z \cup W}(w_Y, w_Z, w_W) =$$
$$= \partial^d_{X \cup Y}(w_X, w_W) \cap \partial^d_{Y \cup W}(w_Y, w_W)$$

and for (19), it is just enough to notice that from inclusion

$$\partial^d_{X \cup Y \cup Z} \subseteq \partial^d_{X \cup Y}$$

one obtains that

$$\partial^d_{X \cup Y \cup Z \cup W} \supseteq \partial^d_{X \cup Y \cup Z} \cap \partial^d_{Y \cup Z \cup W}$$

To prove (20) let us combine premises

$$\partial^d_{X \cup Z \cup W} = \partial^d_{X \cup W} \cap \partial^d_{Z \cup W}$$
$$\partial^d_{X \cup Y \cup Z \cup W} = \partial^d_{X \cup Z \cup W} \cap \partial^d_{Y \cup Z \cup W}$$

as follows

$$\partial^d_{X \cup Y \cup Z \cup W} = \partial^d_{X \cup W} \cap \partial^d_{Z \cup W} \cap \partial^d_{Y \cup Z \cup W} = \partial^d_{X \cup W} \cap \partial^d_{Y \cup Z \cup W}$$

what finishes the proof.

The above result points out some interesting properties of our decomposition criterion, which can be used while reasoning about new objects with decomposed system of decision tables or while searching for optimal decomposition. Given B_1, B_2 satisfying (7), implication (18) enables for some object with values known just on a subset $C_1 \subseteq B_1$ to reason safely with appropriate decision rule corresponding to $\partial^d_{C_1} \cap \partial^d_{B_2}$, unless $(B_1 \backslash C_1) \cap B_2 \neq \emptyset$. Thus, one can see that talking about two-set decomposition we should search for B_1, B_2 with possibly small intersection, what is usually not in contradiction with tending to maximization of function Q defined by (12).

From such a point of view, the rule (20) shows how we can combine our knowledge to obtain the most wanted result. Given, as above, a valid decomposition onto $B_1, B_2 \subseteq A$, if we additionally find a proper subset $C_{12} \subset B_1 \cap B_2$ such that decomposition of B_1 onto $(B_1 \backslash B_2) \cup C_{12}$ and $B_1 \cap B_2$ holds, then we can extend it to decomposition of $B_1 \cup B_2$ onto $(B_1 \backslash B_2) \cup C_{12}$ and B_2, which has smaller intersection than the initial one.

Finally, property (19) states that searching for two-set decomposition satisfying our criterion is in some sense monotonic - it implies that if some proper subset $C_{12} \subset B_1 \cap B_2$ does not satisfy the above conditions, then there is no sense to look through its subsets any more.

After presenting advantages of particular implications proved in Proposition 2, let us conclude this section with another application, to approximation of given decomposition model (regarded as the set of all CI-statements $I_\mathbf{A}^\cap(X,Y/Z)$ valid for decision table $\mathbf{A} = (U, A, d)$) starting from possibly small set of "axioms" already derived from data. To see how it can work, let us derive some new CI-statement from those listed in previous examples.

Example 3. Given $I_\mathbf{A}^\cap(\{a_2\},\{a_4\}/\{a_1,a_3,a_5\})$, together with CI-statements in (16), we obtain

$$I_\mathbf{A}^\cap(\{a_1,a_2,a_4\},\{a_5\}/\{a_3\}) \Rightarrow I_\mathbf{A}^\cap(\{a_4\},\{a_5\}/\{a_1,a_3\})$$
$$I_\mathbf{A}^\cap(\{a_2\},\{a_4\}/\{a_1,a_3,a_5\}) \wedge I_\mathbf{A}^\cap(\{a_4\},\{a_5\}/\{a_1,a_3\})$$
$$\Rightarrow I_\mathbf{A}^\cap(\{a_2,a_5\},\{a_4\}/\{a_1,a_3\})$$
$$I_\mathbf{A}^\cap(\{a_1\},\{a_2,a_4,a_5\}/\{a_3\}) \Rightarrow I_\mathbf{A}^\cap(\{a_1\},\{a_2,a_5\}/\{a_3\})$$
$$I_\mathbf{A}^\cap(\{a_2,a_5\},\{a_4\}/\{a_1,a_3\}) \wedge I_\mathbf{A}^\cap(\{a_1\},\{a_2,a_5\}/\{a_3\})$$
$$\Rightarrow I_\mathbf{A}^\cap(\{a_1,a_4\},\{a_2,a_5\}/\{a_3\})$$

One can see that $I_\mathbf{A}^\cap(\{a_1,a_4\},\{a_2,a_5\}/\{a_3\})$ induces decomposition a little bit worse with respect to decomposition quality, equal to 0.5, than in Example 2, with the same intersection $B_1 \cap B_2 = \{a_3\}$. To justify including this case to our considerations, let us just note that evaluating decomposition with respect to quality function Q or cardinality of intersection surely do not exhaust all possible aspects of optimization. For instance, we can say that just derived CI-statement is more "stable", since cardinalities of attribute subsets remain the same, what was not the case for those from Example 2.

4 Decomposition networks

As we are going to show below, Proposition 2, besides properties analyzed in previous section, implies strong similarity of decomposition models to those known from probability theory. This fact will let us adopt some techniques enabling to store decomposition knowledge expressed by CI-statements in a graphical way. Conditional independence information is going to be encoded with respect to appropriate d-separation criterion in a directed acyclic graph (from now on abbreviated by DAG) spanned over nodes corresponding to conditional attributes. Such an approach turns out to be fruitful for creating an algorithmic framework for searching for optimal decomposition as well as for generalization of two-subset decomposition problem onto some more applicable cases. In fact, the following procedure has been already applied successfully to other fields connected with rough sets (see [16],[17]), what is an additional encouragement to use it for solving decomposition problem as well.

For we are claiming that similarities between decomposition and probabilistic models may be helpful for further considerations, let us recall the notion of probabilistic conditional independence.

Definition 3. ([9]) Given a finite family A of finitely valued random variables,[3] with product distribution P over the set of possible vector values V_A, we say that for pairwise disjoint subsets $X, Y, Z \subseteq A$ X is conditionally independent from Y under Z (denoted by $I_P(X/Z/Y)$) iff there is

$$P(X = w_X/Z = w_Z) = P(X = w_X/Y = w_Y, Z = w_Z) \qquad (23)$$

for each combination of $w_X \in V_X$, $w_Y \in V_Y$ and $w_Z \in V_Z$, with non-zero probability.[4]

The above definition of probabilistic conditional independence is convenient while chaining probabilistic rules of reasoning (generated from conditional probabilistic distributions) in terms of Bayesian-like logic framework (see e.g. [1]), or, from other point of view, in terms of their combination along some probabilistic network $D = (A, E)$, which is a DAG spanned over some finite set of random variables (we come back to this subject in Section 6). On the other hand, the following version is more suitable for comparing with conditional independence introduced in Definition 1 for the needs of decision table decomposition.

Proposition 4. *Given pairwise disjoint subsets $X, Y, Z \subseteq A$, X is conditionally independent from Y under Z iff there is*

$$P(X/Z) \cdot P(Y/Z) = P(X, Y/Z) ^{5} \qquad (25)$$

Proof Simple derivation from definition of conditional probabilistic distribution.

Let us note that the above fact enables to introduce "symmetrical" probabilistic conditional independence, as saying that X and Y are conditionally independent under Z (denoted by $I_P(X, Y/Z)$, equal to both $I_P(X/Z/Y)$ and $I_P(Y/Z/X)$) iff condition (25) is satisfied. Now, it turns out that by considering probabilistic conditional independence in such a form we can formulate its fundamental properties exactly like in case of Proposition 2.

---

[3] Assumptions concerning finite nature of A are taken to simplify the following notions. They are justified in view of rough set applications, where frequency functions $\mu_B : V_B \to [0,1]$ such that

$$\mu_B(w_B) = \frac{|Inf_B^{-1}(w_B)|}{|U|} \qquad (22)$$

are regarded as estimating product distributions over approximated random variables corresponding to considered features (see e.g. [15]). This is also why we are going to denote the set of random variables by A.

[4] Conditional probability, for example $P(X = w_X/Z = w_Z)$, is defined as

$$P(X = w_X, Z = w_Z)/P(Z = w_Z) \qquad (24)$$

if $P(Z = w_Z) > 0$ and zero otherwise.

[5] Just like it was in case of operations on generalized decision functions, we are going to skip universal quantifiers concerning combinations of vector values.

Proposition 5. *Given a product probabilistic distribution P over the set A of discretely valued random variables, we have that for each pairwise disjoint subsets $X, Y, Z, W \subseteq A$, the following implications hold*

$$I_P(X, Y \cup Z/W) \Rightarrow I_P(X, Y/W) \quad (26)$$
$$I_P(X, Y \cup Z/W) \Rightarrow I_P(X, Y/Z \cup W) \quad (27)$$
$$I_P(X, Y/Z \cup W) \wedge I_P(X, Z/W) \Rightarrow I_P(X, Y \cup Z/W) \quad (28)$$

Proof Formulation of the above result can be found e.g. in [9]. The only difference is that it was given for "non-symmetrical" version of probabilistic conditional independence (23). To see how it would look like in such a case, we refer the reader to Proposition 19 in Section 6, where properties of decomposition models are rewritten in terms of appropriate conditional valuations.

It turns out that rules (26), (27) and (28) enabled the researchers in probabilistic theory to interpret directed acyclic graphs of the form $D = (A, E)$ without referring to conditional probabilistic distributions in a straightforward way. In fact, they can be regarded as initial for the analysis of Bayesian networks ([9]), which can be used not only to model probabilistic reasoning but also to encode the knowledge about conditional independence models in graphical way. Numerous applications of that second advantage suggested to generalize DAG representation technique onto the whole class of conditional independence models satisfying conditions analogous to those listed in Propositions 2 or 5. Thus, the following result, preceded by some necessary definitions, is actually a special case of a universal approach.

Definition 6. For $\mathbf{A} = (U, A, d)$ consider a DAG $D = (A, E)$. For any $X, Y, Z \subseteq A$ subset Z is said to d-separate X and Y (denoted by $\langle X, Y/Z \rangle_D$) iff $X \cap Y \subseteq Z$ and there is no undirected path between a node in X and a node in Y, along which (1) every node with converging arrows is in Z or has a descendant[6] in Z and (2) every other node is outside Z ([9]).

Definition 7. By a decomposition network consistent with linear order $A = (a_1, .., a_{|A|})$ over conditional attributes, we understand DAG $D = (A, E)$, where, putting $A_i = \{a_1, .., a_{i-1}\}$, for each $a_i \in A$ the set of parents

$$M_i = \{a \in A_i : (a, a_i) \in E\} \quad (29)$$

is a decomposition boundary of a_i in A_i, i.e. such that equality

$$\partial^d_{A_{i+1}} = \partial^d_{M_i \cup \{a_i\}} \cap \partial^d_{A_i} \quad (30)$$

is satisfied and deleting any attribute from M_i rejects it.[7]

---

[6] By a descendant of $a \in A$ in DAG $D = (A, E)$ we mean any $a' \in A$ such that there is a directed path from a to a'.

[7] In view of correspondence to notions from probability theory, one may suggest to call such an irreducible M_i as Markov-like boundary (compare with Definition 12 in Section 6), but we prefer to use the name "decomposition boundary" to keep the language consistent. For the same purpose we call our DAGs as decomposition - not Bayesian-like - networks.

Proposition 8. *Consider decision table* $\mathbf{A} = (U, A, d)$ *with linearly ordered set* $A = (a_1, .., a_{|A|})$ *and* $D = (A, E)$, *such that*

$$E = \bigcup_{i=1,..,n} \{\langle a, a_i \rangle : a \in M_i\}$$

where, for each $i = 1, .., n$, M_i *is a decomposition boundary of* a_i *in* A_i. *Then, for each* $X, Y, Z \subseteq A$, *we have*

$$\langle X, Y/Z \rangle_D \Rightarrow I_{\mathbf{A}}^{\cap} (X, Y/Z) \tag{31}$$

and deleting any edge from E *destroys this property.*

Proof Although the thesis is formulated for the case of $I_{\mathbf{A}}^{\cap}$, the proof, taken from [19], remains valid for any conditional independence model satisfying implications analogous to those from Propositions 2 or 5, or, in case one wants to keep to original, "non-symmetrical" version of conditional independence - to Corollary 19 in Section 6 (see [9] for further references).

Example 4. Let us consider decomposition network $D = (A, \mathbf{E})$, such that

$$E = \{\langle a_1, a_2 \rangle, \langle a_2, a_3 \rangle, \langle a_3, a_4 \rangle, \langle a_3, a_5 \rangle\}$$

One can check that implication (31) is true, because for all not necessarily disjoint triples satisfying d-separation condition (15) is valid as well. However, such a network encodes only a part of conditional independence information - taking Example 3 as a decomposition model sample, we obtain that d-separation implies only 62.5% of CI-statements gathered there.

5 Searching for minimal decomposition boundaries

Let us recall that initial problem of optimal decomposition was to find subsets $B_1, B_2 \subseteq A$, $B_1 \cup B_2 = A$, which satisfy decomposition condition (7) and such that $Q(B_1, B_2)$, defined by (12), is possibly large. One can see that the complexity of optimization over all possible pairs of such subsets, becomes too huge with respect to the size of given table even for some random search techniques. Thus, a kind of knowledge preprocessing which enables to verify (7) without referring to original decision table seems to be necessary indeed. Proposition 8 provides such a preprocessing method, since it states the concrete tool for constructing DAG representation, which can be used e.g. as a basis for testing decomposition condition for arbitrary triples of attribute subsets, applying (31). In other words, initial design of an appropriate decomposition network sets us free at least from cardinality of the universe. It means that once we have found a decomposition network appropriate for given decision table, further computations do not depend on the number of objects (becoming larger and larger in view of current rough set applications) for they are performed just on the network, being an acyclic graph with the number of nodes equal to the number of conditional attributes.

Going further, we can consider in general not necessarily two-set decomposition

$$\mathbf{D} = B_1, .., B_{|\mathbf{D}|} \tag{32}$$

which, due to tending to the decrease of non-classification risk while reasoning, we are going to evaluate by

$$Q(\mathbf{D}) = \frac{1}{|\mathbf{D}|} \sum_{k=1,..,|\mathbf{D}|} \frac{|V_{B_k}^*|}{|V_{B_k}|} \tag{33}$$

being a generalization of formula (12). For such a decomposition there are many ways of extending two-set condition (7). Probably, the most elegant of them is to require that it holds for any pair $B_{k_1}, B_{k_2} \in \mathbf{D}$ - it would state that given any new object classified with respect to any subfamily of $\mathbf{D}$, we can combine local decision rules without losing information.

Example 5. From the network from Example 4 we can e.g. derive three-set decomposition

$$\mathbf{D} = \{\{a_1, a_2, a_3\}, \{a_3, a_4\}, \{a_3, a_5\}\}$$

with quality $Q(\mathbf{D}) = 0.67$ what is the best result so far.

Obviously, such an approach to multi-set decomposition is not the only possible, maybe too rigorous and thus needed to be applied in approximate form for particular data. In any case, however, for designing any distributed system of combinable decision tables we need an effective technique for verifying two-set decomposition criteria or, as one may prefer to call it, corresponding CI-statements, locally. Given a decomposition network which can be used for such a verification, by its effectiveness we understand (1) average time necessary for checking arbitrary CI-statement, and (2) the percent of statements derivable from a network by implication (31) (to avoid, possibly, a situation from Example 4, where we could search for optimal decomposition from approximately 60% of available knowledge). As about the first point, optimization criterion is very natural - concerning any algorithmic method of implementing d-separation, the speed of computations is related straightly to the cardinality of E. It turns out that, due to taking the advantage of referring to the literature on Bayesian networks (see e.g. [3], [4]), DAGs with smaller number of arrows are supposed to encode (with respect to implication (31)) potentially more information about given decomposition model.

Proposition 8 shows that, given decision table $\mathbf{A} = (U, A, d)$, the problem of searching for graph $D = (A, E)$ with minimal $|E|$ may be formulated on two levels, where (1) for fixed linear order $A = (a_1, .., a_{|A|})$, for each a_i locally, we need an algorithm for finding its minimal (in sense of cardinality) decomposition boundary $M_i \subseteq A_i$, and (2) we have to know how to choose such a linear order that local operations give the best global outcome. As about the first point, we would like to propose the following characterization.

Proposition 9. *Given decision table* $\mathbf{A} = (U, A, d)$ *with linearly ordered* $A = (a_1, .., a_{|A|})$, *for each* $a_i \in A$, $M_i \subseteq A_i$ *is its decomposition boundary iff it is minimal in sense of inclusion, such that attributes from* M_i *discern all pairs of elements of* $V^*_{A_i \cup \{d\}}$ *equal on* d *and different with respect to the value of generalized decision function* $\partial^{a_i}_{A_i \cup \{d\}}$, *i.e., for each such pair* (w_1, w_2) *there is at least one* $a \in M_i$ *satisfying* $w_1^{\downarrow \{a\}} \neq w_2^{\downarrow \{a\}}$.

Lemma 10. *Subset* $M_i \subseteq A_i$ *discerns all pairs of elements of* $V^*_{A_i \cup \{d\}}$ *equal on* d *and different with respect to the value of function* $\partial^{a_i}_{A_i \cup \{d\}}$ *iff it satisfies the following equality, for each* $w \in V^*_{A_i \cup \{d\}}$

$$\partial^{a_i}_{A_i \cup \{d\}}(w) = \partial^{a_i}_{M_i \cup \{d\}}\left(w^{\downarrow M_i \cup \{d\}}\right) \tag{34}$$

Proof of Lemma It is actually an easy fact from decision tables analysis - if M_i does not discern all mentioned pairs, then there exists at least one pair (w_1, w_2) of elements from $V^*_{A_i \cup \{d\}}$ such that $w_1^{\downarrow M_i \cup \{d\}} = w_2^{\downarrow M_i \cup \{d\}}$ (equal to some $w_{12} \in V^*_{M_i \cup \{d\}}$) and $\partial^{a_i}_{A_i \cup \{d\}}(w_1) \neq \partial^{a_i}_{A_i \cup \{d\}}(w_2)$. In such a case, however, $\partial^{a_i}_{M_i \cup \{d\}}(w_{12})$ cannot be equal to both $\partial^{a_i}_{A_i \cup \{d\}}(w_1)$ and $\partial^{a_i}_{A_i \cup \{d\}}(w_2)$ what denies equality (34) for at least one of w_1, w_2.

In opposite direction, given that for some $w \in V^*_{A_i \cup \{d\}}$ equality (34) does not hold, there must exist $w_0 \in V^*_{A_i \cup \{d\}}$ such that $w^{\downarrow M_i \cup \{d\}} = w_0^{\downarrow M_i \cup \{d\}}$ and $\partial^{a_i}_{A_i \cup \{d\}}(w) \neq \partial^{a_i}_{A_i \cup \{d\}}(w_0)$, because if not, then by equality

$$\partial^{a_i}_{M_i \cup \{d\}}\left(w^{\downarrow M_i \cup \{d\}}\right) = \bigcup_{w_i \in V^*_{A_i \cup \{d\}}:\ w_i^{\downarrow M_i \cup \{d\}} = w^{\downarrow M_i \cup \{d\}}} \partial^{a_i}_{A_i \cup \{d\}}(w_i) \tag{35}$$

condition (34) would be satisfied for w, what leads to contradiction.

Proof of Proposition Let us introduce, for arbitrary attribute subset B, occurrence function $Occ_B : V_B \to \{0, 1\}$ such that $Occ_B(w_B) = 1$ iff $w_B \in V^*_B$. Let us fix some $a_i \in A$. We can rewrite boundary condition (30) for given $M_i \subseteq A_i$ as

$$\forall_{w_i \in V_{A_i}} \forall_{v_i \in V_{a_i}} \forall_{v_d \in V_d} \left(Occ(w_i, v_d) \wedge Occ\left(w_i^{\downarrow M_i}, v_i, v_d\right) \Rightarrow Occ(w_i, v_i, v_d)\right) \tag{36}$$

In the same way, condition (34) can be expressed as that for any $w_i \in V_{A_i}$, $v_d \in V_d$ there is

$$Occ(w_i, v_d) \Rightarrow \forall_{v_i \in V_{a_i}}\left(Occ\left(w_i^{\downarrow M_i}, v_i, v_d\right) \Rightarrow Occ(w_i, v_i, v_d)\right) \tag{37}$$

what is logically equal to (36).

Corollary 11. *A remark on the outcome of Proposition 9 is that for fixed ordering over attributes the problem of searching for optimal decomposition network for decision table (and thus - optimal decision table decomposition in general) is easier than in case of information system without distinguished decision, where,* as stated in [17] (see it also for details about what we mean by decomposition of information system), condition (34) took the form of

$$\partial_{A_i}^{a_i}(w) = \partial_{M_i}^{a_i}\left(w^{\downarrow M_i}\right) \tag{38}$$

for each $w \in V_{A_i}^*$. It means that in [17] we had to discern by M_i all pairs of vector values with different values of generalized decision function $\partial_{A_i}^{a_i}$ - the number of such pairs is usually much larger (depending on decision attribute) than in current case, so optimal, so called, reduct networks are expected to have more arrows than decomposition ones.

Example 6. Now we can confess that decomposition network from Example 4 was generated with respect to the above characterization, for linear order $A = (a_1, a_2, a_3, a_4, a_5)$. For illustration, let us show how it was in case of a_4. Below we present decision table $\mathbf{A}_4^d = (U, A_4 \cup \{d\}, a_4)$, which corresponds to generalized decision function $\partial_{A_4 \cup \{d\}}^{a_4}$

$\mathbf{A}_4^d$	a_1	a_2	a_3	d	a_4	$\partial_{A_4 \cup \{d\}}^{a_4}$
u_1	No	No	No	1	No	$\{No\}$
u_2	No	No	Yes	1	No	$\{No\}$
u_3	No	No	Yes	2	No	$\{No\}$
u_4	No	Yes	No	2	Yes	$\{Yes\}$
u_5	No	Yes	No	3	No	$\{No\}$
u_6	Yes	No	Yes	3	No	$\{No\}$

(39)

One can see that the only object (vector value) pair equal on d and different on $\partial_{A_4 \cup \{d\}}^{a_4}$ is (u_3, u_4) and the set of attributes discerning it is equal to $\{a_2, a_3\}$. As a result, a_4 has two possible boundaries - $\{a_2\}$ and $\{a_3\}$. In Example 4 we took that second because it was better (still far from perfection) with respect to our sample of CI-statements taken from Example 3. Obviously, it does not imply that such a choice would be better with respect to the whole conditional independence model. For comparison, let us just mention that choosing $\{a_2\}$ would yield DAG with

$$E = \{\langle a_1, a_2\rangle, \langle a_2, a_3\rangle, \langle a_2, a_4\rangle, \langle a_3, a_5\rangle\}$$

and then decomposition, e.g., onto $\{a_1, a_2\}$, $\{a_2, a_4\}$ and $\{a_2, a_3, a_5\}$.

6 Chaining conditional valuations

We would like to start this section with remark that the reader should not regard it as purely theoretical, irrelevant to the problem of searching for optimal decomposition of decision tables. Indeed, although properties of decomposition networks analyzed below do not seem to have anything in common with application to deriving CI-statements, they will turn out to be of a special importance while considering our optimization problem at the level of searching for appropriate initial ordering over conditional attributes.

So far, talking about directed acyclic graphs and their meaning for decomposition of decision tables, we based on referring of our definition of conditional independence to that known from probability theory in symmetrical form, given by (25). Now, we would like to recall that there is still another way of using DAG representation for modeling dependencies in data. It is connected with so called uncertainty propagation by chaining conditional valuations along ordering of given network.

Definition 12. ([9]) Given linearly ordered, finite family of discrete random variables $A = (a_1, .., a_{|A|})$, by probabilistic network we mean each directed acyclic graph $D = (A, E)$ such that for each $i = 1, .., |A|$, parent set M_i, defined by (29), is a Markov boundary for a_i in A_i, i.e. $I_P(\{a_i\}/M_i/A_i)$ holds unless we delete some attribute from M_i.

Proposition 13. ([9]) For each probabilistic network $D = (A, E)$, constructed consistently with some linear order $A = (a_1, .., a_{|A|})$, we have equality

$$P(A) = \prod_{i=1,..,|A|} P(a_i/M_i) \qquad (40)$$

where, for each $a_i \in A$ being a root of D (i.e. such that M_i is empty) we put $P(a_i/M_i) = P(a_i)$.

One should expect that the analysis of chaining distributions enables to gather much more information than while just verifying conditional independence criterions - besides possibility of handling numerical approximations of such criterions, it could show some new directions for searching for optimal decomposition networks. However, to draw the correspondence between our version of conditional independence and the probabilistic one at such a level, some appropriately defined conditional valuations must be proposed.

Definition 14. By conditional generalized decision function $\partial_{X/Y}^d : V_X \times V_Y \rightarrow 2^{V_d}$ over attribute subset $X \subseteq A$ under attribute subset $Y \subseteq A$, defined by

$$\partial_{X/Y}^d(w_X/w_Y) = \partial_Y^d(w_Y) \setminus \partial_{X \cup Y}^d(w_X, w_Y) \qquad (41)$$

we understand additional amount of certainty which we would obtain by enlarging Y by X while generalized reasoning about decision d. In particular, for $Y = \emptyset$, we put

$$\partial_{X/\emptyset}^d(w_X) = V_d \setminus \partial_X^d(w_X) \qquad (42)$$

to express what information about decision can be obtained from X with respect to completely vague knowledge.

The following result shows the correspondence between Definition 14 and conditional independence based on decomposition criterion.

Proposition 15. *For any $X, Y, Z \subseteq A$, validness of $I_A^\cap(X, Y/Z)$ is equivalent to each of the following:*

$$\partial_{X/Z}^d \cup \partial_{Y/Z}^d = \partial_{X \cup Y/Z}^d \tag{43}$$

$$\partial_{X/Z}^d \supseteq \partial_{X/Y \cup Z}^d \tag{44}$$

$$\partial_{X/Z}^d \cup \partial_{Z/Y}^d \supseteq \partial_{X \cup Z/Y}^d \tag{45}$$

Proof Conditional independence condition (15) can be rewritten as

$$\partial_Z^d \setminus \partial_{X \cup Y \cup Z}^d = \left(\partial_Z^d \setminus \partial_{X \cup Z}^d\right) \cup \left(\partial_Z^d \setminus \partial_{Y \cup Z}^d\right)$$

what implies equivalence with (43), and then, by deleting from both sides elements from $\partial_{Y/Z}^d$, as

$$\partial_{Y \cup Z}^d \setminus \partial_{X \cup Y \cup Z}^d = \partial_{Y \cup Z}^d \setminus \partial_{X \cup Z}^d \tag{46}$$

It shows that $I_A^\cap(X, Y/Z)$ implies (44), because

$$\partial_{Y \cup Z}^d \setminus \partial_{X \cup Z}^d \subseteq \partial_Z^d \setminus \partial_{X \cup Z}^d \tag{47}$$

Moreover, given that we have always equality

$$\partial_{Z/Y}^d \cup \partial_{X/Y \cup Z}^d = \partial_{X \cup Z/Y}^d \tag{48}$$

inclusion (45) becomes to be equivalent to (44), since it implies

$$\partial_{Z/Y}^d \cup \partial_{X/Z}^d \supseteq \partial_{Z/Y}^d \cup \partial_{X/Y \cup Z}^d = \partial_{X \cup Z/Y}^d$$

Opposite implications can be derived from the fact that elements of V_d which deny (15) turn out to deny each of (43), (44) and (45) as well. In case of (43) it is obvious. For the rest, it is enough to consider (44). Now, it is enough to realize that the set of decision values denying equality (46) for some $w \in V_{X \cup Y \cup Z}$ is always contained in $\partial_{X/Y \cup Z}^d(w)$ but disjoint with $\partial_{X/Z}^d(w)$:

$$\left(\partial_{X \cup Z}^d \cap \partial_{Y \cup Z}^d \setminus \partial_{X \cup Y \cup Z}^d\right) \subseteq \left(\partial_{Y \cup Z}^d \setminus \partial_{X \cup Y \cup Z}^d\right) \tag{49}$$

$$\left(\partial_{X \cup Z}^d \cap \partial_{Y \cup Z}^d \setminus \partial_{X \cup Y \cup Z}^d\right) \cap \left(\partial_Z^d \setminus \partial_{X \cup Z}^d\right) = \emptyset \tag{50}$$

Thus, if such a set is non-empty, each of its elements rejects (44).

Although conditions listed above state conditional generalized decision functions as analogous to conditional probability distributions, it would be much more straightforward with equalities instead of inclusions in (44) and (45). Thus, it may be interesting, what should be assumed additionally to obtain them. It turns out that such a condition is in some sense dual to the initial decomposition criterion.

Definition 16. Given decision table $\mathbf{A} = (U, A, d)$, for subsets $X, Y, Z \subseteq A$, we say that X and Y are conditionally independent under Z with respect to equality (denoted by $I_{\mathbf{A}}^{=}(X, Y/Z)$) iff there is $I_{\mathbf{A}}^{\cap}(X, Y/Z)$ and, moreover,

$$\partial_Z^d = \partial_{X \cup Z}^d \cup \partial_{Y \cup Z}^d \tag{51}$$

what means that for each $w_A \in V_A$, there is

$$\partial_Z^d\left(w_A^{\downarrow Z}\right) = \partial_{X \cup Z}^d\left(w_A^{\downarrow X \cup Z}\right) \cup \partial_{Y \cup Z}^d\left(w_A^{\downarrow Y \cup Z}\right) \tag{52}$$

The above requirement is stated as optional, additional to (15) and thus we would like to talk about strong decomposition while satisfying it.

Calling such a notion as conditional independence with respect to equality is justified by the following fact, being complementary to Proposition 15.

Proposition 17. *For any $X, Y, Z \subseteq A$, condition $I_{\mathbf{A}}^{=}(X, Y/Z)$ is equivalent to both of the following:*

$$\partial_{X/Z}^d = \partial_{X/Y \cup Z}^d \tag{53}$$

$$\partial_{X/Z}^d \cup \partial_{Z/Y}^d = \partial_{X \cup Z/Y}^d \tag{54}$$

Proof It is enough to notice that inclusion opposite to (47) is equivalent to (51).

The result stated below can be treated as analogous to Proposition 13.

Proposition 18. *For each decomposition network $D = (A, E)$, we have that*

$$\partial_{A/\emptyset}^d \subseteq \bigcup_{i=1,\ldots,card(A)} \partial_{a_i/M_i}^d \tag{55}$$

where inclusion "$\subseteq$" can be replaced by equality in case of a strong decomposition network, i.e. such that parent sets induce strong decomposition with respect to their children.

Proof It is enough to generalize equality (48) as

$$\partial_{A/\emptyset}^d = \bigcup_{i=1,\ldots,card(A)} \partial_{a_i/A_i}^d \tag{56}$$

and apply Propositions 15 and 17, respectively.

We would like to conclude this section with result drawing the correspondence between ours and probabilistic conditional independence as between properties of conditional valuations defined respectively by (44) and (53), and conditional probability distributions.

Proposition 19. *Let us redefine conditional independence as follows: given decision table* $\mathbf{A} = (U, A, d)$, *for subsets* $X, Y, Z \subseteq A$, *we say that* X *is conditionally independent from* Y *under* Z *with respect to inclusion (denoted by* $I_{\mathbf{A}}^{\subseteq}(X/Z/Y)$) *iff we have condition (44). Similarly, we say that it is the case with respect to equality (denoted by* $I_{\mathbf{A}}^{=}(X/Z/Y)$) *iff (53) takes place. Then, each* $\mathbf{A}$ *treated as conditional independence model is a semi-graphoid with respect to both of the above definitions, i.e., denoting by* $I_{\mathbf{A}}^{*}$ *either* $I_{\mathbf{A}}^{\subseteq}$ *or* $I_{\mathbf{A}}^{=}$, *we have that for any pairwise disjoint* $X, Y, Z, W \subseteq A$

$$I_{\mathbf{A}}^{*}(X/Z/Y) \Rightarrow I_{\mathbf{A}}^{*}(Y/Z/X) \tag{57}$$
$$I_{\mathbf{A}}^{*}(X/W/Y \cup Z) \Rightarrow I_{\mathbf{A}}^{*}(X/W/Y) \tag{58}$$
$$I_{\mathbf{A}}^{*}(X/W/Y \cup Z) \Rightarrow I_{\mathbf{A}}^{*}(X/Z \cup W/Y) \tag{59}$$
$$I_{\mathbf{A}}^{*}(X/Z \cup W/Y) \wedge I_{\mathbf{A}}^{*}(X/W/Z) \Rightarrow I_{\mathbf{A}}^{*}(X/W/Y \cup Z) \tag{60}$$

Proof For, according to Proposition 15, the case of $I_{\mathbf{A}}^{\subseteq}$ is just a reformulation of $I_{\mathbf{A}}^{\cap}$, this part can be treated just as a corollary from Proposition 2, which also states, together with Proposition 17, that to prove the thesis for $I_{\mathbf{A}}^{=}$ it would be enough to show that condition (51) satisfies analogous properties as (15). Proving implications

$$\partial_W^d = \partial_{XUW}^d \cup \partial_{YUZUW}^d \Rightarrow \partial_W^d = \partial_{XUW}^d \cup \partial_{YUW}^d \tag{61}$$
$$\partial_{ZUW}^d = \partial_{XUZUW}^d \cup \partial_{YUZUW}^d \wedge \partial_W^d = \partial_{XUW}^d \cup \partial_{ZUW}^d \tag{62}$$
$$\Rightarrow \partial_W^d = \partial_{XUW}^d \cup \partial_{YUZUW}^d \tag{63}$$

corresponding to (18) and (20), is a simple calculus. However, what is worth a special attention, implication

$$\partial_W^d = \partial_{XUW}^d \cup \partial_{YUZUW}^d \Rightarrow \partial_{ZUW}^d = \partial_{XUZUW}^d \cup \partial_{YUZUW}^d \tag{64}$$

corresponding to (19) does not hold in general. Fortunately, for our purpose it is enough to show that (64) is satisfied if we know that equality (21) (as the head of implication (19) and thus a part of the head of implication (59)) is true. According to (18), mentioned equality implies that

$$\partial_{XUZUW}^d = \partial_{XUW}^d \cap \partial_{ZUW}^d$$

what can be rewritten, due to Proposition 15, as

$$\partial_{ZUW}^d \setminus \partial_{XUZUW}^d \subseteq \partial_W^d \setminus \partial_{XUW}^d$$

Furthermore, the head of considered implication can be rewritten as

$$\partial_W^d \setminus \partial_{XUW}^d \subseteq \partial_{YUZUW}^d \setminus \partial_{XUW}^d$$

Chaining the above inclusions gives that

$$\partial_{ZUW}^d \setminus \partial_{XUZUW}^d \subseteq \partial_{YUZUW}^d \setminus \partial_{XUW}^d$$

so, in particular,
$$\partial_{ZUW}^d \setminus \partial_{XUZUW}^d \subseteq \partial_{YUZUW}^d \setminus \partial_{XUZUW}^d$$
equivalent to
$$\partial_{ZUW}^d = \partial_{XUZUW}^d \cup \partial_{YUZUW}^d$$
what finishes the proof.

7 Deriving optimal decomposition network from data

Proposition 9 proved in Section 5 shows that the problem of finding minimal decomposition boundary is actually comparable to the problem of finding minimal decision reduct (see e.g. [13]). However, although it enables us to apply practically any, appropriately modified, algorithm for searching for minimal reducts (like e.g. presented in [2], [6], [20]), one should realize that we are still at the local level of decomposition network search, when a linear order over conditional attributes is already fixed. Moreover, since such a straightforward correspondence to the minimal reduct problem implies NP-hardness of minimal decomposition boundary problem,[8] we have to be very cautious about the choice of such an ordering.

At the very beginning of the previous section we mentioned that considerations on chaining of conditional valuations might have some impact on choosing linear order over attributes. The idea is to analyze additional decomposition criterion (51) more carefully, with respect to practical applications. First of all, let us argue that it is quite difficult to think about such a criterion as that given two sets $B_1, B_2 \subseteq A$ their intersection separates them. In fact, the intuition seems to be opposite - if for any fixed $w \in V_{B_1 \cap B_2}$ each pair $(w_1, w_2) \in V_{B_1} \times V_{B_2}$ of its extensions (i.e. satisfying $w_i^{\downarrow B_1 \cap B_2} = w$, $i = 1, 2$) sums with respect to GDF-sets to the same subset of V_d, what can be written as

$$\partial_{B_1 \cap B_2}^d (w) = \partial_{B_1}^d (w_1) \cup \partial_{B_2}^d (w_2)$$

then B_1 and B_2 should be treated as strongly dependent rather than independent under $B_1 \cap B_2$. In particular, we obtain that if both w_1 and w_2 do not occur in decision table, then it is also the case for w, what makes (51) very unnatural and rigorous requirement (one can check that there is no CI-statement satisfying it among all listed in Example 3).

The above argumentation suggests that instead of tending to satisfy (51), we should rather search for decomposition not suffering from such irrational dependencies, which can only disturb to satisfy initial decomposition criterion (7). In fact, by rejecting them, we do not lose any useful information - one could

---

[8] The proof of the fact that the problem of finding minimal (in sense of cardinality) decomposition boundary for given attribute is NP-hard is really just a modification of that for minimal reduct problem, where the polynomial equivalence to the problem of finding minimal prime implicants for boolean functions was shown (see [13] for further references).

say that it may be helpful for reasoning with a new object which fits the system only on intersection of decomposition subsets $B_1 \cap B_2$, but decision rule for such a case can be easily derived from those corresponding to local decision tables by the following projection operation, for any $w \in V_{B_1 \cap B_2}$,

$$\partial^d_{B_1 \cap B_2}(w) = \bigcup_{w' \in V_{B_1 \setminus B_2}} \partial^d_{B_1}(w, w') \tag{65}$$

Finally convinced about necessity of running away from condition (51), let us recall that due to Proposition 15 the boundary condition which one has to check for given $M_i \subseteq A_i$ to satisfy $I_\mathbf{A}^\cap(\{a_i\}, A_i/M_i)$ is equal to

$$\partial^d_{a_i/M_i} \supseteq \partial^d_{a_i/A_i} \tag{66}$$

As in case of considering strong boundaries the opposite inclusion would have to hold additionally, the best way of searching for decomposition boundary which satisfies (51) in possibly least degree is to search for subsets B_i with possibly large value sets of conditional generalized decision function $\partial^d_{a_i/M_i}$ (from now abbreviated as CGDF-sets).

Proposition 20. *Given decision table* $\mathbf{A} = (U, A, d)$ *with linearly ordered* $A = (a_1, .., a_{|A|})$, *for each* $a_i \in A$, $M_i \subseteq A_i$ *is its decomposition boundary iff it is minimal in sense of inclusion, such that for each* $C_i \subseteq A_i$ *being its superset (i.e.* $C_i \supseteq M_i$) *there is*

$$\partial^d_{a_i/M_i} \supseteq \partial^d_{a_i/C_i} \tag{67}$$

Proof First of all, let us recall that inclusion (67) corresponds to CI-statement $I_\mathbf{A}^\cap(\{a_i\}, C_i/M_i)$ (or, equivalently, to $I_\mathbf{A}^\cap(\{a_i\}, C_i \setminus M_i/M_i)$ if one wants to stay with pairwise disjoint sets). Thus, if M_i is a decomposition boundary, what corresponds to $I_\mathbf{A}^\cap(\{a_i\}, A_i/M_i)$, we obtain (67) for any C_i by applying (18). Moreover, since M_i is minimal in sense of inclusion, satisfying $I_\mathbf{A}^\cap(\{a_i\}, A_i/M_i)$, we cannot reject any of its elements without losing (67) for some of its supersets - if we take arbitrary proper subset $D_i \subset M_i$, then M_i itself is a superset denying (67) for D_i.

To the opposite direction, given that M_i satisfies (67) for all its supersets, we know in particular, that it the case for superset A_i, so $I_\mathbf{A}^\cap(\{a_i\}, A_i/M_i)$ takes place. Moreover, we know that for each proper subset $D_i \subset M_i$ there is some $C_i \supseteq D_i$ which denies (67), i.e. such that $I_\mathbf{A}^\cap(\{a_i\}, C_i/D_i)$ does not hold. In such a case, however, by implication (18), we obtain that $I_\mathbf{A}^\cap(\{a_i\}, A_i/D_i)$ does not hold as well, what implies that M_i is minimal in sense of inclusion, satisfying $I_\mathbf{A}^\cap(\{a_i\}, A_i/M_i)$.

The above fact shows that tending to subsets with possibly large CGDF-sets is not in contradiction with the process of searching for decomposition boundaries, in sense that it is impossible to find a set maximal with respect to CGDF-sets which is not a decomposition boundary. Still, however, we cannot be sure whether boundaries optimal with respect to such a new criterion turn out to be those minimal in sense of cardinality.

First of all, let us note that checking condition (66) for each particular vector value could be too rigorous for real-life data. Thus, we would like to handle some approximate optimization measure as, e.g., $H_i : 2^{A_i} \to [0, 1]$, such that

$$H_i(M_i) = \frac{1}{|V_d|} \sum_{w_i \in V_{M_i}} \sum_{v_i \in V_{a_i}} \mu_{M_i}(w_i) \mu_{a_i}(v_i) \left| \partial^d_{a_i/M_i}(v_i/w_i) \right| \quad (68)$$

where the frequencies $\mu_{M_i}(w_i)$ and $\mu_{a_i}(v_i)$ of occurrence of w_i and v_i in the universe, defined as by (22), express relative meaning of corresponding indiscernibility classes with respect to the others.

One can see that for fixed ordering $A = (a_1, .., a_{|A|})$, for each $a_i \in A$, function H_i enables to talk about qualities of particular subsets of A_i in numerical way. In fact, H_i can be here regarded as analogous to information functions, like e.g. conditional entropy known from information theory ([5]), which was considered with respect to Bayesian networks ([3]) as well as with respect to rough sets (see e.g. [15]). However, although such an optimization task is easier to handle than in non-approximate case, searching for subsets with respect to their cardinality still remains more attractive in view of effectiveness of decomposition networks, as discussed in Section 5.

At last, we are ready to come back to the global level of decomposition network optimization, i.e. to the question how to state potentially best linear ordering over conditional attributes - we become to be interested in situation, where relatively small decomposition boundaries computed along given ordering still have a chance to keep high values of H_i for corresponding $i = 1, .., |A|$.

To present some heuristics for stating such an ordering, let us introduce (1) function $h : A \times A \to [0, 1]$ defined by

$$h(a_i/a_j) = \frac{1}{|V_d|} \sum_{v_i \in V_{a_i}} \sum_{v_j \in V_{a_j}} \mu_{a_i}(v_i) \mu_{a_j}(v_j) \left| \partial^d_{a_i/a_j}(v_i/v_j) \right| \quad (69)$$

which expresses potential importance of a_j while constructing subsets with high values of formula (68) with respect to a_i, and (2) function $\rho : A \times A \to [0, 1]$ defined by $\rho(a_i, a_j) =$

$$= \frac{1}{|V_d|} \sum_{v_i \in V_{a_i}} \sum_{v_j \in V_{a_j}} \mu_{a_i}(v_i) \mu_{a_j}(v_j) \left| \partial^d_{a_i}(v_i) \cap \partial^d_{a_j}(v_j) \setminus \partial^d_{a_i,a_j}(v_i, v_j) \right| \quad (70)$$

which expresses the correlation between a_i and a_j with respect to decomposition criterion (7). One can see that formula (69) is an analogon of the measure defined by (68). In the same way, formula (70) can be compared with cross entropy or other measures applied to searching for optimal ordering of Bayesian-like networks.

Described below algorithm for searching for initial ordering remains quite naive if compared with methods developed for Bayesian networks (see [4], [12]). On the other hand, it reflects a general strategy of solving the problem of decomposition network optimization clearly enough to conclude this section by stating the directions for further research and implementations.

Algorithm

1. Compute correlation (70) for each pair $a_i, a_j \in A$ and store the pairs in descending order with respect to $\rho(a_i, a_j)$. Optionally, fix some threshold $\alpha \in [0,1]$ and consider only pairs with $\rho \geq \alpha$.
2. Starting from the most correlated pair a_i, a_j, examine which of values $h(a_i/a_j)$, $h(a_j/a_i)$ is higher. If it is $h(a_i/a_j)$, then put a_j as before a_i in ordering being constructed, and otherwise - a_i before a_j. Optionally, fix some threshold $\beta \in [0,1]$ and consider given pair iff there is

$$|h(a_i/a_j) - h(a_j/a_i)| \geq \beta \qquad (71)$$

3. Repeat the above procedure for foregoing pairs along order descending with respect to correlation function ρ, unless the outcome is in contradiction with previous settings of attributes order.
4. Find a linear ordering over A, consistent with the final list of ordered pairs.

Example 7. Let us apply the above heuristics to decision table (10). First of all, setting $\alpha = 0$, we examine correlations between pairs of conditional attributes:

(1) $\rho(a_2, a_3) = 0.33$ (6) $\rho(a_3, a_4) = 0.17$
(2) $\rho(a_3, a_5) = 0.33$ (7) $\rho(a_1, a_5) = 0.13$
(3) $\rho(a_1, a_2) = 0.2$ (8) $\rho(a_2, a_4) = 0.13$
(4) $\rho(a_2, a_5) = 0.19$ (9) $\rho(a_1, a_4) = 0$
(5) $\rho(a_1, a_3) = 0.17$ (10) $\rho(a_4, a_5) = 0$

The second step is to compare values $h(a_i/a_j)$ and $h(a_j/a_i)$ along the above list, to derive an ordering over attributes, by putting $\beta = 0$:

i,j	$h(a_i/a_j)$	$h(a_j/a_i)$	Outcome
2,3	0.44	0.33	$a_2 > a_3$
3,5	0.33	0.44	$a_3 < a_5$
1,2	0.3	0.3	$a_1 \; ? \; a_2$
2,5	0.52	0.48	$a_2 > a_5$
1,3	0.28	0.17	$a_1 > a_3$
3,4	0.17	0.28	$a_3 < a_4$
1,5	0.22	0.22	$a_1 \; ? \; a_5$
2,4	0.22	0.22	$a_2 \; ? \; a_4$
1,4	0.1	0.1	$a_1 \; ? \; a_4$
4,5	0.11	0.11	$a_4 \; ? \; a_5$

Exemplar ordering (for real-life data one can expect that there remains only one ordering possibility) consistent with the last column of the above table is $A = (a_3, a_5, a_1, a_2, a_4)$. After computing minimal decomposition boundaries along such an ordering, we obtain decomposition network $D = (A, E)$, where

$$E = \{\langle a_3, a_1 \rangle, \langle a_3, a_2 \rangle, \langle a_3, a_4 \rangle, \langle a_3, a_5 \rangle\}$$

implies, in particular, decomposition

$$\mathbf{D} = \{\{a_1, a_3\}, \{a_2, a_3\}, \{a_3, a_4\}, \{a_3, a_5\}\}$$

corresponding to decision tables

a_1	a_3	$\partial^d_{a_1,a_3}$		a_2	a_3	$\partial^d_{a_2,a_3}$
No	No	$\{1,2,3\}$		No	No	$\{1\}$
No	Yes	$\{1,2\}$		No	Yes	$\{1,2,3\}$
Yes	Yes	$\{3\}$		Yes	No	$\{2,3\}$

a_3	a_4	$\partial^d_{a_3,a_4}$		a_3	a_5	$\partial^d_{a_3,a_5}$
No	No	$\{1,3\}$		No	No	$\{1,2,3\}$
Yes	No	$\{1,2,3\}$		Yes	Yes	$\{1,3\}$
No	Yes	$\{2\}$		Yes	No	$\{2\}$

with quality $Q(\mathbf{D}) = 0.75$, the best of all presented solutions.

8 Conclusions

We presented the foundations for decision tables decomposition with respect to the generalized decision function rules. Obtained structures of conditional independence models and Bayesian-like decomposition networks are possible to be used by adopting tools developed for both probabilistic network (see e.g. [3], [12]) and rough set (see e.g. [2], [7], [20]) applications.

Although stronger than in our previous research ([18]), suggested decomposition criterion is still definable in rough set framework, having more straightforward advantages with respect to reasoning about new objects, what was illustrated by its properties corresponding not necessarily to Bayesian-like network approach.

The main topics of future research are going to be focused on improving efficiency of presented model of distributive reasoning. Given an information preprocessing technique based on decomposition network search, more attention should be payed to what combinations of CI-statements we should derive for obtaining final decomposition. Also some effort concerning generalization of current model of decomposition, from generalized decision functions to, e.g., rough membership functions (see e.g. [15]), would be worth spending on.

Acknowledgments This work was supported by KBN Research Grant No. 8T11C01011, ESPRIT project CRIT-2 No. 20288 of European Union and Programme de Cooperation Scientifique entre la France et la Pologne, Laboratoire LEIBNIZ, Grenoble, Institute of Telecommunications, Warsaw, project No. 7004.

References

1. Andersen, K.A., Hooker, J.N.: Bayesian Logic. Decision Support Systems 11, (1994) 191-210.
2. Bazan, J.G., Skowron, A., Synak, P.: Discovery of Decision Rules from Experimental Data. Proceedings of the Third International Workshop on Rough Sets and Soft Computing RSSC'94, November 10-12, San Jose University, CA, (1994) 526-535.
3. Bouckaert, R.R.: Properties of Bayesian Belief Network Learning Algorithms. Proceedings of the 10-th Conference on Uncertainty in AI, de Mantarnas R.L., Poole D.(eds.), the University of Washington, Seattle, Morgan Kaufmann, San Francisco, CA, (1994) 102-109.
4. Cooper, F.G., Herskovits, E.: A Bayesian Method for the Induction of Probabilistic Networks from Data. Machine Learning 9, Kluwer Academic Publishers, Boston, (1992) 309-347.
5. Gallager, R.G.: Information Theory and Reliable Communication. John Wiley & Sons, New York (1968).
6. Nguyen, S.H., Nguyen, T.T., Polkowski, L., Skowron, A., Synak, P., Wróblewski, J.: Decision Rules for Large Data Tables. Proceedings of Symposium on Modelling, Analysis and Simulation vol 1, Computational Engeneering in Systems Applications CESA'96, July 9-12, Lille, France, (1996) 942-947.
7. Nguyen, S.H., Polkowski, L., Skowron, A., Synak, P., Wróblewski, J.: Searching for Approximate Description of Decision Classes. Proceedings of the Fourth International Workshop on Rough Sets, Fuzzy Sets and Machine Discovery RSFD'96, November 6-8, Tokyo, Japan, the University of Tokyo, (1996) 153-161.
8. Pawlak, Z.: Rough Sets. Theoretical Aspects of Reasoning about Data. Kluwer Academic Publishers, Dortrecht (1991).
9. Pearl, J.: Probabilistic Reasoning in Intelligent Systems: Networks of Plausible Inference. Morgan Kaufmann (1988).
10. Polkowski, L., Skowron, A.: Rough mereology: a new paradigm for approximate reasoning. Journ. of Approximate Reasoning 15/5 (1996) 333-365.
11. Shenoy, P.P.: Conditional Independence in Valuation-based Systems. International Journal of Approximate Reasoning 10, Elsevier Science Inc. (1994) 203-234.
12. Singh, M., Valtorta, M.: Construction of Bayesian Network Structures from Data: A Brief Survey and an Efficient Algorithm. International Journal of Approximate Reasoning 12, Elsevier Science Inc. (1995) 111-131.
13. Skowron, A., Rauszer, C.: The Discernibility Matrices and Functions. Information Systems in Intelligent Decision Support. Handbook of Applications and Advances of the Rough Sets Theory, Słowiński R.(ed.), Kluwer, Dortrecht (1992) 331-362.
14. Skowron, A., Stepaniuk, J.: Decision Rules based on Discernibility Matrices and Decision Matrices. Proceedings of The Third International Workshop on Rough Sets and Soft Computing RSSC'94, November 10-12, San Jose University, (1994) 62-69.
15. Ślęzak, D.: Approximate Reducts in Decision Tables. IPMU-96 Information Processing and Management of Uncertainty on Knowledge-Based Systems, vol. III, July 1-5, Granada, Spain, Universidad de Granada (1996) 1159-1164.
16. Ślęzak, D.: Tolerance dependency model for decision rules generation. Proceedings of the Fourth International Workshop on Rough Sets, Fuzzy Sets and Machine Discovery RSFD'96, November 6-8, Tokyo, Japan; The University of Tokyo (1996) 131-138.

17. Ślęzak, D.: Rough Set Reduct Networks. Proceedings of Joint Conference of Information Sciences JCIS'97, vol 3, Duke University, Elsevier Publishing Company, (1997) 77-80.
18. Ślęzak, D.: Attribute set decomposition of decision tables. Proceedings of the Fifth European Congress on Intelligent Techniques and Soft Computing EUFIT'97, September 8-12, Aachen, Germany, Verlag Mainz 1 (1997) 236-240.
19. Verma, T.S.: Causal networks: Semantics and expressiveness. Technical Report R-65, Cognitive Systems Laboratory, University of California, Los Angeles (1986).
20. Wróblewski, J.: Finding minimal reducts using genetic algorithms. Proceedings of the Second Annual Joint Conference on Information Sciences, September 28 - October 1, Wrightsville Beach, NC (1995) 186-189.

Conflict Logic with Degrees

Akira Nakamura

Department of Computer Science
Meiji University
Kawasaki, 214, Japan

Abstract. This paper proposes a graded KB-modal monadic predicate logic which has been motivated by Pawlak's conflict analysis. After giving semantics and syntax, we present some properties. Then, the decision problem is discussed about the decidability as well as the undecidability result.

0. Introduction

In [7], Pawlak introduced an interesting concept of relations that is called "conflict". Several years ago, Pawlak proposed the indiscernibility relation in the knowledge representation system, and he established the rough sets theory which is constructed upon this indiscernibility concept. The conflict is a dual concept of this indiscernibility, and is considered as a kind of dissimilarity. The present author proposed modal logics based on this similarity relation, in particular, graded similarities [5]. It is well-known that the systems of modal logics are characterized by relations corresponding to their modal operations. Further, in [4] he has argued a case that the logic contains monadic predicate symbols. That is, information (or knowledge) is considered as "information of a person". For example, a person a has an information that a predicate $P(c)$ is true, but another person b has an information that the same predicate $P(c)$ is not true. Hence, it happens that informations of different persons are not always the same. Here, instead of persons we can consider agents or times. In this case, we have to consider binary relations among persons (or agents).

Along such a research line, we shall consider a modal monadic predicate logic based on conflict relations. Although we can formally extend this logic to the general case that contains n-ary predicate symbols, we treat here only monadic predicate symbols in order to make the meaning of modal predicate logic understandable. In particular, we are focused modal operations with parameters which correspond to the degree of conflicts. After giving semantics and syntax of this logic, we present some properties. Then, the decision problem is argued about the decidability for the special case and the undecidability result. Finally, theoretical problems of axiomatization and applicability of this logic to practical use are remarked.

1. Preliminaries and definitions

Let us consider the following table:

W/A	a	b	c	d	e
1	-	+	+	+	+
2	+	0	-	-	-
3	+	-	-	-	0
4	0	-	-	0	-
5	+	-	-	-	-
6	0	+	-	0	+

This table is called *information system*. Formally, an information system is defined as a pair $S = (W, A)$, where

W is a nonempty, finite set called the *world*,
elements of W will be called *agents*,
A is nonempty, finite set of *attributes*.

For every attribute $x \in A$ a function f_x is a total function from W to V_x, where V_x is the set of values of x.
In the above table, $W = \{1, 2, 3, 4, 5, 6\}$, $A = \{a, b, c, d, e\}$, and $f_a(4) = 0$.

A *conflict system* $C = (W, A)$ is defined as one of information systems. The above table can be considered as a conflict table, that is, W is the set of *agents* and A is the set of issues. In this case, $-, 0, +$ mean *against, neutral, favorable*, respectively. Hence, for example $f_b(2)$ means that the opinion of agent 2 about issue b is neutral.

It is well-known that the equivalence relation can be defined from a given information system. In the conflict case, we define three basic binary relations on the world W: *conflict, neutrality*, and *alliance*. To this end, we first the following auxiliary function:

$$\phi_a(x, y) = \begin{cases} +1, & \text{if } a(x) \cdot a(y) = 1 \quad \text{or} \quad x = y \\ 0, & \text{if } a(x) \cdot a(y) = 0 \quad \text{and} \quad x \neq y \\ -1, & \text{if } a(x) \cdot a(y) = -1. \end{cases}$$

Then, we define as follows:

$$R_a^+ = \{(x, y) | \phi_a(x, y) = 1\},$$
$$R_a^0 = \{(x, y) | \phi_a(x, y) = 0\},$$
$$R_a^- = \{(x, y) | \phi_a(x, y) = -1\}.$$

These three basic relations are called *alliance, neutrality*, and *conflict*. The relation R_a^+ has exactly two equivalence classes: X_a^+ and X_a^-, where $X_a^+ = \{x \in W | a(x) = +1\}$, $X_a^- = \{x \in W | a(x) = -1\}$.

Now, we define a degree of conflict. Let $C = (W, A)$ be a conflict system. Then, $\delta_A(x, y)$ is defined as follows: $\delta_A(x, y) = \{a \in A | a(x) \neq a(y)\}$. This δ_A corresponds to the set of discernible attributes in the rough sets theory [6]. The following properties is obvious.

i) $\delta_A(x,x) = \phi$,
ii) $\delta_A(x,y) = \delta_A(y,x)$,
iii) $\delta_A(x,z) \subseteq \delta_A(x,y) \cup \delta_A(y,z)$.

By making use of this δ_A, a conflict function $\rho_A(x,y)$ is defined as follows:

$$\rho_A(x,y) = \frac{|\delta_A(x,y)|}{|A|}.$$

For the case that it is unnecessary to mention A, A is omitted.
Obviously, $0 \leq \rho(x,y) \leq 1$. This $\rho(x,y)$ is also called the *degree of conflict* between x and y.

Instead of function ρ_A it is better to use the following ρ_A^* which defines conflict between agents more precisely, by assuming that conflict between agents in conflict is greater than conflict between agents which are neutral, i.e.,

$$\rho_A^*(x,y) = \frac{\Sigma_{a \in A} \phi_a^*(x,y)}{|A|},$$

where

$$\phi_a^*(x,y) = \frac{1 - \phi_a(x,y)}{2} = \begin{cases} 0 & \text{if } a(x) \cdot a(y) = 1 \quad \text{or} \quad x = y \\ 0.5 & \text{if } a(x) \cdot a(y) = 0, \\ 1, & \text{if } a(x) \cdot a(y) = -1 \quad \text{and} \quad x \neq y \end{cases}$$

The relation between x and y such that $\rho^*(x,y) = 0$ is nothing but the indiscernible relation.

The following properties are obvious:

1. $\rho^*(x,x) = 0$,
2. $\rho^*(x,y) = \rho^*(y,x)$,
3. $\rho^*(x,z) \leq \rho^*(x,y) + \rho^*(y,z)$.

Also, we have the following interesting properties:

If $\rho^*(x,y) = 1$ and $\rho^*(y,z) = 1$ then $\rho^*(x,z) = 0$.

This means that an enemy of my enemy is my friend.
Thus the $\rho^*(x,y)$ is the distance between x and y. Hereafter, we use again a notation ρ instead of ρ^*. This distance functions can be formulated in a more general way, without referring to the specific issues being debated by agents.

Let W be a nonempty set called the *world* and let ρ be a function $\rho : W^2 \to [0,1]$ satisfying the above 1), 2), and 3) for every $x, y, z \in W$. Obviously, (W, ρ) is a metric space, and $\rho(x,y)$ is the distance between x and y in this space.

In the sequel, any pair (W, ρ) will referred as to a *conflict space* and $\rho(x,y)$ will be called a *degree of conflict* between x and y in the conflict space. The conflict space is considerable as a kind of *fuzzy graphs* [8].

A pair x, y is said to be:

i) *allied*, if $\rho(x,y) < 0.5$,

ii) *in conflict*, if $\rho(x,y) > 0.5$,
iii) *neutral*, if $\rho(x,y) = 0.5$.

By making use of this $\rho(x,y)$, we define a *conflict relation* $C(\lambda)$ as follows:

$$C(\lambda) = \{(x,y)|\rho(x,y) = \lambda\}.$$

Further, C_λ is defined as $\{(x,y)|\rho(x,y) \geq \lambda\}$ and is called λ-*level-conflict degree*.

2. Conflict logic with degree

In this section, we propose a conflict logic based on the conflict space defined in the preceding section. We formally define this logic denoted by *CLD* (which means <u>c</u>onflict <u>l</u>ogic with <u>d</u>egree).

First of all, we define the language of logic *CLD*, namely, a *syntax* of the logic.

Expressions of this language are built from the symbols of the following nonempty, at most denumerable, and pairwise disjoint sets:

i) A set *VAROB* of *objects variables*, denoted by $x, y, x_1, x_2, \ldots$.
ii) A set *PRED* of *monadic predicate letters*, denoted by $p, q, p_1, p_2, \ldots$.
(iii) A set *CONOB* of *object constants*, denoted by $o, o_1, o_2, \ldots$.
(iv) A set $\{\neg, \wedge, \vee, \supset, \equiv\}$ of propositional operations of *negation, conjunction, disjunction, implication*, and *equivalence*, respectively.
(v) A set $\{\exists, \forall\}$ of quantifiers, called *existential quantifier* and *universal quantifier*, respectively.
(vi) Two families $\{[\lambda]\}_{\lambda\in[0,1]}$ and $\{<\lambda>\}_{\lambda\in[0,1]}$ of modal operations, called *necessity operation* and *possibility operation*, respectively.
(vii) A set $\{(,)\}$ of parentheses.

Then, the set *WFF* of well-formed formulas (denoted by wff's) of *CLD* is the least set satisfying the following conditions:

if $x \in VAROB, o \in CONOB, p \in PRED$, then $p(x) \in WFF$ and $p(o) \in WFF$,
if $A, B \in WFF$, then $\neg(F), (F) \wedge (G), (F) \vee (G), (F) \supset (G), (F) \equiv (G) WFF$,
if $A \in WFF$ and $x \in VAROB$, then $\exists x(A), \forall x(A) \in WFF$
if $A \in WFF$ then $[\lambda](A) \in WFF, <\lambda> (A) \in WFF$.

In the above definition, parentheses are omitted in the usual way. Further, as usually we assume that wff's do not contain redundant or overlapping quantifiers. Moreover, we adopt the usual definition of *free* and *bound* variables. A wff without free variables is called a *sentence* and wff's are denoted by $A, B, \ldots, F, \ldots$.

A *semantics* of *CLD* is given as follows:

We take the set *OB* of objects and a mapping $m : CONOB \cup (PRED \times W) \to OB \cup 2^{OB}$ such that $m(o) \in OB$ and $m(p,w) \in 2^{OB}$. This m is called a *meaning function* over the set *OB*. Further, we consider a mapping $v : VAROB \to OB$. This v is called a *valuation* over the set *OB*.

By a *model*, we mean $M = (\mathbf{K}, OB, C_\lambda, m, v, W)$, where

K is a conflict system,
W is the world of a conflict system **K**,
OB is a nonempty (not necessarily finite) set of objects,
C_λ is a λ-level-conflict degree on W that is described in the preceding section,
m is a meaning function over the set OB such that for a propositional variable p $m(p,w)$ is a subset of OB,
v is a valuation function.

In the modal predicate logic, the set OB can change in dependence on at $w \in W$, i.e., OB at w_1 may be different from OB at w_2. (See [2].) But, in this paper we consider the same OB for every $w \in W$. Given a model $M = (\mathbf{K}, OB, C_\lambda, m, v, W)$ and a valuation v over OB and $w \in W$, we say that a wff A is *satisfied* by v, w in M (denoted by $M, v, w \models A$) whenever the following conditions are satisfied:

$M, v, w \models p(x)$ iff $v(x) \in m(p, w)$,
$M, v, w \models p(o)$ iff $m(o) \in m(p, w)$,
$M, v, w \models \neg A$ iff not $M, v, w \models A$,
$M, v, w \models A \wedge B$ iff $M, v, w \models A$ and $M, v, w \models B$,
$M, v, w \models A \vee B$ iff $M, v, w \models A$ or $M, v, w \models B$,
$M, v, w \models A \supset B$ iff $M, v, w \models \neg A \vee B$,
$M, v, w \models A \equiv B$ iff $M, v, w \models (A \supset B) \wedge (A \supset B)$,
$M, v, w \models \exists x A$ iff there is $o \in OB$ such that $M, v_o, w \models A$,
$M, v, w \models \forall x A$ iff for all $o \in OB$ we have $M, v_o, w \models A$,
  where v_o is the valuation over OB such that
  $v_o(x) = o$ and $v_o(y) = v(y)$ for $y \neq x$,
$M, v, w \models \langle \alpha \rangle A$ iff there is a $(w, w') \in C_\alpha$ such that $M, v, w' \models A$,
$M, v, w \models [\alpha] A$ iff for all $(w, w') \in C_\alpha$ we have $M, v, w' \models A$.

A wff A is said to be *true* in a model $M = (\mathbf{K}, OB, C_\lambda, m, v, W)$ (denoted by $\models_M A$) iff for every valuation v over the OB and every $w \in W$ we have $M, v, w \models A$. A wff A is said to be *valid* in the logic CLD (denoted by $\models A$) iff A is true in every model for CLD. A set T of wff's is said to be true in a model M iff every wff $A \in T$ is true in M. A set T is said *satisfiable* iff there is a model M such that T is true in M. Also, we say that a wff A is a *semantical consequence* of a set T (denoted by $T \models A$) iff for every model M the wff A is true in M whenever T is true in M.

We now examine some of properties of this logic.

Proposition 2.1. All valid wff's in the usual monadic predicate logic are also valid in CLD.

Proof: Since the definition of validity contains the usual one, this is easily shown. □

Proposition 2.2. We have the following:

(1) $\models [\alpha]F \supset [\beta]F$, where $\alpha \leq \beta$.
(2) $\models [\alpha](\neg F \vee G) \supset \neg[\alpha]F \vee [\alpha]G$.
(3) $\models [\alpha](\neg F \vee G) \supset \neg\langle\alpha\rangle F \vee \langle\alpha\rangle G$.
(4) $\models [\alpha](F \wedge G) \equiv [\alpha]F \wedge [\alpha]G$.
(5) $\models \langle\alpha\rangle(F \vee G) \equiv \langle\alpha\rangle F \vee \langle\alpha\rangle G$.
(6) $\models \langle\alpha\rangle(F \wedge G) \supset \langle\alpha\rangle F \wedge \langle\alpha\rangle G$.
(7) $\models [\alpha]F \vee [\alpha]G \supset [\alpha](F \vee G)$.

Proof:

(1) Since $\alpha \leq \beta$, $C_\beta \subseteq C_\alpha$.

Hence, $M, v, w \models [\alpha]F \Leftrightarrow$ For all w' such that $(w, w') \in C_\alpha$
we have $M, v, w' \models F$
$\Rightarrow$ For all w' such that $(w, w') \in C_\beta$
we have $M, v, w' \models F$
$\Leftrightarrow M, v, w \models [\beta]F$.

We have this statement for arbitrary M, v, w. Then, we get (1).

(2) $M, v, w \models [\alpha](F \supset G) \Leftrightarrow$ For all w' such that $(w, w') C_\alpha$
we have if $M, v, w' \models F$ then $M, v, w' \models G$
$\Rightarrow$ We have if for all w' such that
$(w, w') \in C_\alpha M, v, w' \models F$ then
for all w' such that $(w, w') C_\alpha M, v, w' \models G$
$\Leftrightarrow M, v, w \models \neg[\alpha]F \vee [\alpha]G$.

We have this statement for arbitrary M, v, w. Then, we get (2).

(3) We have obviously $\models [\alpha](F \supset G) \supset [\alpha](\neg G \supset \neg F)$.

Hence, $\models [\alpha](F \supset G) \supset ([\alpha]\neg G \supset [\alpha]\neg F)$ by (2).
Thus, $\models [\alpha](F \supset G) \supset (\neg[\alpha]\neg F \supset \neg[\alpha]\neg G)$. Therefore, we get (3).
(4)-(7) These formulas are provable in the usual way of modal logic. □

Proposition 2.3. We have the following:

(8) $\models F \supset [\alpha]\langle\beta\rangle F$, where $\alpha \geq \beta$.
(9) From $\models \langle\alpha\rangle F \supset G$, we get $\models F \supset [\alpha]G$.
(10) not $\models F \supset [\alpha]\langle\beta\rangle F$, where $\alpha < \beta$.
(11) not $\models [\alpha]F \supset F$, where α is an arbitrary positive number.
(12) not $\models \langle\alpha\rangle\langle\alpha\rangle F \supset \langle\alpha\rangle F$.

Proof:

(8) Since $\alpha \geq \beta$, $C_\alpha \subseteq C_\beta$. Hence, $M, v, w \models F$ implies $\forall w'((w, w') \in C_\alpha \supset \exists w''((w', w'') \in C_\beta \& M, v, w'' \models F))$.
(9) From $\models <\alpha> F \supset G$, we have $\models [\alpha]\langle\alpha\rangle F \supset [\alpha]G$ and by (8) $\models F \supset [\alpha]G$.

(10) For $\alpha < \beta$, we do not have $C_\alpha \subseteq C_\beta$.
    From this fact, we have the non-validity of $F \supset [\alpha] < \beta > F$.
(11) In general, $(w,w) \notin C_\alpha$. Assume that $\models [\alpha]p(o) \supset p(o)$. Then, for all w' such that $(w,w')C_\alpha M, v, w' \models p(o)$ must imply $M, v, w \models p(o)$. This is not true by considering the following example:

W	a_1	a_2
w_1	+1	+1
w_2	-1	+1

We take OB as $\{o_1, o_2\}$ and write $m(o_i)$ as o_i, for simplicity. Let α be 0.5 and $v(p, w_1) = \{o_1\}$ and $v(p, w_2) = \{o_2\}$.
The no-validity of $[\alpha]F \supset F$ follows from this fact.

(12) Let us consider the following conflict system give by the following model:

W	a_1	a_2	a_3	a_4	a_5
w_1	+1	+1	+1	+1	+1
w_2	+1	+1	-1	-1	-1
w_3	+1	-1	+1	+1	-1

From this table, $\rho(w_1, w_2) = 0.6$, $\rho(w_2, w_3) = 0.6$, $\rho(w_1, w_3) = 0.4$. Then, it is easily known that $M, v, w_1 \models < 0.6 > F$ cannot always get from $M, v, w_1 \models < 0.6 >< 0.6 > F$. □

From Proposition 2.3, we know that CLD is a kind of graded KB-modal system. That is, the relation corresponding to the modal operation is symmetric but neither reflexive nor transitive. This is a reasonable characterization for the conflict relation.

Proposition 2.4. We have the following:

(13) $\models \forall x[\alpha]F \equiv [\alpha]\forall xF$.
(14) $\models \exists x < \alpha > F \equiv < \alpha > \exists xF$.
(15) $\models \exists x[\alpha]F \supset [\alpha]\exists xF$.
(16) $\models < \alpha > \forall xF \supset \forall x < \alpha > F$.

Proof:

(13) From the definition, $\models \forall x[\alpha]F \supset [\alpha]F$. Thus, $\models < \alpha > \forall x[\alpha]F \supset < \alpha > [\alpha]F$. But, $\models < \alpha > [\alpha]F \supset F$. Hence, $\models < \alpha > \forall x[\alpha]F \supset F$ and then $\models < \alpha > \forall x[\alpha]F \supset \forall xF$. Therefore, from (9) $\models \forall x[\alpha]F \supset [\alpha]\forall xF$.
It is easy to show that $\models [\alpha]\forall xF \supset \forall x[\alpha]F$. Hence we have (13).
(14) This is obtained from (13).
(15) and (16) is shown in the usual way. □

The expression (13) which is equivalent to (14) is called *Barcan Formula*. Thus, it is known that Barcan Formula holds in CLD.

Proposition 2.5. We have the following:

(17) not $\models \exists x[\alpha]F \equiv [\alpha]\exists xF$.
(18) not $\models <\alpha> \forall xF \equiv \forall x <\alpha> F$.

Proof:

(17) It is enough to give a counter-model of $\models [\alpha]\exists xp(x) \supset \exists x[\alpha]p(x)$.
Let us consider the following example:

W	a_1	a_2
w_1	+1	+1
w_2	-1	+1
w_3	+1	-1

$OB = \{o_1, o_2\}$ and $v(p, w_2) = \{o_1\}$ and $v(p, w_3) = \{o_2\}$. Then, $M, v, w_1 \models [0.5]\exists xp(x)$ but neither $M, v, w_1 \models [0.5]p(o_1)$ nor $M, v, w_1 \models [0.5]p(o_2)$ is true.

(18) This is shown the similar way as in (17). □

3. Decision problem of CLD

3.1. Decidability case

Now, let us discuss the problem to decide whether or not a given wff of CLD is valid. First, we consider the propositional case which excludes the quantifiers from CLD. This case is denoted by CLD^*. In the propositional CLD^*, $M, v, w \models F$ means that F is *true* at w in the model M.

A method used here is the same as that of [5]. That is, to test the validity of a given wff F we search a counter-model of F; if such a counter-model exists then F is not valid, otherwise F is valid.

Let us describe the procedure below. In the following explanation, $w, w_1, w_2, \ldots$ mean names of rectangles. These $w, w_1, w_1, \ldots$ are considered as agents of conflict system.

Procedure

Beginning
Assign to a given wff F the initial value f at the first rectangle w.

Rules:
1) Rules for propositional operations:

¬t : If ¬G has the assignment t at a rectangle w, then assign
f to G at the same rectangle.

¬f : If ¬G has the assignment f at a rectangle w, then assign
t to G at the same rectangle.

∧t : If $F_1 \wedge F_2$ has an assignment t at a rectangle w, then assign
t to F_1 and F_2 at the same rectangle.

∧f : If $F_1 \wedge F_2$ has an assignment f at a rectangle w, then

    (i) assign f to F_1 at one alternative,
    (ii) assign f to F_2 at another alternative.

∨t : If $F_1 \vee F_2$ has an assignment t at a rectangle w, then

    (i) assign t to F_1 at one alternative,
    (ii) assign t to F_2 at another alternative.

∨f : If $F_1 \vee F_2$ has an assignment f at a rectangle w, then assign
f to F_1 and F_2 at the same rectangle.

2) Rules for modal operations:

[α]t : If [α]G has an assignment t at a rectangle w, then assign t to G
at every rectangle w' such that $(w, w') \in C_\alpha$.

[α]f : If [α]G has an assignment f at a rectangle w,
then start out a new rectangle w' such that
$(w, w') \in C_\alpha$ and assign f to G at this rectangle w'.

It is not necessary to give rules for ⊃, ≡, and < α >, since they are defined in terms of ¬, ∧, ∨, and [α]. Before giving conditions when the procedure terminates, we introduce a notion of closed rectangle. We say that a rectangle w is *closed* iff one of the following conditions is satisfied:

1. some subformula has two assignment t and f in the same rectangle,
2. when applying rule [α]f at w, a rectangle w' introduced from w is closed,
3. all alternatives of w are closed.

Termination:

The procedure terminates when one of the following conditions is satisfied:

(1) all rectangles remain unaffected by any rule,
(2) the first rectangle is closed.

To make clearly understandable the above procedure, let us give simple examples:
In the following rectangles, t and f under symbols mean that the corresponding wff's take values true and false, respectively.

Example 3.1:

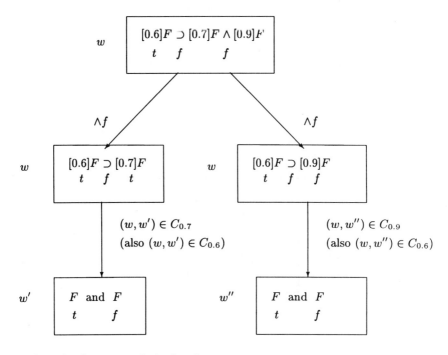

Thus, the first rectangle is closed.

Here, let us give a remark. In the above diagram, two rectangles in the second level are named by the same w at the top level. This is based on the rule $\wedge f$ which does not introduce any "new" rectangle.

Example 3.2:

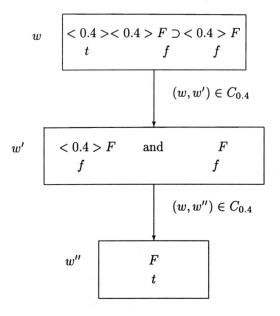

Thus, the first rectangle is not closed.

Example 3.3: We denote [0] by [ ].

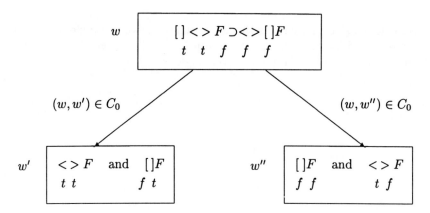

At this stage, we have to introduce a new rectangle w''' for w' since $[\ ]F$ is f at w' and F is t at the same w'. But, it is easily known that an assignment at the new w''' is the same as that at w''. Also, we have to introduce a new rectangle $w^{(4)}$ for w''. But, it is also easily known that an assignment at the new $w^{(4)}$ is the same as that at w'. Then, this process continues infinitely. But, as mentioned below our procedure for rectangle w' stops at this stage.

Consider a sequence of non-closed rectangles $w_1, w_2, w_3, \ldots, w_n, \ldots$, where w_{i+1} is the immediate descendant of w_i. We call the sequence a *path*. By Example 3.3, it is known that a path may be infinite. However, for the following reasons, there must be merely finitely many different assignments to all the formula in all the rectangles in the path:

(i) nothing but subformulas of wff F occur in any rectangle, but there are only finitely many subformulas of F,
(ii) different sets of the subformulas in all the rectangles of the path is less than 2^n, where n is the number of the subformulas.

In a path, w_j is said to be *contained* by w_i if and only if every formula in w_j appears in w_i and has the same assignment as in $w_i (i < j)$. Therefore, if a path is infinite, then there must be a rectangle in the path which contains some of its descendants. This means that there exist infinitely many circles of rectangles in the path. However, the number of rectangles appearing each circle has to be finite.

It is shown by the following reason that there are merely finitely many paths in the procedure.

(1) A new rectangle is created only when the rule $[\alpha]f$ is applied, but modal operations in any rectangle is finite;
(2) Alternatives of a rectangle are introduced only when one of the rules $\vee t, \wedge f$ is applied to a rectangle, while the length of a formula is finite.

Based on the analysis mentioned above, we add the following rule for circle of rectangles:

(3) In a path, if w_j is contained by w_i, then delete w_j.

The w' and w'' of Example 3.3 are this case (3).

Then, we have the following lemmas:

Lemma 3.4. If the rectangle for F is closed, then F is valid.

Proof:
Suppose F is not valid. Then, there exists a counter model M to F, namely,

$$M, w \models \neg F.$$

In this time, from our construction of rectangles we can show that the rectangle w for F is never closed. □

Lemma 3.5. If F is valid, then the rectangle for F is closed.

Proof:
Assume that the rectangle for F is not closed. Then, it is sufficient to show that there is a model M such that $M, w \models \neg F$. If a rectangle construction is finite, this is immediate. When a rectangle construction is infinitely continued, i.e., our procedure enters a circle of rectangles, it is sufficient to consider the infinitely many agents. □

Now, we have the following theorem:

Theorem 3.6. The problem to decide whether or not a given wff of CLD^* is valid is recursively solvable.

Proof:
The procedure mentioned above is finitely completed. Therefore, this theorem follows from Lemma 3.4 and Lemma 3.5. □

4. Undecidability

The undecidability in monadic case of some modal predicate logics is well-known. We prove now the undecidability of decision problem in CLD. This result is shown by making use of the same technique as in [3]. That is, we construct a wff $F\#$ in CLD from a wff F in the ordinary first order predicate logic as follows: $F\#$ is obtained from F by substituting $\langle \alpha \rangle (p^1(x_1) \wedge p^2(x_2) \wedge \ldots \wedge p^n(x_n))$ for n-ary predicate $P(x_1, x_2, \ldots, x_n)$ in F, where $p^i (i = 1, 2, \ldots, n)$ is a new predicate symbol. Then, we have the following theorem:

Theorem 3.7.
F is universally valid in the first order predicate logic if and only if $F\#$ is valid in CLD.

Proof:
From the definition of validity in the first order predicate logic, it is obvious that if F is universally valid then the corresponding $F\#$ is valid in CLD. Thus, we have to prove the converse, in other words, if $\neg F$ is satisfiable then $\neg F\#$ is satisfiable in CLD. Here, according to Löwenheim-Skolem's theorem we can assume that $\neg F$ is satisfiable in the domain of natural numbers. Then, we denote an assignment of n-ary predicate symbol in the model such that $\neg F$ is satisfiable as follows (the general case is similarly discussed by considering some assignment if "true" and "false"):

$$(I) \quad \begin{array}{l} P(1, 1, \ldots, 1, 1, 1) = \text{true}, \\ P(1, 1, \ldots, 1, 1, 2) = \text{false}, \\ P(1, 1, \ldots, 1, 2, 1) = \text{false}, \\ P(1, 1, \ldots, 2, 1, 1) = \text{true}, \\ \quad \vdots \end{array}$$

In this table, the first row, the second row, ... are enumerated by the following rule: Firstly, we enumerate by ordering $P(i_1, i_2, \ldots, i_n)$ according to increasing sum $i_1 + i_2 + \ldots + i_n$ and then lexicographically for those having the same sum.

We consider now an infinite set of worlds $\{w_0, w_1, w_2, \ldots\}$ in which $\rho(w_0, w_i) = \alpha$ for every $i = 1, 2, \ldots$

Then, we consider the following model:

$$\begin{array}{rl} & w_0 \; w_1 \; w_2 \; w_3 \; w_4 \; \ldots\ldots \\ p^1(1) & = (f,\, t,\;\; f,\;\; f,\;\; t,\;\; \ldots\ldots) \\ p^2(1) & = (f,\, t,\;\; f,\;\; f,\;\; t,\;\; \ldots\ldots) \\ & \quad\vdots \\ p^{n-2}(1) & = (f,\, f,\;\; f,\;\; f,\;\; f \;\;\; \ldots\ldots) \\ p^{n-1}(1) & = (f,\, t,\;\; f,\;\; f,\;\; t,\;\; \ldots\ldots) \\ p^n(1) & = (f,\, t,\;\; f,\;\; f,\;\; t,\;\; \ldots\ldots) \\ & \quad\vdots \\ p^{n-2}(2) & = (f,\, f,\;\; f,\;\; f,\;\; t,\;\; \ldots\ldots) \\ & \quad\vdots \end{array}$$

In the above notation, the value of the predicate p^1 for the natural number 1 (i.e., $p^1(1)$) takes the value true at w_1 in which $P(1, 1, \ldots, 1)$ is true. This w_1 corresponds to the first row and the w_4 to the fourth row in which $P(1, 1, \ldots, 2, 1, 1)$ is true, and so on.

By making use of the above model, the value of $<\alpha> (p^1(i_1) \wedge p^2(i_2) \wedge \ldots \wedge p^n(i_n))$ at w_0 is the same as that of $P(i_1, i_2, \ldots, i_n)$ for all $i_1, i_2, \ldots, i_n$.

Therefore, from the satisfiability of F, we get the satisfiability of $F\#$ in CLD.

Thus, we have this theorem. □

From Theorem 3.7, we get the following undecidability theorem:

Theorem 3.8. The decision problem in CLD is unsolvable.

Proof:
From Theorem 3.7 we know that if the decision problem in CLD is solvable then the problem in the first order predicate logic is solvable. This is a contradiction.
□

Remarks
In this paper, we have defined the conflict logic (CLD) with degrees which is a kind of the graded KB-modal monadic predicate logic. We have presented some properties of this logic. Then, we have also presented a decidability result as well as an undecidability. In this paper, we treated only monadic predicate symbols. As mentioned in Introduction, we will be able to extended this logic to a general predicate logic containing n-ary predicate symbols. Since CLD^* is decidable, its axioamtization is theoretically posssible. It will be an interesting problem to give a complete axiomatic system of this general predicate logic. Furthermore, it is meaningful to discuss an embedding problem of this logic to

the usual predicate logic. In practical use, applications of the decidability result of CLD are considered. Let us consider a wff: $[0.8]A \supset [0.6]B \wedge C$. This wff is obviously satisfiable. A wff: $[0.7]A \wedge [0.7]B \supset (A \equiv B)$ is also satisfiable but not valid. Note that $[0]A \wedge [0]B \supset (A \equiv B)$ is a valid wff. The following fact may give us an practical application of satisfiability.

If conditions of conflict relation among various sets of agents are adequately expressed by a wff F and the wff F is not satisfiable (i.e., $\neg F$ is valid), then we have a solution satisfying the conditions.

The conflict is considered as a kind of inequality which satisfies the special conditions. The decidability results for equality system is well-known [1]. By modifying these results to the conflict, we may obtain interesting decision procedures which are useful in practical applications of the conflict system.

Let us consider the meaning of $[\alpha]\forall x p(x)$. We take a set of persons as W. According to the interpretation given in Section 2, this wff means that the sentence $\forall x p(x)$ is approved by a person w only when it is approved for every person w_i who is in conflict by degree α. That is, if at least one person w_i has a contrary opinion then this is denied. It will be considerable to apply this logic to practical uses. Further theoretical progress as well as useful applications of this logic CLD are hoped.

References

1 W. Ackermann: *Solvable Cases of the Decision Problem*, North-Holland, Amsterdam, 1954.
2 G.E. Hughes and M.J. Cresswel: *An Introduction to Modal Logic*, Hethuen, London, 1968.
3 A. Nakamura: *On the undecidability of monadic modal predicate logic*, Zeistchr. f. math. Logik und Grundlagen d. Math., 16, 1970, 257-260.
4 A. Nakamura: On a logic of information for reasoning about knowledge, *Rough Sets, Fuzzy Sets and Knowledge Discovery* (Ed. by W. P Ziarko), Springer-Verlag, Berlin, 1994, 186-195.
5 A. Nakamura: *On a logic based on graded modalities*, IEICE Trans. Inf. and Syst. vol.E76-D, 5, 1993, 527-732.
6 Z. Pawlak: *Rough Sets - Theoretical Aspects of Reasoning about Data*, Kluwer Academic Press, Dordrecht, 1991.
7 Z. Pawlak: *An inquiry into anatomy of conflicts*, to appear in Information Sciences.
8 A. Rosenfeld: *Fuzzy graphs, Fuzzy Sets and their Applications to Cognitive and Decision Processes* (Ed. by L.A. Zadeh, K.S. Fu, K. Tanaka and M. Shimura), Academic Press, NY, 1975, 77-95.

Approximate mathematical morphology. Rough set approach

Lech Polkowski

Institute of Mathematics, Warsaw University of Technology
Pl. Politechniki 1, 00-665 Warsaw, Poland
e-mail: polk@mimuw.edu.pl

Abstract. Mathematical morphology is a well - established discipline whose aim is to provide specific tools, based originally on Minkowski operators in affine spaces, for feature selection in complex objects, primarily patterns and images. Complexity of morphological operations makes it desirable to propose a theoretical scheme of an approximate morphological calculus within a general paradigm of soft computing. We propose a scheme based on ideas of rough set theory. In this scheme, the underlying space of points (e.g. pixels) is partitioned into disjoint cells (classes) by means of some primitive attributes (features) and morphological operations are performed on classes, which allows for compression of data. In our chapter, we discuss topological foundations of morphology, in particular spaces of rough fractals as well as morphological approximate operations and we also point to plausible applications by proving the approximate collage theorem.

keywords: *rough sets, mathematical morphology, hit - or - miss topologies, approximate collage theorem*

1 Introduction

Mathematical morphology is concerned with problems of filtering of complex objects by means of some geometrical operations. We begin with a brief introduction to mathematical morphology (*cf.* [4], [10]) and for purposes of simplicity and applicability we restrict ourselves to the euclidean case in which objects are either subsets of a euclidean space E^n of n - dimensions or subsets of an affinely closed subspace $V \subseteq E^n$ (e.g. V may be the digital space Z^n).

Morphological operations are generated by two binary operations called, respectively, the *Minkowski sum* and the *Minkowski difference*. The Minkowski sum, denoted $\oplus$, is defined for any pair $A, B \subseteq V$ by the formula $A \oplus B = \{x + y : x \in A, y \in B\}$ where $+$ denotes the addition in the space E^n. Similarly, $A \ominus B = \{x \in V : \{x\} \oplus B \subseteq A\}$. We now simplify further our presentation by requiring that the generic set B be symmetric about the origin i.e. $B = -B = \{-x : x \in B\}$.

We fix a set St of *standard objects (structuring objects)*; for an object $B \in St$, and an object $X \subseteq V$, we define the *dilation* $d_B(X)$ of a given object X by the standard object B as the object $d_B(X) = X \oplus B$. Similarly, the *erosion* $e_B(X)$ of X by B is defined via the identity $e_B(X) = X \ominus B$.

Dilations and erosions generate new morphological operations by means of the operation of composition of mappings; in particular, the *opening* $o_B(X)$ of X by B is defined as the composition $d_B(e_B(X))$ i.e. $o_B = d_B \circ e_B$. Similarly, the *closing* $c_B(X)$ of X by B is defined as the composition $e_B(d_B(X))$ i.e. $c_B = e_B \circ d_B$.

It follows immediately from the definitions above that local characterizations of opening, respectively, closing are the following:

(i) $o_B(X) = \{x \in V : \exists y.(x \in \{y\} \oplus B \subseteq X)\}$;

(ii) $c_B(X) = \{x \in V : \forall y.(x \in \{y\} \oplus B \Rightarrow X \cap (\{y\} \oplus B) \neq \emptyset)\}$.

From (i), (ii) it is easily inferred that the operations of morphological opening, respectively, closing, resemble topological operations of interior, respectively, closure (*cf.* [3]) with the difference that the former are defined with respect to a fixed object B while the latter are defined with respect to the whole family St satisfying the requirements for an open basis for topology on V.

Morphological operations can easily be extended to a more general case of functions (*cf.* [12]) (e.g. signals) *viz.* given a function $F : X \longrightarrow E$, where $X \subseteq E^n$, an object in E^{n+1} called the *umbra* of F is associated with F and morphological operations are performed on umbrae; the resulting function is reconstructed then by the standard *envelope* operation. To give details of this approach, we first define the umbra $U(F)$ by letting $U(F) = \{(x,y) \in X \times E : y \leq F(x)\}$. The set St of standard objects consists in this case of functions denoted with the generic symbol $K : domK \longrightarrow E$ where the domain $domK \subseteq E^n$; we assume again the symmetricity of K i.e. $domK = -domK$ and $F(-x) = F(x)$.

The *(functional) envelope* $Env(Y)$ of a subset $Y \subseteq E^{n+1}$ is the function defined by the identity $Env(Y)(x) = \sup\{y \in E : (x,y) \in Y\}$ (observe that the value is $-\infty$ when $\{y \in E : (x,y) \in Y\} = \emptyset\}$). A morphological operation ψ_K (where ψ is, respectively, d, e, o, c) on a given function F is defined via the formula $\psi_K(F) = Env(\psi_{U(K)}U(F))$.

The *theory of rough sets* (cf. [5]) is also concerned with finding a useful set of classifying features for a given class of objects but its language is logical. Objects under consideration form a set U; the description of objects is provided by means of a set A of attributes (features) where each attribute $a \in A$ is represented as a mapping $a : U \longrightarrow V_a$ of the set U into the value set of a.

1.1 Example For instance, in the context of mathematical morphology on the set of subsets of a bounded set $Z \subseteq E^n$, we may adopt $U = Z$, and relate $A = \{a_1, a_2, .., a_n\}$ to a collection $\{P_1, P_2, ..., P_n\}$ of partitions where P_i is the partition of the $i-th$ axis E_i into intervals of the form $(j, j+1]$ (where j is an integer); for $x = (x_1, x_2, ..., x_n) \in U$, we let $a_i(x) = j$ whenever $x_i \in (j, j+1]$.

Objects in the set U are perceived through the attribute set A; in consequence, objects having identical descriptions relative to A are identified into a common class and treated as one generalized object. Formally, this is expressed by means of the relation of indiscernibility $IND(A) = \{(x,y) \in U \times U : a(x) = a(y)$ for each $a \in A\}$. Equivalence classes of this relation form elementary granules of our knowledge: all objects in the equivalence class $[x]_{IND(A)}$ are pairwise indistinguishable.

For a subset $X \subseteq U$, rough set theory offers two approximations in terms of granules of knowledge; the *lower approximation* $A_-X = \{x \in U : [x]_{IND(A)} \subseteq X\}$ consists of objects which are elements of X with certainty.

The *upper approximation* $A^+X = \{x \in U : [x]_{IND(A)} \cap X \neq \emptyset\}$ consists of objects which cannot be excluded from X with certainty. The set X is approximately defined by the pair (A_-X, A^+X) of its approximations. A set X is *rough* in case $A_-X \neq A^+X$. It may be observed that rough set theoretic approximations and morfological approximations are formally related to each other via formulae (i) and (ii).

1.2 Example For the context of Example 1.1, let us select a standard object $B = (0,1]^n$. Since equivalence classes of the relation $IND(A)$ are of the form $Z \cap \prod_{1 \leq i \leq n}(j_i, j_i + 1]$, it follows that (iii) $o_B(X) = A_-X$ for each $X \subseteq Z$; (iv) $c_B(X) = A^+X$ for each $X \subseteq Z$.

Let us observe that approximations A_-X, A^+X may be expressed in topological language *viz.* denoting by $\pi_{IND(A)}$ the partition topology, obtained by taking the set of equivalence classes of $IND(A)$ as its open base, we have: $A_-X = int_{\pi_{IND(A)}} X$ and $A^+X = cl_{\pi_{IND(A)}} X$ where int, respectively, cl are the interior, respectively, the closure operators (*cf.* [11], [13]).

The reader will find a representative picture of the state of art in rough set theory in [9].

Morphological operations are characterized in terms of their topological properties (*cf.* [4]). The prerequisite for approximate morphology is therefore an adequate topological theory. We begin its presentation in the next section, where we outline a general scheme for topologies on families of subsets.

2 Exponential ("hit - or - miss") topologies

The scheme for introducing topologies on a family of sets may be conveniently characterized by the "hit - or -miss" metaphor (*cf.* [8]). The meaning of this metaphor is as follows: for a topological space (X, τ) (*cf.*[3]), and a family A of subsets of X, the proposed way of introducing a topology into A consists in the following: first, select families P,Q of subsets of X such that P is closed under finite unions and then take as an open base for the topology $\tau_{P,Q}(A)$ on A the family of sets
$[P; Q_1, Q_2, ..., Q_k] = \{A \in A : A \cap P = \emptyset; A \cap Q_i \neq \emptyset$ for $i = 1, 2, ..., k\}$
where $P \in P$ and $Q_i \in Q$.

Various choices of families P,Q lead to various topologies. Let us mention briefly the most important cases. We denote, respectively, by G, F, K, the families of all, respectively, open, closed, compact sets in the space (X, τ). We consider usually the following topologies

1. the *exponential topology* $\tau_{F,G}(F)$ (*cf.* [3]);

2. the *hit - or - miss topology* $\tau_{K,G}(F)$ (cf. [4]);
3. the *myope topology* $\tau_{F,G}(K)$ (cf. [4]).

Since our purpose is restricted here to mathematical morphology in Euclidean spaces, we assume that all underlying topological spaces (X, τ) are locally compact metrizable (cf. [3,4]).

The important property of the hit - or - miss topology $\tau_{K,G}(F)$ on the family F of closed sets is its *metrizability* as well as *compactness* (cf. [3,4]); this means a fortiori that this topology can be characterized in terms of convergent sequences (op.cit.) i.e. a closed set $C \in F$ belongs to the closure $\overline{Z}$ of a collection $Z \subseteq F$ if and only if there exists a sequence $(C_n)_n \subseteq Z$ such that $C = \lim_n C_n$; the convergence $C_n \to C$ is in turn characterized in set - theoretic terms (cf. [3]) viz. $C = \lim_n C_n$ if and only if $C = Lim C_n$ which means the conjunction of two conditions:

(i) $C = Li C_n = \{x \in X :$ each neighborhood of x intersects almost every C_n;
(ii) $C = Ls C_n = \{x \in X :$ each neighborhood of x intersects infinitely many C_n.

Important subsets of K in myope topology are the compact ones; the characterization of compact sets in K with respect to myope topology is a simple one viz. a subset $Z \subseteq K$ is compact in myope topology if and only if it is closed in the hit - or - miss topology (i.e. Z with each convergent sequence $(C_n)_n$ contains its limit $Lim C_n$) and there exists a compact set $W \in K$ such that $\cup Z \subseteq W$ (op.cit.[4]). The importance of compact sets lies in the fact that the hit - or - miss topology and myope topology coincide on families of closed subsets of any compact set.

Concerning the explicit form of a metric for hit - or - miss as well as myope topologies, it is useful to recall here the *Hausdorff - Pompeiu metric* D_H (cf. [3], [8]) defined on the family F by the formula:

$$D_H(C, D) = \max\{\sup_{x \in C}\{\inf_{y \in D} d(x, y)\}, \sup_{y \in D}\{\inf_{x \in C} d(x, y)\}$$

where d is a compatible bounded metric on X.

The myope topology is metrizable by the metric D_H restricted to the space K×K (cf. [4]); hence, the hit-or-miss topology is metrizable by the metric D_H on every subspace $Z \subseteq F$ compact in the myope topology (cf. [4]).

We recall that for any compact set $C \subseteq X$, the space K|C endowed with the hit - or - miss (equal in this case to myope) topology is the space of *fractals in C*. The above facts provide therefore a complete description of topologies of spaces of fractals.

In terms of topologies, one may classify morphological operations in terms of their continuity properties. First, we should say what continuity means in case of set - valued mappings. We would give an example only (cf. [4] *for a fuller account*) and so we consider a set - valued mapping $f : K \times K \longrightarrow K$ (examples of such mappings are *d, e, o, c*). We would say that f is *upper - semicontinuous* in the case when for every converging sequence $(C_n, D_n)_n \subseteq K \times K$, the inclusion $Ls f(C_n, D_n) \subseteq f(Lim(C_n, D_n))$ takes place; similarly, f is *lower semi - continuous* when $f(Lim(C_n, D_n)) \subseteq Ls f(C_n, D_n)$ holds. Continuity means both upper - and lower - semi continuity. One shows (cf. [4], [10]) that the mapping d is continuous while e, o, c are upper semi - continuous. The property of upper semicontinuity is given interpretation in terms of stability of a morphological operation (*cf.* [10]).

We now pass to the rough set - theoretic case for which we introduce some topologies being counterparts of the hit - or - miss, respectively, myope topology.

3 Topologies on rough sets

The full account of topological theory of rough sets may be found in [6], [7]. Here we will present some particular cases, notably, we will analyse the countable case as most closely related to applications. To begin with, let us observe, going back to Introduction, that in case of the indiscernibility relation $IND(A)$ on the set U of objects, rough sets may be characterized topologically as those sets X for which $int_{\pi_{IND(A)}} X \neq cl_{\pi_{IND(A)}} X$. We will now consider a more general context in which we are given an information system $A = (U, A)$ where the set A of attributes is countable infinite *i.e.* $A = \{a_n : n = 1, 2, ...\}$; we may and will assume that $IND(a_{n+1}) \subseteq IND(a_n)$ for each n *i.e.* the attribute a_{n+1} induces a finer classification than the attribute a_n.

In this context a natural topology appears: we denote by π_A the topology obtained by taking the union $\cup \{[x]_{IND(a_n)} : x \in U, n = 1, 2, ...\}$ of equivalence classes of all relations $IND(a_n)$ as an open base.

3.1 Example Going back to Example 1.1, for a compact $Z \subseteq E^n$, we may relate the attribute a_m to the partition of Z into the cubes $\prod_{1 \leq i \leq n}(j_i + k_i \cdot 2^{-m}, j_i + (k_i + 1) \cdot 2^{-m}]$ where $j_1, j_2, ..., j_n$ are integers, $k_i = 0, 1, ..., 2^m - 1$ is an integer for each $i \leq n$. Then the topology π_A is finer than the natural euclidean topology on Z.

Topological rough sets

We now return to the information system $A = (U, A)$ and we will call a subset $X \subseteq U$ a $\pi_A - pre - rough$ set in case $int_{\pi_A} X \neq cl_{\pi_A} X$. We introduce on the collection of π_A- pre - rough sets the equivalence relation $\equiv$ defined by the condition: $X \equiv Y$ if and only if $int_{\pi_A} X = int_{\pi_A} Y$ and $cl_{\pi_A} X = cl_{\pi_A} Y$. Equivalence classes of the relation $\equiv$ are called π_A- *rough sets*. These are objects of our study. It will be convenient to call these sets shortly π_A- *sets*.

The first thing to do is to find a convenient characterization of π_A- sets. To this end, given a π_A- set $[X]_\equiv$, we let $Q = cl_{\pi_A} X$, $T = U - int_{\pi_A} X$. We quote the corresponding result and its proof from [7].

3.2 Proposition A given pair (Q, T) of π_A - closed sets does represent a π_A- set $[X]_\equiv$ with $Q = cl_{\pi_A} X$, $T = U - int_{\pi_A} X$ if and only if the following conditions are satisfied:

(R1) $U = Q \cup T$;
(R2) $Q \cap T \neq \emptyset$;
(R3) $Q \cap T$ does not contain a π_A - isolated point.

Proof Only the sufficiency of $(R1 - R3)$ requires a proof. For a given pair (Q, T) satisfying $(R1 - R3)$, we will construct a subset X which will satisfy conditions $Q = cl_{\pi_A} X$, $T = U - int_{\pi_A} X$. We define $A_n = \{[x]_{IND(a_n)} : [x]_{IND(a_n)} \cap P = \emptyset \neq [x]_{IND(a_n)} \cap Q\}$ for $n = 1, 2.....$. We well - order each A_n as $\{R(t, n) : t < \kappa_n\}$.

Inducting on n, we will define points $x(t,n), y(t,n)$ for $t < \kappa_n, n = 1, 2, ..$ which will satisfy the following conditions:
(1) $x(t,n) \in R(t,n) \cap Q$;
(2) $y(t,n) \in R(t,n)$ and $y(t,n) \in R(t,n) - Q$ in case $R(t,n) - Q \neq \emptyset$;
(3) if $x(t',n') \in R(t,n)$ for some $n' < n$ then $x(t,n) = x(t',n')$; similarly, if $y(t',n') \in R(t,n) - Q$ for some $n' < n$ then $y(t,n) = y(t',n')$;
(4) $x(t,n) \notin \{y(t',n') : t' < \kappa_{n'}, n' = 1, 2, ...\}$;
(5) $y(t,n) \notin \{x(t',n') : t' < \kappa_{n'}, n' = 1, 2, ...\}$.

Having this done, we consider the sets $M = \{x(t,n) : t < \kappa_n, n = 1, 2, ...\}$ and $N = \{y(t,n) : t < \kappa_n, n = 1, 2, ...\}$. From (4) and (5), we infer that $M \cap N = \emptyset$.

We now look at the set $X = P \cup M$ where $P = U - T$. We check that X satisfies the following two assertions.

(6) for each $x \in U$ and each n, we have : $[x]_{IND(a_n)} \subseteq X$ if and only if $[x]_{IND(a_n)} \subseteq P$.

Indeed, was $[x]_{IND(a_n)} \subseteq X$ and $[x]_{IND(a_n)} - P \neq \emptyset$ for some x, n, we would have $x(t, n') \in X - P$ with some $n' < n, t < \kappa_n$; it would follow from (2), and (5) that $R(t,n) - X = [x]_{IND(a_n)} - X \neq \emptyset$, a contradiction.

(7) for each $x \in U$ and each n, we have : $[x]_{IND(a_n)} \cap X \neq \emptyset$ if and only if $[x]_{IND(a_n)} \cap Q \neq \emptyset$. Indeed, we have a dychotomy: either $[x]_{IND(a_n)} \cap Q = \emptyset$ or $[x]_{IND(a_n)} = R(t,n)$ for some t, n and then by (1) and (4) we have $[x]_{IND(a_n)} \cap X \neq \emptyset$.

It follows from (6) and (7) that the pair (Q, T) does represent X.

Topologies on π_A-sets

We will introduce topologies of type 'hit - or - miss' on topological rough sets. It is our purpose to control both factors Q, T hence we will 'double' the constructions in Section 2, modifying them to suit the rough case. We will also attempt at controlling the component $Q \cap T$ and thus we will introduce two topologies π and π^* serving this twofold purpose.

We select open sets $W, V \in \pi_A$, natural numbers n_1, n_2, n_3, n_4 and collections $\mathcal{U}, \mathcal{V}$ where $\mathcal{U} \subseteq \{[x]_{IND(a_n)} : x \in U, n \leq n_3\}$, $\mathcal{V} \subseteq \{[x]_{IND(a_n)} : x \in U, n \leq n_4\}$.

For each choice of parameters $W, V, n_1, n_2, n_3, n_4, \mathcal{U}, \mathcal{V}$, we denote by the symbol $[W, V, n_1, n_2, n_3, n_4, \mathcal{U}, \mathcal{V}]$ the collection of all π_A-sets (Q, T) which satisfy the following conditions:

(8) for some $m \geq n_1$, we have $cl_{\pi_{IND(a_m)}} Q \subseteq W$;
(9) for some $m \geq n_2$, we have $cl_{\pi_{IND(a_m)}} T \subseteq V$;
(10) $Q \cap H \neq \emptyset$ for each $H \in \mathcal{U}$;
(11) $T \cap H \neq \emptyset$ for each $H \in \mathcal{V}$.

The topology π is defined as the topology obtained by taking sets of the form $[W, V, n_1, n_2, n_3, n_4, \mathcal{U}, \mathcal{V}]$ as an open base.

We define the topology π^* in a similar way: we add two additional natural numbers n_5, n_6, an open set $G \in \pi_A$, and a collection $\mathcal{W} \subseteq \{[x]_{IND(a_n)} : x \in U, n \leq n_6\}$ and we consider two additional conditions:
(12) for some $m \geq n_5$, we have $cl_{\pi_{IND(a_m)}}(Q \cap T) \subseteq G$;
(13) $(Q \cap T) \cap H \neq \emptyset$ for each $H \in \mathcal{W}$.

We denote by the symbol $[W, V, G, n_1, n_2, n_3, n_4, n_5, n_6, \mathcal{U}, \mathcal{V}, \mathcal{W}]$ the collection of all π_A-sets (Q, T) which satisfy the conditions (8) - (13). The topology

π^* is obtained by taking the sets of the form $[W, V, G, n_1, n_2, n_3, n_4, n_5, n_6, \mathcal{U}, \mathcal{V}, \mathcal{W}]$ as an open base.

Topologies π and π^* are metrizable (cf. [7]); the corresponding metrics are of Hausdorff - Pompeiu type. We will introduce two metrics D and D^* compatible, respectively, with topologies π and π^*.

We begin with the function ρ_n on U, defined for each n, by letting $\rho_n(x,y) = 1$ in case $[x]_{IND(a_n)} \neq [y]_{IND(a_n)}$ and by letting $\rho_n(x,y) = 0$, otherwise.

The function $\rho(x,y) = \sum_{n=1}^{\infty} \rho_n(x,y) \cdot 10^{-n}$ is then a pseudo - metric (meaning that $\rho(x,y) = 0$ does not necessarily imply that $x = y$) compatible with the topology π_A.

We denote by the symbol $\rho(x,C)$ the $\rho-$ distance of x from $C \subseteq U$ i.e.

$$\rho(x,C) = \inf\{\rho(x,y) : y \in C\}.$$

The metric D_H on the space of π_A- closed sets is defined in the standard way:

$$D_H(C,D) = \max\{\max\{\rho(x,C) : x \in D\}, \max\{\rho(x,D) : x \in C\}.$$

The metric D is defined as follows:

$$D((Q,T),(Q',T')) = \max\{D_H(Q,Q'), D_H(T,T')\}.$$

Similarly, the metric D^* is defined as follows:

$$D^*((Q,T),(Q',T')) = \max\{D_H(Q,Q'), D_H(T,T'), D_H(Q \cap T, Q' \cap T')\}.$$

Metrics D, respectively, D^* are compatible with topologies π, respectively, π^*: the proof may be found in [7].

It may be useful to give here procedures **P**, respectively, **P*** for computing the values of D, respectively, D^*.

Procedure P
Input: $(Q,T),(Q',T')$
Output: $D((Q,T),(Q',T'))$
Check: $cl_{\pi_{IND(a_n)}} Q = cl_{\pi_{IND(a_n)}} Q'$ and $cl_{\pi_{IND(a_n)}} T = cl_{\pi_{IND(a_n)}} T'$ for each n.

If yes : 0 else $\frac{1}{9} \cdot 10^{-n+1}$ where n is the first failing either of $cl_{\pi_{IND(a_n)}} Q = cl_{\pi_{IND(a_n)}} Q'$, $cl_{\pi_{IND(a_n)}} T = cl_{\pi_{IND(a_n)}} T'$.

Procedure P*
Input: $(Q,T),(Q',T')$
Output: $D^*((Q,T),(Q',T'))$
Check: $cl_{\pi_{IND(a_n)}} Q = cl_{\pi_{IND(a_n)}} Q'$, $cl_{\pi_{IND(a_n)}} T = cl_{\pi_{IND(a_n)}} T'$ and $cl_{\pi_{IND(a_n)}} Q \cap T = cl_{\pi_{IND(a_n)}} Q' \cap T'$ for each n.

If yes : 0 else $\frac{1}{9} \cdot 10^{-n+1}$ where n is the first failing either of $cl_{\pi_{IND(a_n)}} Q = cl_{\pi_{IND(a_n)}} Q'$, $cl_{\pi_{IND(a_n)}} T = cl_{\pi_{IND(a_n)}} T'$, $cl_{\pi_{IND(a_n)}} Q \cap T = cl_{\pi_{IND(a_n)}} Q' \cap T'$.

It follows that both metrics are discrete - valued.

Completeness properties

The completeness property of a metric space allows for applying fixed - point results of which the *Banach fixed - point theorem* (*cf.* [2]) is an archetype. It would be therefore useful to establish completeness - type properties of our spaces of rough sets. It turns out that in the general case we may establish a property involving both metrics. We would like to recall that a sequence $(x_n)_n$ in a metric space (X, ρ) is called ρ - *fundamental* when $\lim_{n,m} \rho(x_n x_m) = 0$ and (X, ρ) itself is said to be *complete* when every fundamental sequence does converge.

Our study of completeness properties will require a property (C) which we assume about $A = (U, A)$:

(C) for any descending sequence $([x_n]_{IND(a_n)})_n$ of equivalence classes: $\cap [x_n]_{IND(a^n)} \neq \emptyset$.

We denote by the symbol **N** the set of all descending sequences of the form $([x_n]_{IND(a_n)})_n$.

We may offer the following property of π_A- sets (*cf.* [7]).

3.3 Proposition (C) Every D^* - fundamental sequence $((Q_n, T_n))_n$ of π_A- sets is convergent in the metric D.

Proof We may find a strictly increasing sequence $(n_k)_k$ of natural numbers such that:

(i) $D^*((Q_m, T_m), (Q_{m_n}, T_{m_n})) < \frac{1}{n}$ for $m \geq m_n$

and a strictly increasing sequence $(j_n)_n$ of natural numbers such that

(ii) $cl_{\pi_{IND(a_{j_n})}} Q_m = cl_{\pi_{IND(a_{j_n})}} Q_{m_n}$ for $m \geq m_n$;

(iii) $cl_{\pi_{IND(a_{j_n})}} T_m = cl_{\pi_{IND(a_{j_n})}} T_{m_n}$ for $m \geq m_n$;

(iv) $cl_{\pi_{IND(a_{j_n})}} (Q_m \cap T_m) = cl_{\pi_{IND(a_{j_n})}} (Q_{m_n} \cap T_{m_n})$ for $m \geq m_n$.

We define sets:

$Q^* = \cup \{\cap [x_n]_{IND(a_n)} : ([x_n]_{IND(a_n)})_n \in \mathbf{N}$ and $[x_n]_{IND(a_n)} \cap Q_{m_n} \neq \emptyset$ for each $n\}$

and

$T^* = \cup \{\cap [x_n]_{IND(a_n)} : ([x_n]_{IND(a_n)})_n \in \mathbf{N}$ and $[x_n]_{IND(a_n)} \cap T_{m_n} \neq \emptyset$ for each $n\}$.

To prove that (Q^*, T^*) is a π_A- set to which the sequence $((Q_n, T_n))_n$ converges in the metric D, it is sufficient to establish the truth of the following claims (*cf.* [7]).

Claim 1. Q^* and T^* are π_A - closed.

Claim 2. $U = Q^* \cup T^*$.

Claim 3. $Q^* \cap T^* \neq \emptyset$.

Claim 4. $Q^* \cap T^*$ does not contain a π_A - open singleton.

Claim 5. $cl_{\pi_{IND(a_{j_n})}} Q^* = cl_{\pi_{IND(a_{j_n})}} Q_{m_n}$ and $cl_{\pi_{IND(a_{j_n})}} T^* = cl_{\pi_{IND(a_{j_n})}} T_{m_n}$ for each n.

This concludes the proof.

It is known (*cf.* [7]) that neither D nor D^* is a complete metric.

However, we may mention an important case where each attribute a_n induces only finitely many equivalence classes (compare Example 3.1). This is expressed by the condition (F):

(F) the set $\{[x]_{IND(a_n)} : x \in U\}$ is finite for each n.

3. 4 Proposition (C+F) Each sequence $((Q_n, T_n))_n$ of π_A - sets contains a subsequence which is convergent in the metric D.

This proposition will easily follow from Proposition 3. 7, below.

The meaning of 3. 4 is that under (C+F), the metric D makes the space of π_A - sets into a compact metric space.

We now complete our discussion by adding the lacking till now component viz. it may happen that a π_A − set X is definable i.e. $int_{\pi_A} X = cl_{\pi_A} X$ while it is rough with respect to each attribute a_n i.e. $int_{\pi_{IND(a_n)}} X \neq cl_{\pi_{IND(a_n)}} X$. We call such sets *almost π_A− rough sets (almost π_A - sets, shortly)*.

Almost π_A− sets

The theory of these sets parallels our considerations above; it is manifest that such a set X induces a sequence $(Q_n = cl_{\pi_{IND(a_n)}} X, T_n = U - int_{\pi_{IND(a_n)}} X)_n$ and we have to select those sequences which do indeed represent almost π_A - sets. The characterization in question is as follows (cf. [7] for a proof).

Recalling the set **N** defined earlier, we let:

N$(Q) = \{([x_n]_{IND(a_n)})_n \in \mathbf{N} : [x_n]_{IND(a_n)} \cap Q_n \neq \emptyset$, each $n\}$,
N$(T) = \{([x_n]_{IND(a_n)})_n \in \mathbf{N} : [x_n]_{IND(a_n)} \cap T_n \neq \emptyset\}$, each $n\}$,
N$(Q)(y, n') = \{([x_n]_{IND(a_n)})_n \in \mathbf{N}(Q) : [x_{n'}]_{IND(a_{n'})} = [y_{n'}]_{IND(a_{n'})}\}$, for $y \in U$, each n',
N$(T)(y, n') = \{([x_n]_{IND(a_n)})_n \in \mathbf{N}(T) : [x_{n'}]_{IND(a_{n'})} = [y_{n'}]_{IND(a_{n'})}\}$, for $y \in U$, each n'.

3. 5 Proposition A sequence $(Q_n, T_n)_n$ does represent an almost π_A - set if and only if the following conditions hold:

(i) $U = Q_n \cup T_n$ for each n;
(ii) $Q_n \cap T_n \neq \emptyset$ for each n;
(iii) $Q_n \cap T_n$ contains a $\pi_{IND(a_n)}$ - isolated point for no n;
(iv) $cl_{\pi_{IND(a_m)}} Q_n = Q_m$ and $cl_{\pi_{IND(a_m)}} T_n = T_m$ for $m < n$ for each n;
(v) **N**$(Q)(y, n') \neq \emptyset$ implies that: there exists $([x_n]_{IND(a_n)})_n \in \mathbf{N}(Q)(y, n')$ with the property that: $\cap_n [x_n]_{IND(a_n)} \neq \emptyset$; similarly, **N**$(T)(y, n') \neq \emptyset$ implies that: there exists $([x_n]_{IND(a_n)})_n \in \mathbf{N}(T)(y, n')$ with the property that: $\cap_n [x_n]_{IND(a_n)} \neq \emptyset$.

Due to (v) one may suspect that almost π_A− sets would be complete under an appropriate modification of the metric. This is indeed the case.

The metric D' on almost π_A− sets is constructed as follows.

For each n, we define the metric ρ_n as above. Then we define the metric D_{H_n} induced by ρ_n via the formula $D_{H_n}(C, D) = \max\{\max\{\rho_n(x, C) : x \in D\}, \max\{\rho_n(y, D) : y \in C\}\}$ and we induce the metric D'_n on pairs of $\pi_{IND(a_n)}$ closed sets by means of the formula:

$D'_n((Q_n, T_n), (Q'_n, T'_n)) = \max\{D_{H_n}(Q_n, Q'_n), D_{H_n}(T_n, T'_n)\}$.

Finally, the metric D' is defined as follows:

$D'((Q_n, T_n)_n, (Q'_n, T'_n)_n) = \sum_{i=1}^{\infty} D'_n((Q_n, T_n), (Q'_n, T'_n)) \cdot 10^{-n}$.

The procedure for computing D' is analogous to procedures **P** and **P***.

Procedure P'

Input: $(Q_n, T_n)_n, (Q'_n, T'_n)_n$
Output: $D'((Q_n, T_n)_n, (Q'_n, T'_n)_n)$
check: $Q_n = Q'_n$ and $T_n = T'_n$ for each n.

If yes : 0 else $\frac{1}{9} \cdot 10^{-n+1}$ where n is the first that fails either of the equalities.

We have the following result on completeness of almost π_A - sets (cf.[7] for a proof).

3. 6 Proposition The space of almost π_A - sets endowed with the metric D' is complete.

We conclude our discusion in this section with a result showing the important case in which almost π_A - sets and π_A - sets coincide (v. Proposition 3. 4) (cf. [7] for a proof).

3. 7 Proposition Under (C+F) each almost π_A - set is a π_A - set. Moreover, the space of almost π_A - sets with the metric D' is isomorphic to the space of π_A - sets with the metric D.

4 Applications

Topological spaces of rough sets introduced above are either complete or almost complete $e.g.$ complete with respect to two metrics. This opens up a possibility of applications concerned with data (image) compression. We begin with a concise study of this aspect of our theory. In what follows, we apply the context of Example 3. 1.

We should recall that all important applications of a complete context are based either directly or implicitly (via some generalizations) on the renowned *Banach contraction principle* ($cf.$ Théorème 6, par.2 in [1]): given a complete metric space (X, ρ) and a *contracting mapping* $f : X \longrightarrow X$ with a *contraction constant* $c \in (0, 1)$ (i.e. $\rho(f(x), f(y)) \leq c \cdot \rho(x, y))$, there exists a *unique fixed point* x_f (i.e. $f(x_f) = x_f$) of the mapping f.

Moreover, x_f can be found as the limit of any sequence $(a_n)_n$ where

(i) $a_o \in X$ is arbitrary and $a_{n+1} = f(a_n)$ for $n = 0, 1, 2...$.

Even the member of the sequence $(a_n)_n$ giving a desired approximation to the fixed point can be found: we have

(ii) $\rho(a_n, x_f) \leq c^n \cdot (1-c)^{-1} \cdot \rho(a_o, f(a_o))$.

These well - known results have been applied in fractal generation by means of conctracting affine mappings in Euclidean spaces (the *collage theorem, cf.* [2]) and a fortiori in data compression.

We would go a step further; having a sequence of compacta (fractals) $(F_n)_n$ converging in the Hausdorff metric D_E, induced by the Euclidean metric in either *2D* - or *3D* - space, to a limit compactum (fractal) F, we would pose the following question:

Q. what would be the least n such that the equality $'roughification'$ $a_m^+ F_n = 'roughification'$ $a_m^+ F$ eventually holds with a given m?

The practical implication of this question is the following: setting m defines the mesh of the grid by which we approximate our rough compact sets F_n and F. The equality $a_m^+ F_n = a_m^+ F$ implies that the Hausdorff distance between F_n and F is less than $2^{-m+\frac{1}{2}}$ hence for large enough values of m both sets F_n and F are satisfactorily close also with respect to the metric D_E. For computational or transmission reasons, using $a_m^+ F_n$ instead of F_n has an advantage of being more robust and computationally less demanding.

The answer to the question Q. is in the following

4.1 Proposition (the *approximate collage theorem (based on topology of almost rough sets)*) Assume that $(F_n)_n$ is a sequence of compact sets. We assume that sets F_n, F are almost rough (which is *e.g.* the case with all sets having a non - integer fractal dimension). We have

(a) if $\lim_n D'(F_n, F) = 0$ then $\lim_n D_E(F_n, F) = 0$.

Assume now that $f : C \longrightarrow C$ is a contracting mapping on a compact set $C \subseteq E^n$ with a contraction coefficient $c \in (0,1)$ (*e.g.* one selects compact sets $C_o^1,..,C_o^k$ and lets $F_o = \cup_i C_o^i, F_1 = \cup_i f(C_o^i)$ etc.; it is easy to see that the resulting set mapping has also a contraction coefficient c with respect to the Hausdorff metric D_E cf. [2]). Let $K = D_E(F_o, F_1)$.

(b) in order to have $D_E(F_n, F) < \varepsilon$ for a chosen positive threshold ε, it is sufficient to check that $a_m^+ F_n = a_m^+ F$ with $m = \lceil \frac{1}{2} - \log_2 \varepsilon \rceil$ and $n \geq n_o = \lceil \frac{\log_2 [2^{-m+\frac{1}{2}} \cdot K^{-1} \cdot (1-c)]}{\log_2 c} \rceil$.

Proof. Claim (a) is true by the remark made above that the equality $a_m^+ F_n = a_m^+ F$ implies that the Hausdorff distance D_E between F_n and F is less than $2^{-m+\frac{1}{2}}$. The estimate for n_o follows from the remark quoted and the estimate (ii) above.

In the remaining part of this section, we would concern ourselves with an example of a *rough morphological operation*; we take as an example the case of $B-$ *erosion* (*cf.* [7]). To define this operation, we take $B = [x]_{IND(a_n)}$ for some $x \in U$ and n and we assume that B contains no $\pi_{IND(a_n)}$ - open singletons.

We define $B-$ *erosion* e_B on the space of rough sets by letting:

$$e_B((Q,T)) = (Q \cup B, T \cup B).$$

In terms of knowledge engineering, this mapping means that we eventually lose some positive knowledge represented in the granule B - it is transferred to the possible knowledge region.

One can prove (*cf.* [7]).

4.2 Proposition The mapping e_B is continuous with respect to the metric D as well as with respect to the metric D^*.

The reader will find in [7] other results of this type.

References

1. S. Banach, *Sur les opérations dans les ensembles abstraits et leur application aux équations intégrales*, Fundamenta Mathematicae 3 (1922), 133 - 181.

2. S. Barnsley, Fractals everywhere, Academic Press, New York, 1991.
3. K. Kuratowski, Topology, vols.I, II, Academic Press, New York, 1966.
4. G. Matheron, Random Sets and Integral Geometry, Wiley, New York, 1975.
5. Z. Pawlak, Rough Sets: Theoretical Aspects of Reasoning about Data, Kluwer, Dordrecht, 1991.
6. L. Polkowski, *On convergence of rough sets,* in: Intelligent Decision Support. Handbook of Applications and Advances of Rough set Theory, R. Slowinski (ed.), Kluwer, Dordrecht, 1992.
7. L.Polkowski, *Mathematical morphology of rough sets,* Bull. Polish Acad. Ser. Sci.Math. 41(3) (1993), 241 - 273.
8. L. Polkowski, *Hit - or - Miss topology,* in: Encyclopaedia of Mathematics, vol. XI, M. Hazewinkel (ed.), Kluwer, Dordrecht, 1997, 291.
9. L. Polkowski, A. Skowron (eds.), Rough Sets in Knowledge Discovery and Data Mining. Methodology and Applications, Springer (Physica Verlag), 1998.
10. J. Serra, Image Analysis and Mathematical Morphology: Theoretical Advances, Academic Press, New York, 1988.
11. A. Skowron, *On topology in information systems,* Bull. Polish Acad. Ser. Sci. Math. 36 (1988), 477 - 479.
12. R. S. Sternberg, *Grayscale morphology,* Comput. Vision, Graphics&Image Process. 35 (1986), 333 - 355.
13. A. Wiweger, *On topological rough sets,* Bull. Polish Acad. Ser . Sci. Math. 37 (1988), 89 - 93.

Local Approach to Construction of Decision Trees

Mikhail Moshkov

Research Institute for Applied Mathematics and
Cybernetics of Nizhni Novgorod State University
10, Uljanova St., Nizhni Novgorod, 603005, Russia
e-mail: moshkov@nnucnit.unn.ac.ru

In this chapter problems of decision tree construction are considered. Decision trees are widely used in various applications as algorithms for problem solving and as a form of knowledge representation. Decision trees over finite (ordinary) information systems [Pa1, Pa2] are investigated in test theory, rough set theory, theory of questionnaires, in machine learning, etc. The notion of infinite information system is useful in discrete optimization, pattern recognition, computational geometry. But decision trees over infinite information systems are investigated less than those over finite information systems.

In [Mo3, Mo4] two approaches to the study of deterministic decision trees over an arbitrary information system were developed: the local approach where for a problem the decision trees are considered using only attributes from the problem description, and the global one admitting arbitrary attributes from the given information system. Later these approaches were extended on the case of nondeterministic decision trees [Mo5, Mo6].

In this chapter for an arbitrary information system with finite or denumerable set of attributes the local approach to the study of decision trees is considered, and problems of deterministic decision tree construction are discussed in detail.

The chapter consists of six sections. In Sect. 1 basic notions are defined. In Sect. 2 algorithms of decision tree construction for a given decision table are discussed. Sect. 3 contains a criterion of the solvability for the problem of decision tree local optimization. In Sect. 4 cardinality characteristics of decision tables are presented, and in Sect. 5 algorithms of decision table construction are considered. In Sect. 6 algorithms of decision tree construction for a given problem are discussed.

1 Basic Notions

In this section basic notions such as decision table, decision tree, information system and problem are considered, and some relationships among these notions are discussed. Notions of decision table and decision tree are defined irrespective of the notion of information system.

1.1 Decision Tables and Decision Trees

Denote $\omega = \{0, 1, 2, \ldots\}$. Let B be a finite nonempty set with at least two elements and F be a nonempty finite or denumerable set.

An *(B, F)-decision table* is a finite rectangular table filled by elements from B in which columns are labeled by pairwise different elements from F, rows are pairwise different and each row is labeled by a number from ω. It is possible that an (B, F)-decision table has no rows.

We denote by $\mathcal{T}(B, F)$ the set of all (B, F)-decision tables. For a table $T \in \mathcal{T}(B, F)$ we denote by $\mathrm{At}(T)$ the set of all elements from F which are labels of columns of the table T.

An *(B, F)-decision tree* is a finite tree with the root in which each terminal node is assigned a number from ω, each non-terminal node is assigned an element from F, each edge is assigned an element from B. Edges starting in a non-terminal node are assigned pairwise different elements.

Let Γ be an (B, F)-decision tree. We denote by $\mathrm{At}(\Gamma)$ the set of all elements from F which are assigned to non-terminal nodes of Γ. A *complete path* in Γ is an arbitrary sequence $\xi = v_1, d_1, \ldots, v_m, d_m, v_{m+1}$ of nodes and edges of Γ such that v_1 is the root, v_{m+1} is a terminal node, and v_i is the initial and v_{i+1} is the terminal node of the edge d_i for $i = 1, \ldots, m$. The number m is called *the length* of the path ξ. (If the path ξ contains only one node then $m = 0$.)

As time complexity measure we will consider *the depth* of a decision tree which is the maximal length of a complete path in the tree. We denote by $h(\Gamma)$ the depth of a decision tree Γ.

Let T be an (B, F)-decision table, $\mathrm{At}(T) = \{f_1, \ldots, f_n\}$ and for $i = 1, \ldots, n$ let f_i be the label of the i-th column of the table T. Let Γ be an (B, F)-decision tree such that $\mathrm{At}(\Gamma) \subseteq \mathrm{At}(T)$. Let $\xi = v_1, d_1, \ldots, v_m, d_m, v_{m+1}$ be a complete path in Γ and $\bar{\delta} = (\delta_1, \ldots, \delta_n)$ be a row of the table T. We will say that the complete path ξ *accepts* the row $\bar{\delta}$ if either $m = 0$ or $m > 0$ and the following conditions hold: for $i = 1, \ldots, m$ if the node v_i is labeled by an element f_{j_i} then the edge d_i is labeled by the element δ_{j_i}.

We will say that an (B, F)-decision tree Γ is *suitable* for an (B, F)-decision table T if $\mathrm{At}(\Gamma) \subseteq \mathrm{At}(T)$ and for each row $\bar{\delta}$ of T there exists a complete path ξ in Γ which accepts this row, and the terminal node of the path ξ is assigned the number that is the label of the row $\bar{\delta}$.

For an arbitrary decision table $T \in \mathcal{T}(B, F)$ we denote by $h(T)$ the minimal depth of an (B, F)-decision tree which is suitable for the table T.

1.2 Information Systems and Problems

Let A be a nonempty set, B be a finite nonempty set with at least two elements and F be some nonempty set of functions from A to B. Functions from F will be called *attributes* and the triple $I = (A, B, F)$ will be called *an information system*. Further we will assume that F is a finite or a denumerable set.

We will not distinguish attributes and their names. But since we will deal with algorithms of decision table and decision tree construction we will assume that in the most part of considered objects (such as decision tables, decision trees, problems, etc.) names of attributes are used instead of attributes.

We will consider problems over the information system I. A *problem over I* is an arbitrary $(n+1)$-tuple $z = (\nu, f_1, \ldots, f_n)$ where $\nu : B^n \to \omega$ and $f_1, \ldots, f_n$ are pairwise different attributes from F. The number n will be called *the dimension of the problem z* and will be denoted by $\dim z$. Denote $\mathrm{At}(z) = \{f_1, \ldots, f_n\}$. The problem z may be interpreted as a problem of searching for the value $z(a) = \nu(f_1(a), \ldots, f_n(a))$ for an arbitrary $a \in A$. Different problems of pattern recognition, discrete optimization, fault diagnosis and computational geometry can be represented in such form. We denote by $P(I)$ the set of all problems over the information system I.

As algorithms for problem solving we will consider decision trees. Let Γ be an (B,F)-decision tree and $\xi = v_1, d_1, \ldots, v_m, d_m, v_{m+1}$ be a complete path in Γ. Define a subset $A(\xi)$ of the set A associated with ξ. If $m = 0$ then $A(\xi) = A$. Let $m > 0$, and let the attribute f_{j_i} be assigned to the node v_i and let δ_i be the element from B assigned to the edge d_i, $i = 1, \ldots, m$. Then $A(\xi) = \{a : a \in A, f_{j_1}(a) = \delta_1, \ldots, f_{j_m}(a) = \delta_m\}$.

We will say that an (B,F)-decision tree Γ *solves* a problem z over I if for each $a \in A$ there exists a complete path ξ in Γ such that $a \in A(\xi)$ and the terminal node of the path ξ is assigned the number $z(a)$.

For an arbitrary problem z over I we denote by $h_I(z)$ the minimal depth of an (B,F)-decision tree Γ which solves the problem z and for which $\mathrm{At}(\Gamma) \subseteq \mathrm{At}(z)$.

1.3 Relationships among Basic Notions

In this chapter we will discuss the local approach to the study of decision trees where for a problem $z = (\nu, f_1, \ldots, f_n)$ over $I = (A, B, F)$ the decision trees are considered using attributes from the set $\{f_1, \ldots, f_n\}$ only. The investigation of such decision trees is based on the study of the (B,F)-decision table $T(z)$ associated with the problem z. The table $T(z)$ has n columns. The row $(\delta_1, \ldots, \delta_n)$ is contained in the table $T(z)$ if and only if the equation system

$$\{f_1(x) = \delta_1, \ldots, f_n(x) = \delta_n\}$$

is compatible on the set A. This row is labeled by the number $\nu(\delta_1, \ldots, \delta_n)$. For $i = 1, \ldots, n$ the i-th column is labelled by the attribute f_i.

The use of decision tables in investigations of decision trees which solve problems is based on the following statement.

Proposition 1. ([Mo4]) *Let $I = (A, B, F)$ be an information system, z be a problem over I and Γ be an (B,F)-decision tree such that $\mathrm{At}(\Gamma) \subseteq \mathrm{At}(z)$. Then the decision tree Γ solves the problem z if and only if the decision tree Γ is suitable for the decision table $T(z)$.*

Corollary 2. ([Mo4]) *Let I be an information system and z be a problem over I. Then $h_I(z) = h(T(z))$.*

In proofs of results presented in this chapter methods of test theory, the groundwork for which was laid by [CY1, YC1] were used as well as methods of rough set theory created in [Pa1, Pa2, SR1, Sl1]. The mathematical apparatus of rough set theory is a generalization of the mathematical apparatus of test theory to the case of approximate problem statements. For exact statement of problems which will be studied in this chapter mathematical tools of considered theories almost coincide and are based on decision table use.

2 Algorithms of Decision Tree Construction for Decision Tables

In this section algorithms are discussed which for a given decision table construct a decision tree that is suitable for the table. Decision tables and decision trees are considered here as independent objects regardless of the notion of information system.

Let B be a finite nonempty set with at least two elements and F be some nonempty finite or denumerable set. We define the (B,F)-*Decision Tree Construction problem* as follows: for a given (B,F)-decision table T it is required to construct an (B,F)-decision tree with minimal depth which is suitable for the table T. It is not difficult to prove that the (B,F)-Decision Tree Construction problem is solvable and if F is a denumerable set then this problem is NP-hard (see, for example, [Mo4]).

We will not consider exact algorithms for solving of the (B,F)-Decision Tree Construction problem. This section deals with the algorithm U which for a given decision table $T \in \mathcal{T}(B,F)$ constructs the (B,F)-decision tree $U(T)$. The constructed tree is suitable for the decision table T but, possible, has the depth which is greater than $h(T)$. This algorithm resembles algorithms from [Qu1, Qu2] however it was proposed and investigated independently in [Mo1, Mo2].

2.1 Some Definitions

Let $T \in \mathcal{T}(B,F)$. Denote by $L(T)$ the cardinality of the set of numbers which are labels of rows of the table T. Denote by $N(T)$ the number of rows in the table T. Denote by $R(T)$ the number of non-ordered pairs of rows of the table T which are labeled by different numbers.

For any $f_i \in \mathrm{At}(T)$ and $\delta \in B$ denote by $T(f_i, \delta)$ the sub-table of the table T that contains only such rows of the table T which on the intersection with the column labeled by f_i have the element δ. It is clear that $T(f_i, \delta) \in \mathcal{T}(B,F)$.

2.2 Description of Algorithm U

We will use here a description of the algorithm U which is similar to the description from [Mo4]. Let us apply the algorithm U to a table $T \in \mathcal{T}(B,F)$.

1-st step. Construct a tree consisting of a single node w.

Let $L(T) \leq 1$. If $L(T) = 0$ then the node w will be labeled with the number 0. If $L(T) = 1$ then the node w will be labeled with the number which is the label of all rows of the table T. Proceed to the second step.

Let $L(T) \geq 2$. Label the node w with the table T and proceed to the second step.

Suppose $t \geq 1$ steps have already been made. The tree obtained in the step t will be denoted by G.

$(t + 1)$-*th step.* If no one node of the tree G is labeled with a table from $\mathcal{T}(B, F)$ then we denote the tree G by $U(T)$. The algorithm U operation is completed.

Otherwise we choose certain node w in the tree G which is labeled with a table from $\mathcal{T}(B, F)$. Let the node w be labeled with the table K.

If $L(K) = 1$ then replace the table K as the label of the node w with the number which is the label of all rows of the table K and proceed to the $(t+2)$-th step.

Let $L(K) \geq 2$. For every $f_i \in \text{At}(T)$ let $d(f_i) = \max\{R(K(f_i, \sigma)) : \sigma \in B\}$. Let $f_p \in \text{At}(T)$ and $d(f_p) = \min\{d(f_i) : f_i \in \text{At}(T)\}$. Assign the element f_p the node w as label instead of the table K. For every $\delta \in B$ such that the table $K(f_p, \delta)$ has at least one row add the node $w(\delta)$ to the tree G and draw the edge from the node w to the node $w(\delta)$. This edge will be labeled with the element δ while the node $w(\delta)$ will be assigned the table $K(f_p, \delta)$ Proceed to the $(t+2)$-th step.

Proposition 3. ([Mo4]) *For an arbitrary table $T \in \mathcal{T}(B, F)$ the work of the algorithm U is completed in a finite number of steps. The constructed tree $U(T)$ is an (B, F)-decision tree which is suitable for the decision table T.*

2.3 Complexity Bounds for Algorithm U

Denote by $C_U^{step}(T)$ the number of steps made by the algorithm U to construct the decision tree $U(T)$ for a given decision table T.

Consider bounds for the value $C_U^{step}(T)$ which immediately follow from the statement and from the proof of Theorem 2.4.4 of [Mo4].

Theorem 4. *Let $T \in \mathcal{T}(B, F)$. Then the following inequalities hold:*

$$L(T) \leq C_U^{step}(T) \leq 2N(T) + 2 .$$

We will not analyze in detail the algorithm U complexity. Note only that the algorithm U has polynomial time complexity.

2.4 Accuracy Bounds for Algorithm U

The upper bound on the complexity of the decision trees constructed by the algorithm U is considered in the following statement.

Theorem 5. ([Mo2]) *Let $T \in \mathcal{T}(B, F)$. Then the following inequality holds:*

$$h(U(T)) \leq \begin{cases} h(T), & if\ h(T) \leq 1\ ; \\ h(T)(\ln R(T) - \ln h(T) + 1), & if\ h(T) \geq 2\ . \end{cases}$$

The next statement allows us to estimate the quality of the bound from Theorem 5. Denote $\Pi(B, F) = \{(h(T), R(T)) : T \in \mathcal{T}(B, F)\}$.

Theorem 6. ([Mo2]) *Let F be a denumerable set. Then $\Pi(B, F) = \{(0, 0)\} \cup \{(m, r) : m, r \in \omega \setminus \{0\}, m \leq r\}$ and for any pair $(m, r) \in \Pi(B, F)$ there exists a table $T(m, r) \in \mathcal{T}(B, F)$ such that $h(T(m, r)) = m$, $R(T(m, r)) = r$ and the following inequality holds:*

$$h(U(T(m,r))) \geq \begin{cases} m, & if\ m < 2\ or\ r < 3m\ , \\ \lfloor (m-1)(\ln r - \ln 3m) \rfloor + m, & if\ m \geq 2\ and\ r \geq 3m\ . \end{cases}$$

Theorem 6 implies that the bound of Theorem 5 doesn't allow essential improvement in general case. Also from Theorem 6 follows that if F is a denumerable set then there is no function $f : \omega \to \omega$ such that $h(U(T)) \leq f(h(T))$ for any table $T \in \mathcal{T}(B, F)$.

2.5 On Unimprovability of Algorithm U

In this subsection we repeat some reasoning from [Mo11].

Let S be a set of N points and $\mathcal{F} = \{S_1, \ldots, S_m\}$ be a collection of subsets of S. *Set Covering problem* is the problem of selecting as few as possible subsets from $\mathcal{F}$ such that every point in S is contained in at least one of the selected subsets. This problem is NP-hard.

In [LY1] it was proved that for each $0 < c < 1/4$ the Set Covering problem cannot be approximated within factor of $c \log_2 N$ in polynomial time unless $NP \subset DTIME(n^{poly \log_2 n})$. In [Fe1] it was proved that for each $\varepsilon > 0$ the Set Covering problem cannot be approximated within factor of $(1 - \varepsilon) \ln N$ in polynomial time unless $NP \subset DTIME(n^{O(\log_2 \log_2 n)})$.

Let F be a denumerable set. Using mentioned result from [Fe1] it is not difficult to prove that for each $\varepsilon > 0$ the (B, F)-Decision Tree Construction problem cannot be approximated within factor of $(1 - \varepsilon) \ln R(T)$ in polynomial time unless $NP \subset DTIME(n^{O(\log_2 \log_2 n)})$.

Taking into account that the algorithm U has polynomial time complexity and using Theorem 5 we obtain that unless $NP \subset DTIME(n^{O(\log_2 \log_2 n)})$ then the algorithm U is close to unimprovable approximate polynomial algorithms for the (B, F)-Decision Tree Construction problem solving.

3 Problem of Decision Tree Local Optimization

Beginning with this section we will consider the notions of decision table and decision tree only with respect to some information system. In this section a criterion of solvability of the problem of decision tree local optimization is considered and the situation when this problem is unsolvable is discussed.

3.1 Criterion of Solvability of Optimization Problem

Let $I = (A, B, F)$ be an information system and z be a problem over I. Recall that by $h_I(z)$ we denote the minimal depth of an (B, F)-decision tree Γ which solves the problem z and for which $\text{At}(\Gamma) \subseteq \text{At}(z)$.

Now we define two algorithmic problems: the problem of decision tree local optimization $\text{Opt}(I)$ and the problem of compatibility of equation systems $\text{Com}(I)$.

The problem $\text{Opt}(I)$: for a given problem $z \in P(I)$ it is required to find an (B, F)-decision tree Γ which solves the problem z and for which $\text{At}(\Gamma) \subseteq \text{At}(z)$ and $h(\Gamma) = h_I(z)$.

The problem $\text{Com}(I)$: for given attributes $f_1, \ldots, f_n \in F$ and elements $\delta_1, \ldots, \delta_n \in B$ it is required to determine whether the system of equations

$$\{f_1(x) = \delta_1, \ldots, f_n(x) = \delta_n\}$$

is compatible on the set A.

In the following theorem the criterion of the problem $\text{Opt}(I)$ solvability is considered.

Theorem 7. ([Mo4]) *Let I be an information system. Then the problem $\text{Opt}(I)$ is solvable if and only if the problem $\text{Com}(I)$ is solvable.*

Note that the problem $\text{Com}(I)$ is solvable if and only if the following problem is solvable: for a given problem z over I it is required to construct the decision table $T(z)$. So if the problem $\text{Com}(I)$ is solvable then we can use the following way for construction of decision trees: from a problem z over I to the table $T(z)$ and then to a decision tree Γ which is suitable for $T(z)$. From Proposition 1 follows that Γ solves z, and from Corollary 2 follows that if $h(\Gamma) = h(T(z))$ then Γ is a solution of the problem $\text{Opt}(I)$ for z.

3.2 Extensions of Information Systems

All the algorithms of decision tree construction considered in this chapter are applicable only to information systems I for which the problem $\text{Com}(I)$ is solvable. In what follows the notion of an information system extension is introduced which allows to bypass, in some sense, this restriction.

Let $I' = (A', B, F)$ be an information system, $A \subseteq A'$, $A \neq \emptyset$ and $I = (A, B, F)$. The information system I' will be called *an extension* of the information system I Evidently if an (B, F)-decision tree Γ solves a problem z over I' then the decision tree Γ solves the same problem z but over I.

The following statement is a simple consequence of Proposition 3.4.3 from [Mo4].

Proposition 8. *For an arbitrary information system I there exists an extension I' such that the problem $\text{Com}(I')$ is solvable.*

Let I be an information system such that the problem $\mathrm{Com}(I)$ is unsolvable, and I' be an extension of the information system I such that the problem $\mathrm{Com}(I')$ is solvable. Then we can use algorithms of decision tree construction for the information system I' and transfer the obtained results on the information system I.

3.3 Examples

Let $n \in \omega \setminus \{0\}$, $\bar{x} = (x_1, \ldots, x_n)$ and let $\{p_i(\bar{x}) : i \in \omega\}$ be the set of all polynomials which have integer coefficients and depend on the variables $x_1, \ldots, x_n$. Denote $F_n = \{f_i(\bar{x}) : i \in \omega\}$ where for any $i \in \omega$ and $\bar{a} \in \mathbb{R}^n$ let $f_i(\bar{a}) = 0$ if $p_i(\bar{a}) = 0$, and $f_i(\bar{a}) = 1$ if $p_i(\bar{a}) \neq 0$.

Example 1. From results of [Ma1] follows that there exists $m \in \omega$ such that for any $n > m$ for the information system $I_n(\mathbb{Z}) = (\mathbb{Z}^n, \{0, 1\}, F_n)$ the problem $\mathrm{Com}(I_n(\mathbb{Z}))$ is unsolvable.

Example 2. From results of [Ta1] follows that for any $n \in \omega \setminus \{0\}$ for the information system $I_n(\mathbb{R}) = (\mathbb{R}^n, \{0, 1\}, F_n)$ the problem $\mathrm{Com}(I_n(\mathbb{R}))$ is solvable. Obviously the information system $I_n(\mathbb{R})$ is an extension of the information system $I_n(\mathbb{Z})$.

4 Cardinality Characteristics of Decision Tables

Decision tables play the important role in algorithms of decision tree construction. In this section upper and lower bounds on the number of rows in decision tables are considered. These bounds will be used further for estimation of complexity and accuracy of algorithms. The results contained in this section are like those obtained in [Al1, Sa1, Sh1]. The notions used here are similar to the Vapnik-Chervonenkis dimension [BEHW1, VC1].

Let $I = (A, B, F)$ be an information system and $z = (\nu, f_1, \ldots, f_n)$ be a problem over I. Recall that by $N(T(z))$ we denote the number of rows in the table $T(z)$. Now we define the parameter $V(z)$ of the problem z. If $N(T(z)) = 1$ then $V(z) = 0$. Let now $N(T(z)) \geq 2$. Then $V(z)$ is the maximal number $m \in \{1, \ldots, n\}$ for which there exist attributes $f_{i_1}, \ldots, f_{i_m} \in \{f_1, \ldots, f_n\}$ and two-element subsets $B_1, \ldots, B_m$ of the set B such that for any $\delta_1 \in B_1, \ldots, \delta_m \in B_m$ the following system of equations is compatible on the set A:

$$\{f_{i_1}(x) = \delta_1, \ldots, f_{i_m}(x) = \delta_m\} .$$

Theorem 9. ([Mo4]) *Let $I = (A, B, F)$ be an information system, $k = |B|$ and z be a problem over I. Then the following inequalities hold:*

$$2^{V(z)} \leq N(T(z)) \leq (k^2 \dim z)^{V(z)} .$$

Now we consider the behavior of the function $N_I : \omega \setminus \{0\} \to \omega$ defined as follows:

$$N_I(n) = \max\{N(T(z)) : z \in P(I), \dim z \leq n\} .$$

Theorem 10. ([Mo4]) *Let $I = (A, B, F)$ be an information system and $k = |B|$. Then the following statements hold:*
 a) if there exists a constant $c \in \omega$ such that $V(z) \leq c$ for any problem $z \in P(I)$ then $N_I(n) \leq (k^2 n)^c$ for any $n \in \omega \setminus \{0\}$;
 b) if there exists no constant $c \in \omega$ such that $V(z) \leq c$ for any problem $z \in P(I)$ then $N_I(n) \geq 2^n$ for any $n \in \omega \setminus \{0\}$.

Let as consider an example.

Example 3. Define the function sign : $\mathbb{R} \to \{-1, 0, 1\}$ as follows: for any $a \in \mathbb{R}$ let $\text{sign}(a) = -1$ if $a < 0$, $\text{sign}(a) = 0$ if $a = 0$, and $\text{sign}(a) = 1$ if $a > 0$. Let $m, t \in \omega \setminus \{0\}$, $\bar{x} = (x_1, \ldots, x_m)$, $\{p_i(\bar{x}) : i \in \omega\}$ be the set of all polynomials which have integer coefficients and depend on variables $x_1, \ldots, x_m$, and let $\{q_i(\bar{x}) : i \in \omega\}$ be the set of all polynomials of the degree not exceeding t which have integer coefficients and depend on variables $x_1, \ldots, x_m$. Denote $F(m) = \{\text{sign}(p_i(\bar{x})) : i \in \omega\}$ and $F(m, t) = \{\text{sign}(q_i(\bar{x})) : i \in \omega\}$. Define two information systems as follows: $I(m) = (\mathbb{R}^m, \{-1, 0, 1\}, F(m))$ and $I(m, t) = (\mathbb{R}^m, \{-1, 0, 1\}, F(m, t))$. It was shown in [Mo4] that the function V is unbounded above on the set of problems over $I(m)$ but it is bounded above by a constant on the set of problems over $I(m, t)$.

5 Algorithms of Decision Table Construction

This section deals with algorithms of decision table construction and their complexity characteristics.

5.1 Description of Algorithms of Decision Table Construction

Let $I = (A, B, F)$ be an information system for which the problem $\text{Com}(I)$ is solvable and let J be an algorithm which solves the problem $\text{Com}(I)$.

Describe the algorithm WJ [Mo4] which for an arbitrary problem $z \in P(I)$ constructs the decision table $T(z)$. Let $z = (\nu, f_1, \ldots, f_n)$.

1-st step. Construct the tree containing only one node. Label this node with the empty system of equations and proceed to the second step.

Suppose t steps have already been done. Denote by D the labeled finite tree with the root built on the step t.

$(t+1)$-th step. Let every terminal node in the tree D be labeled with an n-tuple from B^n. Define the table $T(z)$ as follows. The set of rows of the table $T(z)$ coincides with the set of n-tuples attached to terminal nodes of the tree D as

labels. A row $(\delta_1, \ldots, \delta_n)$ is labeled by the number $\nu(\delta_1, \ldots, \delta_n)$. For $i = 1, \ldots, n$ the i-th column is labeled by the attribute f_i. The algorithm WJ operation is completed.

Suppose not all terminal nodes in the tree D are labeled with n-tuples from B^n. Choose a terminal node w in the tree D with certain equation system $\Sigma = \{f_1(x) = \delta_1, \ldots, f_r(x) = \delta_r\}$ attached to it as the label (if Σ is the empty system then $r = 0$). Let $r = n$. Replace the system Σ as the label of the node w with the n-tuple $(\delta_1, \ldots, \delta_n)$ and proceed to the $(t+2)$-th step. Let $r < n$. By applying $|B|$ times the algorithm J we construct the set $B(\Sigma)$ which consists of those and only those $\delta \in B$ such that the system of equations $\Sigma \cup \{f_{r+1}(x) = \delta\}$ is compatible on the set A. Erase the label Σ at the node w. For every $\delta \in B(\Sigma)$ add to the tree D a node $w(\delta)$. Draw the edge from the node w to the node $w(\delta)$ and assign the node $w(\delta)$ the system of equations $\Sigma \cup \{f_{r+1}(x) = \delta\}$ as the label. Proceed to the $(t+2)$-th step.

5.2 Complexity Bounds for Algorithm WJ

For $z \in P(I)$ denote by $C_{WJ}^{step}(z)$ the number of steps made by the algorithm WJ to construct the table $T(z)$, and by $C_{WJ}^{call}(z)$ we denote the number of calls of the algorithm J by the algorithm WJ in the process of the table $T(z)$ construction.

Consider bounds for the values $C_{WJ}^{step}(z)$ and $C_{WJ}^{call}(z)$.

Theorem 11. ([Mo4]) *Let $I = (A, B, F)$ be an information system for which the problem $\mathrm{Com}(I)$ is solvable, J be an algorithm which solves the problem $\mathrm{Com}(I)$ and let $k = |B|$. Then the inequalities*

$$2 + N(T(z)) \leq C_{WJ}^{step}(z) \leq 2 + (\dim z + 1)N(T(z)),$$
$$N(T(z)) \leq C_{WJ}^{call}(z) \leq k(\dim z + 1)N(T(z))$$

hold for any problem $z \in P(I)$.

Consider dependencies of values $C_{WJ}^{step}(z)$ and $C_{WJ}^{call}(z)$ on $\dim z$ for problems $z \in P(I)$ in the worst case. To this end define the functions $C_{WJ,I}^{step} : \omega \setminus \{0\} \to \omega$ and $C_{WJ,I}^{call} : \omega \setminus \{0\} \to \omega$ as follows:

$$C_{WJ,I}^{step}(n) = \max\{C_{WJ}^{step}(z) : z \in P(I), \dim z \leq n\},$$
$$C_{WJ,I}^{call}(n) = \max\{C_{WJ}^{call}(z) : z \in P(I), \dim z \leq n\}.$$

The proof of the following theorem is based on bounds of Theorems 10 and 11.

Theorem 12. ([Mo4]) *Let $I = (A, B, F)$ be an information system for which the problem $\mathrm{Com}(I)$ is solvable, J be an algorithm which solves the problem $\mathrm{Com}(I)$ and let $k = |B|$. Then the following statements hold:*

a) if there exists a constant $c \in \omega$ such that $V(z) \leq c$ for any problem $z \in P(I)$ then $C_{WJ,I}^{step}(n) \leq 2 + k^{2c}n^c(n+1)$ and $C_{WJ,I}^{call}(n) \leq k^{2c+1}n^c(n+1)$ for any $n \in \omega \setminus \{0\}$;

b) if there doesn't exist a constant $c \in \omega$ such that $V(z) \leq c$ for any problem $z \in P(I)$ then $C_{WJ,I}^{step}(n) \geq 2^n$ and $C_{WJ,I}^{call}(n) \geq 2^n$ for any $n \in \omega \setminus \{0\}$.

6 Algorithms of Decision Tree Construction for Problems

Let $I = (A, B, F)$ be an information system for which the problem $\mathrm{Com}(I)$ is solvable and J be an algorithm which solves the problem $\mathrm{Com}(I)$.

The problem $\mathrm{Opt}(I)$ may be solve in the following way: for a given problem z over I we construct the decision table $T(z)$ by the algorithm WJ, then using some algorithm that solves the (B, F)-Decision Tree Construction problem we construct an (B, F)-decision tree Γ which is suitable for the table $T(z)$ and for which $h(\Gamma) = h(T(z))$. Using Proposition 1 and Corollary 2 we obtain that Γ is a solution of the problem $\mathrm{Opt}(I)$ for z.

We will not consider more explicitly exact algorithms for the problem $\mathrm{Opt}(I)$ solving. This section is devoted to analysis of the algorithm UJ [Mo4] which for an arbitrary problem $z \in P(I)$ constructs the (B, F)-decision tree $UJ(z)$ with the following properties: the decision tree $UJ(z)$ solves the problem z and $\mathrm{At}(UJ(z)) \subseteq \mathrm{At}(z)$. It is possible that $h(UJ(z)) > h_I(z)$.

The algorithm UJ operates as follows.

For $z \in P(I)$ construct the decision table $T(z)$ by the algorithm WJ. Next construct the decision tree $U(T(z))$ by the algorithm U. Then $UJ(z) = U(T(z))$.

6.1 Complexity Bounds for Algorithm UJ

Complexity parameters of algorithm WJ which for a given problem $z \in P(I)$ constructs the decision table $T(z)$ have been investigated in Sect. 5. The present subsection deals with bounds on the value $C_U^{step}(T(z))$ which is the number of steps made by the algorithm U in the process of the decision tree $U(T(z))$ construction.

Consider a dependence of the value $C_U^{step}(T(z))$ on $\dim z$ for problems $z \in P(I)$ in the worst case. To this end define the function $C_{U,I}^{step} : \omega \setminus \{0\} \to \omega$ as follows:

$$C_{U,I}^{step}(n) = \max\{C_U^{step}(T(z)) : z \in P(I), \dim z \leq n\}\ .$$

The proof of the following theorem is based on bounds from Theorems 4 and 10.

Theorem 13. ([Mo4]) *Let $I = (A, B, F)$ be an information system for which the problem $\mathrm{Com}(I)$ is solvable and let $k = |B|$. Then the following statements hold:*

a) if there exists a constant $c \in \omega$ such that $V(z) \leq c$ for any problem $z \in P(I)$ then $C_{U,I}^{step}(n) \leq 2(k^2 n)^c + 2$ for any $n \in \omega \setminus \{0\}$;

b) if there doesn't exist a constant $c \in \omega$ such that $V(z) \leq c$ for any problem $z \in P(I)$ then $C_{U,I}^{step}(n) \geq 2^n$ for any $n \in \omega \setminus \{0\}$.

6.2 Accuracy Bounds for Algorithm UJ

At first we consider accuracy bounds for the algorithm UJ in the case when there exists a constant $c \in \omega$ such that $V(z) \leq c$ for any problem $z \in P(I)$. The next statement follows from Corollary 2 and Theorems 5 and 9.

Theorem 14. *Let $I = (A, B, F)$ be an information system for which the problem $\text{Com}(I)$ is solvable and there exists a constant $c \in \omega$ such that $V(z) \leq c$ for any problem $z \in P(I)$. Let J be an algorithm which solves the problem $\text{Com}(I)$ and let $k = |B|$. Then for any problem z over the information system I the following inequality holds:*

$$h(UJ(z)) \leq \begin{cases} h_I(z), & \text{if } h_I(z) \leq 1, \\ h_I(z)(2c\ln(k^2 \dim z) - \ln h_I(z) + 1), & \text{if } h_I(z) \geq 2. \end{cases}$$

Now we consider two functions $S_I : \omega \setminus \{0\} \to \omega$ and $S_I^{UJ} : \omega \setminus \{0\} \to \omega$ defined as follows:

$$S_I(n) = \max\{h_I(z) : z \in P(I), \dim z \leq n\},$$
$$S_I^{UJ}(n) = \max\{h(UJ(z)) : z \in P(I), \dim z \leq n\}.$$

Comparison of functions S_I and S_I^{UJ} allows to estimate the accuracy of the algorithm UJ for the information system I.

We will say that the information system $I = (A, B, F)$ satisfies *the condition of reduction* if there exists a number $m \in \omega \setminus \{0\}$ such that for each compatible on A system of equations $\{f_1(x) = \delta_1, \ldots, f_r(x) = \delta_r\}$ where $r \in \omega \setminus \{0\}$, $f_1, \ldots, f_r \in F$ and $\delta_1, \ldots, \delta_r \in B$ there exists a subsystem of this system which has the same set of solutions and contains at most m equations.

We will say that I is *a finite* information system if F is a finite set, and we will say that I is *an infinite* information system if F is an infinite set.

The following theorem is a simple consequence of Theorems 3.3.1 and 3.7.2 and Lemma 3.3.7 from [Mo4].

Theorem 15. *Let $I = (A, B, F)$ be an information system for which the problem $\text{Com}(I)$ is solvable, J be an algorithm which solves the problem $\text{Com}(I)$ and let $k = |B|$. Then the following statements hold:*

a) if I is a finite information system then there exists a constant $c_1 \in \omega \setminus \{0\}$ such that $S_I(n) \leq S_I^{UJ}(n) \leq c_1$ for any $n \in \omega \setminus \{0\}$;

b) if I is an infinite information system which satisfies the condition of reduction then there exist constants $c_2, c_3 \in \omega \setminus \{0\}$ such that $\log_k n \leq S_I(n) \leq S_I^{UJ}(n) \leq c_2 \log_k n + c_3$ for any $n \in \omega \setminus \{0\}$;

c) if I is an infinite information system which does not satisfy the condition of reduction then $S_I(n) = S_I^{UJ}(n) = n$ for any $n \in \omega \setminus \{0\}$.

Example 4. Let us consider the same information systems $I(m)$ and $I(m, t)$ as in Example 3. One can show that $S_{I(m)}(n) = S_{I(m)}^{UJ}(n) = n$ for any $n \in \omega \setminus \{0\}$, $S_{I(1,1)}(n) = \Theta(\log_2 n)$ and $S_{I(1,1)}^{UJ}(n) = \Theta(\log_2 n)$, and if $m > 1$ or $t > 1$ then $S_{I(m,t)}(n) = S_{I(m,t)}^{UJ}(n) = n$ for any $n \in \omega \setminus \{0\}$.

7 Conclusion

We considered some results obtained in the frameworks of the local approach to decision tree investigation and connected with problems of decision tree construction. The detailed discussion of bounds on time complexity of decision trees in the case of the local approach may be found in [Mo4, Mo9].

Problems of the global approach to decision tree investigation are essentially more complicated. Some results in this field relating to bounds on complexity and algorithms for construction of decision trees may be found in [Mo3, Mo4, Mo7, Mo8, Mo10, Mo12].

Acknowledgments

We would like to thank the anonymous referee for useful comments.

This work was partially supported by Russian Foundation of Fundamental Research (grants # 93-01-00488 and # 96-01-00428) and by Russian Federal Purposeful Program "Integration" (project Educational-Research Center "Methods of Discrete Mathematics for New Information Technologies").

References

[Al1] Alexeyev, V.E.: On entropy of two-dimensional fragmentary closed languages. Kombinatorno-algebraicheskiye Metody i ikh Primeneniye. Edited by Al.A. Markov. Gorky University Publishers, Gorky (1987) 5–13 (in Russian)

[BEHW1] Blumer, A., Ehrenfeucht, A., Haussler, D., Warmuth, M.: Learnability and the Vapnik-Chervonenkis dimension. J. ACM **36**(4) (1989) 929–965

[CY1] Chegis, I.A., Yablonskii, S.V.: Logical methods of electric circuit control. Trudy MIAN SSSR **51** (1958) 270–360 (in Russian)

[Fe1] Feige, U.: A threshold of $\ln n$ for approximating set cover (Preliminary version). Proceedings of 28th Annual ACM Symposium on the Theory of Computing (1996) 314–318

[LY1] Lund, C., Yannakakis, M.: On the hardness of approximating minimization problems. J. ACM **41**(5) (1994) 960–981

[Ma1] Matiyasevich, Ju.V.: Diophantinity of enumerable sets. DAN SSSR **191**(2) (1970) 279–382 (in Russian)

[Mo1] Moshkov, M.Ju.: On conditional tests. DAN SSSR **265**(3) (1982) 550–552 (in Russian)

[Mo2] Moshkov, M.Ju.: Conditional tests. Problemy Kybernetiki **40**. Edited by S.V. Yablonskii. Nauka Publishers, Moscow (1983) 131–170 (in Russian)

[Mo3] Moshkov, M.Ju.: Optimization problems for decision trees. Fundamenta Informaticae **21** (1994) 391–401

[Mo4] Moshkov, M.Ju.: Decision Trees. Theory and Applications. Nizhni Novgorod University Publishers, Nizhni Novgorod, 1994 (in Russian)

[Mo5] Moshkov, M.Ju.: Two approaches to investigation of deterministic and nondeterministic decision tree complexity. Proceedings of the World Conference on the Fundamentals of AI. Paris, France (1995)

[Mo6] Moshkov, M.Ju.: Comparative analysis of deterministic and nondeterministic decision tree complexity. Global approach. Fundamenta Informaticae **25** (1996) 201–214

[Mo7] Moshkov, M.Ju.: On the depth of decision trees over infinite information systems. Proceedings of the Congress "Information Processing and Management of Uncertainty in Knowledge-Based Systems". Granada, Spain (1996) 885–886

[Mo8] Moshkov, M.Ju.: On global Shannon functions of two-valued information systems. Proceedings of the Fourth International Workshop on Rough Sets, Fuzzy Sets and Machine Discovery. Tokyo, Japan (1996) 142–143

[Mo9] Moshkov, M.Ju.: Unimprovable upper bounds on complexity of decision trees over information systems. Foundations of Computing and Decision Sciences **21**(4) (1996) 219–231

[Mo10] Moshkov, M.Ju.: On complexity of decision trees over infinite information systems. Proceedings of the Third Joint Conference on Information Sciences. USA, Duke University (1997) 353–354

[Mo11] Moshkov, M.Ju.: Algorithms for constructing of decision trees. Proceedings of the First European Symposium Principles of Data Mining and Knowledge Discovery, Trondheim, Norway, June 1997. Lecture Notes in Artificial Intelligence **1263**. Edited by J. Komorowski and J. Zytkow. Springer Verlag, Berlin (1997) 335–342

[Mo12] Moshkov, M.Ju.: Unimprovable upper bounds on time complexity of decision trees. Fundamenta Informaticae **31** (1997) 157–184

[Pa1] Pawlak, Z.: Information Systems - Theoretical Foundations. PWN, Warsaw, 1981 (in Polish)

[Pa2] Pawlak, Z.: Rough Sets - Theoretical Aspects of Reasoning about Data. Kluwer Academic Publishers, Dordrecht, Boston, London, 1991

[Qu1] Quinlan, J.R.: Discovering rules by induction from large collections of examples. Experts Systems in the Microelectronic Age. Edited by D. Michie. Edinburg University Press (1979)

[Qu2] Quinlan, J.R.: Induction of decision trees. Machine Learning **1**(1) (1986) 81–106

[Sa1] Sauer, N.: On the density of families of sets. J. of Combinatorial Theory (A) **13** (1972) 145–147

[Sh1] Shelah, S.: A combinatorial problem; stability and order for models and theories in infinitary languages. Pacific J. of Mathematics **41** (1972) 241–261

[SR1] Skowron, A., Rauszer, C.: The discernibility matrices and functions in information systems. Intelligent Decision Support. Handbook of Applications and Advances of the Rough Set Theory. Edited by R. Slowinski. Kluwer Academic Publishers, Dordrecht, Boston, London (1992) 331–362

[Sl1] Slowinski, R. (Ed.): Intelligent Decision Support. Handbook of Applications and Advances of the Rough Set Theory. Kluwer Academic Publishers, Dordrecht, Boston, London, 1992

[Ta1] Tarski, A.: Arithmetical classes and types of mathematical systems, Mathematical aspects of arithmetical classes and types, Arithmetical classes and types of Boolean algebras, Arithmetical classes and types of algebraically closed and real closed fields. Bull. Amer. Math. Soc. **55** (1949) 63–64

[VC1] Vapnik, V.N., Chervonenkis, A.Ya.: On the uniform convergence of relative frequencies of events to their probabilities. Theory of Probability and its Applications **16**(2) (1971) 264–280

[YC1] Yablonskii, S.V., Chegis, I.A.: On tests for electric circuits. UMN **10**(4) (1955) 182–184 (in Russian)

Rough-Fuzzy Theory

Shadowed Sets: Bridging Fuzzy and Rough Sets

Witold Pedrycz

Department of Electrical and Computer Engineering,
University of Manitoba, Winnipeg R3T 2N2 Canada
pedrycz@ee.umanitoba.ca

Abstract. The study introduces a new concept of shadowed sets. They can be regarded as a certain operational framework simplifying processing carried out with the aid of fuzzy sets and enhancing interpretation of results derived in such a way. Some conceptual links between this idea and some others known in the literature are established. In particular, it is demonstrated how fuzzy sets can induce shadowed sets. Subsequently, shadowed sets reveal interesting conceptual and algorithmic relationships existing between rough sets and fuzzy sets. It is also studied how different ways Detailed computational aspects of shadowed sets are discussed. Several illustrative examples are provided. Additionally, we re-examine on the ideas of approximate reasoning and mapping of information granules

Keywords. Fuzzy sets, shadowed sets, rough sets, design frameworks of granular computing, three - valued logic, vagueness, decision - making, fuzzy clustering,

1 Introduction

Fuzzy sets are regarded as one of the formal vehicles devoted to capture, represent and processing vagueness [2] [4] [5] and thus allow to cope with diverse phenomena exhibiting unclearly defined boundaries. Do fuzzy sets live up to such expectations? The answer is affirmative yet some outstanding questions remain. When it comes to representing vagueness, fuzzy sets tend to capture this exclusively via membership functions that are mappings from a given universe of discourse to a unit interval containing membership values. As always emphasized in the literature, a membership grade indicates an extent to which a given point in the universe of discourse belongs to a concept we are about to represent. Once the membership function has been established (estimated or defined), the concept is described very precisely as the membership values are exact numerical quantities. This seems to raise a certain dilemma of excessive *precision* in describing *imprecise* phenomena. In fact, this concern has already sparked a lot of debates starting from the very inception of fuzzy sets.

The study is devoted to a certain different model of vagueness called shadowed sets which does not lend itself to precise numerical membership values but relies on basic concepts of truth values (yes - no) and an entire unit interval perceived as a zone of uncertainty.

Fuzzy sets are regarded as an emerging point of the ensuing discussion; it is revealed how shadowed sets can be induced by fuzzy sets by suppressing all detailed numerical information about membership values inherently associated with the latter. First, we introduce a constructive way in which shadowed sets are developed based on fuzzy sets (Section 2). In the sequel (Section 3), we look into the problem of a formal definition and basic operations on shadowed sets. Finally, in Section 4, we discuss several examples highlighting the use of shadowed sets in decision making, clustering, and image analysis.

2 From fuzzy sets to shadowed sets

Consider the problem of representing (or approximating) a fuzzy set by a less precise and rough construct which does not require so much precision and calls for less computational effort. Several ways could be pursued; these have been reported in the existing literature. In particular, consider the following scenarios:

a. Any fuzzy set can be approximated by a certain α - cut [4][5], Figure 1, meaning that all elements that belong to the fuzzy sets to a membership degree less than α are dropped whereas all these with the membership above this threshold are admitted to the resulting set. The appealing question is then the one about optimality of a specific level of the threshold (α).

b. One can restrict to a so-called level fuzzy set [9]. The intent of this construct is to preserve only the most significant membership values and eliminate those lying below the threshold level and therefore being perceived as completely meaningless. This modification reduces an amount of computing, and when tailored to problems of information retrieval allows to filter out a vast number of otherwise irrelevant records resulting as a response to a fuzzy query. Again, the same problem of selecting α remains open - the lower the value of α, the more original information present in A becomes retained.

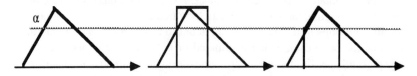

Figure 1. Fuzzy set and its a-cut and corresponding level fuzzy set

Notice that in both these cases a fuzzy set is replaced by another set or a part of the original fuzzy set resulting in a purely numeric structure.

It is interesting to underline that a purely numeric format of representing fuzzy sets has raised some concerns in the past and, in sequel, has triggered a search for some other models that are less numeric and precise. As already mentioned, this trend of questioning excessive precision of fuzzy sets has been motivated by the conceptual shortcoming associated with precise numeric values of membership used to describe vague concepts. The justification of this concern is also on an experimental side. Along this line let us recall several enhancements such as fuzzy sets of type -2 [6], interval valued fuzzy sets [11], and probabilistic sets [3], in particular. The underlying premise is that the grades of membership themselves are rather fuzzy sets, sets, or truncated random variables all being defined in the unit interval. In studies on probabilistic sets [3] (even though they originate on a somewhat different conceptual and computational ground), it has been found that most of uncertainty in the determination of the membership values is associated with those grades situated *around* 0.5. This finding is quite appealing. In contrast, we are usually far more confident about assigning values close 1 (thus counting the elements in) or 0 (therefore making the corresponding element excluded from the concept). On the other hand, the membership values (such as those *around* 0.5) always spark some hesitation and are always more difficult to place on a simple numeric scale. This observation forms a cornerstone of the model to be developed in this study. Consider a fuzzy set A. We elevate some membership values (usually those that are high enough) and reduce those that are viewed as substantially low. The elevation and reduction mechanisms are quite a radical as we go for 1 and 0, respectively. In other words, we can say that by doing that we eliminate (disambiguate) the original concept described by the fuzzy set. As this reduces vagueness we make some extra provisions for maintaining the overall level of vagueness constant by allowing some other regions of the universe of discourse associated with intermediate membership values that are defined in a more relaxed way or, simply, to let them be totally undefined. To do that, rather than attaching there a single specific membership value, we assign a unit interval that could be regarded as a nonnumeric model of membership grade. Figure 2 summarizes the proposed construct - the shadowed set is induced from the fuzzy set by accepting a specific threshold level. Observe that this development produces an effect of vagueness allocation as the regions of vagueness (unit intervals) are allocated to some regions of **X** rather than to the entire space as encountered in fuzzy sets. In essence, we have transformed the fuzzy set to a set with some clearly marked vagueness zones or, put it more descriptively, shadows. Note that the result is a structure that maps **X** to 0, 1, and [0,1]. We call this concept a shadowed set and define it as follows

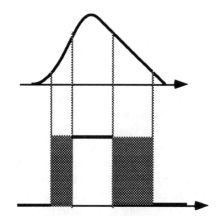

Figure 2. Fuzzy set and induced shadowed set

$$A : X \rightarrow \{ 0, 1, [0,1] \}$$
(1)

The elements of **X** for which $\mathbb{A}$ attains 1 constitute its core while the elements where $\mathbb{A}(x) = [0,1]$ form a shadow of this construct. One can envision some particular cases such as a shadowed set without any core (only shadow available) and shadowed sets with nonexistent shadows.

To proceed with more computational issues, we address a balance of vagueness. As some of the regions come with elevated or reduced membership values (1 and 0), this process should happen at an expense of increased uncertainty in the intermediate membership values, namely an introduction of unit intervals to be distributed across some ranges of **X**. As outlined in Figure 3, we study the areas below the membership function and these need to be balanced by selecting a suitable threshold α meaning that the following relationship holds

$$V = \left| \int_{-\infty}^{a_1} A(x)dx + \int_{a_2}^{+\infty} (1-A(x))dx - \int_{a_1}^{a_2} A(x)dx \right|$$
(2)

i.e.,

$$\Omega_1 + \Omega_2 = \Omega_3$$

In other words, the threshold $\alpha \in [0, 1/2)$ should lead to $V(\alpha) = 0$.

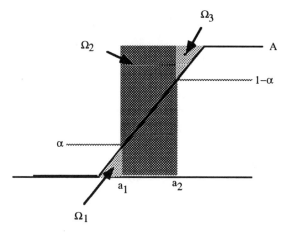

Figure 3. Development of a shadowed set

Shadowed sets exhibit some interesting conceptual links with the existing concepts, especially interval valued sets [11] and rough sets [7]. With respect to the first class, shadowed sets are somewhat subsumed by them. The important operational difference lies in the way in which these concepts have been formed. Interval - valued fuzzy sets are developed independently from fuzzy sets and, by no means are implied by them, whereas shadowed sets, as shown above, can be directly implied (induced) by fuzzy sets. Conceptually, shadowed sets are conceptually close to rough sets even though the mathematical foundations of these latter are very different. In rough sets we distinguish between three regions [7]:
- the regions whose elements are fully accepted (membership value equal 1) and belonging to the concept under discussion,
- the regions whose elements definitely do not belong to the concept
- the regions where membership grade is doubtful - these come in the form of the shadows of the introduced shadowed sets

In this sense shadowed sets narrow down a conceptual and an algorithmic gap between fuzzy sets and rough sets highlighting how these could be directly related. There is some significant difference. In rough sets the approximation space is defined in advance and the equivalence classes are kept fixed [7]. In the concept of shadowed sets these classes are assigned dynamically.

Two points are worth underlining in the setting established so far:
-the proposed concept attempts to capture vagueness in a nonnumeric fashion - we do not commit ourselves to any specific (and precise) membership values over the specific regions of the universe of discourse.
-the factor of vagueness becomes localized in the form of shadows as opposed to the situation existing with fuzzy sets where it is spread across the entire universe of discourse.

For discrete universes of discourse when we are dealing with a collection of membership values rather than continuous functions, (2) involves several sums and emerges in the form

$$V = \left| \sum_{i:\ A(x_i) < \alpha} A(x_i) + \sum_{i:\ A(x_i) > 1-\alpha} (1 - A(x_i)) - \text{card}\{x_i \in X \mid \alpha \leq A(x) \leq 1-\alpha\} \right|$$

(3)

One can think of a certain algorithmic generalization of the shadowed set by admitting two separate thresholds that is α and β, $\alpha < \beta$. This modifies (3) making it a function of two arguments

$$V = \left| \sum_{i:\ A(x_i) < \alpha} A(x_i) + \sum_{i:\ A(x_i) > \beta} (1 - A(x_i)) - \text{card}\{x_i \in X \mid \alpha \leq A(x) \leq \beta\} \right|$$

(4)

This extension, however, does not change the generic underlying idea and will not be pursued here.

Let us discuss computations of the threshold level α for some selected classes of continuous membership functions

1. Triangular membership functions. Assume the membership function of the form

$$y = \begin{cases} \frac{x-a}{b-a} & \text{for } x \in [a, b] \\ 0, & \text{otherwise} \end{cases}$$

For given α the values of a_1 and a_2 are equal

$$a_1 = a + \alpha(b-a)$$

and

$$a_2 = a + (1-\alpha)(b-a)$$

Computing the corresponding areas and solving the resulting quadratic equation with respect to α we derive

$$\alpha = \frac{2^{3/2} - 2}{2} = 0.4142$$

and

$$\alpha = \frac{-2^{3/2} - 2}{2} = -2.414$$

Obviously, the only first root becomes accepted as it satisfies the requirement formulated in (3). Note that the level of α does not depend on the values of "a" and "b". These bounds appear in the formulas describing the shadow itself, namely

$$a_1 = a + 0.414(b-a) = 0.414b - 0.586a$$
$$a_2 = a + 0.586(b-a) = 0.586b - 0.414a$$

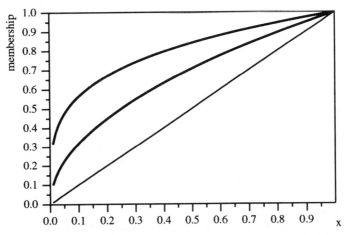

Figure 4. The use of linguistic modifiers: *more or less* applied twice to the original fuzzy set defined by a triangular membership function

We have for the fuzzy set affected twice by the linguistic modifier we get $\alpha = 0.3775$ as opposed to the original threshold being equal to 0.4142.

2. Consider now the nonlinear membership function assuming the form

$$A(x) = \sqrt{\frac{x-a}{b-a}}$$

Here $a_1 = a + \alpha^2(b-a)$ and $a_2 = a + (1-\alpha)^2(b-a)$. Proceeding in the same way as before, the problem gives produces a fifth order polynomial equation - the only root satisfying the imposed requirement is $\alpha = 0.405$.

3. In the optimization of the threshold for the Gaussian membership function, the performance index V regarded as a function of α exhibits a clearly manifesting minimum, Figure 5.

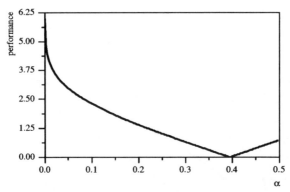

Figure 5. Performance index V for Gaussian membership function; optimal threshold $\alpha = 0.3950$

In fact, one can regard a shadowed set as summarizing a family of fuzzy set whose membership functions are indiscernible with respect to the criterion formulated by (2) or (3). This means that a reconstruction of a fuzzy set from the given shadowed set is not feasible. Obviously, more detailed cases could be discussed when restricting to some specific classes of fuzzy sets. In particular, let us study triangular (or more generally trapezoidal) membership functions. Three characteristic cases emerge as illustrated in Figure 6 - the result is primarily determined by the shadows and core of the shadowed set.

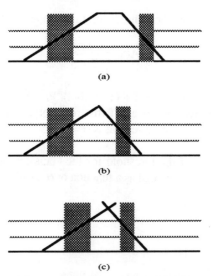

Figure 6. Reconstruction aspects:
(a) broad core with narrow shadows produce a trapezoidal fuzzy set
(b) balanced shadows and core reconstruct a triangular fuzzy set
(c) too broad shadows with a narrow core make a reconstruction impossible

3 Operations on shadowed sets

The basic operations on shadowed sets (union, intersection, and complement) are concisely summarized in Figure 7.

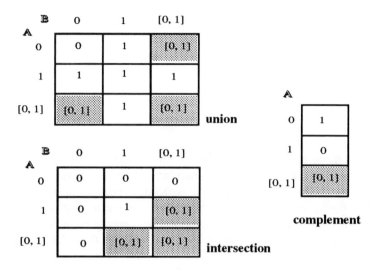

Figure 7. Defining operations on shadowed sets: union, intersection, and complement

The operations on shadowed sets exhibit a number of fundamental properties:

Commutativity
$$A \cup B = B \cup A$$
$$A \cap B = B \cap A$$

Associativity
$$A \cup (B \cup C) = (A \cup B) \cup C = A \cup B \cup C$$
$$A \cap (B \cap C) = (A \cap B) \cap C = A \cap B \cap C$$

Idempotency
$$A \cup A = A$$
$$A \cap A = A$$

Distributivity

$$A \cap (B \cup C) = (A \cap B) \cup (A \cap C)$$
$$A \cup (B \cap C) = (A \cup B) \cap (A \cup C)$$

Boundary conditions

$$A \cup \emptyset = A \qquad A \cup X = X$$
$$A \cap \emptyset = \emptyset \qquad A \cap X = A$$

Involution

$$\overline{\overline{A}} = A$$

Essentially, these operations are isomorphic with the logic connectives encountered in three-valued logic[10], in particular, Lukasiewicz logic. In this comparison the intermediate logical value (1/2) is identified with the [0, 1] interval.

The operations on shadowed sets exhibit an interesting property of information degradation meaning that we end up with less precise piece of information when operating on the shadows. More specifically, we get the following relationships

$$0 \cup [0, 1] = [0, 1]$$

and

$$1 \cap [0, 1] = [0, 1]$$

The operations can also be carried out in a sort of mixed mode embracing both fuzzy sets and shadowed sets. Note that for any a in [0, 1] we have

$$\min([0, 1], a) = [0, a]$$

and

$$\max([0, 1], a) = [a, 1]$$

hence the operations of union and intersection are straightforward. We remark that the result of mixing fuzzy sets and shadowed sets arises in the form of interval valued sets.

In the same way as above, one can develop calculus of shadowed relations by extending the above set operations to higher dimensional constructs. For instance, a shadowed relation R is computed pointwise as

$$(A \times B)(x, y) = A(x) \wedge B(y) = \min(A(x), B(y)) \tag{5}$$

A corresponding illustration is contained in Figure 8.

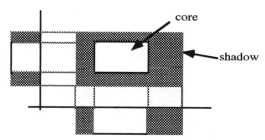

Figure 8. An example of a shadowed set

4 Selected Examples

In this section we are concerned with a series of representative examples illustrating the usefulness of the developed concept. Before proceeding with them, it is instructive to elaborate on the principles of fuzzy computing. As a matter of fact, we should be cognizant of the use of fuzzy sets at the level of granular information. We process such quantities in the framework of fuzzy sets. More importantly, such results are afterwards decoded (defuzzified, as often referred to in the literature) to produce results that are directly used at the numeric level, Figure 9. This type of transformation is omnipresent in fuzzy controllers, fuzzy classifiers, etc. The crucial question about this entire scheme revolves around an associated level of credibility of these final numeric results. How far are these meaningful? How well do they convey message about the level of uncertainty that might have been emerged at the level of available information? To relatively high extent these questions remain unanswered. This is the main point around which shadowed sets have been developed in order to address these essential issues, Figure 10.

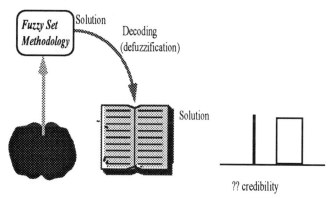

Figure 9. The principles of the use of fuzzy set methodology

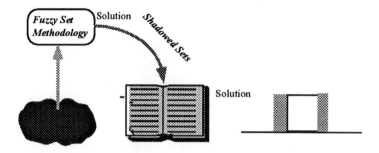

Figure 10. The principles of shadowed sets in shadowed sets

Figure 11 portrays a number of main ways fuzzy sets and shadowed sets interact and get into symbiotic relationships. In general, shadowed sets can serve as a useful back-end interface (where all processing are carried out in the setting of fuzzy sets). The other interesting option comes in the form of the front- as well as back-end realized via shadowed sets. In this case the processing level becomes very much simplified as we are concerned with computing geared into three valued logic. The ensuing computational overhead is thus substantially reduced.

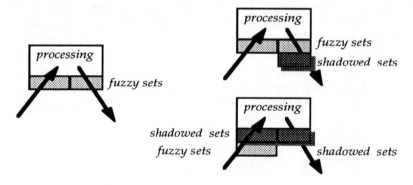

Figure 11. Main conceptual avenues of information processing in the setting of fuzzy sets and shadowed sets

4.1 Fuzzy clustering

Briefly speaking, the aim of fuzzy clustering is to discover meaningful structures (clusters) in data structures in such a way that the patterns close each other are placed in the same group. Fuzzy clustering provides us with an opportunity to assign patterns to the classes with some membership values situated in [0, 1] and in this way help identify potential outliers viz. the object not fully assigned to a single class. With this intent in mind we use the concept of shadowed sets towards cluster analysis. We confine ourselves to so - called objective based fuzzy clustering where the results of clustering are summarized in the form of partition matrices. Fuzzy Isodata (or Fuzzy C - Means) is an example of commonly utilized method belonging to this category, cf. [2]. This resulting partition matrix $U = [u_{ik}]$, $i=1, 2, ..., c$, $k=1, 2, ..., N$ (with c being the number of clusters and N standing for the number of patterns) is nothing but a fuzzy relation. Thus the shadow of the shadowed relation is determined in a usual way by treating V as a function of α

$$V(\alpha) = \left| \sum_{i,k: \, u_{ik} < \alpha} u_{ik} + \sum_{i,k: \, u_{ik} > 1-\alpha} (1 - u_{ik}) - \text{card}\{ (i, k) \,|\, \alpha \leq u_{ik} \leq 1-\alpha \} \right|$$

(6)

and finding α for which $V(\alpha)$ reaches zero.

As an illustrative example, let us consider a set of two - dimensional data (patterns) visualized in Figure 12.

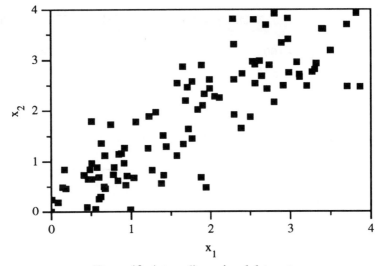

Figure 12. A two-dimensional data set

The clustering method is Fuzzy Isodata and the number of clusters is set to 4. The bar plot of the obtained partition matrix shows a significant number of patterns whose class membership has been evidently split between several categories, Figure 13.

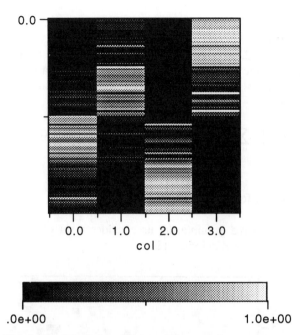

Figure 13. A bar plot of the partition matrix

The plot of V(α), Figure 14, identifies an optimal level of α leading subsequently to the emergence of the shadows of the shadowed sets. This finally identifies patterns that fall under this shadow - they are marked in Figure 15.

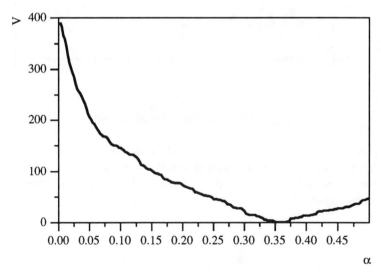

Figure 14. V as a function of α

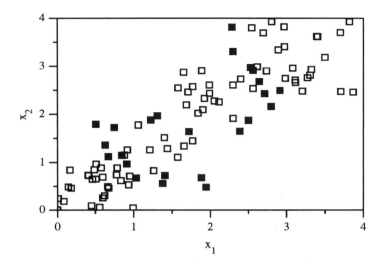

Figure 15. Patterns (black squares) belonging to the shadows of the shadowed sets

4. 2 Image processing

The use of shadowed sets or shadowed relations can be also substantial in image processing especially in the problem of determining boundaries of objects. Consider a blurred circle as in Figure 16. Due to a spectrum of different levels of brightness, this object can be conveniently regarded as a two argument fuzzy relation. The minimization of V leads to $\alpha = 0.395$, Figure 17, and the produced residual structure (shadow of the fuzzy relation) identifying the boundary of the circle is visualized in Figure 18.

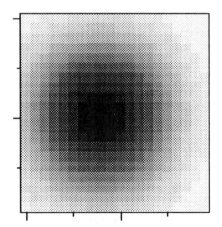

Figure 16. A blurred image of a circle

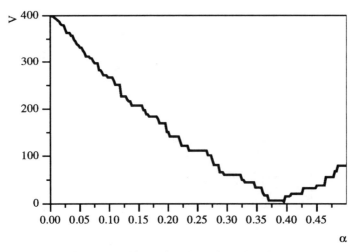

Figure 17. Plot of V as a function of the threshold level

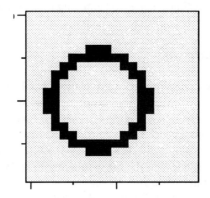

Figure 18. The identified boundary region of the blurred circle

4. 3 Single stage decision - making in presence of fuzzy objectives

The problem of a single stage decision - making has been a favorite example showing the use of fuzzy set technology in this area In the simplest scenario possible, the resulting fuzzy decision is regarded to be a fuzzy set formed as an intersection of several decision objectives (viewed as some goals and constraints) and treated as fuzzy sets defined in the same universe of discourse. For sake of conciseness, let us consider only two objectives (fuzzy sets), namely A and B defined in X. Rather than using these fuzzy sets directly and compute their intersection, we proceed with the corresponding shadowed sets, $\mathbb{A}$ and $\mathbb{B}$, and determine a shadowed set of decision meaning that

$$\mathbb{D} = \mathbb{A} \cap \mathbb{B}$$

Three interesting and qualitatively distinct cases arise

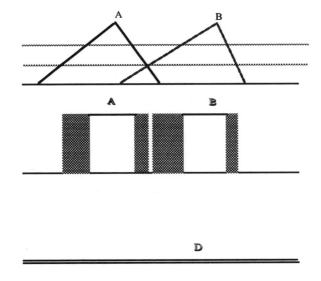

Figure 19. Fuzzy objectives and resulting decision represented as a shadowed set

a. The decision $\mathbb{D}$ is empty; $\mathbb{D}$ assumes zero over **X** (see Figure 19). The analysis worked out with shadowed sets has evidently identified the case where no decision could be made - this is primarily because of strongly conflicting objectives in the decision problem. $\mathbb{A}$ and $\mathbb{B}$ are too distinct to advice making any rational decision. What it occurs when using fuzzy sets is that the resulting fuzzy set of decision D is highly subnormal which could still push us to make a decision by e.g., selecting the modal value of D.

b. In the second case $\mathbb{A}$ and $\mathbb{B}$ are getting closer each other and this results in the form of $\mathbb{D}$ shown in Figure 20. The interpretation is also straightforward: "decide at your own risk" - there is no indication which subset of **X** could be legitimate as the shadowed set does not have any core. Perhaps the only choice worth considering will the region identified by the shadow of $\mathbb{D}$.

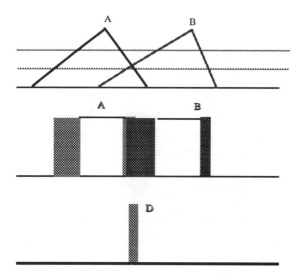

Figure 20. Fuzzy objectives and resulting decision represented as a shadowed set

c. Finally, enough overlap of $\mathbb{A}$ and $\mathbb{B}$ (or less conflicting nature of these) has produced $\mathbb{D}$ visualized in Figure 21 - the core of D identifies potential decision values. Note, however, that further selection between the elements of the core of $\mathbb{D}$ is possible.

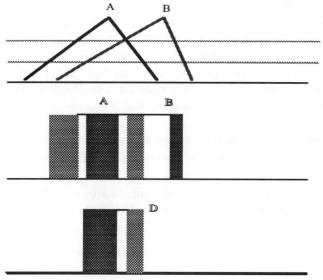

Figure 21. Fuzzy objectives and resulting decision represented as a shadowed set

4.4 Associative memories and rule-based systems with shadowed sets

The standard associative rule - based systems consist of rules assuming the form

$$\text{if } A_k \text{ then } B_k$$

where A_k and B_k are conditions and conclusions defined in **A** and **B**, respectively, k=1, 2, ..., N. The associative memories can be easily placed in the context of shadowed sets. To simplify the entire analysis, let us treat the items to be stored as shadowed sets and assume a Hebbian - like style of learning meaning that the rules are summarized as a union of the above conditions and conclusions producing a fuzzy relation of the form

$$\mathbb{R} = \bigcup_{k=1}^{N} (\mathbb{A}_k \times \mathbb{B}_k)$$

that is

$$\mathbb{R}(x, y) = \bigvee_{k=1}^{N} (\mathbb{A}_k(x) \wedge \mathbb{B}_k(y))$$

for all x in **A** and y in **B**. The inference in completed using the standard max - min composition.

Depending upon a specific distribution of A_k's, several situations are distinguished:

a. Complete coverage of the condition space assuring existence of a justifiable conclusion (shadowed set with a nonempty core) for any input of the associative memory.

b. sufficient coverage of the condition space. Even though the core regions of $\mathbb{R}$ situated in the Cartesian product of the conditions and actions standing in the rules are disjoint they intersect at the level of the shadows of the respective conditions.

c. Insufficient coverage - the shadowed relations produce disjoint regions in **A** $\times$ **B**. This implies insufficient recall capabilities

These three scenarios allow us to analyze any knowledge base with respect to its completeness. Similarly, the inference mechanism (mapping from **A** to **B**) is very much simplified not requiring any tedious numeric computing involving exact membership values. Definitely, the produced results are also quite rough returning shadowed sets rather than fuzzy sets.

Quite closely related are rule-based constructs. The use of shadowed sets helps us establish another useful and practically viable insight into the design of such systems and eventually establish some ways of elimination of deficiencies in these systems. As becomes apparent, a rule-based system is governed by a finite set of rules. From a formal standpoint, they can be regarded as "patches" distributed over experimental data as illustrated in Figure 22.

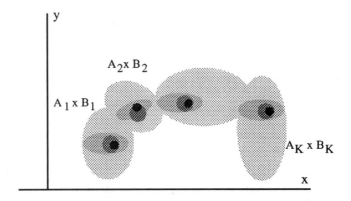

Figure 22. An effect of rule patching realized over experimental data

We are primarily interested in looking into the feasibility of mapping from x to y based on given fuzzy rules (patches). A simple criterion of relevancy of the mapping can be formulated based on the level of coverage of the input space. Here the induced shadowed sets play an important role. We define a union of cores of the shadowed sets

$$\text{core}(\mathbb{A}_1) \cup \text{core}(\mathbb{A}_2) \cup ... \cup \text{core}(\mathbb{A}_N)$$

Then we check for eventual "holes" or shadows in the above construct. They are straightforward indicators of a local lack of relevance of the mapping completed by the rules
There are two possible ways these mapping deficiencies can be alleviated:
- use inputs (X) of lower granularity (assuming normality of X it is sufficient to quantify granularity of fuzzy set by its cardinality)
- decrease granularity of A_k (and B_k, correspondingly) by using linguistic modifiers (*more or less*)

5 Conclusions

We have introduced and studied a new concept of shadowed sets. These constructs are viewed as being induced by fuzzy sets and aimed at less numeric and far more computationally demanding processing of information conveyed by fuzzy sets. This feature could be especially important in all situations where a certain trade-off between numeric precision and computational effort becomes necessary. Shadowed sets enhance and simplify an interpretation of results of processing with fuzzy sets by proposing decision expressed in the language of three - valued logic (that could be interpreted as yes, no, and unknown). As numeric details are suppressed while computing efficiency increased, one can think of shadowed sets as a provider of a quick and dirty approach to computing with fuzzy quantities - if the obtained results are of interest (usually shadowed sets with nonempty cores) then one can resort to detailed yet time consuming computing with fuzzy sets.

Acknowledgments

Support from the Natural Sciences and Engineering Research Council of Canada (NSERC) is gratefully acknowledged.

6 References

1. R. E. Bellman, L. A. Zadeh, Decision making in a fuzzy environment, *Management Sciences*, 17, 1970, 141 - 164.
2. J. C. Bezdek, *Pattern Recognition with Fuzzy Objective Functions*, Plenum Press, N. York, 1981.
3. K. Hirota, Concepts of probabilistic sets, *Fuzzy Sets and Systems*, 5, 1981, 31 - 46.
4. A. Kandel, *Fuzzy Techniques in Pattern Recognition*, J. Wiley, N. York, 1982
5. G. J. Klir, T. A. Folger, *Fuzzy Sets, Uncertainty and Information*, Prentice Hall, Englewood Cliffs, NJ, 1988
6. M. Mizumoto, K. Tanaka, Some properties of fuzzy sets of type 2, *Information and Control*, 31, 1976, 312 - 340.
7. Z. Pawlak, *Rough Sets*, Kluwer Academic Publishers, Dordrecht, 1991.
8. W. Pedrycz, F. Gomide, *An Introduction to Fuzzy Sets: Analysis and Design*, MIT Press, Cambridge, MA, 1998
9. T. Radecki, Level fuzzy sets, *J. Cybernetics*, 7, 1977, 189 - 198.
10. N. Rescher, *Many - Valued Logic*, McGraw Hill, N. York, 1969.
11. R. Sambuc, *Fonctions Φ-flous. Application a l'aide au diagnostic en pathologie thyroidienne*, Ph. D. thesis, Marseille, 1975.

Information Measures for Rough and Fuzzy Sets and Application to Uncertainty in Relational Databases

Theresa Beaubouef[1], Frederick E. Petry[2], and Gurdial Arora[3]

[1] Computer Science Department, Xavier University of Louisiana, New Orleans, LA 70125, USA, email: tbeaubou@mail.xula.edu
[2] Center for Intelligent and Knowledge-Based Systems, Tulane University, New Orleans, LA 70118, USA, email: petry@eecs.tulane.edu
[3] Math Department, Xavier University of Louisiana, New Orleans, LA 70125, USA, email: gdial@mail.xula.edu

1 Introduction

Uncertainty pervades every aspect of daily living. It is the norm, not the exception, and any system that attempts to be realistic must also incorporate uncertainty management. In recent years researchers have accepted this fact. The problem which must then be addressed is how to model the uncertainty and how to measure it in terms of the application in question.

In communication theory, Shannon [30] introduced the concept of entropy which was used to characterize the information content of signals. Since then, variations of these information theoretic measures have been successfully applied to applications in many diverse fields. In particular, the representation of uncertain information by entropy measures has been applied to all areas of databases, including fuzzy database querying [9], data allocation [16], and classification in rule-based systems [27].

In fuzzy set theory the representation of uncertain information measures has been extensively studied [12,20]. So this paper relates the concepts of information theory to rough sets and compares these information theoretic measures to established rough set metrics of uncertainty. The measures are then applied to the rough relational database model [5]. Information content of both stored relational schemas and rough relations are expressed as types of rough entropy. We then discuss some new fuzzy information measures for application to fuzzy rough databases.

2 Rough Sets and Fuzzy Rough Sets

Rough set theory, introduced by Pawlak and discussed in greater detail in [17,26], is a technique for dealing with uncertainty and for identifying cause-effect relationships in databases as a form of database learning [32]. Rough sets involve the following:

U is the *universe*, which cannot be empty,
R is the *indiscernibility relation*, or equivalence relation,
$A = (U,R)$, an ordered pair, is called an *approximation space*,
$[x]_R$ denotes the equivalence class of R containing x, for any element x of U,
elementary sets in A - the equivalence classes of R,
definable set in A - any finite union of elementary sets in A.

Given an approximation space defined on some universe U which has an equivalence relation R imposed upon it, U is partitioned into equivalence classes called elementary sets which may be used to define other sets in A. A rough set X, where $X \subseteq U$, X can be defined in terms of the definable sets in A by the following:

lower approximation of X in A is the set $\underline{R}X = \{x \in U \mid [x]_R \subseteq X\}$
upper approximation of X in A is the set $\overline{R}X = \{x \in U \mid [x]_R \cap X \neq \emptyset\}$.

$POS_R(X) = \underline{R}X$ denotes the R-positive region of X, or those elements which certainly belong to the rough set. The R-negative region of X, $NEG_R(X) = U - \overline{R}X$, contains elements which do not belong to the rough set and the boundary or R-borderline region of X, $BN_R(X) = \overline{R}X - \underline{R}X$, contains those elements which may or may not belong to the set. X is R-definable if and only if $\underline{R}X = \overline{R}X$. Otherwise, $\underline{R}X \neq \overline{R}X$ and X is rough with respect to R. A *rough set in A* is the group of subsets of U with the same upper and lower approximations.

Because there are advantages to both fuzzy set and rough set theories, several researchers have studied various ways of combining the two theories [13,14,23]. Others have investigated the interrelations between the two theories [10,25,32]. Fuzzy sets and rough sets are not equivalent, but complementary.

It has been shown in [32] that rough sets can be expressed by a fuzzy membership function $\mu \longrightarrow \{0, 0.5, 1\}$ to represent the negative, boundary, and positive regions. In this model, all elements of the lower approximation, or positive region, have a membership value of one. Those elements of the boundary region are assigned a membership value of 0.5. Elements not belonging to the rough set have a membership value of zero. Rough set definitions of union and intersection were modified so that the fuzzy model would satisfy all the properties of rough sets [25]. This allowed a rough set to be expressed as a fuzzy set.

We integrate fuzziness into the rough set model in order to quantify levels of roughness in boundary region areas through the use of fuzzy membership values. Therefore, we do not require membership values of elements of the boundary region to equal 0.5, but allow them to range from zero to one, noninclusive. Additionally, the union and intersection operators for fuzzy rough sets are comparable to those for ordinary fuzzy sets, where MIN and MAX are used to obtain membership values of redundant elements.

Let U be a *universe*, X a rough set in U.

Definition. A *fuzzy rough set* Y in U is a membership function $\mu_Y(x)$ which associates a grade of membership from the interval [0,1] with every element of U where

$\mu_Y(\underline{R}X) = 1$, $\mu_Y(U - \overline{R}X) = 0$, and $0 < \mu_Y(\overline{R}X - \underline{R}X) < 1$.

Definition. The *union* of two fuzzy rough sets A and B is a fuzzy rough set C where C = {x | x ∈ A OR x ∈ B}, where $\mu_C(x) = MAX[\mu_A(x), \mu_B(x)]$.

Definition. The *intersection* of two fuzzy rough sets A and B is a fuzzy rough set C = {x | x ∈ A AND x ∈ B}, where $\mu_C(x) = MIN[\mu_A(x), \mu_B(x)]$.

3 The Rough Relational Database Model

The rough relational database model [5] is an extension of the standard relational database model of Codd [11]. It captures all the essential features of rough sets theory including indiscernibility of elements denoted by equivalence classes and lower and upper approximation regions for defining sets which are indefinable in terms of the indiscernibility.

Every attribute domain is partitioned by some equivalence relation designated by the database designer or user. Within each domain, those values which are considered indiscernible belong to an equivalence class. This information is used by the query mechanism to retrieve information based on equivalence with the class to which the value belongs rather than equality, resulting in less critical wording of queries.

Recall is also improved in the rough relational database because rough relations provide *possible* matches to the query in addition to the *certain* matches which are obtained in the standard relational database. This is accomplished by using set containment in addition to equality of attributes in the calculation of lower and upper approximation regions of the query result.

The rough relational database has several features in common with the ordinary relational database. Both models represent data as a collection of *relations* containing *tuples*. These relations are sets. The tuples of a relation are its elements, and like elements of sets in general, are unordered and nonduplicated. A tuple t_i takes the form $(d_{i1}, d_{i2}, ..., d_{im})$, where d_{ij} is a *domain value* of a particular *domain set* D_j. In the ordinary relational database, $d_{ij} \in D_j$. In the rough database, however, as in other non-first normal form extensions to the relational model [21,29], $d_{ij} \subset D_j$, and although it is not required that d_{ij} be a singleton, $d_{ij} \neq \emptyset$. Let $\mathcal{P}(D_i)$ denote the powerset(D_i) - ∅.

Definition. A *rough relation* R is a subset of the set cross product $\mathcal{P}(D_1) \times \mathcal{P}(D_2) \times \cdots \times \mathcal{P}(D_m)$.

A rough tuple t is any member of R, which implies that it is also a member of $\mathcal{P}(D_1) \times \mathcal{P}(D_2) \times \cdots \times \mathcal{P}(D_m)$. If t_i is some arbitrary tuple, then $t_i = (d_{i1}, d_{i2}, ..., d_{im})$ where $d_{ij} \subseteq D_j$. A tuple in this model differs from that of ordinary databases in that the tuple components may be sets of domain values rather than single values. The set braces are omitted from singletons for notational simplicity.

Let $[d_{xy}]$ denote the equivalence class to which d_{xy} belongs. When d_{xy} is a set of values, the equivalence class is formed by taking the union of equivalence classes of members of the set; if $d_{xy} = \{c_1, c_2, ..., c_n\}$, then $[d_{xy}] = [c_1] \cup [c_2] \cup ... \cup [c_n]$.

Definition. Tuples $t_i = (d_{i1}, d_{i2}, ..., d_{im})$ and $t_k = (d_{k1}, d_{k2}, ..., d_{km})$ are *redundant* if $[d_{ij}] = [d_{kj}]$ for all $j = 1,..., m$.

In the rough relational database, redundant tuples are removed in the merging process since duplicates are not allowed in sets, the structure upon which the relational model is based.

There are two basic types of relational operators. The first type arises from the fact that relations are considered sets of tuples. Therefore, operations which can be applied to sets also apply to relations. The most useful of these for database purposes are *set difference*, *union*, and *intersection*. Operators which do not come from set theory, but which are useful for retrieval of relational data are *select*, *project*, and *join*.

In the rough relational database, relations are rough sets as opposed to ordinary sets. Therefore, new rough operators (-, ∪, ∩, ,σ, π, ⋈), which are comparable to the standard relational operators, must be developed for the rough relational database. Moreover, a mechanism must exist within the database to mark tuples of a rough relation as belonging to the lower or upper approximation of that rough relation. Properties of the rough relational operators can be found in [5].

4 The Fuzzy Rough Relational Database Model

The fuzzy rough relational database is an extension of the rough relational database. A tuple t_i takes the form $(d_{i1}, d_{i2}, ..., d_{im}, d_{i\mu})$, where d_{ij} is a *domain value* of a particular *domain set* D_j and $d_{i\mu} \in D_\mu$, where D_μ is the interval [0,1], the domain for fuzzy membership values. In the ordinary relational database, $d_{ij} \in D_j$. In the fuzzy rough relational database, except for the fuzzy membership value, however, $d_{ij} \subset D_j$, and although d_{ij} is not restricted to be a singleton, $d_{ij} \neq \emptyset$. Let $\mathcal{P}(D_i)$ denote any non-null member of the powerset of D_i.

Definition. A *fuzzy rough relation* R is a subset of the set cross product $\mathcal{P}(D_1) \times \mathcal{P}(D_2) \times \cdots \times \mathcal{P}(D_m) \times D_\mu$.

For a specific relation, R, membership is determined semantically. A *fuzzy rough tuple* t is any member of R. If t_i is some arbitrary tuple, then $t_i = (d_{i1}, d_{i2}, ..., d_{im}, d_{i\mu})$ where $d_{ij} \subseteq D_j$ and $d_{i\mu} \in D_\mu$.

Definition. An *interpretation* $\alpha = (a_1, a_2, ..., a_m, a_\mu)$ of a fuzzy rough tuple $t_i = (d_{i1}, d_{i2}, ..., d_{im}, d_{i\mu})$ is any value assignment such that $a_j \in d_{ij}$ for all j.

The interpretation space is the cross product $D_1 \times D_2 \times \cdots \times D_m \times D_\mu$, but is limited for a given relation R to the set of those tuples which are valid according to the underlying semantics of R. In an ordinary relational database, because domain values are atomic, there is only one possible interpretation for each tuple t_i. Moreover, the interpretation of t_i is equivalent to the tuple t_i. In the fuzzy rough relational database, this is not always the case.

Let $[d_{xy}]$ denote the equivalence class to which d_{xy} belongs. When d_{xy} is a set of values, the equivalence class is formed by taking the union of equivalence classes of members of the set; if $d_{xy} = \{c_1, c_2, ..., c_n\}$, then $[d_{xy}] = [c_1] \cup [c_2] \cup ... \cup [c_n]$.

Definition. Tuples $t_i = (d_{i1}, d_{i2}, ..., d_{in}, d_{i\mu})$ and $t_k = (d_{k1}, d_{k2}, ..., d_{kn}, d_{k\mu})$ are *redundant* if $[d_{ij}] = [d_{kj}]$ for all $j = 1, ..., n$.

If a relation contains only those tuples of a lower approximation, i.e., those tuples having a μ value equal to one, the interpretation α of a tuple is unique. This follows immediately from the definition of redundancy. In fuzzy rough relations, there are no redundant tuples. The merging process used in relational database operations removes duplicate tuples since duplicates are not allowed in sets, the structure upon which the relational model is based.

Tuples may be redundant in all values except μ. As in the union of fuzzy rough sets where the maximum membership value of an element is retained, it is the convention of the fuzzy rough relational database to retain the tuple having the higher μ value when removing redundant tuples during merging. If we are supplied with identical data from two sources, one certain and the other uncertain, we would want to retain the data that is certain, avoiding loss of information.

5 Information-Theoretic Measures and Rough Sets

Rough set theory [26] inherently models two types of uncertainty. The first type of uncertainty arises from the indiscernibility relation which is imposed on the universe, partitioning all values into a finite set of equivalence classes. If every equivalence class contains only one value, then there is no loss of information caused by the partitioning. In any coarser partitioning, however, there are fewer classes, and each class will contain a larger number of members. Our knowledge, or information, about a particular value decreases as the granularity of the partitioning becomes coarser.

Uncertainty is also modeled through the approximation regions of rough sets where elements of the lower approximation region have total participation in the rough set and those of the upper approximation region have uncertain participation in the rough set. Equivalently, the lower approximation is the *certain* region and the boundary area of the upper approximation region is the *possible* region.

Pawlak [26] discusses two numerical characterizations of imprecision of a rough set X: *accuracy* and *roughness*. Accuracy, which is simply the ratio of the number of elements in the lower approximation of X, $\underline{R}X$, to the number of elements in the upper approximation of the rough set X, $\overline{R}X$, measures the degree of completeness of knowledge about the given rough set X. It is defined as a ratio of the two set cardinalities as follows:

$$\alpha_R(X) = \text{card}(\underline{R}X) / \text{card}(\overline{R}X), \quad \text{where } 0 \leq \alpha_R(X) \leq 1.$$

The second measure, roughness, represents the degree of incompleteness of knowledge about the rough set. It is calculated by subtracting the accuracy from 1:
$\rho_R(X) = 1 - \alpha_R(X)$.

These measures require knowledge of the number of elements in each of the approximation regions and are good metrics for uncertainty as it arises from the boundary region, implicitly taking into account equivalence classes as they belong wholly or partially to the set. However, accuracy and roughness measures do not necessarily provide us with information on the uncertainty related to the granularity of the indiscernibility relation for those values which are totally included in the lower approximation region. For example,

Let the rough set X be defined as follows: X = {A11, A12, A21, A22, B11, C1} with lower and upper approximation regions defined as

$$\underline{R}X = \{A11, A12, A21, A22\} \text{ and } \overline{R}X = \{A11, A12, A21, A22, B11, B12, B13, C1, C2\}$$

These approximation regions may result from one of several partitionings. Consider, for example, the following indiscernibility relations:

$A_1 = \{[A11, A12, A21, A22], [B11, B12, B13], [C1, C2]\}$,
$A_2 = \{[A11, A12], [A21, A22], [B11, B12, B13], [C1, C2]\}$,
$A_3 = \{[A11], [A12], [A21], [A22], [B11, B12, B13], [C1, C2]\}$.

All three of the above partitionings result in the same upper and lower approximation regions for the given set X, and hence the same accuracy measure (4/9 = .444) since only those classes belonging to the lower approximation region were re-partitioned. It is obvious, however, that there is more uncertainty in A_1 than in A_2, and more uncertainty in A_2 than in A_3. Therefore, a more comprehensive measure of uncertainty is needed.

We derive such a measure from techniques used for measuring entropy in classical information theory. Countless variations of the classical entropy have been developed, each tailored for a particular application domain or for measuring a particular type of uncertainty. Our rough entropy is defined such that we may apply it to rough databases. We define the entropy of a rough set X as follows:

Definition: The *rough entropy* $E_r(X)$ of a rough set X is calculated by

$E_r(X) = -(\rho_R(X)) [\sum Q_i \log(P_i)]$ for i = 1,... n equivalence classes.

The term $\rho_R(X)$ denotes the roughness of the set X. The second term is the summation of the probabilities for each equivalence class belonging either wholly or in part to the rough set X. There is no ordering associated with individual class members. Therefore the probability of any one value of the class being named is the reciprocal of the number of elements in the class. If c_i is the cardinality of, or number of elements in, equivalence class i and all members of a given equivalence class are equally probable, $P_i = 1/c_i$ represents the probability of one of the values in class i. Q_i denotes the probability of equivalence class i within the universe. Q_i is computed by taking the number of elements in class i and dividing by the total number of elements in all equivalence classes combined. The entropy of the sample rough set X, $E_r(X)$, is given for each of the possible indiscernibility relations A_1, A_2, and A_3:

A_1: $-(5/9)[(4/9)\log(1/4) + (3/9)\log(1/3) + (2/9)\log(1/2)] = .274$

A_2: $-(5/9)[(2/9)\log(1/2) + (2/9)\log(1/2) + (3/9)\log(1/3) + (2/9)\log(1/2)] = .20$
A_3: $-(5/9)[(1/9)\log(1) + (1/9)\log(1) + (1/9)\log(1) + (1/9)\log(1) + (3/9)\log(1/3) + (2/9)\log(1/2)] = .048$

From the above calculations it is clear that although each of the partitionings results in identical roughness measures, the entropy decreases as the classes become smaller through finer partitionings.

6 Entropy and the Rough Relational Database

The basic concepts of rough sets and their information-theoretic measures carries over to the rough relational database model [4,5]. Recall that in the rough relational database all domains are partitioned into equivalence classes and relations are not restricted to first normal form. We therefore have a type of rough set for each attribute of a relation. This results in a rough relation, since any tuple having a value for an attribute which belongs to the boundary region of its domain is a tuple belonging to the boundary region of the rough relation.

There are two things to consider when measuring uncertainty in databases: uncertainty or entropy of a rough relation that exists in a database at some given time and the entropy of a relation schema for an existing relation or query result. We must consider both since the approximation regions only come about by set values for attributes in given tuples. Without the extension of a database containing actual values, we only know about indiscernibility of attributes. We cannot consider the approximation regions.

We define the entropy for a rough relation schema as follows:

Definition. The *rough schema entropy* for a rough relation schema S is

$E_s(S) = -\sum_j [\sum Q_i \log(P_i)]$ for $i = 1,\ldots n_j$; $j = 1,\ldots, m$

where there are n equivalence classes of domain j, and m attributes in the schema $R(A_1, A_2, \ldots, A_m)$.

This is similar to the definition of entropy for rough sets without factoring in roughness since there are no elements in the boundary region (lower approximation = upper approximation). However, because a relation is a cross product among the domains, we must take the sum of all these entropies to obtain the entropy of the schema. The schema entropy provides a measure of the uncertainty inherent in the definition of the rough relation schema taking into account the partitioning of the domains on which the attributes of the schema are defined.

We extend the schema entropy $E_s(S)$ to define the entropy of an actual rough relation instance $E_R(R)$ of some database D by multiplying each term in the product by the roughness of the rough set of values for the domain of that given attribute.

Definition. The *rough relation entropy* of a particular extension of a schema is

$E_R(R) = -\sum_j D\rho_j(R) [\sum DQ_i \log(DP_i)]$ for $i = 1,\ldots n_j$; $j = 1,\ldots, m$

where $D\rho_j(R)$ represents a type of database roughness for the rough set of values of the domain for attribute j of the relation, m is the number of attributes in the database relation, and n is the number of equivalence classes for a given domain for the database.

We obtain the $D\rho_j(R)$ values by letting the non-singleton domain values represent elements of the boundary region, computing the original rough set accuracy and subtracting it from one to obtain the roughness. DQ_i is the probability of a tuple in the database relation having a value from class i, and DP_i is the probability of a value for class i occurring in the database relation out of all the values which are given. These concepts are explained in more detail in the ensuing example where actual calculations of rough schema and relation entropies are illustrated.

7 Example

Soil scientists classify soils based on several properties. We will consider two such properties: the basic overall color of the soil sample and the average size of soil particles in the sample. Often there are skilled technicians who record data about the samples, any two of which may differ in the actual values used for classification. A rough set approach allows equivalence of values to be defined in the indiscernibility relation. Consider the following domains with equivalence classes denoted by [ ], which may be defined by the scientist:

COLOR = {[black,ebony], [brown,tan,sienna],[white], [gray], [orange]}
PARTICLE-SIZE = {[big,large], [huge, enormous], [medium], [small, little, tiny]}

The soil sample information is entered into a database relation of the following schema:

SOIL(BIN-NO, COLOR, PARTICLE-SIZE)

where BIN-NO, the key, represents the soil sample number based on the bin in which it is stored, COLOR is the general color of the soil, and PARTICLE-SIZE describes the average size of soil particles for the sample. The schema entropy $E_s(SOIL)$ is computed as

$$E_s(SOIL) = -\Sigma_j [\Sigma Q_i \log(P_i)] \quad \text{for } i = 1,\ldots,n; j = \text{BIN-NO, COLOR, PARTICLE-SIZE.}$$

There are three partial sums in this computation, one for each attribute j. Because there is no equivalence relation defined for BIN-NO, all values are singletons ($P_i = 1$) which cause this term to be zero in the above equation. The other partial sums are computed as follows:

COLOR: $-[(2/8)\log(½) + (3/8)\log(⅓) + (1/8)\log(1) + (1/8)\log(1) + (1/8)\log(1)]$ = .254
PARTICLE-SIZE: $-[(2/8)\log(½) + (2/8)\log(½) + (1/8)\log(1) + (3/8)\log(⅓)] = .329$

Hence, the entropy of relation schema SOIL, $E_s(SOIL) = 0 + .254 + .329 = .583$.

Let us now compute the entropy of the database relation instance SAMPLE-114, found in Table 1.

Table 1. SAMPLE-114

BIN-NO	COLOR	PARTICLE-SIZE
P21	brown	medium
P22	{black,tan}	large
P23	gray	{medium,small}
T01	black	tiny
T04	{gray,brown}	large

We first calculate the terms $D\rho_{COLOR}(\text{SAMPLE-114}) = 4/7$ and $D\rho_{PARTICLE\text{-}SIZE}(\text{SAMPLE-114}) = 2/6$. This is accomplished by taking the number of attribute values for the particular attribute j which make up a non-singleton entry and dividing by the total number of values. The result is a type of roughness measure for the relation instance since it is a ratio of *possible* values to *all* values, *certain* and *possible* combined. For attribute COLOR, there are four values which belong to non-singleton sets: black, tan, gray, and brown. In addition, there are three singleton values: brown, gray, and black. Therefore, the roughness for attribute COLOR is equal to four divided by seven. Alternatively, we could have calculated the accuracy by dividing the number of singleton values by the total (3/7) and subtracting this number from one. DQ_i represents a type of probability that some value from equivalence class i will be present in a tuple. Actual value occurrences in the relations are used in calculating the percentages DP_i.

The rough relation entropy of the instance of SAMPLE-114 shown in the table is calculated as follows:

$$E_R(\text{SAMPLE-114}) = -(4/7)[(2/5)\log(2/7) + (3/5)\log(3/7) + 0 + (2/5)\log(2/7) + 0]$$
$$- (2/6)[(2/5\log(2/6) + 0 + (2/5)\log(2/6) + (2/5)\log(2/6)] = .56$$

The factor (4/7) is derived from values for COLOR; four of the seven values which appear in the relation for COLOR appear as part of a non-singleton attribute value. Likewise, (2/6) is derived from values for PARTICLE-SIZE; two of the six values, medium and small in the third tuple, are part of a non-singleton attribute value.

The logarithmic terms within the first set of brackets are derived from the equivalence classes for COLOR. The first term, for example, is computed from the class [black, ebony]. Two of the five tuples (2/5) contain a member of this class as a value; two of the seven (2/7) values under COLOR contain a member of this class. The second term is calculated similarly for class [brown, tan, sienna] and the fourth for class [gray]. Because there are no tuples containing a color from the class [white] or the class [orange], the third and fifth terms are zero. The logarithmic terms within the second set of brackets are calculated from the equivalence classes for PARTICLE-SIZE: [big, large], [huge, enormous], [medium], [small, little, tiny]. The second of these terms is zero since no value from the class [huge, enormous] appears in the relation.

The example illustrates how all parts of the relation, instance values for all attributes as well as the indiscernibility, take part in the calculation of rough relation entropy. The greater the number of non-singleton attribute values, the greater the entropy. Additionally, the granularity of the partitioning affects both rough schema entropy and rough relation entropy, with finer partitionings resulting in lower entropies.

8 New Fuzzy Information Measures

Several measures for fuzzy sets have been proposed. Each one of the measures has its advantages and disadvantages. It is to be noted that measures of fuzziness estimate the average fuzzy uncertainty in fuzzy sets in some well defined sense. Recently Pal and Bezdek [24] reviewed several measures of fuzziness for discrete fuzzy sets and introduced new multiplicative and additive classes. Also Bhandari and Pal [6] reviewed in their paper fuzzy measures and introduced a new measure of information similar to the Renyi's [28] probabilistic entropy of order α. In [2], a review of some of the existing measures of fuzziness was given and some new measures of fuzziness based on generalized entropies were defined. In this section we discuss those measures that we plan to apply to fuzzy rough databases in our ongoing research.

Let X be a discrete random variable that takes values in U with probabilities $P = \{p_1, \ldots, p_n\}$. Let $P_n(U)$ denote the set of all fuzzy subsets of the universe U. We can define a measure of fuzziness for discrete fuzzy set as follows:

Definition. A *measure of fuzziness* for a discrete fuzzy subset A is a mapping $H_A: P_n(U) \longrightarrow R^+$.

Let A and B be two fuzzy subsets of U. Ebanks [E83] proposed that any measure of fuzziness should satisfy the following properties:

(P1) H(A) is *minimum* if and only if A is a crisp set: H(A) is 0 if and only if $\mu_A(x) = 0$ or 1 for all x.
(P2) H(A) is *maximum* if and only if A is the most fuzzy set. In other words, H(A) is maximum if and only if $\mu_A(x) = .5$ for all x.
(P3) $H(A) \geq H(A^*)$, where A^* is a sharpened version of A.
(P4) $H(A) = H(1-A)$ where $\mu_{1-A}(x) = 1 - \mu_A(x)$ for all x.
(P5) $H(A \cup B) + H(A \cap B) = H(A) + H(B)$.

Pal and Bezdek [24] also considered these five (desirable) requirements to define new multiplicative and additive classes of fuzzy uncertainty measures.

Deluca and Termini [12] introduced the fuzzy entropy which is based on the concept of fuzziness without reference to probabilities. They defined the entropy measure of a fuzzy subset A as

$$H_{DT}(A) = (1/(n \ln 2)) \sum S_n(\mu_A(x_i))$$

where $S_n(\mu_A(x_i)) = -\sum \mu_A(x_i)\ln\mu_A(x_i) - (1 - \mu_A(x_i))\ln(1 - \mu_A(x_i))$.
The function $S_n(\mu_A(x_i))$ is called Shannon's information function due to its similarity to Shannon's function in the probabilistic form. The entropy defined by Deluca and Termini satisfies the five properties mentioned above.

Renyi [28] in 1961 introduced the entropy of order α of a probability distribution $P = \{p_1, \ldots, p_n\}$ associated with a discrete random variable X as

$H^\alpha(P) = (1/\alpha)\ln(\sum p_i^\alpha)$, $\alpha > 0$, $\alpha \neq 1$.

Based on this definition, Bhandari and Pal [6] defined a new measure of fuzziness for fuzzy sets as:

Definition. The *measure of fuzziness of order* α is defined as

$$H_{BP}(\alpha:A) = (K/(1-\alpha)) \sum \log(\mu_A^\alpha(x_i) + (1 - \mu_A(x_i))^\alpha)$$

where $\alpha > 0$, $\alpha \neq 1$, and K is a constant. This measure satisfies properties (P1) through (P4).

In the literature of information theory, there exist other generalized measures of information. One reason to consider alternative measures of uncertainty is to have the flexibility which may be necessary in a variety of applications. Different measures of entropy lead to a unique model for each situation. Another reason to consider generalized measures of information is that some probability distributions can be obtained by maximizing Shannon entropy but with complicated and artificial constraints. Thus if we have at our disposal a variety of measures, then we can obtain a variety of models and the model that is closest to observations will emerge as the satisfactory one. In the literature, some of the generalized measures have been applied successfully to different fields such as marketing and accounting [19]. We introduce new measures and determine if they satisfy all five properties listed above.

Havrda and Charvat [18] proposed the following measure of uncertainty of type β as

$$H_\beta(P) = A_\beta(\sum p_i^\beta - 1), \quad \beta > 0, B \neq 1 \quad \text{where } A_\beta = (2^{1-\beta} - 1)^{-1}.$$

A new measure of fuzziness based on the entropy of type β is defined as follows:

Definition. The *measure of fuzziness of type* β is

$H_\beta(A) = A_\beta K(\sum(\mu_A^\beta(x_i) + (1 - \mu_A(x_i))^\beta - 1)$ where K is a normalizing constant ($K > 0$).

Now we will verify if this measure satisfies the desirable properties listed in the previous section.

(P1) $H^\beta(A) = 0$ if and only if A is a non-fuzzy set.

Proof. Case (i): Let A be a non-fuzzy set. This means that $\mu_A(x_i) = 0$ or 1 for all i. Therefore $\mu_A^\beta(x_i) + (1 - \mu_A(x_i))^\beta - 1 = 0$ for all i. Thus $H^\beta(A) = 0$.

Case (ii): Let us assume that $H^\beta(A) = 0$. Now consider

$$f(\mu_A(x_i)) = A_\beta(\mu_A^\beta(x_i) + (1 - \mu_A(x_i))^\beta - 1)$$

for all i. Differentiating $f(\mu_A(x_i))$ with respect to $\mu_A(x_i)$, we get

$f'(\mu_A(x_i)) = A_\beta(\beta\mu_A^{\beta-1}(x_i) + \beta(1 - \mu_A(x_i))^{\beta-1}(-1)) = A_\beta\beta(\mu_A^{\beta-1}(x_i) + (1 - \mu_A(x_i))^{\beta-1}(-1))$.

Let $0 \leq \mu_A(x_i) < 0.5$. Then for $\beta > 1$, we can verify that $f(\mu_A(x_i)) \geq 0$. Also, for $\beta < 1$, we can verify that $f(\mu_A(x_i)) \geq 0$. This implies that $f(\mu_A(x_i))$ is an increasing function in $0 \leq \mu_A(x_i) < 0.5$. From symmetry, we can prove that $f(\mu_A(x_i))$ is a decreasing function in $0.5 \leq f(\mu_A(x_i)) \leq 1$. Also $f(0) = f(1) = 0$. This completes the proof.

Observation: From above, we can conclude that $f(\mu_A(x_i))$ is a concave function over $0 \leq f(\mu_A(x_i)) \leq 1$. Thus $H^\beta(A)$ is a concave function.

(P2) $H^\beta(A)$ is maximum if and only if A is the most fuzzy set.
Proof. We know that $f'(\mu_A(x_i)) = A_\beta(\beta\mu_A^{\beta-1}(x_i) + \beta(1 - \mu_A(x_i))^{\beta-1} (-1))$.
Set $f'(\mu_A(x_i)) = 0$. This implies that $\mu_A^{\beta-1}(x_i) - (1 - \mu_A(x_i))^{\beta-1} = 0$ or $\mu_A^{\beta-1}(x_i) = (1 - \mu_A(x_i))^{\beta-1}$ or $\mu_A(x_i) = 1 - \mu_A(x_i)$ or $\mu_A(x_i) = .5$.
Also $f''(\mu_A(x_i)) = A_\beta\beta(\beta-1)(\mu_A^{\beta-2}(x_i) + (1 - \mu_A(x_i))^{\beta-2})$.
This implies that $f(0.5) < 0$. This and the concavity of the function imply that $H^\beta(A)$ is maximum if and only if A is the most fuzzy set. We can easily see that the ymaximum value of the function when the set is most fuzzy and the minimum value when the set is least fuzzy are independent of the parameter β.

(P3) Sharpening reduces the value of the entropy.
Proof. Let A^* be a sharpened version of A. This means that
(i) if $\mu_A(x) < 0.5$, then $\mu_{A^*}(x) \leq \mu_A(x)$, or (ii) $\mu_{A^*}(x) \geq \mu_A(x)$, otherwise.
Since $f(\mu_A(x_i))$ is increasing in $0 \leq \mu_A(x_i) < 0.5$ and is decreasing in $0.5 \leq f(\mu_A(x_i)) \leq 1$ for each i, therefore $f(\mu_{A^*}(x_i)) \leq f(\mu_A(x_i))$. This completes the proof.

(P4) $H^\beta(A) = H^\beta(1 - A)$. It is obvious from the definition.
(P5) $H^\beta(A \cup B) + H(A \cap B) = H(A) + H(B)$. The proof follows easily.

Aczel and Daroczy [1] generalized the Renyi entropy of order α which can be written in the form:
$$H^{\alpha,\beta}(P) = ((1-\alpha)/(1-\beta)) H^\alpha(P) + ((1-\beta)/(\alpha-\beta)) H^\beta(P).$$
From this relation we can see that the entropy of order α can be expressed in terms of Renyi's entropy. Using this idea, we can define the fuzziness measure of order (α,β) in terms of fuzzy measure of order β and fuzzy measure of order α as follows:

Definition. The *fuzzy measure of order (α,β)* of a fuzzy set A can be defined as
$$H^{\alpha,\beta}(A,P) = ((1-\alpha)/(1-\beta)) H_{BP}(\alpha:A) + ((1-\beta)/(\alpha-\beta)) H_{BP}(\beta:A).$$

It can be easily verified that this measure satisfies the five properties listed in the previous section under the condition that $\beta < 1 < \alpha$ or $\alpha < 1 < \beta$.

There also exists a generalization of the entropy of type β which involves two parameters and is called the entropy of type (α,β) for a probability distribution associated with a discrete random variable X [1]. It is defined as
$$H_{\alpha,\beta}(P) = (2^{1-\alpha} - 2^{1-\beta})^{-1}\sum(p_i^\alpha - p_i^\beta), \quad \alpha,\beta > 0, \quad \alpha \neq \beta.$$
It is also easy to see that the entropy of type (α,β) can be written in terms of the entropy of type α and entropy of type β in the following way:
$$H_{\alpha,\beta}(P) = ((2^{1-\alpha} - 1)/(2^{1-\alpha} - 2^{1-\beta})) H_\alpha(P) + ((2^{1-\beta} - 1)/(2^{1-\beta} - 2^{1-\alpha})) H_\beta(P).$$
Using this idea, we can also define the fuzziness measure of type (α,β) in the following way:

Definition. The *amount of fuzziness measure of type (α,β) of a fuzzy set A* is defined as
$$H_{\alpha,\beta}(P) = ((2^{1-\alpha} - 1)/(2^{1-\alpha} - 2^{1-\beta})) H_\alpha(P) + ((2^{1-\beta} - 1)/(2^{1-\beta} - 2^{1-\alpha})) H_\beta(P).$$

It is easy to check the five properties under suitable conditions on parameters α and β listed previously.

9 Conclusion

Information theoretic measures again prove to be a useful metric for quantifying information content. In rough sets and the rough relational database, this is especially useful since in ordinary rough sets Pawlak's measure of roughness does not seem to capture the information content as precisely as our rough entropy measure.

In rough relational databases, knowledge about entropy can either guide the database user toward less uncertain data or act as a measure of the uncertainty of a data set or relation. As rough relations become larger in terms of the number of tuples or attributes, the automatic calculation of some measure of entropy becomes a necessity. Our rough relation entropy measure fulfills this need.

In fuzzy database research, fuzzy information measures have been used to determine how well a fuzzy query differentiates among possible responses [8,22] and for database security measures [7]. We expect to develop equivalent results for rough database querying. Furthermore, we are studying combined new rough/fuzzy information measures in our fuzzy rough database model [3] for data mining applications.

References

[1] J. Aczel and Z. Daroczy, *On Measures of Information and their Characterization*, Academic Press, New York, 1975.

[2] G. Arora, F. Petry, and T. Beaubouef, "New Information Measures for Fuzzy Sets," *Seventh IFSA World Congress*, Prague, June 25-29, 1997.

[3] T. Beaubouef and F.E. Petry, "Fuzzy Set Quantification of Roughness in a Rough Relational Database Model," *Proc. Third IEEE International Conference on Fuzzy Systems*, Orlando, Florida, pp. 172-177, 1994.

[4] T. Beaubouef, F. Petry, and G. Arora, "Information-Theoretic Measures of Uncertainty for Rough Sets and Rough Relational Databases," *Fifth International Workshop on Rough Sets and Soft Computing (RSSC'97)*, Research Triangle Park, NC, March 1997.

[5] T. Beaubouef, F. Petry, and B. Buckles, "Extension of the Relational Database and its Algebra with Rough Set Techniques," *Computational Intelligence*, vol. 11, no. 2, pp. 233-245, 1995.

[6] D. Bhandari and N.R. Pal, "Some New Information Measures for Fuzzy Sets," *Information Sciences*, vol. 67, pp. 209-228, 1993.

[7] B. Buckles and F. Petry, "Security and Fuzzy Databases," *Proceedings of the 1982 IEEE International Conference on Cybernetics and Society*, pp. 622-625, 1982.

[8] B.P Buckles and F. Petry, "Uncertainty models in information and database systems," *Journal of Information Science*, vol. 11, pp. 77-87, 1985.

[9] B. Buckles and F. Petry, "Information-Theoretical Characterization of Fuzzy Relational Databases," *IEEE Trans. on Systems, Man, and Cybernetics*, vol. SMC-13, no. 1, Jan/Feb, 1983.

[10] S. Chanas and D. Kuchta, "Further remarks on the relation between rough and fuzzy sets," *Fuzzy Sets and Systems*, vol. 47, pp. 391-394, 1992.
[11] E.F. Codd, "A relational model of data for large shared data banks," *Communications of the ACM*, vol. 13, pp. 377-387, 1970.
[12] A. de Luca and S. Termini, "A Definition of a Nonprobabilistic Entropy in the Setting of Fuzzy Set Theory," *Information and Control*, vol. 20, pp. 301-312, 1972.
[13] D. Dubois and H. Prade, "Twofold Fuzzy Sets and Rough Sets--Some Issues in Knowledge Representation," *Fuzzy Sets and Systems*, vol. 23, pp. 3-18, 1987.
[14] D. Dubois and H. Prade, "Putting Rough Sets and Fuzzy Sets Together," *Intelligent Decision Support: Handbook of Applications and Advances of the Rough Sets Theory*, (R. Slowinski, ed.). Boston: Kluwer Academic Publishers, 1992.
[15] B. Ebanks, "On Measures of Fuzziness and their Representations," *J. Math. Anal. Appl.*, vol. 94, pp. 24-37, 1983.
[16] K.T. Fung and C.M. Lam, "The Database Entropy Concept and its Application to the Data Allocation Problem," *INFOR*, vol. 18, no. 4, pp. 354-363, November, 1980.
[17] J. Grzymala-Busse, *Managing Uncertainty in Expert Systems*, Kluwer Academic Publishers, Boston, 1991.
[18] J.H. Havrda and F. Charvat, "Quantification Methods of Classification Processes: Concepts of Structural α Entropy," *Kybernetica*, vol.3, pp. 149-172, 1967.
[19] J. Kapur and H. Kesavan, *Entropy Optimization Principles with Applications*, Academic Press, New York, 1992.
[20] G.J. Klir and T.A. Folger, *Fuzzy Sets, Uncertainty, and Information*, Englewood Cliffs, 1988.
[21] A. Makinouchi, "A consideration on normal form of not-necessarily normalized relation in the relational data model," *Proceedings of the Third International Conference on Very Large Databases*, pp. 447-453, 1977.
[22] J.M. Morrissey, "Imprecise Information and Uncertainty in Information Systems," *ACM Trans. on Information Systems*, vol. 8, no. 2, pp. 159-180, 1990.
[23] S. Nanda and S. Majumdar, "Fuzzy rough sets," *Fuzzy Sets and Systems*, vol. 45, pp. 157-160, 1992.
[24] N.R. Pal and James C. Bezdek, "Measuring Fuzzy Entropy," *IEEE Trans. Fuzzy Systems*, vol. 2, No. 2, pp. 107-118, 1994.
[25] Z. Pawlak, "Rough Sets and Fuzzy Sets," *Fuzzy Sets and Systems*, vol. 17, pp. 99-102, 1985.
[26] Z. Pawlak, *Rough Sets: Theoretical Aspects of Reasoning About Data*, Kluwer Academic Publishers, Norwell, MA, 1991.
[27] J.R. Quinlan, "Induction of Decision Trees," *Machine Learning*, vol. 1, pp. 81-106, 1986.
[28] A. Renyi, "On Measures of Entropy and Information," *Proc. 4th Berkeley Symp. Maths. Stat. Prob.*, vol. 1, pp. 547-561, 1961.
[29] M.A. Roth, H.F. Korth, and D.S. Batory, "SQL/NF: A query language for non-1NF databases," *Information Systems*, vol. 12, pp. 99-114, 1987.

[30] C.L. Shannon, "The Mathematical Theory of Communication," *Bell System Technical Journal* **27**, 1948.

[31] R. Slowinski, "A Generalization of the Indiscernibility Relation for Rough Sets Analysis of Quantitative Information," *First International Wrokshop on Rough Sets: State of the Art and Perspectives*, Poland, September, 1992.

[32] M. Wygralak, "Rough Sets and Fuzzy Sets--Some Remarks on Interrelations," *Fuzzy Sets and Systems*, vol. 29, pp. 241-243, 1989.

Soft Concept Analysis

Robert E. Kent

Washington State University
Pullman WA 99164, USA

1 Overview

In this chapter we discuss soft concept analysis[6, 5], a study which identifies an enriched notion of *conceptual scale* as developed in formal concept analysis [2] with an enriched notion of *linguistic variable* as discussed in fuzzy logic [17]. The identification *enriched conceptual scale* ≡ *enriched linguistic variable* was made in a previous paper [5]. In this chapter we offer further arguments for the importance of this identification by discussing the philosophy, spirit, and practical application of conceptual scaling to the discovery, conceptual analysis, interpretation, and categorization of networked information resources. We argue that a linguistic variable, which has been defined at just the right generalization of valued categories [10], provides a natural definition for the process of soft conceptual scaling. This enrichment using valued categories models the relation of indiscernability [6, 11, 12], a notion of central importance in rough set theory [13]. At a more fundamental level for soft concept analysis, it also models the derivation of formal concepts[4], a process of central importance in formal concept analysis [15]. Soft concept analysis is synonymous with enriched concept analysis. From one viewpoint, the study of soft concept analysis that is initiated here extends formal concept analysis to soft computational structures. From another viewpoint, soft concept analysis provides a natural foundation for soft computation by unifying and explaining notions from soft computation in terms of suitably generalized notions from formal concept analysis, rough set theory and fuzzy set theory.

2 Networked Information Resources

An information management software system for the World-Wide Web called WAVE (Web Analysis and Visualization Environment) [1] is currently under development [9]. WAVE is a third generation World-Wide Web tool used for conceptual navigation and discovery over a universe of networked information resources.

---

[1] Accessible at the Web address http://wave.eecs.wsu.edu/.

Figure 1 is a diagram of the architecture of the WAVE system. This consists of three major components (the digital object store, the metadata object store, and the conceptual space), and three processes (metadata abstraction, conceptual scaling, conceptual browsing) which connect the components. The digital object store represents the information space for a community on the World-Wide Web as stored in various Web document collections or online databases. The metadata object store represents information abstracted from the digital object store. The process of metadata abstraction includes extraction from raw HTML[2] documents or translation from annotated XML[3] files, both done by Web robots. The WAVE system represents its metadata in a markup language (an XML application) called Ontology Markup Language (OML)[4]. OML is a semantic data model — an extended form of the entity-relationship model of database theory. OML represents information in terms of abstract objects and relations between those objects. The type structure of this information is specified in OML by an ontology, which consists of a generalization-specialization hierarchy or taxonomy of categories (object types) and relational schemata (relation types) between those categories.

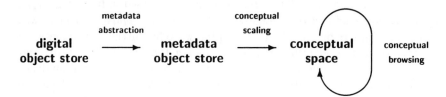

Fig. 1. The WAVE system architecture

The conceptual space is a form of conceptual knowledge representation of the original information. It organizes the ontological information of the metadata object store in terms of the formal concepts of formal concept analysis. The conceptual space represents information conceptually scaled from the metadata object store. The process of conceptual scaling applies conceptual scales to the metadata captured in the ontological categories and relations of OML, thereby transforming it into a framework suitable for the concept lattices of formal concept analysis[15]. An extension of OML called Conceptual Knowledge Markup Language (CKML)[5] specifies in the WAVE system the conceptual scales

---

[2] HyperText Markup Language, the current lingua franca of the World-Wide Web.
[3] eXtensible Markup Language, a data format for structured document interchange on the World-Wide Web; XML is a metalanguage used to define markup languages, which are called XML applications; see http://www.w3.org/XML/.
[4] See http://wave.eecs.wsu.edu/WAVE/Ontologies/OML/OML-DTD.html; Ontology Markup Language (OML) owes much to pioneering efforts of the SHOE initiative (see http://www.cs.umd.edu/projects/plus/SHOE/) at the University of Maryland at College Park.
[5] See http://wave.eecs.wsu.edu/WAVE/Ontologies/CKML/CKML-DTD.html; CKML

used in the scaling process. The conceptual space component is essentially a concept lattice with a naming facility for bookmarking favorite formal concepts. Bookmarked concepts are called conceptual views.

The WAVE system has a client/server architecture. The metadata object store component, the conceptual scaling process, and the conceptual space construction, all reside or take place on the WAVE server. The WAVE system has an online client software interface that allows the user to download the conceptual space and browse over its lattice of formal concepts. The WAVE system is a state transition system with formal concepts playing the role of states and the concept lattice functioning as a state space. The conceptual browsing process provides state transitioning over the concept lattice by either of two methods: direct transition to named concepts by mouse click; and definition of next conceptual state by use of lattice meets and joins. The process of conceptual browsing is dual mode: extensional and intentional. The extensional browsing mode, which visibly browses globally over conceptual views and attribute generated concepts, defines the next conceptual state in terms of the meets and joins of a collection of views restricted to the meets of a collection of attributes. Neighboring concepts, neither above or below the current concept, are compared to the current conceptual state in terms of the objects common to their extents. The intentional browsing mode is dual.

Fig. 2. The WAVE conceptual interface

---

follows the philosophy and practice of conceptual knowledge processing, a principled approach to knowledge representation and data analysis[15].

> Figure 2 is an image of the WAVE system client interface taken during conceptual browsing over a movie information space. The extensional browsing mode is being used here (as indicated by the depressed inverted tree symbol button). The three panes in Figure 2 are: the definition pane on the upper right, which lists elements that are being used in the definition of the current concept; the global pane on the left, which is a global display of all view and attribute concepts in the conceptual space; and the local pane on the bottom right, which is a local display of both the extent of the current concept and views more specific (below) the current concept. Named formal concepts are visible as explicit entries in the client interface: objects, attributes and conceptual views. The kind column indicates whether the lattice element is an object, attribute or view. The type column lists the scope category for an object, the conceptual scale of an attribute, and the defining/owning agent for a conceptual view. The relation column gives an "Equivalent" label for a lattice element which labels the current conceptual state, gives an "Intent" ("Extent") label for an attribute (object) which in the intent (extent) of the current state, gives an "Ancestor" ("Descendant") label for a view which above (below) the current state in the lattice order, and shows a "Similar" label for a lattice element which is off to the side — neither above nor below the current concept. Finally, the similarity column displays the extensional similarity between a lattice element and the current conceptual state; in extensional mode elements above the current state have maximal similarity. The current conceptual state in Figure 2 is the lattice meet (the depressed big M symbol button) of (1) the "Western" attribute defined by a nominal conceptual scaling of a composite description function consisting of just the binary relation "genre" between movie ontology categories "Movie" and "Genre" ("Western" is defined in CKML by the query "What movies have western genre?") and (2) the "Recent Movie" attribute defined by an ordinal conceptual scaling of the simple description function "year" from the category "Movie" to the datatype "Date" ("Recent Movie" is defined in CKML by the query "What movies appeared in year $\geq$ 1990?"). The current concept in Figure 2 does not have a name; that is, it is not labeled as a conceptual view. If the current concept was a view, that view name would appear in the global pane with an "Equivalent" value in the relation column. Anonymous concepts such as this can be given a name by using them to define a new conceptual view (the bright dot symbol button) — conceptual views are named formal concepts. The extent of the current conceptual state in Figure 2 is the collection of three movies listed in the local pane. The similarity column in the global pane shows that the current concept has one object in common with the attribute "Suitable for US teenagers". Extensional similarity, which provides the user with a sideways neighboring dimension within the concept lattice, is defined as the extent cardinality of the lattice meet of two concepts.

The analogies in Table 1 link the various major components and processes in the architecture of WAVE with the traditional library, and illustrate the fact that the WAVE system is a digital library. Interpretation of resource descriptions, via conceptual scaling or faceted analysis, plays a central role in the WAVE system. At the present time, the WAVE system conceptually analyzes, interprets, and categorizes resources, such as Web textual and image documents, in a crisp or hard fashion. However, using ideas developed in this chapter, an excellent approach for the extension to an enriched WAVE system is quite clear. The following short list of conceptually scalable attributes indicates that notions of approximation are very important for networked information resources: the visible size of textual documents in pages or some other meaningful unit; the concept extent cardinality as a count of equivalent instances of resources; similarity measures between Web documents based upon numbers of common attributes; relative scores for search engine keyword search; the cost of resources; the duration of play for audio/video data; the critical review of resources; etc. As a particular example of some importance, the co-occurrence matrices of semantic retrieval[1] effectively represent networked information as weighted or fuzzy formal contexts. An enriched WAVE system will allow the user to define according to his own judgement various enriched interpretations of networked resource information.

The Traditional Library	The WAVE System
holdings	digital object store
catalogs and indexes	metadata object store
classification scheme Dewey, LC, Colon, etc.	concept space
cataloging practices	metadata abstraction conceptual scaling
reference librarian	conceptual browsing

Table 1. Analogies

3 The Process of Conceptual Scaling

The application of facets in the theory of library classification was first tested and developed by Ranganathan in his Colon classification system. Faceted analysis and classification provides a flexible means to classify complex, multi-concept subjects[14]. Complex subjects are divided into their component, single-concept subjects. Single-concept subjects are called *isolates*. Faceted analysis examines the literature of an area of knowledge and identifies its isolates. A *facet* is the sum total of isolates formed by the division of a subject by one characteristic of division. Some examples of facets in musical literature are: composer, instrument, form, etc. Isolates within facets are known as foci.

faceted analysis	conceptual scaling	linguistic variable use
facet	conceptual scale	linguistic variable
isolate/foci	scale attribute	linguistic value

Table 2. Identities

Comparing these ideas to ontologically structured metadata[5], facets are identified with conceptual scales or linguistic variables and are often associated with a composite description function, and isolate/foci are identified with scale attributes or linguistic values. A composite description function in the ontology may consist of one function with primitive image values, or a binary relation connecting two categories of objects composable with such a function, or something more complex. These observations are recorded in Table 2. In conceptual knowledge processing each facet is computed by a conceptual scale. A conceptual scale is an active filter or lens through which information is interpreted. Faceted analysis is conceptual scaling. It involves four steps.

1. Gather ontologically structured metadata.
2. Identify conceptual scales of interest and specify attributes within them.
3. Specify the structure of conceptual scales.
4. Apply the conceptual scales to the metadata, producing a composable vector of facets which constitutes the conceptual space.

As stated in the following principles of interpretation and classification, our use of information involves our interpretion of it.

- **Information use is interpretation:** Humans use information by representing it in conceptual structures. Such conceptual structures are constructable. This construction is partly interpretive and partly automatic. The design of the conceptual constructors, although governed by principles, requires interactive advice from human experts. The application of the conceptual constructors, the actual categorization of information and the construction of conceptual classes, can be automatic.
- **Interpretation involves classification:** Interpretation defines (implicit) conceptual structures. Since explicit categories are special conceptual structures (tree structures for single inheritance or directed acyclic graphs for multi-inheritance), specified classification can help define interpretation.
- **Conceptual classification is general:** Most conceptual classification structures are hierarchical tree structures. But tree structures are special cases of lattice structures — just add a bottom node which is the meet for all pairs of categories; the join of two categories is their most specific ancestor.
- **Conceptual classification is composable:** Conceptual classification schemes can be composed, using summing operations (such as apposition) and/or producting operations, from a small set of primitive conceptual structures called *conceptual scales*.
- **Conceptual generation is polar:** Each collection of instances (objects) generates a category (conceptual class). The smaller the collection of objects, the more prototypical and exemplary they are of the category that they generate.

scale type	mathematical structure	intuitive idea
nominal	set	partition/separateness
ordinal	(often) total order	ranking
interordinal	partial order of intervals	betweenness
hierarchical	tree structure	nesting
metrical	generalized metric space	similarity

Table 3. Conceptual scales

For conceptual knowledge representation to be useful, we need to develop some practical guides for conceptual scaling. From the user's standpoint, there must be a *purpose* in mind and an intended *use* for the information. It would be good to write these down explicitly. The information itself is usually concerned with *entities*, although entity tuples might be appropriate. In traditionally crisp or hard conceptual knowledge representation, in order to form a base for conceptual knowledge we must ask *true-false questions* about the information. We can compile these, for purpose of interactive management, into a database of natural language queries. The bottom level of this query database forms coherent components which we call *conceptual scales*. This partitions the queries. A conceptual scale is associated with a composite description function. As listed in Table 3, conceptual scales can themselves be typed according to purpose or use, and mathematical structure [2]. Mathematical types of scales represent intuitive ideas of design.

The process of conceptual scaling as depicted in Figure 3 consists of the interpretation of ontologically structured metadata. From a technical standpoint as depicted in Figure 5, conceptual scaling is the conversion of a composite description function to a facet. But more importantly from both philosophical and practical standpoints, the development of conceptual scales is an act of interpretation which defines views of the information along a variety of information dimensions called facets. These facets form a spectrum of interpretation/classification, from the very particular and often ad hoc, through the more pragmatic and utilitarian, to the very general and scientific.

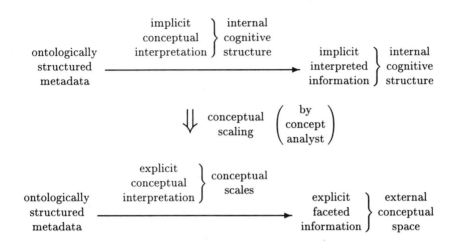

Fig. 3. The process of conceptual scaling

There are three constituents in the development of facets by means of conceptual scales: the abstract conceptual scale, the concrete conceptual scale, and the process of conceptual scaling.

– [**Abstract conceptual scale**] The first constituent is the development of a conceptual scale associated with a composite description function. This involves the linguistic analysis of this information dimension.
  1. The creation and choice of scale attributes or linguistic values (terms) — words or phrases meaningful for this particular information dimension. It is very important to observe that terms form a spectrum, from terms used for very individualistic and ad hoc interpretation/classification to terms with a common and accepted meaning in science and society.

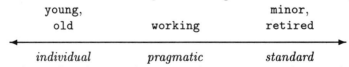

  2. The analysis of implicational structure between terms. As an aid in the explication of conceptual scaling we will use our adaptation of an example of people's age developed in an unpublished report by Karl Erich Wolff. The abstract **Age** conceptual scale is represented (equivalently) in Figure 4 as either a basis of implications, a lattice, or a one-valued formal context. The Person ontology in Table 6 specifies the abstract **Age**

conceptual scale in CKML. This particular abstract Age conceptual scale has the form of a biordinal scale. The total set of implications can be generated from the basis of implications by use of the following inference rules.

$$\text{transitive: } \frac{X \Rightarrow Y, Y \Rightarrow Z}{X \Rightarrow Z} \quad \text{projective: } \frac{X \supseteq Y}{X \Rightarrow Y} \quad \text{additive: } \frac{X \Rightarrow Y, X \Rightarrow Z}{X \Rightarrow (Y \cup Z)}$$

- [Concrete conceptual scale] The second constituent is the development of a concrete conceptual scale over the natural numbers primitive data type $D = \{0, 1, 2, \cdots, \} = \aleph$. This involves the binding of the abstract conceptual scale with logical queries.
  1. Assignment of logical query formula to the terms of the abstract Age conceptual scale as in Table 4. The logical query formulas must satisfy the constraints listed in Table 5 specified by the implicational basis of the abstract Age conceptual scale.

$$\left. \begin{array}{rl} \texttt{minor} \longmapsto \phi_{\text{minor}} & = (x \leq 18)? \\ \texttt{young} \longmapsto \phi_{\text{young}} & = (x < 40)? \\ \texttt{working} \longmapsto \phi_{\text{working}} & = (x \leq 65)? \\ \texttt{retired} \longmapsto \phi_{\text{retired}} & = (x > 65)? \\ \texttt{old} \longmapsto \phi_{\text{old}} & = (x \geq 80)? \end{array} \right\} \text{assignment}$$

Table 4. Conceptual scale assignment

$$\left. \begin{array}{r} \phi_{\text{minor}} \subseteq \phi_{\text{young}} \\ \phi_{\text{young}} \subseteq \phi_{\text{working}} \\ \phi_{\text{old}} \subseteq \phi_{\text{retired}} \\ \phi_{\text{working}} \cap \phi_{\text{retired}} = \emptyset \end{array} \right\} \begin{array}{l} \text{implicational} \\ \text{constraint} \\ \text{satisfaction} \end{array}$$

Table 5. Conceptual scale constraints

  2. Calculation of conceptual contingents (distinguishing characteristics). Recall the structure of the usual dictionary definition in terms of superordinate concept and distinguishing characteristics. For example, the definiton of the category of trees given as follows.

     tree *noun*: a woody perennial plant having a single usually elongated main stem generally with few or no branches on its lower part

     Here the category of trees has the category of plants as its immediate superordinate, and has the distinguishing characteristics: woody, perennial, one-branching-stem. The concrete Age conceptual scale is represented as

a lattice in the center of Figure 5 by calculating concept contingents γ_n for each concept n via the formula
$$\gamma_n \stackrel{\mathrm{df}}{=} \phi_n \wedge (\bigwedge_i \neg \phi_i),$$
where i ranges over all children nodes of n. Together with the Person ontology, the Person collection in Table 6 specifies the concrete Age conceptual scale in CKML by the assignment of queries to terms of the Age scale in the Person ontology.

- [**Conceptual scaling process**] The third constituent is the evaluation of logical queries with respect to ontologically structured metadata. Facet interpretation of metadata involves contingent query evaluation. Consider the People category, described on the left side in Figure 5, which might be part of a questionaire. Missing values, so-called database nulls, are represented by the question mark "?". The Age dimension of the People data is interpreted in terms of the Age conceptual scale by evaluating the concept contingent logical query. These evaluations interpret the information as a single Age facet, which can be visualized as the concept lattice on the right side in Figure 5.

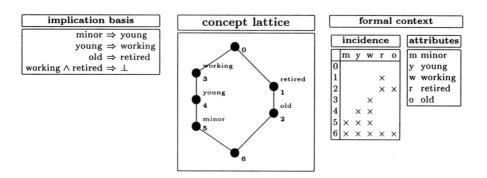

Fig. 4. Abstract conceptual scale: Age (3 forms)

4 Valuated Enrichment

Indiscernibility, a central concept in rough set theory, is traditionally treated as a hard relationship — either two objects are indiscernible or they are not. In order to define and develop a soft theory of rough sets, it would seem quite appropriate, if not necessary, to define and develop a soft or graded version of indiscernibility. We review this approach[5] by using ideas from the theory of valuated categories[10].

```
<ONTOLOGY NAME="Person" VERSION="1.0">
    ...
    <CATEGORY NAME="Person"/>
    ...
    <FNSCHEMA NAME="age"
        ARGTYPE="Person"
        IMAGETYPE="Integer"/>
    ...
    <SCALE CATEGORY="Person" NAME="Age">
        <TERM NAME="Young"/>
        <TERM NAME="Old"/>
        <TERM NAME="Working"/>
        <TERM NAME="Minor"/>
        <TERM NAME="Retired"/>
        <IMPLICATION>
            <IF><TERM NAME="Minor"/></IF>
            <THEN><TERM NAME="Young"/></THEN>
        </IMPLICATION>
        <IMPLICATION>
            <IF><TERM NAME="Young"/></IF>
            <THEN><TERM NAME="Working"/></THEN>
        </IMPLICATION>
        <IMPLICATION>
            <IF><TERM NAME="Old"/></IF>
            <THEN><TERM NAME="Retired"/></THEN>
        </IMPLICATION>
        <IMPLICATION>
            <IF><TERM NAME="Working"/>
                <TERM NAME="Retired"/></IF>
        </IMPLICATION>
    </SCALE>
    ...
</ONTOLOGY>
```

```
<COLLECTION KIND="attribute" SCOPE="People">
    <USES ONTOLOGY="Person" VERSION="1.0"/>
    ...
    <ATTRIBUTE SCALE="Age" KEY="Minor">
        <QUERY VARIABLE="person" CATEGORY="People"/>
            <FN2REL NAME="age" ORDER="less-equal">
                <ARGUMENT VALUE="person"/>
                <ARGUMENT VALUE="18"/>
            </FN2REL>
        </QUERY>
    </ATTRIBUTE>
    ...
</COLLECTION>
```

Table 6. Ontology and attribute collection in CKML

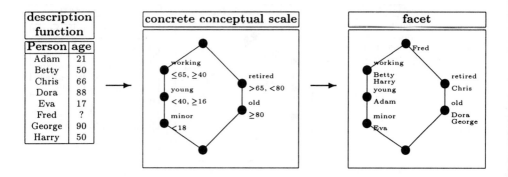

Fig. 5. Conceptual scaling process: Age

An *approximation space* [13] is traditionally defined as a pair $\mathcal{G} = \langle G, E \rangle$ consisting of a set of objects or entities G and an equivalence relation $E \subseteq G \times G$ called indiscernibility. Two objects $g_1, g_2 \in G$ are *indiscernible* when $g_1 E g_2$; that is, when $E(g_1, g_2) = \text{true}$. Equivalently, an approximation space (function version) is a triple $\langle G, \phi, D \rangle$, where G is a set of objects, D is a set (hard, crisp and unenriched!) of values, and $G \xrightarrow{\phi} D$ is a (not necessarily surjective) function called a *description function*. The description function ϕ represents a certain amount of knowledge about the objects in G. Two objects $g_1, g_2 \in G$ are *indiscernible* when the procedure ϕ cannot distinguish between them, $\phi(g_1) = \phi(g_2)$; that is, when $\text{Eq}_D(\phi(g_1), \phi(g_2)) = \text{true}$. To enriched rough set notions, we allow grades of indiscernibility by assuming that D has **V**-enriched structure on it, where $\mathbf{V} = \langle V, \preceq, \otimes, \Rightarrow, e \rangle$ is a closed preorder; that is, we assume that D is an approximation **V**-space.

A *closed preorder* [10] $\mathbf{V} = \langle V, \preceq, \otimes, \Rightarrow, e \rangle$ consist of the following data and axioms. $\langle V, \preceq, \otimes, e \rangle$ is a symmetric monoidal preorder, with $\langle V, \preceq \rangle$ a preorder and $\langle V, \otimes, e \rangle$ a commutative monoid, where the binary operation $\otimes: V \times V \to V$, called **V**-*composition*, is monotonic (if both $u \preceq u'$ and $v \preceq v'$ then $u \otimes v \preceq u' \otimes v'$), and symmetric or commutative ($a \otimes b = b \otimes a$ for all elements $a, b \in V$), and satisfies the closure axiom: the monotonic **V**-composition function $(\,) \otimes b: V \to V$ has a specified right adjoint $b \Rightarrow (\,): V \to V$ for each element $b \in B$ called **V**-*implication*, hence satisfying the equivalence $a \otimes b \preceq c$ iff $a \preceq b \Rightarrow c$ for any triple of elements $a, b, c \in V$. We list some important closed preorders which can be used in soft concept analysis for the interpretation in linguistic variables.

booleans $\langle 2 = \{0, 1\}, \leq, \wedge, \to, 1 \rangle$, where 0 is **false**, 1 is **true**, $\leq$ is the usual order on truth-values, $\wedge$ is the truth-table for **and**, and $\to$ is the truth-table for **implies**. This defines the hard context of traditional set theory and logic.

fuzzy truth-values $\langle [0, 1], \leq, \wedge, \to, 1 \rangle$ where 0 is **false**, 1 is **true**, $0 \leq r \leq 1$ is some grade of truth-value between **false** and **true**, $\leq$ is the usual order on fuzzy truth-values in the interval, $\wedge$ is the minimum operation representing the interval truth-table for the fuzzy **and**, and $\to$ is the operation (defined by $r \to s = 1$ if $r \leq s$, or s otherwise) representing the interval truth-table for the fuzzy **implies**. This defines a soft context for fuzzy set theory and logic.

reals $\Re = \langle \Re = [0, \infty], \geq, +, \dot{-}, 0 \rangle$, $\geq$ is the usual downward ordering on the nonegative real numbers $\Re$ (regarded as quantitative truth-values), $+$ is sum, and $\dot{-}$ is the operation (defined by $s \dot{-} r = 0$ if $r \geq s$, or $s - r$ otherwise) representing the truth-table for the metrical **difference**. This defines the soft context of metric spaces.

In soft concept analysis the operation of **V**-implication is used for at least three different purposes: (1) the enriched lower approximation operator uses **V**-implication; (2) **V**-implication is sometimes used in the queries which define enriched conceptual scales; and (3) the operation of derivation, which defines the notion of an enriched formal concept, is a special case of enriched relational residuation, which itself is defined using **V**-implication. The values in the closed preorder V are regarded as being a set of generalized truth values.

While enriched approximation spaces are the appropriate abstraction of indiscernibility and our main concern in this paper, it seems that these approximation spaces are best defined in terms of an asymmetric generalization called simply an enriched space. A pair $\mathcal{X} = \langle X, \mu \rangle$ consisting of a set X and a function $\mu: X \times X \to \mathbf{V}$ is called a **V**-*enriched space* or **V**-*space* when it satisfies the reflexivity (zero law) $e \preceq \mu(x, x)$ for all $x \in X$, and the transitivity (triangle axiom) $\mu(x_1, x_2) \otimes \mu(x_2, x_3) \preceq \mu(x_1, x_3)$ for all $x_1, x_2, x_3 \in X$. The function μ, called a *metric*, represents a distance or measure of agreement between the elements of X. We can interpret μ to be either an enriched preordering, a generalized distance function, a similarity measure, or a gradation. When $\mathbf{V} = 2$, the crisp boolean case, a **V**-space $\mathcal{X}$ is precisely a preorder $\mathcal{X} = \langle X, \preceq \rangle$ with order characteristic function $\preceq: X \times X \to 2$. When $\mathbf{V} = \Re$, the metric topology case, a **V**-space $\mathcal{X}$ is (generalize) metric space $\mathcal{X} = \langle X, \delta \rangle$ with distance function $\delta: X \times X \to \Re$. When $\mathbf{V} = [0, 1]$, the fuzzy case, a **V**-space $\mathcal{X}$ is a fuzzy space $\mathcal{X} = \langle X, \mu \rangle$ with similarity measure $\mu: X \times X \to [0, 1]$.

Any **V**-space $\mathcal{X} = \langle X, \mu \rangle$ has a *dual* or *opposite* **V**-space $\mathcal{X}^{\mathrm{op}} = \langle X, \mu^{\mathrm{op}} \rangle$, where $\mu^{\mathrm{op}}(x_1, x_2) = \mu(x_2, x_1)$ is the dual or opposite metric. In general our metrics are asymmetrical: $\mu(x_1, x_2) \neq \mu(x_2, x_1)$. A **V**-*enriched approximation space* or *approximation* **V**-*space* is defined to be a symmetrical **V**-space. Here the metric μ, called an *indiscernibility measure*, is a **V**-enriched equivalence relation on X satisfying reflexivity, transitivity and symmetry $\mu(x_2, x_1) = \mu(x_1, x_2)$ for all $x_1, x_2 \in X$. Any **V**-space $\mathcal{X} = \langle X, \mu \rangle$ can be symmetrized and made into an approximation space, by defining the metric $\mu^{\mathrm{sym}}(x_1, x_2) = \mu(x_1, x_2) \otimes \mu^{\mathrm{op}}(x_1, x_2)$.

A **V**-*map* $f: \mathcal{X} \to \mathcal{Y}$ between two **V**-spaces $\mathcal{X} = \langle X, \mu \rangle$ and $\mathcal{Y} = \langle Y, \nu \rangle$ is a function $f: X \to Y$ that preserves measure by satisfying the condition $\mu(x_1, x_2) \preceq \nu(f(x_1), f(x_2))$ for all $x_1, x_2 \in X$. When $\mathbf{V} = 2$, the crisp boolean case, a **V**-map $f: \mathcal{X} \to \mathcal{Y}$ is precisely a monotonic function. When $\mathbf{V} = \Re$, the metric topology case, a **V**-map $f: \mathcal{X} \to \mathcal{Y}$ is precisely a contraction. When $\mathbf{V} = [0,1]$, the fuzzy case, a **V**-map $f: \mathcal{X} \to \mathcal{Y}$ is a fuzzy measure preserving function.

Each element $x \in X$ of a **V**-space $\mathcal{X} = \langle X, \mu \rangle$ can be represented as the **V**-predicate $\mathrm{y}(x) = \mu(x, -)$ over $\mathcal{X}$ where $\mathrm{y}(x)(x') = \mu(x, x')$ for each element $x' \in X$. The function $\mathrm{y}_X: X \to V^X$, which is called the *Yoneda embedding*, is a **V**-isometry $\mathrm{y}_{\mathcal{X}}: \mathcal{X}^{\mathrm{op}} \to \mathbf{V}^{\mathcal{X}}$. Composition of (the opposite of) a **V**-map $f: \mathcal{X} \to \mathcal{Y}$ on the right with the Yoneda embedding $\mathrm{y}_{\mathcal{Y}}: \mathcal{Y}^{\mathrm{op}} \to \mathbf{V}^{\mathcal{Y}}$, resulting in the **V**-map $f_*: \mathcal{X}^{\mathrm{op}} \to \mathbf{V}^{\mathcal{Y}}$, allows us to generalize the concept of a **V**-map. Such a generalized **V**-map, equivalent to a **V**-map $\mathcal{X}^{\mathrm{op}} \times \mathcal{Y} \xrightarrow{\tau} \mathbf{V}$, may be regarded to be a **V**-*enriched relation* or **V**-*relation* from $\mathcal{X}$ to $\mathcal{Y}$. It is denoted by $\mathcal{X} \xrightarrow{\tau} \mathcal{Y}$, with $\tau(x, y)$ an element of **V** interpreted as the "truth-value of the τ-relatedness of x to y" [10].

A pair of **V**-relations $\mathcal{X} \xrightarrow{\sigma} \mathcal{Y}$ and $\mathcal{Y} \xrightarrow{\tau} \mathcal{Z}$ can be composed, yielding the **V**-relation $\mathcal{X} \xrightarrow{\sigma \circ \tau} \mathcal{Z}$ called *composition*, and defined to be the supremum (iterated disjunction) $(\sigma \circ \tau)(x, z) = \bigvee_{y \in \mathcal{Y}} (\sigma(x, y) \otimes \rho(y, z))$. An enriched relation $\mathcal{X} \xrightarrow{\tau} \mathcal{Y}$ is closed with respect to the metrics on both left and right: $\mu \circ \tau \preceq \tau$ and $\tau \circ \nu \preceq \tau$. Relational composition has a right adjoint called residuation. The *residuation* of a pair of **V**-relations $\mathcal{X} \xrightarrow{\sigma} \mathcal{Y}$ and $\mathcal{X} \xrightarrow{\rho} \mathcal{Z}$, denoted by the **V**-relation $\mathcal{Y} \xrightarrow{\sigma \setminus \rho} \mathcal{Z}$, is defined to be the infimum (iterated conjunction) $(\sigma \setminus \rho)(y, z) = \bigwedge_{x \in \mathcal{X}} (\sigma(x, y) \Rightarrow \rho(x, z))$.

As mentioned above, every **V**-map $\mathcal{X} \xrightarrow{f} \mathcal{Y}$ determines a **V**-relation $\mathcal{X} \xrightarrow{f_*} \mathcal{Y}$ defined by $f_* = f^{\mathrm{op}} \cdot \mathrm{y}_{\mathcal{Y}}$, or on elements by $f_*(x, y) = \nu(f(x), y)$. In particular, the Yoneda embedding becomes the relation $\mathcal{X} \xrightarrow{\mu} \mathcal{X}$. Dually every **V**-map $\mathcal{X} \xrightarrow{f} \mathcal{Y}$ also determines a **V**-relation $\mathcal{Y} \xrightarrow{f^*} \mathcal{X}$ in the opposite direction defined by $f^* = \mathrm{y}_{\mathcal{Y}} \cdot \mathbf{V}^f$, or on elements by $f^*(y, x) = \nu(y, f(x))$.

The *power* **V**-*space* $\mathbf{V}^{\mathcal{X}}$ of all **V**-valued **V**-maps on $\mathcal{X}$ is an **V**-space with metric $\phi \Rightarrow \psi = \bigwedge_{x \in X} (\phi(x) \Rightarrow \psi(x))$. We interpret an element of $\mathbf{V}^{\mathcal{X}}$, a **V**-map $\phi: \mathcal{X} \longrightarrow \mathbf{V}$, to be a **V**-enriched subset, which satisfies the internal pointwise metric constraint $\mu: \mu(x_1, x_2) \preceq \phi(x_1) \Rightarrow \phi(x_2)$ for all $x_1, x_2 \in X$; or equivalently, by the $\otimes$ - $\Rightarrow$ adjointness, the constraint $\phi(x_1) \otimes \mu(x_1, x_2) \preceq \phi(x_2)$ for all $x_1, x_2 \in X$. Such a characteristic function $\phi: \mathcal{X} \to \mathbf{V}$, which is constrained by the metric on $\mathcal{X}$, is called a **V**-*predicate* or *enriched predicate* in $\mathcal{X}$. Using terminology from rough set theory, it can also be called a **V**-*definable subset* in $\mathcal{X}$.

When **V** = 2, the crisp boolean case, for an approximation space $\mathcal{X} = \langle X, E \rangle$, a **V**-predicate $\phi: \mathcal{X} \to 2$ satisfying the constraint $\phi(x_1) \wedge E(x_1, x_2) \leq \phi(x_2)$ for all $x_1, x_2 \in X$ is precisely a definable subset in X. When **V** = $\Re$, the metric topology case, for an **V**-space $\mathcal{X} = \langle X, \delta \rangle$, a **V**-predicate $\phi: \mathcal{X} \to \Re$ satisfying the constraint $\phi(x_1) + \delta(x_1, x_2) \geq \phi(x_2)$ for all $x_1, x_2 \in X$ is called a *closed subset* of $\mathcal{X}$ — closed w.r.t. the distance function δ. When **V** = [0, 1], the fuzzy case, a **V**-predicate $\phi: \mathcal{X} \to [0, 1]$ satisfying the constraint $\phi(x_1) \wedge \mu(x_1, x_2) \leq \phi(x_2)$ for all $x_1, x_2 \in X$, or equivalently, the constraint "equal below points of similarity" $\mu(x_1, x_2) \leq \phi(x_1), \phi(x_2)$ or $\phi(x_1) = \phi(x_2) \leq \mu(x_1, x_2)$.

5 Enriched Conceptual Scales

We describe enriched conceptual scales ($\equiv$ enriched linguistic variables) in terms of a use-case scenario. We start with a collection of objects $\mathcal{G} = \langle G, \gamma \rangle$. We assume that some observations have been made or some experimental measurements have been done, resulting in the production of some "raw" data $\mathcal{D} = \langle D, \delta \rangle$. This data is associated with the objects by a map called a *description function* $\mathcal{G} \xrightarrow{\phi} \mathcal{D}$. Both objects and data can be enriched as approximation spaces for benefit of flexibility by using soft structures. We will use enriched conceptual scales in order (1) to interpret this data and (2) to provide a facet of it which is meaningful to the user. The creation of enriched conceptual scales is an act of interpretation.

Mathematically, the notion of an enriched attribute ($\equiv$ linguistic value) is represented here by the notion of an enriched predicate. An *enriched attribute* over data domain $\mathcal{D} = \langle D, \delta \rangle$ is an enriched predicate in $\mathbf{V}^{\mathcal{D}}$. An *enriched conceptual scale* ($\equiv$ enriched linguistic variable) [17, 2, 5] over data domain $\mathcal{D} = \langle D, \delta \rangle$ is a collection $\sigma = \{\sigma_m \in \mathbf{V}^{\mathcal{D}} \mid m \in M\}$ of enriched attributes over $\mathcal{D}$, indexed by a collection of attribute symbols or terms M. In the crisp case, **V** = 2, the assignments are part of a concrete conceptual scale (see Table 4). Using functional notation we can write this as the **V**-map $\sigma: \mathcal{M} \to \mathbf{V}^{\mathcal{D}}$, where we have enriched the attributes to an **V**-space $\mathcal{M} = \langle M, \mu \rangle$. In the crisp boolean case, **V** = 2, the metric μ represents the implication basis order in an abstract conceptual scale (see Figure 4). An enriched conceptual scale can be represented as the relation $\mathcal{M} \xrightarrow{\sigma} \mathcal{D}$ where $\sigma(m, d) \stackrel{\mathrm{df}}{=} \sigma(m)(d)$. In the crisp boolean case, **V** = 2, closure of this enriched relation with respect to the term metric $\mu \circ \sigma \preceq \sigma$ represents constraint satisfaction in the concrete conceptual scale (see Table 5). The four parts of a enriched conceptual scale can be interpreted as follows.

1. $\mathcal{D}$ gives its data scope or range,
2. **V** represents our interpretation style,
3. $\mathcal{M}$ gives attributes of the enriched conceptual scale
4. σ assigns enriched predicates to terms.

These are listed in order of volatility — of these four, $\mathcal{D}$ varies slowest (it is given to us), whereas σ is most volatile. The standard way to combine two enriched conceptual scales is used in the apposition of formal contexts [2]. Given two enriched conceptual scales (with no apparent relationships) $\mathcal{M}_0 \xrightarrow{\sigma_0} \mathcal{D}_0$ and $\mathcal{M}_1 \xrightarrow{\sigma_1} \mathcal{D}_1$, the *apposition* enriched conceptual scale $\mathcal{M}_0 \oplus \mathcal{M}_1 \xrightarrow{\sigma_0 | \sigma_1} \mathcal{D}_0 \otimes \mathcal{D}_1$ from the unconstrained sum space of terms to the tensor product space of data, is defined by $\sigma_0|\sigma_1\,(m_0,(d_0,d_1)) \stackrel{\mathrm{df}}{=} \sigma_0(m_0,d_0)$ and $\sigma_0|\sigma_1\,(m_1,(d_0,d_1)) \stackrel{\mathrm{df}}{=} \sigma_1(m_1,d_1)$.

Using the Age example discussed above for conceptual scaling, we can provide a crisp interpretation of people's age description using the boolean closed poset

$$\mathcal{M} \xrightarrow{\sigma} \mathcal{D} = \left[\begin{array}{l} \mathcal{G} \;=\; \text{Person} \\ \phi \;=\; \text{age description function} \\ \mathcal{D} \;=\; \aleph = \{0,1,2,\ldots\} \\ \mathbf{V} \;=\; \mathbf{2} = \text{crisp boolean closed poset} \\ \mathcal{M} \;=\; \{\text{``minor''}, \text{``young''}, \text{``working''}, \text{``retired''}, \text{``old''}\} \\ \sigma(\text{``young''})(d) \;=\; \begin{cases} 1, & \text{if } 0 \leq d < 40 \\ 0, & \text{if } 40 \leq d \end{cases} \\ \qquad\qquad\qquad \text{etc.} \end{array} \right]$$

or we can provide a fuzzy interpretation of people's age description using the fuzzy truth-values closed poset and fuzzy predicates assigned to terms, such as

$$\left[\begin{array}{l} \sigma(\text{``young''})(d) \;=\; \begin{cases} 1, & \text{if } 0 \leq d \leq 20 \\ -\frac{1}{20}d + 2, & \text{if } 20 \leq d \leq 40 \\ 0, & \text{if } 40 \leq d \end{cases} \\ \qquad\qquad\qquad \text{etc.} \end{array} \right]$$

We use an enriched conceptual scale to interpret the meaning of the metadata — the description function ϕ. This interpretation, called *simple enriched conceptual scaling*, applies the conceptual scale σ by composing it with the description function metadata ϕ, resulting in the facet $\iota \stackrel{\mathrm{df}}{=} \phi_* \circ \sigma^{\mathrm{op}}$. This takes the form of a **V**-relation $\mathcal{G} \xrightarrow{\iota} \mathcal{M}$ called an enriched formal context. In terms of elements this definition is $\iota(g,m) = \tilde{\sigma}(\phi(g))(m) = \sigma(m,\phi(g)) = \sigma^{op}(\phi(g),m)$. It is important to observe that simple enriched conceptual scaling is synonymous with the notion of *granulation* in fuzzy set theory.

Composite description functions within the metadata allow for the definition of richer and more complex attributes in composite conceptual scales. Such *composite enriched conceptual scaling* provides for automatic translation of natural language specifications[16] of conceptual scales. For example, consider the composite description function consisting of a binary "membership" relation between the categories "Person" and "Social Organization" and the simple description function "age" from the category "Person" to the datatype of natural numbers $\aleph$. Define the crisp conceptual scale attribute "youth organization" by the query "What social organizations have only young members?" This can be expressed mathematically as $s \in$ "Social Organization" and for all $p \in$ "Person" we have $\mathrm{member}(p,s) \Rightarrow \mathrm{age}(p, \text{``young''})$. Using residuation, this can be expressed as $(s, \text{``young''}) \in (\psi \backslash (\phi_* \circ \sigma^{\mathrm{op}}))$, where

$$\begin{array}{rl} \mathcal{G}' \;=\; & \text{Social Organization} \\ \mathcal{M} \xrightarrow{\sigma} \mathcal{D} \;=\; & \text{simple age conceptual scale} \\ \mathcal{G} \xrightarrow{\psi} \mathcal{G}' \;=\; & \text{membership relation} \\ \psi \backslash (\phi_* \circ \sigma^{\mathrm{op}}) \colon \mathcal{G}' \to \mathcal{M} \;=\; & \text{resulting facet} \end{array}$$

This works quite well in the case of composite description functions consisting of simple description functions and binary relations, and many examples of composite description functions can be modeled using relational composition and residuation. However, it becomes more complicated in the case of n-ary relations. Consider the context of corporate information. Specifically, information about the revenue rank of corporations on a scale 0 to 100. For investors, an interesting query is: "Has the revenue rank of the corporation been universally high recently?" The processing here involves multiple ordinal scales — a Date ordinal scale and an ordinal scale on the subrange [0, 100] of natural numbers. The metadata here is principally concerned with the ternary "revenue rank" relation $\rho \subseteq \mathcal{C} \times \mathcal{D} \times [0, 100]$, where $\mathcal{C}$ represents the category of all corporations, and $\mathcal{D}$ is the date datatype. Here we need to define the two attributes: "high" within a suitable conceptual scale on values [0, 100], and "recently" within a suitable conceptual scale on dates $\mathcal{D}$. For example, crisp specifications might be that "high" means any rank above 80 and "recently" means within the last 10 years. Whatever the interpretations for these two terms, the central question is how to handle the ternary relation ρ. One method would be to use the process of reification, which is used for interoperability between metadata standards [8]. Mathematically, the relation ρ is replaced by its three projection functions. Even after this conversion from the n-ary to the binary case, there yet remains the question of how to combine the scales with the metadata. There appear to be several answers to this question.

6 Summary and Future Work

Conceptual scales are now being used for the interpretation, classification and organization of networked information resources [8]. An enriched notion of conceptual scale would provide for a more flexible approach in the interpretation of networked information. Enriched conceptual scales equivalent to enriched linguistic variables unify ideas from formal concept analysis, rough set theory, and fuzzy set theory.

Future work could possibly include any of the following initiatives. The basic theorem [15] of formal concept analysis should be developed in the enriched context. Also in the area of formal concept analysis, the basic operations on conceptual scales, such as sums, products, and the important apposition operation, should be further developed in the enriched setting. It may be profitable to investigate possible connections, such as fixpoints, between enriched concept spaces and generalized metric spaces [10, 3]. From the standpoint of soft computation, there needs to be a closer integration of the enriched notions discussed in this paper with the rough notions of formal concept analysis, such as the rough formal concept [7]. There should also be an integration of soft computation ideas, such as discussed in this paper and elsewhere, into current standards (Resource Description Framework, Conceptual Knowledge Markup Language, Conceptual Graph Markup Language, Knowledge Interchange Format and Conceptual Graph Interchange Format) for conceptual knowledge representation and ontological modeling [8].

References

1. H. Chen, P. Hsu, R. Orwig, L. Hoopes, and J.F. Nunamaker. Automatic concept classification of text from electronic meetings. *Communications of the ACM*, 37:56–73, October 1994.
2. B. Ganter and R. Wille. Conceptual scaling. In F. Roberts, editor, *Applications of Combinatorics and Graph Theory in the Biological and Social Sciences*, pages 139–167. Springer, New York, 1989. Preprint No. 1174 (1988), Technische Hochschule Darmstadt, Darmstadt, Germany.
3. R.E. Kent. The metric closure powerspace construction. In M. Main, A. Melton, M. Mislove, and D. Schmidt, editors, *Mathematical Foundations of Programming Semantics, 3rd Workshop*, pages 173–199. Tulane University, New Orleans, Springer-Verlag, 1987. Lecture Notes in Computer Science, Vol. 298.
4. R.E. Kent. Conceptual collectives. Technical report, Department of Computer and Information Science, University of Arkansas at Little Rock, 1992.
5. R.E. Kent. Enriched interpretation. In *Third International Workshop on Rough Sets and Soft Computing (RSSC'94)*, San Jose, California, USA, 1994.
6. R.E. Kent. Rough concept analysis. In W. Ziarko, editor, *Rough Sets, Fuzzy Sets and Knowledge Discovery*, pages 248–255. Springer-Verlag, August 1994.
7. R.E. Kent. Rough concept analysis: A synthesis of rough sets and formal concept analysis. *Fundamenta Informatica*, 27(2,3):169–181, August 1996.
8. R.E. Kent. Organizing conceptual knowledge online: Metadata interoperability and faceted classification. In *Fifth International Conference of the International Society of Knowledge Organization (ISKO'98)*, Lille, France, August 1998. To be distributed in the forthcoming issue of Knowledge Organization.
9. R.E. Kent and C. Neuss. Creating a web analysis and visualization environment. *Computer Networks and ISDN Systems*, 28:107–117, 1995.
10. F.W. Lawvere. Metric spaces, generalized logic, and closed categories. In *Seminario Mathematico E. Fisico*, volume 43, pages 135–166. Rendiconti, Milan, 1973.
11. P. Pagliani. A modal relation algebra for generalized approximation spaces. In *Rough Sets, Fuzzy Sets and Machine Discovery*, 1996.
12. P. Pagliani. Information gaps as communication needs: A new semantic foundation for some non-classical logics. *Journal of Logic, Language, and Information*, 6:63–99, 1997.
13. Z. Pawlak. Rough sets. *International Journal of Information and Computer Science*, 11:341–356, 1982.
14. J. Rowley. *Organising Knowledge: An Introduction to Information Retrieval*. Gower Publishing Company Ltd., Hants, England, 1987.
15. R. Wille. Restructuring lattice theory: An approach based on hierarchies of concepts. In I. Rival, editor, *Ordered Sets*, pages 445–470. Reidel, Dordrecht-Boston, 1982.
16. W.A. Woods. Understanding subsumption and taxonomy: A framework for progress. In J. Sowa, editor, *Principles of Semantic Networks: Explorations in the Representation of Knowledge*, pages 45–94. Morgan Kaufmann, San Mateo, California, 1991.
17. L. Zadeh. The concept of a linguistic variable and its application to approximate reasoning. *Information Science*, 8:199–249, 1975.

Rough Set Application

A Rough Set Framework for Mining Propositional Default Rules

Torulf Mollestad[1] *and Jan Komorowski*[2]

[1] Computas AS, PO box 444, 1301 Sandvika, Norway
[2] Department of Computer and Information Science, Norwegian University of Science and Technology (NTNU), 7034 Trondheim, Norway

O sancta simplicitas (J. Huss)

1 Introduction

As the amount of information in the world is steadily increasing, there is a growing demand for tools for analysing the information with the aim of finding patterns in terms of implicit dependencies in data. Realising that much of the collected data will not be handled or even seen by human beings, systems that are able to generate pragmatic summaries from large quantities of information will be of increasing importance in the future. Although several statistical techniques for data analysis were developed long ago, advanced techniques for intelligent data analysis are not yet mature. As a result, there is a growing gap between data generating and data understanding. At the same time, there is a growing realisation and expectation that data, intelligently analysed and presented, will be a valuable resource to be used for a competitive advantage.

The concept of *knowledge discovery* [22] has recently been brought to the attention of the business community. One main reason for this is that there is a general recognition of the untapped value in larger databases. Knowledge discovery is the nontrivial extraction of implicit, previously unknown, and potentially useful information from data. Knowledge discovery is thus a form of machine learning which discovers interesting knowledge and represents the information in a high-level language. If the underlying source of information is a database, the term *data mining* is used to denominate the process of automatic extraction of information. The information is then represented in terms of rules reflecting the intra-dependencies in the data. Applications of such rules include customer behaviour analysis in a supermarket or banking environment, telecommunications alarm diagnosis and prediction, etc. Such rules would typically emulate an expert's reasoning process when *classifying* objects.

A great deal of work done in data mining has focused on the generation of rules that cover the situation where the training data is entirely consistent, i.e. all objects that are indiscernible are classified equally. In these cases, *definite* rules that map all objects into the same decision class may be generated. There is in many cases, however, a clear need to be able to reason also in the presence of inconsistencies. Different experts may disagree on the classification of one particular object in which case it is desirable to assign different *trust* to the respective conclusions. Also, if objects are classified inconsistently, we still want to be able to generate rules that reflect the *normal* situation. Such normalcy rules typically sanction a particular conclusion given some information, whereas additional knowledge may invalidate previous conclusions.

In this work we look at the problems related to incompleteness, uncertainty and inconsistency in data represented in tabular form. We particularly wish to be able to generate classification rules that can handle these common phenomena. More specifically, we investigate how information systems and rough sets [18, 19] can be applied to the problem of generating *default rules* [30, 25, 24] from a set of low-level data since such rules enable us to express *common* relationships. The conclusions that are drawn from a default theory rely on the soundness of a set of assumptions that themselves may be disproven when new knowledge is made available. This generalises the notion of decision rules and provides a framework under which more, potentially interesting, statistical information may be extracted. Much of this information would be lost if the knowledge extraction process were restricted to generating definite rules only. The contention of this paper is that generating of default rules from databases provides a framework that is suited to handling many of the problems described above (see, for instance, [14, 15, 16]). By using the rough set approach, we are able to emulate an expert's decision process by generating a set of decision rules. In combination with default reasoning, we generate rules that cover the most general patterns in the data. The synthesis of such normalcy rules will not be prevented by noise, i.e. by abnormal objects.

The input to the knowledge extraction process is a set of *example objects* and information about their properties. In order to learn rules from a set of examples, we assume the existence of an *expert* who has complete knowledge of the domain. The expert is able to classify the elements of the universe in the sense that he can make decisions wrt. some restricted set of *decision properties*. In doing this, he identifies a set of concepts – the classes of the classification. The task of the *learner* is to learn the expert's knowledge. He/she can do so by trying to find the characteristic features of each concept and to describe the expert's concepts in terms of attributes that are available to the learner. The task of learning is thus the problem of expressing the expert's basic concept in terms of the learner's basic concepts.

In Sect. 2 and 3, we give an introduction to default reasoning and rough sets theory, respectively. The following section presents the principle behind definite rule generation and it gives a small example that will be referred to throughout the rest of the paper. The problem of default rule generation using rough sets is investigated in Sect. 5. Section 6 presents a case study. References to related work are given in Sect. 7. In the last section we give a summary of work done and point to some questions that give directions for future research.

2 Non-monotonic and Default Reasoning

The purpose of this work is to investigate the relationship between observations about the world (represented in terms of a set of objects in an information system and their individual properties/features) and "common–sense" rules that reflect normal dependencies between different properties of objects.

The notion of default reasoning is omnipresent in common-sense and is manifested in all reasoning from inconsistent and incomplete information. A good introduction to the field is given by Brewka [4]. The default logic of Reiter [30] formalises the notion of default rules. The framework uses a non-monotonic reasoning strategy to draw conclusions from defaults that are consistent with the intended model. Default logic seems to be a non-monotonic formalism that can be quite easily adapted to the problem of explanation finding. It is possible to interpret defaults as predefined hypotheses and reasoning with them as a special way of logical theory formation [30, 25, 24].

David Poole proposes an abductive approach to non-monotonic reasoning which corresponds to Reiter's default logic. Poole's *Theorist* [25, 26, 24] is a simple framework for default reasoning, providing one possible implementation of Reiter's formalism. In Poole's original formalisation, a default theory consists of two sets of first order formulae, *facts* $\mathcal{F}$ and *defaults* Δ, which together represent a *Theorist framework*. The facts are closed formulae that are taken to be true in the domain, whereas defaults may be seen as possible hypotheses which *may* be used as premises in a logical argument. They may be applied in order to explain propositions provided that consistency is maintained. Hence, it is possible to select a set $\mathcal{D} \subseteq \Delta$ from the defaults. The combination of $\mathcal{F}$ and $\mathcal{D}$ constitutes what is called a *Theorist scenario*.

For a given *Theorist* framework there generally exist a number of different scenarios each of which explains a different set of propositions. *Theorist* relies on a backward-chaining theorem prover to collect the assumptions that will be needed to explain a given set of observations and to prove their consistency. If new facts are added to the theory, it may be that the goal can no longer be explained because the defaults used are inconsistent with the new facts. In its standard form, Poole's framework is capable of manipulating first order expressions. The presentation below will, however, be limited to propositional theories. The more general definitions, covering the first order case, can be found in [4], among others. For a comparison of *Theorist* with Reiter's logic, see [25].

Example 2.1 The birds Tweety and Woody is the most common example of the inadequacy of monotonic reasoning from consistent facts that illustrates the need for non-monotonic reasoning capabilities. Assume that we know that Tweety is a penguin, and that Woody is (some kind of) a bird. Consider the framework $(\mathcal{F}, \Delta)$, defined as follows (for reasons of tradition, the rules are here written with the consequent on the left hand side):

$$\mathcal{F} = \{\neg can\_fly \leftarrow penguin\}$$

$$\Delta = \{can\_fly \leftarrow bird\}$$

If the proposition *bird* is known to be true about an object, as in Woody's case, the theory $\mathcal{F} \cup \Delta$ explains *can_fly*. For Tweety, *penguin* is known to hold also, therefore, Δ contains no element which may be consistently used to extend the theory and it cannot be assumed that Tweety flies. □

Recall that only a subset of the defaults have to be generally selected in order to make a *Theorist* scenario. Hence, it may be that a number of different *extensions* to a given set of facts $\mathcal{F}$ and defaults Δ exist. In other words, the framework may have different and mutually inconsistent entailments. This phenomenon is referred to in the literature as the *multiple extensions problem* and arises when there exist several different combinations of defaults that are consistent with the given facts. One possible remedy for this situation is described below.

Brewka [3, 4] defines a generalisation of Poole's framework such that defaults of many different *priority levels* may exist. When defaults are in conflict, the defaults of higher priority take precedence and the others are deemed inconsistent. Brewka describes a multiple-layered framework where a linear ordering has been imposed on different sets of rules. The ordering of layers reflects the relative reliability of the rules. Mutually inconsistent rules are still allowed within any given hypothesis set and multiple extensions may still exist.

Example 2.2 Assuming that abnormal flying penguins exist, the following level default theory may be constructed:

$$T_1 = \{ \ bird \leftarrow penguin$$
$$penguin \leftarrow air\_penguin$$
$$fly \leftarrow air\_penguin \quad \}$$
$$T_2 = \{ \ \neg fly \leftarrow penguin \quad \}$$
$$T_3 = \{ \ fly \leftarrow bird \quad \}$$

Note here that the conjecture that birds fly is overturned by the higher priority default that penguins do not. The latter sentence, in turn, is overridden by the statement that one particular kind of penguin, i.e. the "air penguin" does indeed fly. Hence, if nothing is known about Woody except that he is a bird, he is assumed to fly. If more specific or credible information is made available, i.e. that he is a penguin, the default no longer applies. If even more specific information is obtained, that he is indeed an *air_penguin*, he flies once again. □

3 Overview of the Rough Set theory

The rough set approach was designed as a tool to deal with uncertain or vague knowledge. It has shown to provide a successful theoretical basis for the solution of many problems within knowledge discovery. There also exist a few implementations of tools supporting rough set analysis that make the theory practical. For example, in all our experimental work we use now the ROSETTA toolkit [38] that provides a comprehensive set of data analysis tools and an advanced graphical user interface. It runs under Windows 95/NT. There exists a public version of the software [31].

The notion of classification is central to the approach, i.e. the ability to discern between objects and to reason about partitions of the universe. In rough sets, objects are perceived only through the information that is available about them, that is, through their values for a predetermined set of attributes. In the case of inexact information, one has to be able to give and reason about *rough classifications* of objects.

In this work we investigate how rough sets and information systems [18, 19] can be applied to the problem of generating default rules from a set of low-level data given in terms of a set of *objects* and information about their properties. As our starting point data about some domain has been collected and represented in a tabular form called an *information system*. It contains knowledge about a set of *objects* in terms of a predefined set of *attributes* (or properties). The attribute values are given for each object in the domain.

Definition 3.1 (*Information System*). An *information system* (hence abbreviation IS) is an ordered pair $\mathcal{A} = (U, A)$ where U is a non-empty, finite set called the *universe*, A is a non-empty, finite set of *attributes*. Each attribute $a \in A$ is a total function $a : U \to V_a$, where V_a is the set of values of a, called the *range* of a. The elements of the universe will in the following be referred to as *objects*. A *decision system* (abbreviated DS) $\mathcal{A} = (U, (C, D))$ is an information system for which the attributes are separated into disjoint sets of *condition* attributes C and *decision* attributes D ($C \cap D = \emptyset$). ■

Definition 3.2 (*Indiscernibility Relation*). Let $\mathcal{A} = (U, A)$ be an information system. Every subset of attributes $B \subseteq A$ defines an equivalence relation $\text{IND}(B)$, called an *indiscernibility relation*, defined as follows:

$$\text{IND}(B) = \{(x, y) \in U^2 : a(x) = a(y) \text{ for every } a \in B\}$$

■

Definition 3.3 (*Equivalence Class*). Let $\mathcal{A} = (U, A)$ be an information system, and let $B \subseteq A$. Then, for any $x \in U$,

$$[x]_B = \{y \in U : (x, y) \in \text{IND}(B)\}$$

■

Throughout the paper, the notation $|\,.\,|$ denotes the function that returns the cardinality of the argument set.

The set $U/\text{IND}(B)$ is the set of all equivalence classes in the relation $\text{IND}(B)$. For a given B, the number of equivalence classes is $n_B = |U/\text{IND}(B)|$. When it is clear from the context which subset B is concerned the index will be dropped. It is then convenient to enumerate these classes by $E_1, \ldots, E_n$.

It is obvious that for any object $x \in U$ the equivalence class $[x]_B$ contains the objects that are indiscernible from x using attributes B. These equivalence classes will be referred to as *object classes* or simply *classes*. The intuition behind the notion of the indiscernibility relation is that selecting a set of attributes $B \subseteq A$ effectively defines a complete partitioning of the universe into sets of objects that cannot be distinguished using only the attributes in B. In the following, the lowest level of perception is the individual classes in the equivalence relation $\text{IND}(B)$ while individual objects are not considered. We will reason about the properties of the classes; by a slight overloading of the notation we extend each attribute function $a \in A$ to be defined on the quotient set, i.e. $a : U/\text{IND}(B) \to V_a$.

Definition 3.4 (*Discernibility Matrix*). Let $\mathcal{A} = (U, A)$, $n = |U/\text{IND}(A)|$ and $E_1, \ldots, E_n \in U/\text{IND}(A)$ be the corresponding partitioning of U. The *discernibility matrix* of $\mathcal{A}$ is $M_{[A]} = \{m_{[A]}(E_i, E_j)\}_{n \times n}$, where

$$m_{[A]}(E_i, E_j) = \{a \in A : a(E_i) \neq a(E_j)\} \text{ for } i, j = 1, \ldots, n$$

∎

In other words, each entry $m_{[A]}(E_i, E_j)$ in the discernibility matrix is the set of attributes from A that discern between object classes $E_i, E_j \in U/\text{IND}(A)$.

Let $\tilde{a}$ be a unique Boolean variable associated with the corresponding attribute $a \in A$. To each element $m_{[A]}(E_i, E_j)$ of the discernibility matrix there corresponds a set $\tilde{m}_{[A]}(E_i, E_j) = \{\tilde{a} : a \in m_{[A]}(E_i, E_j)\}$. We put $\bigvee \tilde{S}$, where $\tilde{S}$ is a set of Boolean variables, to refer to a disjunction taken over $\tilde{S}$. If $\tilde{S} = \{\tilde{a}_1, \tilde{a}_2, \ldots, \tilde{a}_m\}$, then $\bigvee \tilde{S} = \tilde{a}_1 \vee \tilde{a}_2 \vee \ldots \vee \tilde{a}_m$. Two *discernibility functions* are defined as follows.

Definition 3.5 (*Discernibility Function*). Given an information system $\mathcal{A} = (U, A)$, let $m = |A|$ and $n = |U/\text{IND}(A)|$. The *discernibility function* of $\mathcal{A}$ over attributes $a_1, \ldots, a_m \in A$ is a Boolean function of m variables

$$f_{[A]}(a_1, \ldots, a_m) = \bigwedge \{\bigvee \tilde{m}_{[A]}(E_i, E_j) : 1 \leq j < i \leq n, m_{[A]}(E_i, E_j) \neq \emptyset\}$$

∎

Definition 3.6 (*Discernibility Function for a Class*). Given an information system $\mathcal{A} = (U, A)$, let $m = |A|$, $n = |U/\text{IND}(A)|$ and $1 \leq k \leq n$. The *discernibility function for a class* E_k of $\mathcal{A}$ over attributes $a_1, \ldots, a_m \in A$ is a Boolean function of m variables

$$f_{[A]}(E_k)(a_1, \ldots, a_m) = \bigwedge \{\bigvee \tilde{m}_{[A]}(E_k, E_j) : 1 \leq j \leq n, m_{[A]}(E_k, E_j) \neq \emptyset\}$$

∎

This function is also called k-*relative discernibility function*. When it will be clear from the context we shall write $f(B)$ for $f_{[B]}(a_1, \ldots, a_m)$ and $f(E, B)$ for $f_{[B]}(E)(a_1, \ldots, a_m)$, where $B \subseteq A$ in $\mathcal{A} = (U, A)$.

Definition 3.7 (*Dispensability*). Given $\mathcal{A} = (U, A)$, an attribute a is said to be *dispensable* or *superfluous* in $B \subseteq A$ if $\text{IND}(B) = \text{IND}(B - \{a\})$, otherwise the attribute is *indispensable* in B. If all attributes $a \in B$ are indispensable in B, then B is called *orthogonal*. ∎

Definition 3.8 (*Reduct*). Given $\mathcal{A} = (U, A)$, let $B \subseteq A$. A reduct of B is a set of attributes $B' \subseteq B$ such that all attributes $a \in B - B'$ are dispensable, and $\text{IND}(B') = \text{IND}(B)$. The set of reducts of B is denoted $red(B)$. ∎

Before continuing the discussion, we recall some basic notions from standard Boolean reasoning. A *prime implicant* of a Boolean expression is an expression which is simpler than the original, but the truth of which implies the truth of the original expression. Informally, the set of prime implicants of a logical expression is the set of disjuncts in the function written in a disjunctive normal form. Given a Boolean function f, p_imp will denote a unary function which returns the set of prime implicants of function f. In turn, the set of prime implicants of the discernibility function $f(B)$ determines the reducts of B [34]. A reduct (there may be several of them) is thus a subset of the attributes A by which all classes discernible by the original IS may still be kept separate.

Definition 3.9 (*Reduct for a Class*). For a given $\mathcal{A} = (U, A)$ and $B \subseteq A$ a *reduct for a class* $E \in U/\text{IND}(B)$ is a prime implicant of the discernibility function $f_{[B]}(E)$. The set of reducts for E and attributes B is:

$$red(E, B) = \{\{a \in B : \tilde{a} \in r\} : r \in p\_imp(f_{[B]}(E))\}$$

∎

Definition 3.10 (*Lower and Upper Approximation, Boundary Region*). Given an IS $\mathcal{A} = (U, A)$, let $X \subseteq U$ be a set of objects and $B \subseteq A$ be a selected set of attributes. The B-*lower approximation* $\underline{B}X$ and the B-*upper approximation* $\overline{B}X$ of X with reference to attributes B (defining an equivalence relation on U) are

$$\underline{B}X = \{x \in U : [x]_B \subseteq X\}$$
$$\overline{B}X = \{x \in U : [x]_B \cap X \neq \emptyset\}$$

The region $\text{BN}_B(X) = \overline{B}X - \underline{B}X$ is called the B-*boundary (region)* of X. For any $X \subseteq U$, $B \subseteq A$, $\underline{B}X \subseteq X \subseteq \overline{B}X$. ∎

The lower approximation $\underline{B}X$ is the set of elements from U that can *with certainty* be classified as elements of X, according to the attribute set B. Correspondingly, the set $\overline{B}X$ contains the set of objects that may possibly be classified as elements of X. The boundary region contains elements that neither can be classified as being definitely within nor definitely outside X, again using the attributes B. If $\text{BN}_B(X) \neq \emptyset$, the set is rough, otherwise the set is crisp.

The degree of membership is formulated in terms of the *rough membership function*.

Definition 3.11 (*Rough Membership Function*). Given $\mathcal{A} = (U, A)$, let $E, X \subseteq U$. The *rough membership function* $\mu(E, X)$ of E in X is defined as:

$$\mu(E, X) = \frac{|E \cap X|}{|E|}$$

∎

Obviously, $0 \leq \mu(E, X) \leq 1$

In the following we will give some definitions that apply to decision systems, as specified in Def. 3.1. First of all, the notion of determinism (equivalently, consistency) has to be defined in terms of the set of condition and decision attributes.

Definition 3.12 (*Deterministic DS*). A decision system $\mathcal{A} = (U, (C, D))$ is *deterministic* (equivalently, *consistent*), if

for all $E \in U/\text{IND}(C)$, *there exists an* $X \in U/\text{IND}(D)$ *such that* $E \subseteq X$

Otherwise, the system is said to be *indeterministic* (equivalently, *inconsistent*).

∎

It is easy to see that a decision system is indeterministic in situations where objects that cannot be distinguished from each other (by attributes C) are classified differently by the expert (D). Skowron and Grzymała-Busse [33] showed that a knowledge system may always be split into two distinct parts: one which is totally deterministic and the other one indeterministic. The notion of determinism is related to the upper and lower approximations of sets; the objects contained in the lower approximation of some decision class are exactly those for which consistent decision rules may be generated. For objects contained in the boundary region of several classes no such consistent decision can be made. In the following, we will use the notation $\delta_C(E)$ to denote the set of possible decisions for a given class $E \in U/\text{IND}(C)$. In other words, $\delta_C(E)$ identifies the union of decision values for individual objects in E.

In computing the discernibility matrix and corresponding discernibility function, a simplification may be made in the case of decision tables. If we are interested in performing classification (into D) on the basis of values for the condition attributes C, it is not necessary to distinguish between classes that are mapped into the same decision class X. Hence, the items in the discernibility matrix that record the differences between these classes need not be considered. When this simplification is performed, in the following we will refer to indiscernibility *modulo* decision attributes D. The notion is formally defined by a slight alteration of the definition of the discernibility matrix, as follows:

Definition 3.13 (*Discernibility Matrix Modulo D*). Given a decision system $\mathcal{A} = (U, (C, D))$ let $n = |U/\text{IND}(C)|$. The *discernibility matrix modulo D* of $\mathcal{A}$ is

$$M_{[C,D]} = \{m_{[C,D]}(E_i, E_j) : 1 \leq i \leq n, 1 \leq j \leq n\}, \text{where}$$

$$m_{[C,D]}(E_i, E_j) = \begin{cases} m_{[C]}(E_i, E_j) = \{a \in C : a(E_i) \neq a(E_j)\} & \text{if } \delta_C(E_i) \neq \delta_C(E_j) \\ \emptyset & \text{otherwise} \end{cases}$$

∎

The entry $m_{[C,D]}(E_i, E_j)$ in the discernibility matrix is the set of attributes from C that discern object classes $E_i, E_j \in U/\text{IND}(C)$ if $\delta_C(E_i) \neq \delta_C(E_j)$. In analogy with the definition of the discernibility function (Def. 3.5), we define the discernibility function modulo D as follows.

Definition 3.14 (*Discernibility Function Modulo D*). Given a decision system $\mathcal{A} = (U, (C, D))$ the *discernibility function modulo D of $\mathcal{A}$* is

$$f_{[C,D]} = \bigwedge_{i,j \in \{1,..,n\}} \bigvee \tilde{m}_{[C,D]}(E_i, E_j)$$

where $n = |U/\text{IND}(C)|$. ∎

4 Extraction of Knowledge from Information Systems

In this section we present a simple example that will help clarify the concepts that have been introduced. It will also serve as an illustration throughout the rest of the paper.

Example 4.1 The information system displayed in Tab. 4.1 resulted from having observed a total of one hundred objects that were classified according to condition attributes $C = \{a, b, c\}$. The expert classification is represented as a set of decision attributes $D = \{d\}$.

	a b c	d
E_1	1 2 3	1 (50×)
E_2	1 2 1	2 (5×)
E_3	2 2 3	2 (30×)
E_4	2 3 3	2 (10×)
$E_{5,1}$	3 5 1	3 (4×)
$E_{5,2}$	3 5 1	4 (1×)

Table 1. An Example Decision System

The partition of the universe induced by the condition attributes contains five classes. The objects that have the same values for the attributes $C = \{a, b, c\}$

(indiscernible objects) are represented by their respective equivalence classes $E_1, .., E_5$. For instance, the class E_1 contains 50 objects, characterised by their values for the condition attributes, namely $(a = 1), (b = 2), (c = 3)$. There are four decision classes: X_1 through X_4. The class E_5 is shown split into two disjoint sets of objects $E_5 = E_{5,1} \cup E_{5,2}$ which reflects the two different decisions $d = 3$ (for $E_{5,1}$) and $d = 4$ ($E_{5,2}$). No unique decision may be made for the objects in E_5 since the system is indeterministic wrt. the objects in this class.

Since this system is a decision table, we do not have to distinguish between classes that are mapped into the same decision class. In the example (with $C = \{a, b, c\}$, $D = \{d\}$), this simplification may be applied to the classes E_2, E_3 and E_4 which are all mapped into decision $d = 2$. Taking the decision attribute into account in the manner described above gives the discernibility matrix $M_{[C,\{d\}]}$, displayed in Tab. 4.1. On the right side of the table we have shown the discernibility functions for the respective classes, i.e. the sets of attributes that are needed to discern one particular class E_i from all other classes. For instance, in order to be able to separate the objects in class E_1 from any object class that maps into another decision, we need to make use of both attributes a and c.

	E_1	E_2	E_3	E_4	E_5	
E_1	×	c	a	ab	abc	ac
E_2	c	×	×	×	ab	$c(a \vee b)$
E_3	a	×	×	×	abc	a
E_4	ab	×	×	×	abc	$a \vee b$
E_5	abc	ab	abc	abc	×	$a \vee b$

Table 2. The Discernibility Matrix Modulo Decision Attribute d

Only two of the three attributes are needed to distinguish the classes when the decision mapping is taken into account since the discernibility function for the system modulo d is $f_{[C,\{d\}]} = ca(a \vee b)(a \vee b \vee c) = ca$. □

4.1 Definite Rule Generation

Several authors have worked on applying the rough set approach to the problem of decision rules generation. They designed algorithms to extract information from a set of low-level data by building propositional rules that cover the available input knowledge; see, for instance, [5, 36, 33, 6, 10, 9, 32, 21]. A typical learning task is to define an expert's knowledge (the classification of objects wrt. (hence abbreviation wrt.) the decision attribute set D) in terms of the properties available to the learner, i.e. in terms of the attributes contained in the set C. The result of the learning process is a decision algorithm, called an *oracle*, which is able to emulate the expert's decision procedure when applied to new objects. We introduce now some concepts and definitions that will be referred to later.

Definition 4.1 (*Descriptor* [32]). Given information system $\mathcal{A} = (U, A)$, let $B \subseteq A$. The set $\mathcal{F}_=(B, V_B)$ of *descriptors* (atomic formulae) over attributes B and values $V_B = \bigcup_{a \in B} V_a$ is

$$\mathcal{F}_=(B, V_B) = \{(a = v_a) : a \in B, \ v_a \in V_a\}$$

∎

A descriptor is an atomic propositional formula of the form $(a = v)$, where $a \in B$, $v \in V_a$. In the following we will also use a_v as a syntactic variant. Descriptors may be used as building blocks in conjunctive formulae. The set of all possible conjunctive formulae that may be generated over attributes B and their values is denoted $\mathcal{F}_\wedge(B, V_B)$ The most interesting conjunctive formulae are the so called *class descriptions*.

Definition 4.2 (*Class Description*). Given an information system $\mathcal{A} = (U, A)$, let $B \subseteq A$. A *class description* $Des(E, B) \in \mathcal{F}_\wedge(B, V_B)$ of a class $E \in U/\text{IND}(B)$, attributes B, is a conjunctive formula over B:

$$Des(E, B) = \bigwedge_{a \in B} (a = a(E))$$

∎

A decision rule is an implication between conjunctive formulae:

Definition 4.3 (*Decision Rule*). Given a decision system $\mathcal{A} = (U, (C, D))$, the set $\mathcal{F}_\to(C, D, V_C, V_D)$ of *decision rules* for $\mathcal{A}$ is

$$\mathcal{F}_\to(C, D, V_C, V_D) = \{\tau \to \tau' : \tau \in \mathcal{F}_\wedge(C, V_C), \ \tau' \in \mathcal{F}_\wedge(D, V_D)\}$$

∎

Definition 4.4 (*Semantics of Descriptors and Class Descriptions*). Given an information system $\mathcal{A} = (U, A)$, let $a \in A$. The semantics $[\![\tau]\!]_\mathcal{A}$ of a formula τ is defined inductively:

$\tau \equiv (a = v) \in \mathcal{F}_=(\{a\}, V_{\{a\}}): [\![\tau]\!]_\mathcal{A} = \{x \in U : a(x) = v\}$

$\tau \equiv (\tau' \wedge \tau'') \in \mathcal{F}_\wedge(A, V_A): [\![\tau]\!]_\mathcal{A} = [\![(\tau' \wedge \tau'')]\!]_\mathcal{A} = [\![\tau']\!]_\mathcal{A} \cap [\![\tau'']\!]_\mathcal{A}$

∎

Definition 4.5 (*Semantics of Decision Rules*). Given a decision system $\mathcal{A} = (U, (C, D))$, let $(\tau \to \tau') \in \mathcal{F}_\to(C, D, V_C, V_D)$ be a decision rule. Then

$$[\![(\tau \to \tau')]\!]_\mathcal{A} = (U - [\![\tau]\!]_\mathcal{A}) \cup [\![\tau']\!]_\mathcal{A}$$

∎

A decision rule is *valid* in a decision system if each of the objects that satisfies (i.e. is in the extension of) the antecedent is in the extension of the consequent. This property may be modelled in terms of the lower approximation of the respective decision classes according to the available condition attributes. Given a decision system $\mathcal{A} = (U, (C, D))$ and using the lower approximation $\underline{C}X$ of each $X \in U/\text{IND}(D)$, decision rules may be generated for those object classes $E \in U/\text{IND}(C)$ that are totally contained in X, i.e. the classes can be classified with full certainty into that decision class.

As mentioned above, the rules are defined as mappings from a class description over the set of condition attributes into a description over the decision attribute(s). Correspondingly, definite rules are generated according to the schema below, if all objects of a class E with full certainty have the decision $Des(X, D)$, i.e., if E is fully contained in X.

$$Des(E, C) \to Des(X, D), \text{ where } E \subseteq X$$

In defining the rules, the *minimal* class description of each class is used, applying the reducts for the class, i.e. the reducts corresponding to the prime implicants of the discernibility function.

Example 4.1 (Continued) Consider the example discernibility matrix given in Tab. 4.1. The minimal set of attributes needed to distinguish E_1 from all other classes is ac. Hence, one rule is generated for this class mapping the description/value trace over the reduct $(a_1 c_3)$ into the decision class d_1. There are two minimal class descriptions of the E_2 class: $a_1 c_1$ and $b_2 c_1$. They correspond to each of the two disjuncts of the discernibility function $f(E_2, C)$. Note again that the subscripts identify the values of respective attributes. From the table it can easily be verified that the objects contained in classes E_1 through E_4 are all mapped deterministically into one of the decision classes. Each of these classes is contained in the lower approximation $\underline{C}X_j$ for some decision class X_j, $j \in \{1, 2\}$. The following five (definite) rules are generated applying the minimal class descriptions.

$$\begin{array}{ll} E_1: & a_1 c_3 \to d_1 \\ E_2: & a_1 c_1 \to d_2 \\ & b_2 c_1 \to d_2 \\ E_3, E_4: & a_2 \to d_2 \\ E_4: & b_3 \to d_2 \end{array}$$

Since there is an ambiguity for the class E_5 wrt. the decision d, no rule can be generated. All objects in the class are contained in the boundary region of both classes X_3 and X_4. The fraction γ_C of instances for which deterministic decision rules can be learned may be used as a metric on the quality of the learning process. By this simple definition, the quality of the learning (quality of approximation) described above would be:

$$\gamma_C = \frac{|\bigcup_{i=1}^{4} \underline{C}X_i|}{|U|} = \frac{|\underline{C}X_1 \cup \underline{C}X_2|}{|U|} = \frac{|E_1 \cup E_2 \cup E_3 \cup E_4|}{|U|} = \frac{95}{100} = 0.95$$

□
The above result seems to be very good indeed. In many cases, however, the structure of the learning data is such that only a small fraction – if any – of the objects may be classified with certainty. In this case, steps should be taken to make sure that the information which is available is really put to use. In the following section one way of doing exactly this will be considered.

Example 4.1 (Continued) Assume that one new object, x_{101}, has been observed, and the decision table is updated accordingly, i.e. the object is added to the decision table. The rule base should now be updated to comprise the existence of the new object.

	a b c	d
x_{101}	1 2 3	4 (1×)

According to its values for the condition attributes, the object is a member of the E_1 class. The introduction of the object, however, causes an indeterminacy. Since all 51 objects in E_1 are indiscernible, it can no longer be stated with full certainty what the decision is, it can be either 1 or 4. In this case, it can be easily verified that, if learning is restricted to generating definite rules, the quality of the learning has fallen to $45/101 \approx 0.45$, no longer so impressive a result. The indeterminacy for E_1 is caused by one single object out of the total of 51. This object represents an *exception* to a general rule and this knowledge should be taken into account.
□

5 Generation of Default Rules

The framework described above enabled the generation of rules that cover the *deterministic* situation, i.e. where, for a decision system $\mathcal{A} = (U, (C, D))$, all objects of a particular class $E \in U/\text{IND}(C)$ are mapped into the same decision class $X \in U/\text{IND}(D)$. The example given in the last section does, however, demonstrate the fact that it is not always possible to obtain a set of deterministic rules by which all objects can be classified with full certainty. More importantly, in general it is not even desirable to generate such a set of rules. Definite rules over a large number of condition attributes are typically very specific, as they have to identify the (small) classes that are defined over the entire set of attributes. Even if they do cover all the objects in the training information system, such rules may prove completely insufficient when used on other input data.

It is very often important to be able to handle inconsistencies in the data. From such data we may still be able to extract a lot of interesting information, specifically knowledge that reflects the *most common* or *normal* situation. A strict requirement on the absolute correctness of the rules has proven insufficient in many real- world applications. Computer systems – as well as people – are often under pressure to make a decision under strict time constraints, and the ability to reason in absence of knowledge is a great advantage. To reason in this way, we may choose to believe in some rule provided that the evidence supporting it is strong enough. Furthermore, results were obtained [27] that suggest that simplified, "uncertain" rules in general prove better than the original ones when applied to new cases.

Finding such simple and natural expressions for regularities in databases is useful in various analysis or prediction tasks. We wish to obtain a set of rules that model the *general characteristics* of the data and that are less susceptible to noise. In the following section we define a first generalisation that establishes a framework for the generation of *default* rules. We also show how default reasoning may be used as a framework for representing and reasoning about indeterministic information systems.

5.1 Handling the Indeterministic Situation

There are various reasons of uncertainty in information systems. It may stem from uncertainty in the collection of data (e.g. due to noise), or from uncertainty in the knowledge itself. Inconsistencies arise when different objects that are described by the same values of conditions, i.e. the objects that are indiscernible by these attributes, are classified differently by the expert. In the case where the decisions of several experts are represented, disagreement wrt. the decision is another source of inconsistency. In the latter case, it would be of great value to be able to assign relative weights to the differing opinions of the experts, i.e. to represent the degree of trust that we have in the respective conclusions.

Ultimately, the reason that indeterminacy arises in information systems is the lack of ability to represent – in the particular information system – some attributes that differentiate the objects in the "real" world. If an expert performs different classifications of two objects that are indiscernible to the learner, it is because the expert is able to distinguish them according to some attribute or attributes that are not present in the system. It may be, however, that these attributes are significant only in exceptional situations.

As suggested earlier, default reasoning provides a natural framework for reasoning about uncertainty. Default rules account for common dependencies between objects and their properties allowing conclusions to be drawn about objects for which no deterministic rules can be applied. A default rule is acceptable if there exists a decision that dominates, to a certain high extent, the set of objects characterised by the rule. Using the concepts of information systems we can say that a great majority of the cases that match the antecedent of the rule does indeed also match the consequent. The question of what constitutes sufficient evidence wrt. the dominance of one decision over another is, however, highly debatable. A pragmatic choice has been made in this work. It involves the setting of a *threshold value* μ_{tr} which reflects the degree of trust desired for the rules.

We start from some fundamental concepts. (Recall that the terminology $\mu([\tau]_\mathcal{A}, [\tau']_\mathcal{A})$ refers to the degree of membership of the class identified by formula τ in the class $[\tau']_\mathcal{A}$.)

Definition 5.1 (*Default Rule*). Let a decision system $\mathcal{A} = (U, (C, D))$ and a threshold value μ_{tr} be given. A *default rule* for $\mathcal{A}$ wrt. the threshold $0 < \mu_{tr} \leq 1$ is any decision rule $(\tau \to \tau')$, where $\mu([\tau]_\mathcal{A}, [\tau']_\mathcal{A}) \geq \mu_{tr}$. ∎

As indicated, the acceptance of a default rule $\tau \to \tau'$ relies on whether the degree of membership (i.e. confidence or trust) of $[\tau]_\mathcal{A}$ in the class $[\tau']_\mathcal{A}$ exceeds the threshold value μ_{tr}. The threshold may be a preset constant or a parametrised function. It reflects the degree of confidence that the user wishes to have in the rules [16]. Obviously, if for some class E there exists no decision class X such that $\mu(E, X) \geq \mu_{tr}$, then no default rules should be generated for that particular class. The default rules that are accepted according to the above standard, may be added to the rule base (i.e. the set of definite rules) to accommodate the indeterministic part of the decision. In the deterministic situation $\mu([\tau]_\mathcal{A}, [\tau']_\mathcal{A}) = 1$, since the class described by the formula τ is in its entirety contained in $[\tau']_\mathcal{A}$. Hence, by the above definition, any definite rule is a default rule, too. Note also that if the threshold value is set above 0.5, at most one rule may be generated for each class, whereas a threshold lower or equal to 0.5 may result in the generation of mutually inconsistent rules. For instance, this will happen in the case where exactly 50% of the objects are mapped into one decision and the remaining ones into the other decision.

Although default rules may be generated from a decision system following the same pattern as definite rules, there is one major difference. In Def. 5.1 it is no longer imperative that the condition class be totally contained in the decision class. The degree of membership, reflected in the value $\mu(E, X)$ (Def. 3.11) should rather be high enough for the rule to be accepted as a default, i.e. $\mu(E, X) \geq \mu_{tr}$.

Example 4.1 (Continued) Referring again to the original example, we now aim at characterising the data represented in the class E_5 for which no deterministic rules could be generated. The computation of the membership functions of the two possible decision classes yields $\mu(E_5, X_3) = 4/5 = 0.8$ and $\mu(E_5, X_4) = 1/5 = 0.2$. Depending on the value of the preset threshold μ_{tr}, default rules may be generated for one, two or even zero alternative classifications. Assuming that the threshold value lies below 0.8 (but above 0.2), exactly one decision class will prevail. Two default rules are generated for the class (more precisely, for each of the two minimal class descriptions of E_5): $a_3 \to d_3$ and $b_5 \to d_3$. □

Which rules will be generated in the process depends upon the setting of μ_{tr}. It is again a result of the desired confidence that the user wishes to have in the generated rules. In a wider perspective, the quality of a given rule may be estimated not only according to its degree of confidence, as defined above, but also according to its generality and compactness. A rule is considered *good* if it is trusted to the required degree, *general* enough to be interesting when classification of new cases is to be done and sufficiently *compact*. The latter involves usability not only for mechanical classification of new cases. One example use can be education, i.e. paraphrasing the knowledge in a form which is easy to grasp for a human being and where irrelevant details have been eliminated. The aspect of generality will be the center of attention in the following.

5.2 Finding More General Patterns

In the last section it was explained that a number of default rules may be generated to cover the indeterministic subpart of a given decision system. Such rules could be added to the set of definite rules that themselves cover the deterministic part of the system. Default rules are generated in the cases where $\mu(E, X) \geq \mu_{tr}$, for condition class $E \in U/\mathrm{IND}(C)$ and decision class $X \in U/\mathrm{IND}(D)$. We are, however, often interested in finding more general patterns in any (deterministic or non-deterministic) information system. We propose to do this by a framework which is able to generate and reason about classes for which no unique decision can be made. Sets may, in fact, be constructed that cover more objects (and hence have a simpler description) by forming unions of the classes induced by the condition attributes C. Rules that map these composed classes into the decision that dominates the set may be subsequently generated. The rules that result have generally at least two substantial advantages compared to deterministic rules. They are usually *simpler* in structure because they cover larger classes defined over fewer conjuncts. The attributes that discern the objects contained in these compound classes are no longer considered relevant. Secondly, even though the rules might not be entirely correct wrt. the training data, they will in many cases prove to be better when handling unseen cases. The quality of the system's classification of future events will in general greatly depend on its ability to *generalise* over the knowledge.

At the core of the approach is the idea of *creating* indeterminacy in information systems in order to synthesise rules that cover the majority of the cases. There are in principle three ways of *creating* an indeterminacy in an information system [16]. We consider here generation of indeterminacy through selecting *projections* over the condition attributes allowing certain attributes to be excluded from consideration. More concretely, we define the following operation on decision systems:

$$remC((U,(C,D)), C_{Cut}) = (U, (C - C_{Cut}, D))$$

The $remC$ operation creates a simplified decision system where the attributes contained in $C_{Pr} = C - C_{Cut}$ are left for consideration. Each of these simplified systems will be called a *variant* of $(U,(C,D))$. In doing this, we effectively join equivalence classes over the condition attributes. These classes may be mapped into different decisions. Each new class $E_{(k,C_{Pr})} \in U/\mathrm{IND}(C_{Pr})$, where k ranges over the newly generated equivalence classes, is constructed as a union of those $E_i, E_j \in U/\mathrm{IND}(C)$ that originally were discernible only by attributes that were removed in the projection, i.e. $m_{[C]}(E_i, E_j) \subseteq C_{Cut}$. In other words:

$$E_{(k,C_{Pr})} = \bigcup \{E_i \cup E_j \mid E_i, E_j \in U/\mathrm{IND}(C), m_{[C]}(E_i, E_j) \subseteq C_{Cut}\}$$

It follows that classes in the equivalence relation defined by the condition attributes C are "glued" together by selecting suitable projections C_{Cut} of C. At the same time it is ensured that for the resulting object class one particular decision remains the dominating one. For the new, compound classes, default decision rules may be constructed. Also, a *block* of a particular default rule is generated for each class that was glued together when making the projection and for which the classification is inconsistent with the rule. We note that a block is a fact that may be applied to effectively prevent the application of a default.

We consider now the problem of selecting between possible projections. So far, a projection operation $remC$ has been defined and the imperative that a new indeterminacy be created was mentioned. It should be clear that, if no new indeterminacy results from a given projection, no new rules can be generated. Recall now that the rules were generated using the minimal possible description. This means that there is no need to separate two classes that are mapped into the same generalised decision. The gluing of these classes is futile, since no truly new information can be obtained. In order for new and interesting (default) rules to be found, new classes have to be constructed. These classes may only result from an indeterminacy gluing as defined below. Any two object classes $E_i, E_j \in U/\text{IND}(C)$ may be glued together by making a projection which excludes all the attributes represented by $m_{[C]}(E_i, E_j) = \{a \in C : a(E_i) \neq a(E_j)\}$.

Definition 5.2 (*Indeterminacy Gluing*). Let $C_{Cut} \subseteq C$ and $C_{Pr} = C - C_{Cut}$ in a decision system $\mathcal{A} = (U, (C, D))$. $remC(\mathcal{A}, C_{Cut})$ is an *indeterminacy gluing* if there exist object classes $E_i, E_j \in U/\text{IND}(C)$ such that $\delta_C(E_i) \neq \delta_C(E_j)$ and $m_{[C]}(E_i, E_j) \subseteq C_{Cut}$. ∎

The selection of projections that will potentially lead to interesting rules is dictated by the content of the discernibility matrix. The consequences of removing a set of attributes may moreover be studied by syntactically removing all the corresponding Boolean variables from the discernibility matrix. The cells in the matrix that become empty signify exactly the classes that are glued together by performing the projection. Consider now an information system having discernibility function $f_{[C,D]}$ (Def. 3.14). We modify the definition as follows.

Definition 5.3 (*Discernibility Factor Set Modulo D*). Given a decision system $\mathcal{A} = (U, (C, D))$. The sets $\Phi^0_{[C,D]}, \Phi_{[C,D]} \subseteq 2^C$ are defined as follows:

$$\Phi^0_{[C,D]} = \{m_{[C]}(E_i, E_j) : E_i, E_j \in U/\text{IND}(C),\ \delta_C(E_i) \neq \delta_C(E_j)\}$$

$$\Phi_{[C,D]} = \{\varphi \in \Phi^0_{[C,D]} : \text{there exists no } \varphi' \in \Phi^0_{[C,D]} \text{ such that } \varphi' \subset \varphi\}$$

The set $\Phi^0_{[C,D]}$ will be called the *unsimplified discernibility factor set modulo D* (UDFS) of $\mathcal{A}$ modulo D, whereas $\Phi_{[C,D]}$ is called the *(simplified) discernibility factor set* (SDFS) of $\mathcal{A}$ modulo D. Each set $\varphi \in \Phi^0_{[C,D]}$ is called a *factor* of $\Phi^0_{[C,D]}$. ∎

We contend that the set of factors $\Phi_{[C,D]}$ is identical to the set of (in set theoretic terms) *minimal* projections C_{Pr} such that $remC(\mathcal{A}, C_{Cut})$ is an indeterminacy gluing. Moreover, the factor φ uniquely identifies the variant information system $\mathcal{A}' = (U, (C-\varphi, D))$. The only immediate way of creating indeterminacies in the system is to remove either of the factors φ in the simplified discernibility factor set.

Theorem 5.1 (*Removal of a Factor yields Indeterminacy*). Given a decision system $\mathcal{A} = (U, (C, D))$. Let $\Phi_{[C,D]}$ be the discernibility factor set modulo D for $\mathcal{A}$. Then, a projection $remC(\mathcal{A}, C_{Cut})$ for $C_{Cut} \subseteq C$ is an indeterminacy gluing if and only if $C_{Cut} \supseteq \varphi$ for some $\varphi \in \Phi_{[C,D]}$.
Proof: See [16]. ∎

For instance, for a system having SDFS, say, $\Phi_{[C,D]} = \{\{a\}, \{b, c\}\}$ ($f_{[C,D]} = a(b \lor c)$) there are two immediate ways of creating an indeterminacy: by removing either a or both attributes b and c.

Each projection $remC(C, \varphi)$ results in a "new" indeterministic information system defined over the condition attributes $C - \varphi$. From each of these systems, a recursive call is made to the rule generating procedure that will eventually generate still more simplified rules. The projection operation is defined over subsets of the condition attributes C. The recursive process is thus defined on the lattice over the powerset 2^C. Using the properties of the discernibility factor set described above, we may significantly reduce the search in the lattice. For instance, in the example IS having discernibility function ac, we find that two immediately interesting projections are $C - \{a\} = \{b, c\}$ and $C - \{c\} = \{a, b\}$. Each of these projections is made, respective rules are generated and recursive calls are made initiating further search from each of the subsystems. The search for the example system is shown in Sect. 5.3.

We will now explain the rule generation procedure whereupon an outline of the algorithm will be given. Assume that a variant of a decision system $\mathcal{A}' = (U, (C_{Pr}, D))$ has been constructed through a projection which removed a set of attributes C_{Cut}. As Def. 5.1 states, a rule is an allowable default if the conditional probability of the consequent given the truth of the antecedent is higher than a threshold. In other words, if a class $E_{(C_{Pr})} \in U/\text{IND}(C_{Pr})$ has a membership degree in any $X \in U/\text{IND}(D)$ which is higher than the threshold ($\mu(E_{(C_{Pr})}, X) \geq \mu_{tr}$), then it is safe to generate a default rule. The new object class $E_{(C_{Pr})}$ may be characterised by its minimal description $Des(E_{(C_{Pr})}, MinDes)$, where $MinDes$ is a prime implicant of the discernibility function $f_{[C_{Pr},D]}(E_{(C_{Pr})})$ (see Def. 3.14). Since there may exist several minimal descriptions of a given class, each class $E_{(C_{Pr})}$ obviously may be represented by any number of rules. The set of default rules constitutes a default decision algorithm and is constructed as follows:

Definition 5.4 (*Default Decision Algorithm*). Given a decision system $\mathcal{A} = (U, (C, D))$, let $C_{Cut} \subseteq C$ and $C_{Pr} = C - C_{Cut}$. Let also $\mathcal{A}' = remC(\mathcal{A}, C_{Cut}) =$

$(U, (C_{Pr}, D))$. Then, for $0 < \mu_{tr} \leq 1$:

$$DRules(\mathcal{A}') = \bigcup \{ Des(E_{(C_{Pr})}, redE) \to Des(X, D) :$$
$$E_{(C_{Pr})} \in U/\text{IND}(C_{Pr}),$$
$$redE \in red(E_{(C_{Pr})}, C_{Pr})$$
$$X \in U/\text{IND}(D),$$
$$\mu(E_{(C_{Pr})}, X) \geq \mu_{tr} \}$$

The set $DRules(\mathcal{A}')$ is called a *default decision algorithm* for $\mathcal{A}'$. ∎

Correspondingly, a set of blocks $DBlocks(\mathcal{A}')$ is associated with the variant $\mathcal{A}' = remC(\mathcal{A}, C_{Cut}) = (U, (C_{Pr}, D))$. It covers the objects that constitute exceptions to the default rule, i.e. that are not mapped into the dominating decision. For each default $Des(E_{(C_{Pr})}, redE) \to Des(X, D)$, a block is generated for each class $E \in E_{(C_{Pr})}/\text{IND}(C_{Cut})$ such that $E \cap X = \emptyset$. For the default $\tau_1 \to \tau_2 \in DRules(\mathcal{A}')$, a block takes the form $Des(E, C_{Cut}) \to \neg(\tau_1 \to \tau_2)$.

We now give an outline of an algorithm for extracting rules that express dependencies in an information system. A more procedural definition of this algorithm can be found in [16].

- **Input:**
  1. A decision system $\mathcal{A} = (U, (C, D))$
  2. A threshold value μ_{tr}, which controls the desired confidence in the rules that are generated.
- **Output:**
  1. The default decision algorithms (Def. 5.4) for variants $\mathcal{A}'$ of $\mathcal{A}$

- **Step 1:** Selection of projections $C_{Pr} = C - C_{Cut}$ over the attributes. To each such projection over attributes corresponds a projection over an information system. The projections are selected so that new indeterminacies result, i.e. each projection is an indeterminacy gluing.
- **Step 2:** By each projection C_{Cut}, a variant $\mathcal{A}' = remC(\mathcal{A}, C_{Cut})$ of $\mathcal{A}$ is created. Since the projection is an indeterminacy gluing, classes are glued together into more compound object classes $E_{(k,C_{Pr})} \in U/\text{IND}(C_{Pr})$, $k = 1, .., |U/\text{IND}(C_{Pr})|$.
- **Step 3:** Rules are generated for each compound class $E_{(k,C_{Pr})}$ ensuring that one particular decision X remains the dominating one by checking that

$$\mu(E_{(k,C_{Pr})}, X) = \frac{|E_{(k,C_{Pr})} \cap X|}{|E_{(k,C_{Pr})}|} \geq \mu_{tr}$$

- **Step 4:** Blocks to the rules for class $E_{(k,C_{Pr})}$ are generated for those classes $E \in E_{(k,C_{Pr})}/\text{IND}(C_{Cut})$ which are now contained in the compound class, but which map into a decision other than the dominating one, i.e. $E \cap X = \emptyset$.
- **Step 5:** Iterate on each variant system $\mathcal{A}' = remC(\mathcal{A}, C_{Cut})$

At this point it is appropriate to describe the space in which these operations are defined. We are computing in a space of possible variants of an information system, i.e. the set of systems that may result from performing any projection on the original system. The projection operation $remC(\mathcal{A}, C_{Cut})$ is a mapping between information systems defined over subsets of the condition attribute set C. It follows that the iterative process of attribute removal is defined on the lattice $(2^C, \cup, \cap)$ over the powerset 2^C. Corresponding to the projection operations done on attribute sets, the projections on information systems define a space over the poset $2^\mathcal{A}$, i.e. a *lattice* over information systems. It is obvious at this point that a search over all possible variants of a decision system is an insurmountable task in general. A selection has to be made from the set of possible projections from a given lattice node. Some initial measures and heuristics for reducing the search were defined in [16].

5.3 The Example

Let us now consider the original example once again, applying the algorithm that was explained in the previous section.

Example 4.1 (Continued) The algorithm is initially called with the original decision system as the argument, that is, over attributes $\{a, b, c\} - \emptyset$. No attributes have been removed in the initial call. At this (top) level in the lattice, a number of definite rules are generated along with default rules that cover the indeterministic part of the IS. (Recall that a definite rule is simply a specialised default rule.) As explained earlier, we are able to generate five definite rules that cover the available "certain" knowledge. The only class which is indeterministic wrt. the decision is E_5. The corresponding degrees of membership are $\mu(E_5, X_3) = 4/5$ and $\mu(E_5, X_4) = 1/5$. If a new object is seen that matches the description of E_5 (i.e. either of the reducts a or b), the prior classification will intuitively be preferred.

In this example, we illustrate the total set of rules that may be generated in the process. To allow this, the threshold value has been set to zero ($\mu_{tr} = 0$) in order to accept any possible rule. The following default rules are extracted (each rule is shown with its corresponding degree of membership value $\mu(E_i, X_j)$ for $i = 1, .., 5$, $j = 1, .., 4$):

$$
remC(\mathcal{A}, \emptyset): \begin{array}{rll} E_1: & a_1 c_3 \to d_1 & |50/50 \\ E_2: & a_1 c_1 \to d_2 & |5/5 \\ E_2: & b_2 c_1 \to d_2 & |5/5 \\ E_3, E_4: & a_2 \to d_2 & |40/40 \\ E_4: & b_3 \to d_2 & |10/10 \\ E_5: & a_3 \to d_3 & |4/5 \\ E_5: & a_3 \to d_4 & |1/5 \\ E_5: & b_5 \to d_3 & |4/5 \\ E_5: & b_5 \to d_4 & |1/5 \end{array}
$$

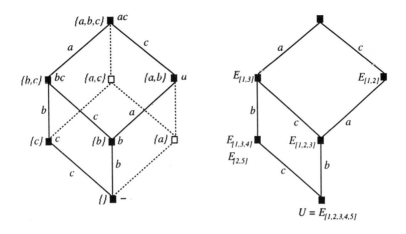

Fig. 1. The Search in the Lattice for Ex. 4.1

As explained earlier, the factors of the discernibility function for each variant of the system will define which projections may allow new and interesting rules to be obtained. The simplified discernibility factor set for the original system is $\{\{a\}, \{c\}\}$ which means that the attribute b is immaterial for the task of discerning the objects. New patterns may only be found through removal of either one of attributes a or c (or supersets of these). The search in the lattice is shown in Fig. 5.3, where rules are generated at nodes over condition attribute sets $\{a, b, c\}, \{a, b\}, \{b, c\}, \{c\}, \{b\}$ and $\emptyset$. These are marked black in the figure. On the right side the new classes that are generated at the respective nodes are shown.

The resulting discernibility factor sets for the constructed nodes are displayed in the table below, showing the minimal set of attributes needed to distinguish the classes that now exist (also displayed). In other words, if any factor is removed in its entirety, there is a change in the discernibility relation.

$$\begin{aligned}
\Phi_{[\{a,b,c\},\{d\}]} &= \{\{a\},\{c\}\} & \{E_1, E_2, E_3, E_4, E_5\} \\
\Phi_{[\{b,c\},\{d\}]} &= \{\{b\},\{c\}\} & \{E_{[1,3]}, E_2, E_4, E_5\} \\
\Phi_{[\{a,b\},\{d\}]} &= \{\{a\}\} & \{E_{[1,2]}, E_3, E_4, E_5\} \\
\Phi_{[\{c\},\{d\}]} &= \{\{c\}\} & \{E_{[1,2,3]}, E_4, E_5\} \\
\Phi_{[\{b\},\{d\}]} &= \{\{b\}\} & \{E_{[1,3,4]}, E_{[2,5]}\} \\
\Phi_{[\emptyset,\{d\}]} &= \emptyset & \{U\}
\end{aligned}$$

For the projections $C - \{a\}$ and $C - \{c\}$, respectively, the following sets of new default rules are generated:

$remC(\mathcal{A}, \{a\})$: $E_{[1,3]}: b_2 c_3 \to d_1 |_{50/80}$
$\phantom{remC(\mathcal{A}, \{a\}):}\ E_{[1,3]}: b_2 c_3 \to d_2 |_{30/80}$

$remC(\mathcal{A}, \{c\})$: $E_{[1,2]}: a_1 \to d_1 |_{50/55}$
$\phantom{remC(\mathcal{A}, \{c\}):}\ E_{[1,2]}: a_1 \to d_2 |_{5/55}$

Continuing to remove attributes from the two obtained systems implies that new compound classes are formed and new default rules may be generated. The system over $C - \{a\}$ has discernibility function bc, i.e. SDFS is $\{\{b\},\{c\}\}$). Further removal of attributes b or c yields two new sets of default rules:

$$remC(\mathcal{A},\{a,b\}): \begin{array}{l} E_{[1,3,4]}: c_3 \to d_1|_{50/90} \\ E_{[1,3,4]}: c_3 \to d_2|_{40/90} \\ E_{[2,5]}: c_1 \to d_2|_{5/10} \\ E_{[2,5]}: c_1 \to d_3|_{4/10} \\ E_{[2,5]}: c_1 \to d_4|_{1/10} \end{array} \qquad remC(\mathcal{A},\{a,c\}): \begin{array}{l} E_{[1,2,3]}: b_2 \to d_1|_{50/85} \\ E_{[1,2,3]}: b_2 \to d_2|_{35/85} \end{array}$$

If the attribute c is removed starting from the system over $C - \{a,b\}$, there is obviously nothing that enables us to distinguish the different classes any more. The rules generated at the bottom level of the lattice reflect simply the distribution of values for the decision attribute d, i.e. $U/\text{IND}(D)$ (the sets of objects that are classified into the respective decisions). The following rules are generated:

$$remC(\mathcal{A},\{a,b,c\}): \begin{array}{l} E_{[1,2,3,4,5]}: T \to d_1|_{50/100} \\ E_{[1,2,3,4,5]}: T \to d_2|_{45/100} \\ E_{[1,2,3,4,5]}: T \to d_3|_{4/100} \\ E_{[1,2,3,4,5]}: T \to d_4|_{1/100} \end{array}$$

It should be noted that the unconstructed nodes, over attributes $\{a,c\}$ and $\{a\}$ respectively, also have associated rule sets. These rule sets will normally be defined in terms of the rule sets for the closest constructed ancestor. The collection of rules shown in Tab. 5.3 covers the example information system. However, only the rules with the highest confidence are shown. Note that some of the rules, for instance the second default rule, have a high confidence, but a very low support.

Definite Rules:
$a_1 c_3 \to d_1$
$a_1 c_1 \to d_2$
$b_2 c_1 \to d_2$
$a_2 \to d_2$
$b_3 \to d_2$

Defaults:
$a_1 \to d_1|_{50/55=0.91}$
$a_3 \to d_3|_{4/5=0.80}$
$b_5 \to d_3|_{4/5=0.80}$
$b_2 c_3 \to d_1|_{50/80=0.62}$
$b_2 \to d_1|_{50/85=0.59}$
$c_3 \to d_1|_{50/90=0.56}$
$c_1 \to d_2|_{5/10=0.50}$
$T \to d_1|_{50/100=0.50}$

Table 3. The Rules, Sorted by Membership Degree

Observe that each node in the lattice is identified by the set of attributes that it represents. The path taken down the lattice in order to reach the node is of no importance in this respect. There is, therefore, no reason to recompute the matrix and rules for $M_{[\{b\},\{d\}]}$ reached through path $\{a,b,c\} \supset \{a,b\} \supset \{b\}$ as shown in Fig. 5.3, if the node has already been reached through path $\{a,b,c\} \supset \{b,c\} \supset \{b\}$.

Let us consider again the first (highest priority) default rule:

$$a_1 \to d_1$$

The rule was output at level $C - \{c\}$ in the lattice which resulted from the gluing of classes E_1 and E_2. Any counter-example to this rule may be defined in terms of the attributes that were removed in the last projection, that is c in this case. The value of this attribute is thus recorded for those objects that are contained in $[\![a_1]\!]_A = E_1 \cup E_2$ and that map into any decision other than d_1. This concerns the objects in the class E_2 for which the value for c is 1. A fact is generated that blocks the default in this case. We notice, however, that other objects could well exist in the information system that match the description $c = 1$ but still map into decision $d = 1$. Rule $c_1 \to \neg(a_1 \to d_1)$ is therefore generated. It may block an application of the default rather than the conclusion itself. When classifying new objects, the default rule may be applied to objects for which the value of a is 1, *unless* the value for c is 1 for that same object. □

In [16], it is argued that the quality of classification – in terms of wrongly classified objects – will decrease monotonically in downward traversal of the lattice, provided that the decision algorithms are tested on the training set itself. In other words, under the latter restriction, the number of objects that are correctly classified can never decrease if more attributes are taken into account. The value of machine learning, however, obviously relies on the ability of the decision algorithm to handle new, and previously unseen examples. Assume that such a set of objects has been submitted for classification. These objects might not be completely specified, rather the information may be provided in terms of some subset of the total set of conditional attributes. In this light, information given in terms of a set of attributes dictates which particular default decision algorithm should be used, namely the one which accounts for the available knowledge. That algorithm reflects exactly the power of classification represented in the known attributes. If additional information about the objects is obtained later, the decision algorithm for the lattice element identified by the new set of available attributes can be applied, i.e. more *specific* rule sets may be used. Using the terminology of decision systems, we may say that the default decision algorithm for a system $\mathcal{A}' = (U, (C', D))$ is more specific than the decision algorithm for any $\mathcal{A}'' = (U, (C'', D))$ if $C' \supset C''$.

A basic contention of this approach is that degree of specificity, as used above, is related to *prioritised* information and lends itself naturally to a framework such as described in Sect. 2. A system performing classification of new cases may do this based only on a small number of attributes using the proper default decision algorithm. However, if new information becomes available in terms of attribute values, the system may consult a more specific default decision algorithm generated from a decision system higher up in a lattice structure. In other words, we are considering an *upward traversal* in the lattice of decision systems emulating the decision procedure of an expert. Upward traversal in the lattice reflects a process of gathering increasing amounts of information, the default decision algorithm used is the theory which is best suited to handling the information at hand. Obviously, in total ignorance of the properties of the new object, the only classification which can be made is totally uninformed, i.e. the default decision algorithm for the bottom element of the lattice is applied.

The discussion above should not be interpreted as saying that more specific decision algorithms are always better. When it comes to classifying new cases, the opposite is often the case. There is no guarantee that adding more information will yield a better overall classification quality. The experiment reported in Sect. 6 gives one out of many examples of this. Again, due to over-fitting, the rules at higher levels of the lattice may be all too specific with many of the rules not matching any of the new objects. It may happen that a number of the objects are not classified at all by the rules. Yet, and this is quite important: for each single object in $\mathcal{A} = (U, (C, D))$, if there is a *conflict* between the classification made by the default decision algorithm at level $C' \subseteq C$ and the classification at level $C'' \subset C'$ (i.e. lower in the lattice), then the more specific algorithm takes priority. It makes sense to assume that all knowledge given about a particular object is relevant to some degree. *if* a given rule matches the object closer than another (lower-level) rule does, then the former would normally be supported. For those objects for which no conflicts arise, however, the original classification may be upheld.

Example 4.1 (Continued) Starting in the uninformed situation, i.e. at the bottom level of the lattice, we find that the most common decision is $d = 1$, made for 50 of the training objects. If knowledge becomes available wrt., say, the attribute c, a more accurate/specific classification may be made by using the rules for $\mathcal{A}' = remC(\mathcal{A}, \{a, b\})$. The process of information collection can, hence, be seen as making an upward traversal on the arc labelled c. Provided that the observed value for the attribute c is 3 or 1, the original hypothesis is strengthened or falsified, respectively. (Note that c_1 (i.e. $c = 1$) implies that the decision is d_2, d_3 or d_4. If, however, the new object has a value for c that is neither 1 nor 3, the currently considered rule set (i.e. $DRules(\mathcal{A}')$) may not be applied. At this point nothing may be done except to rely on the crude classification provided at the bottom node, and await new information. □

Again, the classification process described above may be modelled in terms of a prioritised default framework. In Brewka's *Prioritized Theorist*, reviewed in Sect. 2, rules are ordered in different layers, according to a measure of priority/specificity. Facts as such do not exist, rather they are replaced by defaults that may override any other default in the system.

6 A Case Study

The algorithm has been implemented twice. The first version resulted in the RGEN system [11, 2]. A number of experiments have been performed using case sets that were taken from the *UCI Repository Of Machine Learning Databases and Domain Theories*, moderated by Patrick M. Murphy. That list presently contains 110 different databases and domain theories and can be found at [17]. We will briefly report on one of the case studies here which the interested reader may consult in [16].

The case study is the well-known *Australian Credit Approval* data base [28, 29]. It registers information of credit card applications and the decision made in each case. In order to protect the confidentiality of the data, the attributes were changed to meaningless symbols. The data set offers a good mixture of attributes: continuous, nominal with small numbers of values and nominal with larger numbers of values. There are also a few missing values.

The original data set contains information about a total number of 690 customers, registered over 14 conditional attributes (6 numerical and 8 categorical (unordered)), and one decision (accept/reject). From the total number of cases, 307 objects (44.5%) were classified as 1 (accept), whereas 383 (55.5%) were classified 0 (reject). Before applying the algorithm for default rules generation to the Australian credit data set, the decision system was scaled, i.e. the numeric values were mapped into intervals. Following the scaling, the decision system was split into two parts, a training set containing a third, i.e. 227, of the objects, and a test set comprising the rest of the objects. Then the RGEN algorithm was applied to the data and a number of default decision algorithms were generated and tested. As can be expected, the performance of the respective default decision algorithms on the training set itself was generally very good. About 36 percent of the rule sets classified these objects to a degree of 90% or better, whereas 65 percent of the decision algorithms gave correct classification at least to the degree 85%. The algorithm for the top node was able to classify with one hundred percent certainty. However, using eight (out of the 14) attributes, the correctness rate could come as high as 99.56 (one out of 227 objects was misclassified).

Ultimately however, the criterion for quality of a default decision algorithm is determined by its performance on new cases. The default decision algorithms were therefore applied to the test set, which (as could be expected) gave results of varying quality. There were totally 34 nodes found for which the corresponding decision algorithm was able to classify to a degree 85.0% or better. It should be recalled at this point that the training itself was done on basis of only 227 objects. The default decision algorithm which gave the best result was defined over attributes 1, 3, 7 and 8 (A2, A4, A8 and A9), and gave correct classification for 399 (86.18%) of the 463 objects in the test table. The default decision algorithm for this node contained 57 rules. So from all the nodes, only one was found which could do a better job with the test system than the top node. The size, and thereby the efficiency and understandability of the default decision algorithms should, however, also be taken into account. With the scaled version of the decision system, the top node (overall 14 condition attributes) produced a default decision algorithm containing in excess of 20,000 rules. For the same test table, the decision algorithm for attribute 7 (A8), containing the two rules shown below, gave a classification accuracy of 85.10%, i.e. only 0.65% lower than the result obtained using the complete knowledge of attributes. The corresponding numbers for the original (training) set were 100% and 86%, respectively.

```
RBaseNo: 18
Attrib: (7)
1     : [0.913793,116  ]  (A8=0)=>(Dec=0)
2     : [0.810811,111  ]  (A8=1)=>(Dec=1)
```

The numbers associated with each of the rules show the confidence/trust and the support/generality of the rule, respectively. The first rule matched 116 of the 227 objects. From these 91,4% were classified into decision class 0.

There is a very apparent gain in restricting the attention to smaller subsets of the original set of condition attributes. The quality of classification does not suffer from this simplification so equally good results may be obtained using drastically smaller rule sets. This shows that decisions can be made by means of much more efficient decision algorithms that are at the same time much more *understandable* to a human being. In this light, default decision algorithms may serve also a *pedagogical* purpose. The upward traversal in the lattice mirrors a process of knowledge refinement, where increasingly specific knowledge is introduced in a controlled way and reflected in increasingly specific rules. Through development of suitable presentation tools, the process of knowledge collection can be simulated by a user and the gain in terms of decision making can be studied.

7 Related Work

Hu, Cercone and Han [10] describe an algorithm that treats tuples that occur rarely in the training data as *noise* not allowing for generation of rules from such tuples. This is accomplished by calculating a simple frequency ratio for each tuple and filtering out the exceptional tuples using a noise filter threshold. In our example, the ratio for the object class $E_{5,2}$ would be $1/100 = 0.01$, and one might consider cutting out the tuple from the information system. (Incidentally, this would have solved our problem with the indeterminacy of the decision attribute for the E_5 class). In *boundary region thinning* [32], objects corresponding to small values of the probability distribution wrt. the decision are treated as *abnormal* or *noisy* objects.

> *The generalised decision for a given information vector can then be modified by removing from it the decision values corresponding to these small values of the probability distribution. The decision rules generated for the modified generalised decision can give better quality of classification of new, yet unseen objects.* ([32])

In [5], Grzymała–Busse describes the process of knowledge acquisition under uncertainty, using the rough set approach. The inconsistencies present in the original information system are not corrected; rather they are used to generate *unsafe* rules that cover a greater set of the objects represented in the information system. The framework produces rules that are classified into *certain* and *possible* depending on whether the respective classes are mapped deterministically into a decision class or not. The author suggests propagating the possible rules in parallel with the propagation of the certain rules effectively using two different production systems that are run concurrently. In [6, 7, 9, 8], Grzymala–Busse describes the system LERS, which builds on the same ideas. The algorithm looks for regularities in a data set and induces certain and possible decision rules.

Yasdi [36] presents an approach which allows generation of rules from non-deterministic situations. The rules are built up according to the schema

$$\{Des(E) \rightarrow Des(X): E \cap X \neq \emptyset\}$$

where E and X are the learner's and the oracle's concept partitions, respectively. The approach effectively uses the upper approximations of sets, creating indeterministic/inconsistent rules. Yasdi's paper provides no way of dealing with such inconsistencies in the generated rules and there is no reference to information which could imply that one decision be *preferred* over others in situations of conflict.

8 Summary and Future Research

We have addressed what we see as a problem in the data mining field, namely the ideal that the rules that are extracted should be *correct*. Correctness (i.e. consistency) in the limited training set does not guarantee that the rules behave well in new situations. It is contended that default rules provide a powerful tool for representing common characteristics of a set of data. Situations of limited knowledge are emulated via controlled introduction of indeterminacy. Through an iterative process of removing conditional attributes, a number of variants of an original decision system are obtained. Some of these combinations, generally much smaller than the original set of condition attributes, yield simple decision algorithms that perform remarkably well on unseen objects. We suggest that rough set theory may be applied to solve the problem of generation of default decision rules from low-level data. A theory for default rule extraction has been developed and an algorithm has been constructed and tested on different real-life data sets. The experiments that have been made suggest that the method is well capable of competing with data mining systems currently in use.

Apart from the specific issues tackled, we foresee several interesting topics and questions for future investigation, the most important of which are indicated below. This study suggests that rough sets are beneficial to the research in these areas.

We plan to continue the work in several directions.

- Further work ought to be done on the problem of heuristics development, testing and possibly improving the proposed methods for search limitation through selection of projections. This problem needs to be solved, in order for the framework to scale to larger-size decision systems.
- The implementation of the algorithm should be developed further, including improvements of the rule generation procedure. A new version of the algorithm which incorporates the proposed methods for search limitation should be obtained.

- It should be possible to assign *costs* to attributes, effectively favouring certain attributes over others, and thereby influencing the selection from the set of possible projections from a given node. Such cost measures could model the trust in the data for a given attribute due to the presumed importance of the attribute or the ease by which values for the attribute can be obtained.
- The framework should be extended to encompass interesting default rules from general information systems, i.e. relaxing the restriction that there exist a predefined set of decision attributes. The system ought to recognise as a result interesting patterns between any attribute sets in the system.

At the moment of revising this chapter, we can report that a new implementation of the algorithm presented here has been made in the ROSETTA system [38] and several new experiments were performed. The results reported by Jenssen, Kommorowski and Øhrn in [12] further corroborate the applicability of the method presented here.

Acknowledgments

We have enjoyed and profited from the co-operation with Andrzej Skowron and Lech Polkowski, Warsaw, Poland, while working on this project. In addition to resources acknowledged below, Torulf Mollestad has received support from the Polish State Committee on Research while visiting Warsaw University. His main support came as a PhD scholarship from the Department of Computer and Information Science, NTNU. This research has been also supported in part by the European Union 4th Framework Telematics project CARDIASSIST, by the Norwegian Research Council grant #74467/410 and by the Norwegian Research Council grant for Cooperation with Central Europe.

Several of the students contributed to this work: Ø. T. Aasheim, H.G. Solheim, J.P. Hjulstad, T.-K. Jenssen, S. Vinterbo and A. Øhrn. We thank them all.

Last but not least, we appreciate the thorough comments from an anonymous reviewer as well as the proofreading by Anna Komorowska.

References

1. A. Aamodt and J. Komorowski, (Eds.) (1995), *Proc. Fifth Scandinavian Conference on Artificial Intelligence*, Trondheim, Norway, May, IOS Press.
2. Ø. T. Aasheim and H.G. Solheim (1996), *Rough sets as a framework for data mining.* Technical report, The Norwegian Univ. of Science and Technology, Trondheim.
3. G. Brewka (1989), *Preferred subtheories: An Extended Logical Framework for Default Reasoning*, pp. 1043–1048. In: Proc. of IJCAI-89, pp. 1043–1048, Morgan Kaufmann.
4. G. Brewka (1991), *Nonmonotonic Reasoning: Logical Foundations of Commonsense.* Cambridge University Press.

5. J.W. Grzymala-Busse (1988), *Knowledge Acquisition under Uncertainty – A Rough Set Approach*. J. of Intelligent and Robotic Systems, 1:3–16.
6. J.W. Grzymala-Busse (1992), *LERS – A System for Learning from Examples based on Rough Sets*. In: Słowinski [35], pp. 3–18.
7. J.W. Grzymala-Busse (1993), *ESEP: An Expert System for Environmental Protection*. In: Ziarko [37], pp. 466–473.
8. J.W. Grzymala-Busse (1994), *Managing Uncertainty in Machine Learning from Examples*. In: *Proc. of the III Workshop on Intelligent Information Systems*, pp. 70–84, Wigry, June, Poland.
9. D. M. Grzymała-Busse and J.W. Grzymała-Busse (1993), *Comparison of Machine Learning and Knowledge Acquisition Methods of Rule Induction Based on Rough Sets*. In: Ziarko [37], pp. 282–289.
10. X. Hu, N. Cercone, and J. Han (1993), *An Attribute-Oriented Rough Set approach for Knowledge Discovery in Databases*. In: Ziarko [37] pp. 90–99.
11. J.P. Hjulstad (1996) *Mining Propositional Default Rules*. Master's thesis, The Norwegian Univ. of Science and Technology, Trondheim, Norway.
12. T.-K. Jensen, J. Komorowski and A. Øhrn (1998), *Improving Mollestad's Algorithm for Default Knowledge Discovery*, to appear in the Proc. of 1st International Conference on Rough Set and Soft Computing, RSCTC'98, A. Skowron and L. Polkowski, (Eds.) Warsaw, June 1998, Springer Verlag, LNAI.
13. J. Komorowski and Z.W. Ras, (Eds.) (1993), *Proc. Seventh International Symposium on Methodologies for Intelligent Systems, ISMIS'93*, Trondheim, Norway, June, Springer Verlag, LNAI Vol. 689.
14. T. Mollestad (1995), *Learning Propositional Default Rules using the Rough Set Approach*. In: Aamodt and Komorowski [1], pp. 208–219.
15. T. Mollestad and A. Skowron (1996), *A Rough Set Framework for Data Mining of Propositional Default Rules* In: Pawlak and Ras [20] pp. 448–457. Full version available at http://www.idi.ntnu.no
16. T. Mollestad (1997) *A Rough Set Approach to Data Mining: Extracting a Logic of Default Rules from Data*. PhD thesis 1997:1, The Norwegian Univ. of Science and Technology, Trondheim, Norway, February.
17. P.M. Murphy (1997), *UCI Repository Of Machine Learning Databases and Domain Theories*, 1997. At: http://www.ics.uci.edu/~mlearn/MLRepository.html
18. Z. Pawlak (1982) *Rough Sets*. Inter. J. of Information and Computer Science, 11(5):341–356.
19. Z. Pawlak (1991), *Rough Sets – Theoretical Aspects of Reasoning about Data*. Kluwer Academic Publishers.
20. Z. Pawlak and Z. Ras, (Eds.), (1996), *Proc. Ninth International Symposium on Methodologies for Intelligent Systems, ISMIS'96*, June, Springer Verlag, LNAI.
21. Z. Pawlak and A. Skowron (1993), *A Rough Set Approach to Decision Rules Generation*. Technical Report, University of Warsaw.
22. G. Piatetsky-Shapiro and W.J. Frawley, Eds. (1991), *Knowledge Discovery in Databases*. AAAI/MIT Press.
23. L. Polkowski and A. Skowron (Eds.), (1998), *Rough Sets in Knowledge Discovery*, Physica Verlag.
24. D. Poole, R. Goebel, and R. Aleliunas (1986), *Theorist: A Logical Reasoning System for Defaults and Diagnosis*. In N.J. Cercone and G. McCalla, (Eds.), The Knowledge Frontier: Essays in the Representation of Knowledge, pp. 331–352. Springer-Verlag, New York.

25. D. Poole (1988), *A Logical Framework for Default Reasoning.* J. Artificial Intelligence, 36:27–47.
26. D. Poole (1989), *Explanation and Prediction: An Architecture for Default and Abductive Reasoning.* J. Computational Intelligence, 5(2):97–110.
27. J.R. Quinlan (1986), *Induction of Decision Trees.* J. Machine learning, 1:81–106.
28. J.R. Quinlan (1987), *Simplifying decision trees.* International J. Man–Machine Studies, 27:221–234, December.
29. J.R. Quinlan (1992), *C4.5: Programs for Machine Learning.* Morgan Kaufmann.
30. R. Reiter (1980), *A Logic for Default Reasoning.* J. Computational Intelligence, 13:81–132.
31. The ROSETTA WWW homepage, http://www.idi.ntnu.no/~aleks/rosetta/
32. A. Skowron (1993) *Boolean Reasoning for Decision Rules generation.* In: Komorowski and Ras [13] pp. 295–305.
33. A. Skowron and J. Grzymała-Busse. *From Rough Set Theory to Evidence Theory.* John Wiley & Sons, 1994. Also: ICS Research Report 8/91, Warsaw University of Technology.
34. A. Skowron and C. Rauszer (1992), *The Discernibility Matrices and Functions in Information Systems.* In: Słowinski [35] pp. 331–362.
35. R. Słowinski, (Ed.), (1992) *Intelligent Decision Support – Handbook of Applications and Advances of the Rough Sets Theory*, Dordrecht, Kluwer Academic Publishers.
36. R. Yasdi (1991), *Learning Classification Rules from Databases in the Context of Knowledge-Acquisition and -Representation.* IEEE Transactions on Knowledge and Data Engineering, 3:293–306, September.
37. W. P. Ziarko, (Ed.) (1993), *Rough Sets, Fuzzy Sets and Knowledge Discovery,* Proc. of the International Workshop on Rough Sets and Knowledge Discovery (RSKD'93), Banff, Alberta, Canada, Springer Verlag.
38. A. Øhrn, J. Komorowski, A. Skowron, P. Synak (1998), *The Design and Implementation of a Knowledge Discovery Toolkit Based on Rough Sets - The ROSETTA System*, To appear in: Polkowski and Skowron [23]. 24 pp.

Power System Security Analysis based on Rough Classification

Germano Lambert-Torres[1], *Ronaldo Rossi*[2], *José Antônio Jardini*[3], *Alexandre P. Alves da Silva*[1], *Victor Hugo Quintana*[4]

[1] Escola Federal de Engenharia de Itajubá,
Grupo de Engenharia de Sistemas — Instituto de Engenharia Elétrica,
Av. BPS 1303, Itajubá, MG, 37500–000, Brazil,
E-mail: germano@iee.efei.br
[2] Faculdade de Engenharia de Guaratinguetá, FEG—UNESP
[3] Universidade de São Paulo — USP
[4] University of Waterloo

Abstract: This paper presents an example of a systematic approach to help knowledge engineers during the extraction process of facts and rules of a set of examples for power system operation problems. The described approach tries to reduce the number of examples, offering a more compact set of examples to the user. This approach is based on Rough Set Theory.

Keywords: Power System Operation, Rough Set, Knowledge Engineering, Decision Rules.

1 Introduction

Artificial intelligence techniques have been used as an alternative to solve problems that traditional computer approaches can not handle easily. Among these techniques, those that use deductive systems have been extensively used to accomplish tasks such as diagnosis, design, control, planning, and management. Expert systems and case-based reasoning are examples of deductive systems.

The knowledge acquisition process is a typical task during the construction of deductive systems. During this process, an expert accurately describes to the knowledge engineer all steps of the problem-solving for a specific problem. On the other hand, the knowledge engineer tries to capture this information and express it in facts-and-rules form, following the expert's reasoning. However, this acquisition process is one of the most important drawbacks of this approach, because getting the expert knowledge is not a simple task. Usually, experts have difficulty to put in words many of their analyses and reasoning. In many cases, this leads to knowledge bases that may contain two different problems: (a) superfluous details about specific points, producing bases bigger than necessary; and, (b) lack of information in other points, generating problems during the inference process.

One of the most common way to reduce the problems described above is to generate the knowledge base from a set of examples selected by the expert. Manipulating these examples, the knowledge engineer can extract facts and rules to build the knowledge base. In this approach, the number of interviews during the acquisition process is reduced, but the problem of superfluous details remains. Usually, in this approach, the problem is the superfluous number of input variables and some examples without importance.

2 Overview of Power System Operation

The operation of a power system is intrinsically complex due to the high degree of uncertainty and the large number of variables involved. The various supervision and control actions require the presence of an operator, who must be capable of efficiently responding to the most diverse requests, by handling various types of data and information [1].

These data and information come from measurements of the system or from computational processes. The size of the current database in a power control center has increased a lot in the last years due to the use of telecommunication. The system operator needs to know the current state of the system and some forecasted position, such as load forecasting, maintenance scheduling, and so on in order to take a control action (switching, changing taps and voltage levels, and so on).

One of the most important operator tasks is to determine the current operational state of the system. To accomplish this task, the operator receives many data from into the system. These data can be discrete (e.g., status of circuit breakers) or analogical (e.g., real power flow in a specific transmission line). By handling these data, the operator tries to built an image of the operation point. Figure 1 shows a representation of this process.

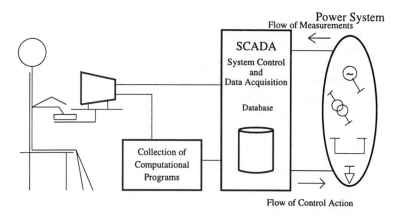

Fig. 1. Pictorial Representation of a Power System Control Center.

The analysis performed by the operator tries to make a classification of the operational mode in one of the three state: normal, emergency, and restorative. In the first state, normal, all loads are supplied and all measurements are inside of the nominal rates. In the second state, emergency, all loads continue to be supplied but some of the measurements are outside the nominal rates. For the restorative operational state, some loads are not supplied, i.e., there was a load shedding process [2].

Even when the operation state is normal, the operator needs to analyze the system security. This analysis is made according to possible contingencies that could affect the power system. Loss of a transmission line, shut down of a power plant or an increase of the load are some contingencies that can occur during the operation. An example of safe or unsafe points is shown in Figure 2. It shows the same contingency for two different operation points. For the operation point A, the contingency produces an abnormal operation point; while for the operation point B, the system continues in the normal state. Thus, the point A is an unsafe operation point and point B is a safe one.

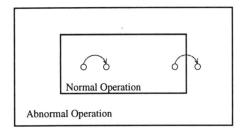

Fig. 2. Operational State of a Power System and Changing of Operation Point.

The operator's difficulty is to deal with all available data. A huge number of measurements must be manipulated by clustering and suppression of information. This allows the operator to visualize the current state of the system. The manipulation of all data/information is not an easy task.

3 Description of the Problem

The main purpose of the illustrative example that follows is to help the understanding of the rough set theory fundamental concepts. The idea is to transform a set of examples in a set of rules that represent the operational state of a power system. Some assumptions and simplifications are made to allow a better understanding of each step of the formulation without loss of generality. In fact, the data used in this paper comes from a Brazilian electric utility [3].

Consider a control center database composed by a set of measurements, such as the one shown in Table 1. The operational state of the hypothetical power

system depends on four elements: status of circuit-breaker A, transmission lines B and C, and voltage of bus D. Moreover, Table 1 contains the attributes represented by the set A, B, C, D and the corresponding decision S, where:
 - the status of circuit-breaker A is defined by 0 (close) or 1 (open);
 - the values of transmission lines B and C are percentages of real power flows according to their maximum capacities, in [%]; and,
 - the bus voltage D is expressed as a fraction of the rated voltage.

The classification of each state is made according to an expert (usually, a senior operator/engineer), and four possible outputs can be selected for the power system operational state: safe (S) or unsafe levels 1,2 and 3 (U1, U2, U3, respectively).

Observing the above set of examples, it is really hard to conclude that the condition of transmission line B is not necessary in the classification process. Notice that, this attribute is a dispensable one, as shown later. Even in this very small database it is very hard to reach a conclusion. For real control center database, usually with hundreds important attributes and thousands of examples, it could be impossible to take a reliable control action.

4 Presentation of the Algorithm

Before the presentation of the algorithm, two major concepts in Rough Set theory, reduct and core, will be defined. These concepts are important in the knowledge base reduction.

Let R be a family of equivalence relations. The reduct of R, RED(R), is defined as a reduced set of relations which conserve the same inductive classification of set R. The core of R, CORE(R), is the set of relations which appear in all reduct of R, i.e., the set of all indispensable relations to characterize the relation R.

The main idea behind the knowledge base reduction is a simplification of a set of examples [5, 6]. This can be obtained by the following procedure:
 a) calculate the core of the problem;
 b) eliminate or substitute a variable by another one; and
 c) redefine the problem using new basic categories.

The algorithm for the reduction of a decision table can be described using algebraic developments or based on logical relations [4, 7]. On the other hand, one of the most common approaches to get knowledge from an expert is by examples. The algorithm that provides the reduction of conditions has been proposed in [8, 9], and can be represented by the following steps:

Step 1: Transform continuous values in ranges.
Step 2: Eliminate identical attributes.
Step 3: Eliminate identical examples.
Step 4: Eliminate dispensable attributes.
Step 5: Compute the core of the decision table.
Step 6: Compose a table with reduct set.
Step 7: Merge possible examples and compose the final set of rules.

Table 1. Reduced Control Center Database

U	Attributes				S
	A	B	C	D	
1	0	57	82	1.07	U2
2	0	37	32	0.97	U1
3	1	0	87	0.95	U3
4	1	72	30	1.07	U3
5	0	28	39	1.02	U1
6	0	42	82	1.07	U2
7	0	52	59	1.01	S
8	1	62	67	1.04	U3
9	0	57	45	0.99	S
10	0	45	58	1.00	S
11	0	32	47	0.94	S
12	0	0	57	1.08	U2
13	1	58	87	0.95	U3
14	0	58	56	1.07	U2
15	0	25	57	1.03	S
16	0	56	54	1.08	U2
17	1	59	72	1.08	U3
18	0	32	0	0.93	U1
19	0	32	45	0.94	S
20	1	72	67	0.96	U3
21	0	57	45	1.01	S
22	0	32	45	0.94	S
23	0	29	43	1.08	U2
24	1	0	72	0.95	U3
25	1	57	79	1.07	U3
26	0	31	43	0.99	S
27	0	32	42	0.94	S
28	0	17	32	0.92	U1
29	0	23	22	0.95	U1
30	0	23	57	0.90	S

5 Description of the Knowledge Base Reduction

This section describes step-by-step the application of the algorithm presented in Section 4.

Step 1:

The first step of the algorithm is to redefine the value of each attribute according to a certain metric. In this illustrative example, typical ranges in power system operation are used:
- real power values:
under 40% of nominal capacity = low (L)
between 40% and 60% = medium (M)
above 60% of nominal capacity = high (H)
- bus voltage values:
under 0.95 pu = low (L)
between 0.95 and 1.05 = normal (N)
above 1.05 = high (H)

The status of circuit-breakers are maintained because the values 0 and 1 are normalized already. Using these new definitions, each example of Table 1 can be rewritten, given the examples shown in Table 2.

Notice: using these ranges, an incomplete set of examples has been generated (the complete search space is composed by 54 examples). Many problems can be analyzed by considering only an incomplete set of cases. In power systems, it is not probable to get all possible combinations of intervals. Thus, the produced set of examples in a real power system is always incomplete.

Step 2:

This next step verifies if any attribute can be eliminated by repetition. In Table 2, it can be verified that it does not occur.

Step 3:

This step verifies identical examples. In this case, many examples are identical (for instance, examples 12-23 and 11-19-22-27-30). The identical examples are merged, and the resultant set of examples is shown in Table 3.

Step 4:

This step verifies if the decision table contains only indispensable attributes. This task can be accomplished eliminating step-by-step each attribute and verifying if the table still gives the correct classification. For example, if attribute B is eliminated, the table continues to give the correct classification. Therefore, B is a dispensable attribute for this decision table (Table 4).

However, when the attribute A is eliminated (Table 5), it can be verified that the examples 7 and 8 have the same set of attributes but they give different classifications. In this case, the attribute A is indispensable for the decision table. After considering the elimination of each attribute, B is dispensable for the decision table. Table 6 presents the resultant set of examples.

Step 5:

Using Table 6, the core of the set of examples is computed. This can be done eliminating each attribute step-by-step, and verifying if the decision table continues to give the correct answer (i.e., it continues to be consistent). This procedure

Table 2. Database with Ranges

U	Attributes				S
	A	B	C	D	
1	0	M	H	H	U2
2	0	L	L	N	U1
3	1	L	H	N	U3
4	1	H	L	H	U3
5	0	L	L	N	U1
6	0	M	H	H	U2
7	0	M	M	N	S
8	1	H	H	N	U3
9	0	M	M	N	S
10	0	M	M	N	S
11	0	L	M	L	S
12	0	L	M	H	U2
13	1	M	H	N	U3
14	0	M	M	H	U2
15	0	L	M	N	S
16	0	M	M	H	U2
17	1	M	H	H	U3
18	0	L	L	L	U1
19	0	L	M	L	S
20	1	H	H	N	U3
21	0	M	M	N	S
22	0	L	M	L	S
23	0	L	M	H	U2
24	1	L	H	N	U3
25	1	M	H	H	U3
26	0	L	M	N	S
27	0	L	M	L	S
28	0	L	L	L	U1
29	0	L	L	N	U1
30	0	L	M	L	S

Table 3. Resultant Set of Examples.

U	Attributes				S	U old
	A	B	C	D		
1	0	L	M	H	U2	12,23
2	1	M	H	N	U3	13
3	1	H	H	N	U3	8,20
4	0	L	M	L	S	11,19,22,27,30
5	0	L	L	L	U1	18,28
6	0	M	M	N	S	7,9,10,21
7	0	M	H	H	U2	1,6
8	1	M	H	H	U3	17,25
9	0	L	L	N	U1	2,5,29
10	0	L	M	N	S	15,26
11	1	L	H	N	U3	3,24
12	0	M	M	H	U2	14,16
13	1	H	L	H	U3	4

will be illustrated for example U1. When the attribute D is eliminated, it can be verified that the decision table becomes inconsistent because the examples 1, 3, and 5 have the same attributes (A equals to 0, and C equals to M) and different decisions (L2 and S, respectively). However, when the attribute A is eliminated, the decision table continues to be consistent. Table 7 shows the core of the decision table.

Step 6:

This step computes the reduced set of relations that conserve the same inductive classification of the original set of examples. Table 8 contains the reduction of each example.

Step 7:

According to Table 8, the knowledge existent in Table 1 can be expressed by the following set of rules:

Rule 1: If C is M and (D is B or D is N) then S is Safe.

Rule 2: If C is L and (A is 0 or D is N or D is B) then S is Unsafe level 1.

Rule 3: If (A is 0 and (C is P or D is A)) or (C is M and D is A) then S is Unsafe level 2.

Rule 4: If (A is 1) or (C is P and D is N) or (C is L and D is A) then S is Unsafe level3.

or, using a complete rule formulation, in natural language:

Rule 1:

If (the power flow in transmission line C is between 40% and 60%) and (the

Table 4. Verification of the Dispensability of Attribute B.

U	Attributes			S
	A	C	D	
1	0	M	H	U2
2	1	H	N	U3
3	1	H	N	U3
4	0	M	L	S
5	0	L	L	U1
6	0	M	N	S
7	0	H	H	U2
8	1	H	H	U3
9	0	L	N	U1
10	0	M	N	S
11	1	H	N	U3
12	0	M	H	U2
13	1	L	H	U3

voltage on bus D is below 1.05) then the classification of the current state of the system is safe.

Rule 2:

If (the power flow in transmission line C is below 40%) and (the voltage on bus D is below 1.05) then the classification of the current state of the system is unsafe level 1.

If (the power flow in transmission line C is below 40then the classification of the current state of the system is unsafe level 1.

Rule 3:

If (the circuit-breaker A is closed) and (the power flow in transmission line C is above 60%) then the classification of the current state of the system is unsafe level 2.

If (the circuit-breaker A is closed) and (the voltage on bus D is above 1.05) then the classification of the current state of the system is unsafe level 2.

If (the power flow in transmission line C is between 40% and 60%) and (the voltage on bus D is above 1.05) then the classification of the current state of the system is unsafe level 2.

Rule 4:

If (the circuit-breaker A is opened) then the classification of the current state of the system is unsafe level 3.

If (the power flow in transmission line C is above 60%) and (the voltage on bus D is between 0.95 and 1.05) then the classification of the current state of

Table 5. Verification of the Indispensability of Attribute A.

U	Attributes			S
	B	C	D	
1	L	M	H	U2
2	M	H	N	U3
3	H	H	N	U3
4	L	M	L	S
5	L	L	L	U1
6	M	M	N	S
7	M	H	H	U2
8	M	H	H	U3
9	L	L	N	U1
10	L	M	N	S
11	L	H	N	U3
12	M	M	H	U2
13	H	L	H	U3

Table 6. Set of Examples with Indispensable Attributes.

U	Attributes			S	U old
	A	C	D		
1	0	M	H	U2	1,12
2	1	H	N	U3	2,3,11
3	0	M	L	S	4
4	0	L	L	U1	5
5	0	M	N	S	6,10
6	0	H	H	U2	7
7	1	H	H	U3	8
8	0	L	N	U1	9
9	1	L	H	U3	13

the system is unsafe level 3.

If (the power flow in transmission line C is below 40%) and (the voltage on bus D is above 1.05) then the classification of the current state of the system is unsafe level 3.

Table 7. Core of the Set of Examples.

U	Attributes			S
	A	C	D	
1	-	-	H	U2
2	-	-	-	U3
3	-	M	L	S
4	-	L	-	U1
5	-	M	N	S
6	0	-	-	U2
7	1	-	-	U3
8	-	L	-	U1
9	-	-	-	U3

Table 8. Reduction of the Set of Examples.

U	Attributes			S
	A	C	D	
1A	0	-	H	U2
1B	-	M	H	U2
2A	1	-	-	U3
2B	-	H	N	U3
3	-	M	L	S
4A	0	L	-	U1
4B	-	L	L	U1
5	-	M	N	S
6A	0	H	-	U2
6B	0	-	H	U2
7	1	-	-	U3
8A	0	L	-	U1
8B	-	L	N	U1
9A	-	L	H	U3
9B	1	-	-	U3

6 Conclusions

The knowledge acquisition process is the most difficult task during the construction of knowledge based systems, such as expert systems and case-based reasoning. Usually, the experts have difficulty to explain to the knowledge engineers how they solve a given problem.

This paper presents a systematic approach to transform examples in a reduced set of rules. This approach uses Rough Set theory and concepts of core and reduction of knowledge. An example for power system control centers has been developed. For the sake of clarity, a reduced database is used in the illustrative example. However, the same methodology is applicable to real databases.

Acknowledgments
The authors would like to thank the CNPq, FAPEMIG and FINEP/RECOPE for the financial support to develop this project.

References

1. Valiquette B., Lambert-Torres G., Mukhedkar D., An Expert Sy stem Based Diagnosis and Advisor Tool for Teaching Power System Operation Emerge ncy Control Strategies. *IEEE Transactions on Power Systems*, Vol.6, No.3, 1991, 1315–1322.
2. Lambert-Torres G. et al., A Fuzzy Knowledge-Based System for Bus Load Forecasting. In: Marks II R.J. (ed.) *Fuzzy Logic Technology and Applications*. IEEE Press, 1994, 221-228.
3. Santos, C.C., Diagnosis and Characterization of Power System Operational State. M.Sc. Thesis, EFEI, Brazil, 1996 (in Portuguese).
4. Pawlak Z., Rough Sets. *International Journal of Information and Computer Sciences*, 11, 1982, 341–356.
5. Pawlak Z., *Rough Sets - Theoretical Aspects of Reasoning about Data*. Kluwer Academic Publishers, 1991.
6. Pawlak Z., Rough Classification. *International Journal on Man–Machine Studies*, 20, 1984, 469–483.
7. Slowinski R., Stefanowski J., Rough Classification in Incomplete Information Systems. *Mathematical and Computing Modeling*, Vol. 12, No. 10/11, 1989, 1347–1357.
8. Slowinski R., *Intelligent Decision Support*. Kluwer Academic Publishers, 1992.
9. Skowrow A., Synthesis of Adaptive Decision Systems from Experimental Data. *Proc. Fifth Scandinavian Conference on Artificial Intelligence*, Trondheim, Norway, IOS Press, May 29-31, 1995, 220–238.

Rough Sets and Principal Component Analysis and Their Applications in Data Model Building and Classification

Roman W. Swiniarski

San Diego State University, San Diego, California 92182-7720, U.S.A.

Abstract The method of an application of principal component analysis (PCA) and rough sets for feature extraction and selection is discussed. The case study of texture images classification is provided as an illustration of the method.

1 Introduction

Dimensionality reduction of patterns and data set is still one of most demanding issues in data mining and knowledge discovery. This problem relates to feature extraction, reduction and selection for data sets with high-dimensional patterns. In this chapter we discuss two methods of feature extraction, reduction and selection from high dimensional pattern data set:

1. Principal component analysis (PCA) with optimal Karhunen-Loéve transformation.
2. Rough sets based feature reduction.

We show how to combine PCA transformation with rough set feature selection/reduction based on minimum concept description paradigm. First, a short introduction to PCA and Karhunen-Loéve transformation, with application to feature extraction/reduction, is presented. Then an introduction to rough sets with feature reduction mechanism, is provided. Then we discuss combination of PCA-rough sets sequence for feature extraction/selection. We illustrate of presented technique by a case study of texture recognition.

2 Principal Component Analysis. Feature Extraction and Reduction Based on the Karhunen-Loéve Transformation

Reduction of pattern dimensionality possibly may improve the recognition process by considering only compact, the most important data representation, possibly with uncorrelated elements retaining maximum information about the original data and with possible better generalization abilities.

The best reduced pattern space, and thus amount of information preserved, may be different for example for the data compression and reconstruction task, than those for the pattern recognition task.

Reduction of the original pattern dimensionality generally refers to a transformation of original $n-dimensional$ patterns (vectors in the pattern space) into other $m-dimensional$ feature patterns (with $m < n$). The pattern reduction (feature extraction) can be considered as a nonlinear transformation

$$\mathbf{y} = \mathbf{F}_t(\mathbf{x}) \qquad (1)$$

of $n-dimensional$ original patterns $\mathbf{x}$ into $m-dimensional$ transformed patterns $\mathbf{y}$ (vectors in the $m-dimensional$ feature space).

Transforming the original patterns into the feature vectors has more general meaning and does not only correspond to the pattern dimensionality reduction. The general goal of obtaining the feature vectors is to represent data in the best form for a given processing goal.

2.1 Principal component analysis (PCA)

The most popular, statistical method of feature extraction and data compression is based on a linear transformation of the original patterns is the Karhunen-Loéve transformation (KLT) related to statistical principal component analysis (PCA) method (Duda, and Hart, 1973; Oja, 1989; Bishop, 1995; Diamantras, and Kung, 1996). This linear transformation is based on statistical characteristic of a given data set represented by data patterns covariance matrix, its eigenvalues and corresponding eigenvectors.

Principal component analysis (PCA) is a statistical data analysis technique which determines an optimal linear transformation

$$\mathbf{y} = \mathbf{W}\mathbf{x} \qquad (2)$$

of a real-valued $n-dimension$ random data pattern $\mathbf{x}$ into another $m-dimensional$ ($m \leq n$) transformed vector $\mathbf{y}$. The $m \times n$ fixed linear transformation matrix $\mathbf{W}$ is designed optimal (from the point of view maximal information retaining) by exploring statistical correlations among elements of the original patterns and by finding possibly reduced compact data representation retaining maximum nonredundant and uncorrelated intrinsic informations of the original data.

Depending on the nature of an original one can obtain a substantial reduction of feature vector dimensionality $m << n$ comparing with a dimensionality of original data patterns. First, having the optimal transformation matrix $\mathbf{W}$ determined, one can reduce of decorrelated feature vector dimension and use reduced feature vectors for classification. Second, all original $n-dimensional$ data patterns can be optimaly transformed to data patterns in the feature space with lower dimensionality. This means that the original data would be compressed with the minimal information loss when data will be reconstructed (maximally preserving information contents of the original data).

A linear transformation of an original data pattern can be also interpreted as a projection of original patterns into $m - dimensional$ feature space with orthonormal basis providing decorrelation of feature vector elements.

The PCA is an unsupervised learning method from data. It means that the PCA does not use knowledge about associated with a pattern class, but only discovers correlation amongs patterns and their elements, as well as ordered intrinsic directions where the data patterns change most (with maximum variance).

Despite a fact that PCA is an unsupervised method, it can be also apply for extraction/reduction of features for data set containing patterns labeled by corresponding classes with classification being a processing goal. Here, PCA is applied solely to patterns, to provide transformation and reduction of original patterns to reduced features spaces. After Karhunen-Loéve transformation has been computed, the projected patterns will have the same class assignments as these in the original data set.

2.2 Statistical characteristic of data required by PCA. A data covariance matrix and its eigenvalues and eigenvectors.

Let us consider data objects characterized by n-dimensional patterns $\mathbf{x} \in \mathbf{R}^n$ in $n - dimensional$ pattern space whose elements take real values $x_i \in \mathbf{R}$. We assume that our knowledge about a domain is represented as limited size sample of N random patterns $\mathbf{x}^i$ gathered as an unlabeled training data set T

$$T = \{\mathbf{x}^1, \mathbf{x}^2, \cdots, \mathbf{x}^N\} \tag{3}$$

A whole training set data will be represented as a $N \times n$ data pattern matrix

$$\mathbf{X} = \begin{bmatrix} \mathbf{x}^1 \\ \mathbf{x}^2 \\ \cdots \\ \mathbf{x}^N \end{bmatrix} \tag{4}$$

where each row contains one pattern. If a data set contains patterns labeled by classes, for unsupervised PCA analysis we need to extract from this data set only patterns.

The data can be characterized by the second order statistics, namely by the $n - dimensional$ mean vector $\boldsymbol{\mu}$ and the square $n \times n$ dimensional *covariance* matrix

$$\mathbf{R}_x = \boldsymbol{\Sigma} = E[(\mathbf{x} - \boldsymbol{\mu})(\mathbf{x} - \boldsymbol{\mu})^T] \tag{5}$$

The true values $\boldsymbol{\mu}$ and $\mathbf{R}_x$ for the mean vector and the covariance matrix are in practice rather not available. Our knowledge about pattern generation are contained in a given data set T containing finite number N of patterns $\{\mathbf{x}^1, \mathbf{x}^2, \cdots, \mathbf{x}^N\}$. Hence, one can find estimates for the mean

$$\hat{\boldsymbol{\mu}} = \frac{1}{N} \sum_{i=1}^{N} \mathbf{x}^i \qquad (6)$$

and the covariance matrix (unbiased estimate)

$$\hat{\mathbf{R}}_x = \frac{1}{N-1} \sum_{i=1}^{N} (\mathbf{x}^i - \boldsymbol{\mu})(\mathbf{x}^i - \boldsymbol{\mu})^T \qquad (7)$$

based on a given limited sample.

The intrinsic characteristic of a given data $\mathbf{X}$ can be found as a set of n eigenvalues λ_i and the corresponding eigenvectors $\mathbf{e}^i$ by solving so called the *eigenvalue problem*

$$\mathbf{R}_x \mathbf{e}^i = \lambda_i \mathbf{e}^i, \quad i = 1, 2, \cdots, n \qquad (8)$$

In the PCA the orthonormal eigenvectors are considered. This means that the eigenvectors are orthogonal $(\mathbf{e}^i)^T \mathbf{e}^j = 0$ $(i, j = 1, 2, \cdots, n, i \neq j)$ with the unit length.

It is essential in the PCA that the eigenvalues of the matrix $\mathbf{R}_x$ are arranged in decreasing order

$$\lambda_1 \geq \lambda_2 \geq \cdots \lambda_n \geq 0 \qquad (9)$$

with $\lambda_1 = \lambda_{max}$. The corresponding orthonormal eigenvectors $\mathbf{e}^i$ are composed into a square $n \times n$ matrix

$$\mathbf{E} = [\mathbf{e}^1, \mathbf{e}^2, \cdots, \mathbf{e}^n] \qquad (10)$$

with *ith* column representing one eigenvector $\mathbf{e}^i$ corresponding to the eigenvalue λ_i. The arrangement of eigenvalues and corresponding eigenvectors in descending order is essential for data dimensionality reduction, when only the first m principal components of projected feature vectors (carrying most information) corresponding to the first m dominant eigenvalues will be considered.

2.3 The PCA and resulting optimal Kurhunen-Loéve transformation

The goal of the principal is to find of the optimal linear transformation $\mathbf{y} = \mathbf{W}\mathbf{x}$ of the original *n-dimensional* data patterns $\mathbf{x}$ into *m-dimensional* feature vectors $\mathbf{y}$, possibly with lower dimensionality ($m < n$). In the PCA analysis it is desired that:

- The optimal transformation will be orthogonal (with orthonormal basis).
- The transformed feature vector $\mathbf{y}$ elements will be uncorrelated (statistically independent).

- The orthonormal basis of the linear projections will show in decreasing order the orthogonal intrinsic directions in data along which data changes (variances) are maximal.
- The pattern reconstruction error will be minimal in the mean least squares sense.

For the orthonormal linear transformation $\mathbf{y} = \mathbf{W}\mathbf{x}$, with $m \times n$ dimensional orthonormal tranformation matrix $\mathbf{W}$, an estimate of the reconstructed pattern is $\hat{\mathbf{x}} = \mathbf{W}^{-1}\mathbf{y}$. Since for the orthonormal matrices we have $\mathbf{W}^{-1} = \mathbf{W}^T$, thus

$$\hat{\mathbf{x}} = \mathbf{W}^{-1}\mathbf{y} = \mathbf{W}^T\mathbf{y} = \mathbf{W}^T\mathbf{W}\mathbf{x} \tag{11}$$

The criterion for the optimal, PCA based, linear transformation is selected in order to guarantee obtaining a minimum of reconstruction error metric. The mean least square reconstruction error based criterion will have a following form

$$J(T) = E[||\mathbf{x} - \hat{\mathbf{x}}||^2] \tag{12}$$

with more practical computation version

$$J(T) = \frac{1}{2}\sum_i^N ||\mathbf{x}^i - \hat{\mathbf{x}}^i||^2 = \frac{1}{2}\sum_i^N \sum_j^n (x_j^i - \hat{x}_j^i)^2 \tag{13}$$

PCA seeks the optimal transformation matrix $\mathbf{W}$ guaranteeing minimization of the mean square error criterion $J(T)$ for a given data set T.

The optimal PCA transformation can be found in the following way. Let for a given data set (a training set) $T = \{\mathbf{x}^1, \mathbf{x}^2, \cdots, \mathbf{x}^N\}$, containing N *n-dimensional* zero-mean randomly generated patterns $\mathbf{x} \in \mathbf{R}^n$ with real-valued elements, a matrix $\mathbf{R}_x \in \mathbf{R}^{n \times n}$ be a symmetric, real-valued $n \times n$ covariance matrix. Let the eigenvalues of the covariance matrix $\mathbf{R}_x$ are arranged in the decreasing order $\lambda_1 \geq \lambda_2 \geq \cdots \lambda_n \geq 0$ (with $\lambda_1 = \lambda_{max}$). Let us ssume that the corresponding orthonormal eigenvectors (orthogonal with unit length $||\mathbf{e}|| = 1$) $\mathbf{e}^1, \mathbf{e}^2, \cdots, \mathbf{e}^n$ compose the $n \times n$ orthonormal matrix

$$\mathbf{E} = [\mathbf{e}^1, \mathbf{e}^2, \cdots, \mathbf{e}^n] \tag{14}$$

with columns being orthonormal eigenvectors. Then the optimal linear transformation

$$\mathbf{y} = \hat{\mathbf{W}}\mathbf{x} \tag{15}$$

is provided for the $m \times n$ optimal transformation matrix $\hat{\mathbf{W}}$, denoted also by $\mathbf{W}_{KL}$, (under the constraints $\mathbf{W}\mathbf{W}^T = \mathbf{I}$)

$$\hat{\mathbf{W}} = \begin{bmatrix} \mathbf{e}^1 \\ \mathbf{e}^2 \\ \cdots \\ \mathbf{e}^m \end{bmatrix} \tag{16}$$

composed with m rows being the first m orthonormal eigenvectors of the original data covariance matrix $\mathbf{R}_x$.

The optimal matrix $\hat{\mathbf{W}}$ transforms the original *n-dimensional* patterns $\mathbf{x}$ into *m-dimensional* ($m \leq n$) feature patterns minimizing the mean least square reconstruction error (a criterion $J(T)$). This is equivalent here to maximizing the variance of projection. The proof of the PCA method is given in (Diamentras, and Kung, 1996).

2.4 Properties and interpretation of the optimal Karhunen-Loéve linear transformation

The optimal KLT transformation guarantees the minimum of reconstruction error in mean least squares sense, with the minimal value

$$\min J(T) = \sum_{i=m+1}^{n} \lambda_i \qquad (17)$$

The minimum value of reconstruction error is equal to the sum of the trailing $n - m$ eigenvalues $\lambda_{m+1}, \lambda_{m+2}, \cdots, \lambda_n$ of the covariance matrix $\mathbf{R}_x$, where m is possibly reduced length of the projected feature vector $\mathbf{y}$ ($m \leq n$).

Simultaneously the transformation guarantees the maximum of the projected feature vector variance, with the maximum value

$$\max J_{variance}(T) = \sum_{i=1}^{m} \lambda_i \qquad (18)$$

equal to the sum of the first m eigenvalues of $\mathbf{R}_x$.

The orthonormal eigenvectors $\mathbf{e}^1, \mathbf{e}^2, \cdots, \mathbf{e}^n$ (rows of the optimal transformation matrix $\hat{\mathbf{W}}$), corresponding to the arranged in the descending order eigenvalues $\lambda_1, \lambda_2, \cdots, \lambda_n$, of the data covariance matrix $\mathbf{R}_x$, are called the *principal eigenvectors*. They show orthogonal directions (in descending order corresponding to the principal eigenvectors and eigenvalues) in the pattern space where data changes maximally (with maximal variance) and signals have most energy.

2.5 Optimal KLT transformation (projection) of the original patterns

Having the optimal KLT transformation matrix $\hat{\mathbf{W}}$ defined, the optimaly transform a given *n-dimensional* original pattern $\mathbf{x}$ into *m-dimensional* optimal feature pattern $\mathbf{y}$ is given by by $\mathbf{y} = \hat{\mathbf{W}}\mathbf{x}$. The inverse transformation can be obtained from

$$\hat{\mathbf{x}} = \hat{\mathbf{W}}^{-1}\mathbf{y} = \hat{\mathbf{W}}^T \mathbf{y} = \sum_{i=1}^{m} y_i \mathbf{e}^i \qquad (19)$$

The optimal transform of a whole original data set is given by the formula

$$\mathbf{Y} = (\hat{\mathbf{W}}\mathbf{X}^T)^T = \mathbf{X}\hat{\mathbf{W}}^T \qquad (20)$$

2.6 Dimensionality reduction

The PCA can be effectively used for the feature extraction and the dimensionality reduction of transformed optimaly feature vectors by Karhunen-Loéve transformation. Instead of a whole $n-dimensional$ original data patterns **x** one can form the $m-dimensional$ ($m \leq n$) feature vector $\mathbf{y} = [y_1, y_2, \cdots, y_m]^T$ containing only the first m most dominant principal components of **x**. We can also say that the m principal components y_i are the *most expressive* features of a data set. PCA gives way to reduce data representation (compress) by choosing the feature vector with lower dimensionality but also provides that reduced size feature vectors will represent data in new feature space, where feature vector elements are uncorrelated and placed along the orthogonal directions of principal components with maximal variances. This might be also possibly desired for better classifiers design for some types of data. However, the PCA does not consider in a transformation criterion improvement of classification in principal component space. This mean that obtained through PCA most expressive features are well suited for data representation (model) and compression. However, they are might not be good for classification (discriminating classes). The open question remains, which principal component select as the best for a given processing goal. One of possible methods (criteria) for (data representation oriented) selection of a dimension of a reduced feature vector **y** is to choose a minimal number of the first m most dominant principal components $y_1, y_2, \cdots, y_m$ of **x** for which the mean square reconstruction error is less than heuristically set the error threshold ϵ. Other, more practical method may assume selection the minimal number of the first m most dominant principal components for which a percentage V of a sum of unused eigenvalues of a sum of all eigenvalues

$$V = \frac{\sum_{i=m+1}^{n} \lambda_i}{\sum_{i=1}^{n} \lambda_i} 100\% \tag{21}$$

is less then a defined threshold ζ: $P < \zeta$.

One can try to use principal components for classification. However, the selection of the best principal components for a classification purpose is the other task.

3 Introduction to rough sets

The rough sets theory has been developed for knowledge discovery in databases and experimental data sets (Pawlak, 1982; Pawlak 1991; Skowron 1990; Skowron, 1995). This theory provides a powerful foundation to reveal and discover important structures in data and to classify objects.

An attribute-oriented rough sets technique reduces the computational complexity of learning processes and eliminates the unimportant or irrelevant attributes so that the knowledge discovery in database or in experimental data sets can be efficiently learned. Using rough sets has been shown to be very effective for revealing relationships within imprecise data, discovering dependencies among objects and attributes, evaluating the classificatory importance of attributes, removing data redundancies and thus reducing an information systems, and generating the decision rules.

The idea of rough sets is based on equivalence relations which partition a data set into equivalence classes, and consists of the approximation of a set by a pair of sets, called lower and upper approximation. The lower approximation of a set of objects (a concept) contains all objects that, based on the knowledge of a given sets of attributes, can be classified as certainly belonging to the concept. The upper approximation of a set contains all objects that based on the knowledge of a given set of attributes cannot be classified categorically as not belonging to the concept. A *rough set* is defined as an approximation of a set, defined as pair of sets: the upper and lower approximation of a set.

Rough sets provide a quantitative, numerical measure of classification approximation. The rough sets theory based on the concept of an *upper* and a *lower approximation* of a set, the *approximation space* and deterministic models of sets. By processing information based on the rough sets theory, the classification of objects in the information system can be discovered.

Rough sets allow to determine for a given information system the most important attributes from a classificatory point of view, and thus attribute and data set reduction. A reduct is the essential part of an information system (related to a subset of attributes) which can discern all objects discernible by the original information system. A core is a common part of all reducts. Core and reduct are fundamental rough sets concepts allowing for knowledge reduction.

3.1 Information System

An information system is consider to be the data arranged in the tabular way. An **information system** is composed of a 4-tuple as follows:

$$S = < U, Q, V, f > \tag{22}$$

where
U – the **universe**, a finite set of N objects $\{x_1, x_2, ..., x_N\}$ (a nonempty set),
Q – a finite set of n **attributes** $\{q_1, q_2, \cdots, q_n\}$ (a nonempty set),
$V = \cup_{q \in Q} V_q$ where V_q is a **domain (value)** of the attribute q,
$f : U \times Q \to V$ – the total **decision function** called the information function such that $f(x, q) \in V_q$ for every $q \in Q$, $x \in U$.

Any pair (q, v) for $q \in Q, v \in Vq$ is called the **descriptor** in an information system S. An information system consists of a finite data table, in which the *columns* are labeled by **attributes**, the *rows* by **objects** and the *entry* in column q and row x has the **value** f(x,q). Each row in the table describes the information about some object in S. Any nonempty set of objects X is called a **concept** in S.

3.2 Indiscernibility Relation

Let $S = <U, Q, V, f>$ be an information system, $A \subseteq Q$ be a subset of attributes and $x, y \in U$ are objects.
Then, objects x and y are **indiscernible** by the set of attributes A in S (denoted by $x\tilde{A}y$) iff $f(x, a) = f(y, a)$ for every $a \in A$.
For any subset of attributes $A \subseteq Q$ the $IND(A)$ (denoted by $\tilde{A}$) is an **equivalence relation** on universe U and is called an **indiscerniblility relation**. The indiscernibility relation $IND(A)$ is defined as follows:

$$IND(A) = \{(x, y) \in U \times U : \text{ for all } a \in A, f(x, a) = f(y, a)\} \quad (23)$$

If the pair of objects (x, y) belongs to the relation $IND(A)$ $((x, y) \in IND(A))$ then objects x and y are called **indiscernible** with respect to A.

The binary indiscernibility relation $IND(A)$ splits the universe U into family of equivalence classes $\{X_1, X_2, ..., X_r\}$. The family of all equivalence classes $\{X_1, X_2, ..., X_r\}$ generates a *partition* of U and it is denoted by A^*. The family of equivalence classes A^* is also referred as **classification** and also denoted by $U/IND(A)$.
The equivalence classes X_i, $i = 1, 2, ..., r$ of the relation $IND(A)$ are called A-**elementary sets** in an information system S. The $[x]_A$, defined as

$$[x]_A = \{y \in U : xIND(A)y \text{ or } x\tilde{A}y\} \quad (24)$$

denotes an A-elementary set (an equivalence class) including an object x. The elementary sets X_i are blocks of the partition of U and represent the smallest discernible group of objects and they are also called **A-basic knowledge**.

For a given information system S a given subset of attributes $A \subseteq Q$ generates the indiscernibility relation $IND(A)$ (equivalence relation). An ordered pair $AS = (U, IND(A))$ is called an **approximation space**. Any finite union of elementary sets in AS is called a *definable set* or *composed set* in AS. The Q-elementary sets are called **atoms** of an information system S. $Des_A(X)$ denotes the description of A-elementary set $X \in A^*$ (an equivalence class) and it is defined as follows:

$$Des_A(X) = \{(a, b) : f(x, a) = b, \forall x \in X, a \in A\} \quad (25)$$

3.3 Decision Tables

An information systems can be designed as a **decision table** if the attribute set Q is divided into *two disjoint* sets which are **condition** attribute set C and decision attribute set D, i.e. $C \cup D = Q$ and $C \cap D = \emptyset$. We can consider a specific type of an information system S as a decision table denoted as follows:

$$DT = <U, \ C \cup D, \ V, \ f> \qquad (26)$$

where
C - is a set of **condition** attributes (a nonempty set, inputs, input pattern's elements),
D - is a set of **decision** attributes (a nonempty set, decisions, actions, classes),
$V = \bigcup_{q \in C \cup D} V_q$, where V_q is the set of **domain** (value) of attribute $q \in Q$,
$f : U \times (C \cup D) \to V$ - is a total **decision function** (information function, decision rule in DT) such that $f(x, q) \in V_q$ for every $q \in Q$ and $x \in V$.

A decision table is also denoted by $(U, C \cup D)$ or by DT_C where C denotes the condition attribute set.

3.4 Approximation of Sets. Approximation space.

Some subsets (classes, categories) of objects in an information system can be only roughly (approximately) defined. The idea of **rough sets** consists of the approximation of a set by a pair of sets, called *lower* and *upper* approximation of this set.

A given subset of attributes $A \subseteq Q$ determines the approximation space $AS = (U, IND(A))$ in S. For a given $A \subseteq Q$ and $X \subseteq U$ (a concept X), the A-**lower approximation** $\underline{A}X$ of set X in AS and the A-**upper approximation** $\bar{A}X$ of set X in AS are defined as follows:

$$\underline{A}X = \{x \in U : [x]_A \subseteq X\} = \bigcup\{Y \in A^* : Y \subseteq X\} \qquad (27)$$

$$\bar{A}X = \{x \in U : [x]_A \cap X \neq \emptyset\} = \bigcup\{Y \in A^* : Y \cap X \neq \emptyset\} \qquad (28)$$

We say also that $\underline{A}X$ and $\bar{A}X$ are A-**lower** and A-**upper** approximation of a concept X in AS. The lower approximation $\underline{A}X$ of set X is the union of all those elementary sets (a composed set of elementary sets) each of which is contained by X. For any $x \in \underline{A}X$, it is certain, on the basis on knowledge from A, that x belongs to X. The upper approximation $\bar{A}X$ of set X is the union of those elementary sets each of which has non-empty intersection with X. For any $x \in \bar{A}X$, we can only say, on the basis on knowledge from A, that x can possibly belong to X.

The **accuracy** of an approximation of set X by the set of attributes A (shortly *accuracy* of X) is defined as follows:

$$\alpha_A(X) = \frac{\text{card } \underline{A}X}{\text{card } \bar{A}X} \qquad (29)$$

3.5 Dependency and Reduction of Attributes.

Some attributes in an information system may be redundant and can be eliminated without loosing any essential classificatory information in a given information systems (with a given set of objects). The process of finding a smaller set of attributes than original one, with the same classificatory power as the original set is called *attribute reduction*. As a result the original larger information system may be reduced to a smaller system.

Rough sets allow us to determine for a given information system the most important attributes from a classificatory point of view. A reduct is the essential part of an information system (related to a subset of attributes) which can discern all objects discernible by the original information system. A core is a common part of all reducts.

Rough sets determine a degree of attributes dependency and their significance. In the indiscernibility relation, dependency of attributes is one of the important features of information systems.

Given an information system $S = <U, Q, V, f>$, with condition and decision attributes $Q = C \cup D$, for a given set of condition attributes $A \subset C$ we can define the A positive region $POS_A(D)$ in the relation $IND(D)$, as

$$POS_A(D) = \bigcup \{\underline{A}X | X \in IND(D)\} \tag{30}$$

The positive region $POS_A(D)$ contains all objects in U which can be classified without error (ideally) into distinct classes defined by $IND(D)$ based only on information in the relation $IND(A)$.

The definition of the positive region can be formed for any two subsets of attributes $A, B \in Q$ in the information system S. We know that the subset of attributes $B \in Q$ defines the indiscernibility relation $IND(B)$ and thus the classification B^* ($U/IND(B)$) with respect to the subset A. The **A-positive region** of B is defined as

$$POS_A(B) = \bigcup_{X \in B^*} \underline{A}X \tag{31}$$

The A-positive region of B contains all objects that by using attributes A, can be certainly classified to one of distinct classes of the classification B^*.

The cardinality (size) of the A-positive region of B is used to define a measure (a degree) $\gamma_A(B)$ of dependency of the set of attributes B on A:

$$\gamma_A(B) = \frac{card(POS_A(B))}{card(U)} \tag{32}$$

We say that the set of attributes B depend on the set of attributes A in a degree $\gamma_A(B)$.

Let us consider the information system $S = <U, Q, V, f>$ and two sets sets of attributes $A, B \subseteq Q$. Set of attributes B **depends** (is **dependent**) on a set A in S, and denoted by $A \rightarrow B$ iff an equivalence relation satisfies

$IND(A) \subseteq IND(B)$. A sets A and B are **independent** in S iff neither $A \rightarrow B$ nor $B \rightarrow A$ hold. A set B is dependent in **degree** k on the set A in S, and this is denoted as follows:

$$A \xrightarrow{k} B, \ 0 \le k \le 1, \text{ if } k = \gamma_A(B) \tag{33}$$

The **measure of significance (coefficient of significance)** of the attribute $a \in A$ from the set A with respect to the classification B^* $(U/IND(B))$ generated by a set B is defined as follows:

$$\mu_{A,B}(a) = \frac{card\,(POS_A(B)) - card\,(POS_{A-\{a\}}(B))}{card\,U} \tag{34}$$

The significance of the attribute a in the set $A \subseteq Q$ computed with the respect to the original classification Q^* generated by whole set of attributes Q from the information system S is denoted as

$$\mu_A(a) = \mu_{A,Q}(a) \tag{35}$$

We can check the properties of an attribute set A in an information system $S = <U, Q, V, f>$ as follows:

1. A set $A \subset Q$ is **dependent** in S iff $\exists B \subset A$ such that $IND(B) = IND(A)$ (i.e. $\alpha_B(X) = \alpha_A(X)$).
2. A set $A \subset Q$ is **independent** in S iff $\forall \ B \subset A$, $IND(B) \supset IND(A)$ (i.e. $\alpha_B(X) < \alpha_A(X)$).
3. A set $A \subset Q$ is **superfluous** in Q iff $IND(Q-A) = IND(Q)$ (i.e. $\alpha_{Q-A}(X) = \alpha_Q(X)$).
4. A set $A \subseteq Q$ is a **reduct** of Q in S iff $Q - A$ is *superfluous* in Q and A is *dependent* in S.

A given information system may have many different reducts. A given information system may have many different reducts. If for a given information system S a subset $A \subset Q$ is a reduct, then the corresponding information system $S' = <U, A, V, f'>$, with the attribute set equal to a reduct A, is called a **reduced system** (where f' is the restriction of a function f to a set $U \times A$).

Reduct and Core

From the dependent properties of attributes, we can find a *reduced* set of the attributes, by removing *superfluous* attributes, without loses in classification power of the reduced information system. For an information system S and a subset of attributes $A \subseteq Q$ an attribute $a \in A$ is called **dispensable** in the set A if $IND(A) = IND(A - \{a\})$ (it means that indiscernibility relations generated by sets A and $A - \{a\}$ are identical). Otherwise a parameter a is **indispensable** in A. The dispensable attribute does not improve the classification of the original information system S. Indispensable attribute carry the essential information about objects of an information system, and cannot by removed without changing the classificatory power of the original system.

The set of all indispensable attributes in the set $A \subseteq Q$ is called a **core** of A in

S. and it is denoted by $CORE(A)$. The core contains all attributes that cannot be removed from the set A without decreasing the original classification A^* accuracy.

Now, let us consider two subsets of attributes $A, B \subseteq Q$ in S. An attribute a is called B-**dispensable** (indispensable with respect to B) in the set A if $POS_A(B) = POS_{A-\{a\}}(B)$. Otherwise the attribute a is B-**indispensable**. If every attribute of A is B-indispensable, than A is indispensable with respect to B. The set of all B-indispensable attributes from the set A is called B-**relative core** (or B-**core**) of A and denoted by $CORE_B(A)$:

$$CORE_B(A) = \{a \in A : POS_A(B) \neq POS_{A-\{a\}}(B)\} \quad (36)$$

The set $A \subseteq Q$ is called **orthogonal** if all its attributes are indispensable. A proper subset $E \subset A$ is defined as a **reduct** set of A in S if E is an orthogonal and preserves the classification generated by A. Thus reduct set of A denoted by $RED(A)$ is defined as

$$E = RED(A) \iff (E \subset A, IND(E) = IND(A), E \text{ orthogonal}) \quad (37)$$

E is a reduct of A (i.e. $E = RED(A)$) if E is a minimal set of attributes that discerns all objects in S discernible by the whole set A, and cannot be further reduced. In general, more then one reduct of A can be identified.

The intersection of all reducts family of A is a core of A

$$CORE(A) = \bigcap RED(A) \quad (38)$$

Relative reduct

Similarly we can define the **relative reduct** related to two sets of attributes $A, B \subseteq Q$ in S. The set A is called **B-orthogonal** if all attributes of A are B-indispensable. Any B-orthogonal proper subset of A is called a **B-reduct** of A and it is denoted by $RED_B(A)$

$$E = RED_B(A) \iff (E \subset A, POS_E(B) = POS_A(B), E is B-orthogonal) \quad (39)$$

A B-reduct $RED_B(A)$ of A is a minimal set of attributes in A that discern all objects in S discernible by the whole set A and cannot be further reduced. All B-reducts (family) are denoted by $RED_B^F(A)$. Depending on data character, more then one relative B-reduct of A can be identified. The intersection of all B-reducts family of A is a relative B-core of A

$$CORE_B(A) = \bigcap RED_B(A) \quad (40)$$

3.6 Dynamic reducts

Skowron (Skowron, at al, 1994) has proposed a novel technique of choosing dynamic reducts from a given decision table, which guarantee better generalization of decision rules. Selection of dynamic reducts belongs to open loop (filter) techniques for feature (attribute) selection, where no feedback of predictor performance is considered.

If $DT_C =< U, C \cup D, V, f >= (U, C \cup \{d\})$ denotes the decision table and $RED(DT_C, d)$ the set of relative reducts i.e. the family of all minimal conditional attribute sets sufficient to discern between objects from different decision classes.

If $DT_C = (U, C \cup \{d\})$ is a decision table then any system $DT'_C = (U', C \cup \{d\})$ such that $U' \subseteq U$ is called a subtable of DT_C. Let $DT_C = (U, C \cup \{d\})$ be a decision table and F be a family of subtables DT'_C of DT_C. By $DR(DT_C, F)$ we denote the set of relative reducts

$$RED(DT_C, d) \cap \bigcap_{DT'_C \in F} RED(DT'_C, d)$$

Any element of $DR(DT_C, F)$ is called an F-*dynamic reduct of* DT_C.
From the definition of dynamic reducts it follows that a relative reduct of DT_C is dynamic if it is also a reduct of all subtables from a given family F. This notion can be sometimes too much restrictive so one can apply also a more general notion of dynamic reducts (Bazan, et al, 1994; Skowron, et al, 1994; Bazan, 1998). They are called (F, ε)-*dynamic reducts*, where $\varepsilon \geq 0$. The set $DR_\varepsilon(DT_C, F)$ of all (F, ε)-*dynamic reducts* is defined by

$$DR_\varepsilon(DT_C, F) =$$

$$\left\{ R \in RED(DT_C, d) : \frac{card\{DT'_C \in F : R \in RED(DT'_C, d)\}}{card\{F\}} \geq 1 - \varepsilon \right\}$$

Let us consider a reduct $R \in RED(DT'_C, d)$. As a measure of a dynamic reduct R robustness (*relatively to* F) the *stability coefficient* $\kappa(C, F)$ is defined as a proportion of number N_R of subtables from the family F, for which reduct R appears and a cardinality of a family F (a number of subtables)

$$\kappa(C, F) = \frac{N_R}{card\{F\}} \tag{41}$$

The stability coefficient may be calculated be sampling the decision table, for example using leave-one-out method (Bazan, Skowron, Synak, 1994).

4 Relevance of features. Feature selection.

4.1 Optimal feature selection - a problem statement

Let us ssume that a data set T_{all} is given (containing N_{all} cases), constituted with n-feature patterns **x** (labeled or unlabeled by classes), sometime accompanied by a prior knowledge about a domain. Let all n features of pattern (a pattern vector elements x_i ($i = 1, 2, \cdots, n$) form a whole original feature set $X_{all} = \{x_1, x_2, \cdots, x_n\}$. An optimal feature selection is a process of finding, for a given type of a predictor, a subset $X_{opt} = \{x_{1,opt}, x_{2,opt}, \cdots, x_{m,opt}\}$ containing $m \leq n$ features from the set of all original features $X_{opt} \subseteq X_{all}$, which guarantee accomplishment of a processing goal while minimizing a defined feature selection criterion (a feature goodness criterion) $J_{feature}(X_{feature\_subset})$. An optimal feature set will depend on type of designed predictor. A solution of optimal feature selection does not need to be unique. Different subsets of original features may guarantee the same goal accomplishment with the same performance measure.

4.2 Relevance of features

Feature relevance idea can help in selection of the best feature subset for a given prediction task. Relevance of a feature can be understood as its ability to contribute to improving predictor performance. There have been few attempts (both deterministic and probabilistic) in machine learning to define feature relevance (Almuallim, Dietterich, 1991; Gennari, et al, 1989; John, Kohavi, Pfleger, 1994; Pawlak, 1991). Let us assume a labeled data set T with N cases (**x**, **target**), containig n-feature patterns **x** and associated **targets**. For classification a **target** is a categorical class target c_{target} (a concept c) with values from set of l discrete classes $\{c_1, c_2, \cdots, c_l\}$. In (Almuallim, Dietterich, 1991) the following definition of deterministic relevance was proposed for Boolean features in noise-free data sets for classification task.

Definition 1 A feature x_i is *relevant to a class c* (a concept c) if x_i appears in every Boolean formula that represents c, and *irrelevant* otherwise.

In (Gennari, et al, 1989), more general definition of relevance for multi-valued features using probabilistic assessments.

Definition 2 A feature x_i is *relevant* if there exists some value of that feature a_{x_i} and a predictor output **y** value $\mathbf{a_y}$ (generally a vector) for which $P(x_i = a_{x_i}) > 0$ such

$$P(\mathbf{y} = \mathbf{a_y} | x_i = a_{x_i}) \neq P(\mathbf{y} = \mathbf{a_y}) \tag{42}$$

According to this definition a feature x_i is relevant if knowledge of its value can change the estimates of **y**, or in another words, if an output vector **y** is conditionally dependent of x_i. In (John, Kohavi, Pfleger, 1994), modification of above definition was proposed. Let us denote a vector of features $\mathbf{v}_i = (x_1, x_2, \cdots, x_{i-1}, x_{i+1}, \cdots, x_n)^T$ (with its values denoted by $\mathbf{a_{v_i}}$) obtained from an original feature vector **x** by removing x_i feature.

Definition 3 A feature x_i is *relevant* if there exists some value of that feature a_{x_i} and a predictor output **y** value $\mathbf{a_y}$ (generally a vector) for which $P(x_i = a_{x_i}) > 0$ such

$$P(\mathbf{y} = \mathbf{a_y}, \mathbf{v}_i = \mathbf{a_{v_i}} | x_i = a_{x_i}) \neq P(\mathbf{y} = \mathbf{a_y}, \mathbf{v}_i = \mathbf{a_{v_i}}) \tag{43}$$

According to this definition a feature x_i is relevant if probability of a **target** (given all features) can change if we remove knowledge about a value of that feature. In (John, Kohavi, Pfleger, 1994) more precise definitions of so called strong and weak relevance were introduced.

Definition 4 A feature x_i is *strongly relevant* if there exists some value of that feature a_{x_i}, a predictor output **y** value $\mathbf{a_y}$ and a value $\mathbf{a_{v_i}}$ of a vector $\mathbf{v}_i$ for which $P(x_1 = a_{x_i}, \mathbf{v}_i = \mathbf{a_{v_i}}) > 0$ such that

$$P(\mathbf{y} = \mathbf{a_y} | \mathbf{v}_i = \mathbf{a_{v_i}}, x_i = a_{x_i}) \neq P(\mathbf{y} = \mathbf{a_y} | \mathbf{v}_i = \mathbf{a_{v_i}}) \tag{44}$$

Strong relevance indicated that a feature is indespensable, which means that its removal from a feature vector will decrease prediction accuracy.

Definition 5 A feature x_i is *weakly relevant* if it is not strongly relevant, and there exists some subset of features (forming a vector $\mathbf{z}_i$) from a set of features forming a patterns $\mathbf{v}_i$, for which there exists: some value of that feature a_{x_i}, a predictor output value $\mathbf{a_y}$, and a value $\mathbf{a_{z_i}}$ of a vector $\mathbf{z}_i$, for which $P(x_i = a_{x_i}, \mathbf{z}_i = \mathbf{a_{z_i}}) > 0$ such that

$$P(\mathbf{y} = \mathbf{a_y} | \mathbf{z}_i = \mathbf{a_{z_i}}, x_i = a_{x_i}) \neq P(\mathbf{y} = \mathbf{a_y} | \mathbf{z}_i = \mathbf{a_{z_i}}) \tag{45}$$

Weak relevance indicates that a feature might be dispensable, however, sometime (in companion of some other features) may improve prediction accuracy.

A feature is *relevant* if it is either *strongly relevant* or *weakly relevant*, otherwise it is *irrelevant*. By definition irrelevant feature will never contribute to prediction accuracy, hence can be removed.

The theory of rough sets (Pawlak, 1991; Skowron, 1990) defines deterministic strong and week relevance for discrete features and discrete targets. For a given data set a set of all strongly relevant features forms a *core*. A minimal set of features satisfactory to describe concepts in a given data set, including a core and possibly some weakly relevant features, form a *reduct*. A core is an intersection of reducts.

It has been shown that in some predictors design (John, Kohavi, Pfleger, 1994) feature relevance (even strong relevance) does not imply that that that feature must be in an optimal feature subset. Relevance, although helping in feature assessing, does not necessary should contribute to optimal predictor design with generalization ability.

4.3 Feature selection methods and algorithms

The existing feature selection methods, depending on feature selection criterion used, contain two main streams (Duda, Hart, 1973; Fukunaga, 1990; Bishop, 1995; John, Kohavi, Pfleger, 1994; Pregenzer, 1997)

- open loop (filter, preset bias, front end) methods
- closed loop (wrapper, classifier feedback) methods

The *open loop* methods, called also the *filter*, *preset bias*, or the *front end* methods do not consider effect of selected features on a whole processing algorithm performance (for example a classifier) performance, since feature selection criterion does not involve a whole predictor evaluation for reduced data set containing patterns with selected feature subsets only. Instead, they select for example these features for which resulting reduced data set has maximal between class separability, defined usually based on between class and between class covariances (or scatter matrices) and their combinations.

The *closed loop* methods (John, et al, 1994; Wettschereck, et al, 1996), called also the *wrapper, performance bias*, or the *classifier feedback* methods are based on feature selection using a predictor performance as a criterion of feature subset selection. A selected feature subset goodness is evaluated using as a criterion $J_{feature} = J_{predictor}$ a performance evaluation $J_{predictor}$ of a whole prediction algorithm for reduced data set containing patterns with the selected features as pattern's elements. The selection algorithm is a "wrapper" around the prediction algorithm. The closed loop methods might provided better selection of feature subset, since they based on the ultimated goal and criterion of optimal feature selection: provide best prediction. A prediction algorithm used in closed loop feature selection could be the final predictor PR (providing the best selection, $J_{feature} = J$), or for computational feasibility reasons a simpler predictor $PR_{feature}$ used only for wrapped feature selection.

Commonly, a procedure for optimal feature selection contains:

- A feature selection criterion $J_{feature}$, allowing to judge whether one subset of features is better than another (evaluation method).
- A systematic search procedure through candidate subsets of features, including initial state of search and stopping criteria of search.

We will discuss open loop method of feature selection.

Criteria based on minimum concept description

A open loop type criteria of feature selection based on minimum construction paradigm were studied earlier (Almuallim, Dietterich, 1991) in machine learning and in statistics for discrete features noise free data sets. One of straightforward techniques of best feature selection could be choosing a minimal feature subset that fully describes all concepts (for example classes in classification) in a given data set (Almuallim, Dietterich, 1991; Pawlak,1991; Doak, 1992; Kononenko, 1994) Here a criterion of feature selection could be defined as Boolean function

$J_{feature}(X_{feature})$ with value 1 if a feature subset $X_{feature}$ is satisfactory to describe all concepts in a data set, otherwise having a value 0. The final selection would based on choosing a minimal subset for which a criterion gives value 1.

This idea of open loop feature selection, with minimum concept criterion, can be extended by using concept of reduct defined by theory of rough sets (Pawlak, 1991; Skowron, 1990). A reduct is a minimal set of attributes that describes all concepts in a data set. A data set may have many reducts. If we will use definition of the above open loop feature selection criterion, we will see that for each reduct (defining a subset of attributes $X_{feature,reduct}$) we have maximum value of the criterion $J_{feature}(X_{feature,reduct})$. Following a paradigm of the minimum concept description, we can select a minimum length reduct as the best feature subset. However, the minimal reduct is good for ideal situations where a given data set fully represents a domain of interest. For real life situations, and limited size data sets, other reduct (generally other feature subsbset) might be better for generalizing prediction. A selection of robust (generalizing) reduct, as a best open loop feature subset, can be supported by introduced by Skowron an ideal of dynamic reduct (Skowron, et al. 1994) described in the next section.

5 Application of rough sets' reducts for selection of discriminatory features from principal components of Karhunen-Loéve transformation

The PCA, with resulting linear Karhunen-Loéve projection, provides feature extraction and reduction optimal from the point of view of minimizing the reconstruction error. However, PCA does not guarantee that selected first principal components, as a feature vector, will be adequate for classification. Nevertheless, the projection of high dimensional patterns into lower dimensional orthogonal principal components feature vectors might help for some data types to provide better classification.

In many applications of PCA an arbitrary number of the first dominant principal components is selected as a feature vector. However, these methods do not cope with the selection of the most discriminative features well suitable for classification task. Even assuming that the Karhunen-Loéve projection can help in classification, and can be used as a first step in the feature extraction/selection procedure, still an open question remains: "Which principal components to choose for classification"?

One of possibilities for selecting features from principal components is to apply rough sets theory (Pawlak, 1991; Skowron, 1990). Specifically, defined in rough sets computation of a reduct can be used for selection some of principal components being a reduct. Thus these principal component will describe all concepts in a data set. For a suboptimal solution one can choose the minimal length reduct or dynamic reduct as selected set of principal components forming a selected, final feature vector.

The following steps can be proposed for PCA and rough sets based procedure for feature selection. Rough sets assume that a processed data set contains patterns labeled by associated classes, with the discrete values of its elements (attributes, features). We know that PCA is predisposed to transform optimally patterns with real-valued features (elements). Thus after realizing the Karhunen-Loéve transformation, resulting projected patterns features must be discretized by some adequate procedure. The resulting discrete attribute valued data set (an information system) can be processed using rough sets methods.

Let us assume that we are given a limited size data set T, containing N cases labeled by associated classes

$$T = \{(\mathbf{x}^1, c_{target}^1)(\mathbf{x}^2, c_{target}^2), \cdots, (\mathbf{x}^N, c_{target}^N)\} \quad (46)$$

Each case $(\mathbf{x}^i, c_{target}^i)$ $(i = 1, 2, \cdots, N)$ is constituted with a $n - dimensional$ real-valued pattern $\mathbf{x} \in \mathbf{R}^n$ with corresponding categorical target class c_{target}^i. We assume that a data set T contains N_i ($\sum_i^l N_i = N$) cases from each categorical class c_i, with the total number of classes denoted by l.

Since PCA is an unsupervised method, first, from the original, class labeled data set T, a pattern part is isolated as $N \times n$ data pattern matrix

$$\mathbf{X} = \begin{bmatrix} \mathbf{x}^1 \\ \mathbf{x}^2 \\ \cdots \\ \mathbf{x}^N \end{bmatrix} \quad (47)$$

which each row contains one pattern. The PCA procedure is applied for extracted pattern matrix $\mathbf{X}$, with resulting full size $n \times n$ optimal Karhuen-Loéve matrix $\mathbf{W}_{KL}$ (where n is a length of an original pattern $\mathbf{x}$). Now, according to the designer decision, the number $m \leq n$ of first dominant principal components has to be selected. Then the reduced $m \times n$ Karhunen-Loéve matrix $\mathbf{W}_{KL}$, containing only first m rows of a full size matrix $\mathbf{W}$, is constructed. Applying the matrix $\mathbf{W}_{KL}$ the original $n - dimensional$ pattern $\mathbf{x}$ can be projected using transformation $\mathbf{y} = \mathbf{W}_{KL}\mathbf{x}$, into the reduced $m - dimensional$ patterns $\mathbf{y}$ in the principal components space. The entire projeced $N \times m$ matrix $\mathbf{Y}$ of patterns can be obtained by the formula $\mathbf{Y} = \mathbf{X}\mathbf{W}_{KL}^T$.

At this stage, the reduced, projected data set, represented by Y (with real-valued attributes), has to be discretized. As a result, the discrete attribute data set represented by the $N \times m$ matrix Y_d is computed. Then, the patterns from $\mathbf{Y}_d$ are labeled by the corresponding target classes from the original data set T. It forms a decision table DT_m with $m - dimensional$ principal component related patterns. From the decision table DT_m one can compute the selected reduct $X_{feature,reduct}$ of size r (for example minimal length or dynamic reduct) as a final selected attribute set. Here a reduct computation is pure feature selection procedure. Selected attributes (being a reduct) are some of elements of projected principal components vector **y**.

Once selected attribute set has been found (as a selected reduct), the final discrete attribute decision table $DT_{f,d}$ is composed. It consist of these columns from the discrete matrix $\mathbf{Y}_d$ which are included in the selected feature set $X_{feature,reduct}$. Each pattern in $DT_{f,d}$ is labeled by the corresponding target class. Similarly one can obtained a real-valued resulting reduced decision table $DT_{f,r}$ extracting (and adequately labeling by classes) these columns from the real-valed projected matrix **Y** which are included in the selected feature set $X_{feature,reduct}$. Both resulting reduced decision tables can be used for a classifier design.

Algorithm: Feature extraction/selection using PCA and rough sets

Given: A $N-case$ data set T containing $n-dimensional$ patterns, with real-valued attributes, labeled by l associated classes
$\{(\mathbf{x}^1, c_{target}^1), (\mathbf{x}^2, c_{target}^2), \cdots, (\mathbf{x}^N, c_{target}^N)\}$.

1. Isolate from the original class labeled data set T, a pattern part as $N \times n$ data pattern matrix **X**.
2. Compute for the matrix **X** the covariance matrix $\mathbf{R}_x$.
3. Compute for the matrix $\mathbf{R}_x$ the eigenvalues and corresponding eigenvectors, and arrange them in descending order.
4. Select the reduced dimension $m \leq n$ of a feature vector in principal components space using defined selection method, which may base on judgement of the ordered values of computed eigenvalues.
5. Compute the optimal $m \times n$ Karhunen-Loéve transform matrix $\mathbf{W}_{KL}$ based on eigenvectors of $\mathbf{R}_x$.
6. Transform original patterns from **X** into $m-dimensional$ feature vectors in the principal component space by formula $\mathbf{y} = \mathbf{W}_{KL}\mathbf{x}$ for a single pattern, or formula $\mathbf{Y} = \mathbf{X}\mathbf{W}_{KL}$ for a whole set of patterns (where **Y** is $N \times m$ matrix).
7. Discretize the patterns in **Y** with resulting matrix $\mathbf{Y}_d$.
8. Compose the decision table DT_m constituted with the patterns from the matrix $\mathbf{Y}_d$ with the corresponding classes from the original data set T.
9. Compute a selected reduct from the decision table DT_m treated as a selected set of features $X_{feature,reduct}$ describing all concepts in DT_m.
10. Compose the final (reduced) discrete attribute decision table $DT_{f,d}$ containing these columns from the projected discrete matrix $\mathbf{Y}_d$ which correspond to the selected feature set $X_{feature,reduct}$. Label patterns by corresponding classes from the original data set T.
11. Compose the final (reduced) real-valued attribute decision table $DT_{f,r}$ containing these columns from the projected discrete matrix $\mathbf{Y}_d$ which correspond to the selected feature set $X_{feature,reduct}$. Label patterns by corresponding classes from the original data set T.

The results of discussed method of feature extraction/selection depend on a data set type and three designer decisions:

1. Selection of dimension $m \leq n$ of projected pattern in the principal component space.
2. Discretization method (and resulting quantization) of projected data.
3. Selection of a reduct.

First, for the selected dimension m, the applied quantization method may lead to the inconsistent decision table DT_m. Then, a designer should return to the discretization step and select other discretization. Even, if for all possible discretization attempts a reduct cannot be found, then a return is realized to the stage of selecting a dimension m of reduced feature vector $\mathbf{y}$ in principal component space. It means that possibly the projected vector does not contain satisfactory set of features. In this situation a design procedure should provide the next iteration with selected larger value of m. If for $m = n$ a reduct cannot be found, a data set is not classifiable in precise deterministic sense.

Lastly, selection of reduct will impact an ability of a designed classifier to generalize prediction for unseen objects.

6 Numerical Experiments - Feature Extraction, Selection and Classification for Texture Images

The objective of this case study was to design a neural network classifier of texture images using feature extraction and selection techniques. Two feature extraction and reduction techniques from texture images have been investigated:

1. Principal Component Analysis (PCA) with Karhunen-Loéve optimal transformation/projection.
2. Principal Component Analysis (PCA) followed by rough sets method of feature selection/reduction.

The rough sets (Pawlak, 1991; Skowron, 1990) methodology and minimum concept description paradigm for selection/reduction of a final feature vector was used for 2-nd method of feature extraction/selection. Two types of neural network based texture classifiers were designed: the error backpropagation neural network and radial basis neural network. Design and testing were supported by designed Matlab programs. The rough sets processing was provided by the RoughFuzzyLab system (Swiniarski, 1995).

6.1 Raw data preprocessing, feature extraction and pattern forming

The error back propagation and radial basis neural network classifiers were trained and tested to classify images of bark, knit and sand (constituting three classes). The training and test sets were built from the three master images each for separate texture. The original images were in the PCX format containing 512×512 pixels per image. In order to define a recognition objects, sub-windows were then extracted from these original images and grouped into training and testing set images (with 100 training images and 100 test images for all classes). Recognition experiments were performed for different size of texture subwindow representing a texture object for recognition. We present results for 5×5, 10×10 and 15×15 subwindow sizes. Raw image patterns from each subwindow were stored as $n \times n$ two-dimensional raw matrix representing texture recognition objects. Here, each pixel of a subwindow is considered as one row pattern feature. Each raw pattern was scaled to the range of $[0, 1]$ by dividing each element by 255, a maximum value on the gray scale.

The first feature extraction method was based on principal component analysis (PCA) of a given data set with an optimal Karhunen-Loéve transformation. PCA was applied to $N = 100$ training cases data set with original raw pattern of size $n \times n$ ($n = 5, 10, 15$) labeled by categorical classes for textures. The resulting optimal Karhunen-Loéve transformation was used to project the original $n \times n$ raw pattern into principal components feature space of the same pattern size $n \times n$ for 5×5 and 10×10 subwindows, and reduced principal component space with 149 most dominant principal components for 15×15 subwindow. The resulting PCA feature vectors were used for classifiers design.

In the second approach a sequence of PCA feature extraction followed by rough sets based feature selection was examined. The rough sets method (Pawlak, 1992; Skowron, 1990), supported by the paradigm of the minimum concept description for feature selection, was used to select minimum length reduct treated as a selected final feature vector. For 5×5 subwindow, with fully dimensional $n \times n$ PCA transformation vector, the 6 elements from PCA features were selected as a reduct and as a final selected feature vector. Similar selection was obtained for 10×10 subwindow. For 15×15 subwindow roughs sets were applied for the reduced PCA feature vector comprising first 149 principal components of projected patterns space (from total number of 235 principal components). The rough sets selected 8 element reduct as a final feature vector used for classifiers design.

Discussion of the numerical results

As the texture images classifiers the error backpropagation and radial basis neural networks were designed and trained.

The summary of numerical experiments are presented in Tables 1, 2, and 3. Through numerous trials, it was determined that all both feature extraction and reduction methods are effective in capturing the image characteristics without significant lost in prediction accuracy. The rough sets minimal reduct, as a selected feature subset, provides reduction of final feature vector size, allowing simpler construction of classifiers. Each reduction techniques offer distinct features which will perform differently under different neural networks types. Therefore, the precision of the final prediction may vary with the employment of different reduction techniques and neural networks. The PCA technique showed to be sensitive to both the input subwindow size and the content of the image. One important factor of the sensitivity analysis is to determine the image size requirement. From heuristic numerical experiments it has been determined that an input image of size 10 × 10, representing a texture object, is sufficient for texture recognition. On the other hand, the summary tables show a slight downward trend for prediction rate with input size of 15 × 15.

Table 1 Accuracy of classification - % of correctly classified textures

BP -back propagation neural network
RBF - radial basis neural network

size	Features PCA		Features PCA +rough sets	
	BP	RBF	BP	RBF
5 × 5	92	92	93	93
10 × 10	93	94	94	94
15 × 15	89	88	89	91

Table 2 Summary of pattern sizes - original and reduced

Image size	Features PCA		Features PCA +rough sets	
	BP	RBF	BP	RBF
5 × 5	25 × 1	25 × 1	6 × 1	6 × 1
10 × 10	100 × 1	100 × 1	6 × 1	6 × 1
15 × 15	149 × 1	149 × 1	8 × 1	8 × 1

Table 3 Neurons distribution in neural networks classifiers

size	Features PCA		Features PCA +rough sets	
	BP	RBF	BP	RBF
5 × 5	10	200	10	200
10 × 10	14	200	14	200
15 × 15	20	200	20	200

An observation from the performance point of view shows, that the back propagation neural network is slightly less efficient compared to the radial basis network for this specific texture recognition task with proposed feature extraction and reduction methods.

References

1. Almuallim, H., and Dietterich, T.G. (1991). Learning with many irrelevant features. In Proceedings of the Ninth National Conference on Artificial Intelligence, pp. 547-552. Menlo Park, CA: AAAI Press.
2. Bazan, J. (1998). A Comparison of Dynamic and non–Dynamic Methods, In: L.Polkowski, A. Skowron (eds.), Rough Sets in Knowledge Discovery, Physica Verlag, Heidelberg 1998 (to appear)
3. Bishop, Ch. M. (1995). *Neural Networks for Pattern Recognition*. New York. Oxford Press.

4. Diamentras, K. I. and S. Y. Kung. (1996). *Principal Component Neural Networks. Theory and Applications*. New York, John Wiley, 1996.
5. Duda, R. O., and Hart, P. E. (1973). *Pattern Recognition and Scene Analysis*. New York. John Wiley.
6. Fukunaga, K. (1990). Introduction to Statistical Pattern Recognition. Academic Press, San Diego.
7. Fisher, R. A. (1936). The use of multiple measurements in taxonomy problems. Annals of Eugenics, 7, pp. 179- 188.
8. Gennari, J. H., Langley, P., and Fisher, D. (1989). Models of incremental concept formation. *Artificial Intelligence*, Vol. 40 , pp. 11-61.
9. John, G., Kohavi, R., and Pfleger, K. (1994). Irrelevant features and the subset selection problem. In Machine Learning: Proceedings of the Eleventh International Conference (ICML-94), pp. 121-129, New Brunswick, NJ: Morgan Kaufmann.
10. Pearson, K., (1901). On lines and planes of closest fit to systems of points in space. *Philosophical Magazine*, 2, pp. 559-572.
11. Karhunen, K. (1947). Uber lineare methoden in der Wahrscheinlichkeitsrechnung. *Annales Acedemiae Scientiarum Fennicae, Series AI: Mathematica-Physica* , 37, pp. 3-79.
12. Lewler, E. L., and D. E. Wood. (1966). Branch and bound methods: a survey. *Operational Research*, 149, No. 4.
13. Loéve, M. (1963). *Probability theory*. 3rd ed. Van Nostrand, New York.
14. Oja, E. (1989). Neural networks, principal components, and subspaces. *Int. J. Neural Systems*, Vo. 1, April 1989, pp. 61-68.
15. Narendra, P. M., and K. Fukunaga. (1977). A branch and bound algorithm for feature subset selection. *Transactions IEEE. Computers*, C-26, pp. 917-922.
16. D. L. Swets, and J. J. Weng. (1996). Using discriminant eigenfeatures for image retrieval. IEEE Trans. on Pattern Recognition and Machine Intelligence. Vol. 10, no. 8. August, pp. 831-836.
17. Bazan, J., Skowron, A., and Synak, P., (1994), "Dynamic reducts as a tool for extracting laws from decision tables", *Proc. of the Symp. on Methodologies for Intelligent Systems*, Charlotte, NC, October 16-19, 1994, Lecture Notes in Artificial Intelligence Vol. 869, Springer-Verlag, 346-355; see also *ICS Research Report* 43/94, Warsaw University of Technology.
18. Duentsch Ivo, and Guenther Gediga, (1995), "Rough set dependency analysis in evaluation studies: an application in the study of repeated heart attacks", *Informatics Research Reports*, University of Ulster, Vol. 10, 25–30.
19. Grzymała-Busse, J.W., (1995). "Rough Sets", *Advances in Imaging and Electrons Physics*, 94, to appear.
20. Lenarcik, A., and Piasta, Z., (1997), "Probabilistic rough classifiers with mixture of discrete and continuous attributes", in: T.Y. Lin, N. Cercone (eds.), *Rough Sets and Data Mining. Analysis for Imprecise Data*, Kluwer Academic Publishers, Boston, London, Dordrecht, 373–383.
21. Lin, T., and Chen, R., (1997), "Finding reducts in very large databases", in: P.P. Wang (ed.), *Joint Conference of Information Sciences*, March 1-5, Duke University, Vol. 3, 350–352.
22. Nguyen, T., and R. Swiniarski, A. Skowron, J. Bazan, K. Thyagarajan. (1994) "Application of Rough Sets, Neural Networks and Maximum Likelihood for Texture Classification Based on Singular Value Decomposition". Proceedings of "Third International Workshop on Rough Sets and Soft Computing'. San Jose, U.S.A., Nov. 10-12, 1994.

23. Pawlak, Z. (1982). Rough sets. International Journal of Computer Sciences, 11, pp. 341-356.
24. Pawlak, Z. (1991). Rough sets, Theoretical aspects of reasoning about data, Kluwer, Dordrecht 1991.
25. Pawlak, Z., and Skowron A., (1994), "Rough membership functions", in: R.R Yaeger, M. Fedrizzi and J. Kacprzyk (eds.), *Advances in the Dempster Shafer Theory of Evidence*, John Wiley & Sons, Inc., New York, Chichester, Brisbane, Toronto, Singapore, 251–271.
26. Pawlak, Z., Wong, S.K.M. and Ziarko, W. (1988). "Rough sets: probabilistic versus deterministic approach", *International Journal of Man-Machine Studies*, 29, 81–85. Jose, California, USA, November 10-12, 65–68.
27. Piasta, Z., Lenarcik, A., and Tsumoto S., (1996), "Machine discovery in databases with probabilistic rough classifiers", in: S. Tsumoto, S. Kobayashi, T. Yokomori, H. Tanaka and A. Nakamura (eds.), *The fourth International Workshop on Rough Sets, Fuzzy Sets, and Machnine Discovery, PROCEEDINGS (RS96FD)*, November 6-8, The University of Tokyo, 353–359.
28. Slowinski, R. (ed.) (1992). Intelligent decision support. Handbook of applications and advances of rough sets theory. Kluwer, Dordrecht.
29. Skowron, A. (1990).The rough sets theory and evidence theory. Fundamenta Informaticae vol.13 pp.245-262.
30. Skowron, A., Rauszer C. The discernibility matrices and functions in information systems. In (Slowinski R. ed.) *Decision Support by Experience - Application of the Rough Sets Theory*, Kluwer, 1992, pp. 331-362.
31. Skowron, A. (1995). Synthesis of adaptive decision systems from experimental data. In: A. Aamodt, J. Komorowski (eds.), Proceedings of the Fifth Scandinavian Conference on Artificial Intelligence (SCAI'95), May 29–31, 1995, Trondheim, Norway, IOS Press, Amsterdam (1995) 220–238
32. Skowron, A., and Stepaniuk, J., (1995), "Decision rules based on discernibility matrices and decision matrices", in: T.Y. Lin, and A.M. Wildberger (eds.), *Soft Computing*, Simulation Councils, Inc., San Diego, 6–9.
33. Stepaniuk, J., and Skowron, A., (1997), "Information reduction based on constructive neighborhood system", in: P.P. Wang (ed.), *Joint Conference of Information Sciences*, March 1-5, Duke University, Vol. 3, 158–160.
34. Swiniarski, R. 1995. RoughFuzzyLab. Software package developed at San Diego State University. San Diego.
35. Swiniarski, R. (1993). Introduction to Rough Sets. Materials of The International Short Course Neural Networks. Fuzzy and Rough Systems. Theory and Applications. April 2, 1993. San Diego State University.
36. Swiniarski, R., (1996a), "Rough sets for intelligent data mining, knowledge discovering and designing of an expert systems for on-line prediction of volleyball game progress", in: S. Tsumoto, S. Kobayashi, T. Yokomori, H. Tanaka and A. Nakamura (eds.), *The fourth International Workshop on Rough Sets, Fuzzy Sets, and Machnine Discovery, PROCEEDINGS (RS96FD)*, November 6-8, The University of Tokyo, 413–418.
37. Swiniarski, R., (1996b), "Rough sets expert system for robust texture classification based on 2D fast Fourier transformation spectral features", in: S. Tsumoto, S. Kobayashi, T. Yokomori, H. Tanaka and A. Nakamura (eds.), *The fourth International Workshop on Rough Sets, Fuzzy Sets, and Machnine Discovery, PROCEEDINGS (RS96FD)*, November 6-8, The University of Tokyo, 419–425.

38. Wasilewska, A., (1997), "RS logics and algebras" - in: A. Skowron (ed.), *Rough Set Theory and its Applications in various fields*, Soft Computing, Springer Verlag.
39. Yao, Y.Y., Wong, S.K.M., and Lin, T.Y., (1997), "A review of rough set models", in: T.Y. Lin, N. Cercone (eds.), *Rough Sets and Data Mining. Analysis for Imprecise Data*, Kluwer Academic Publishers, Boston, London, Dordrecht, 47–75.
40. Ziarko, W., (1993a), "Analysis of uncertain information in the framework of variable precision rough sets", *Foundations of Computing and Decision Sciences* 18/3-4 381–396.
41. Ziarko, W., (1993c), "Variable Precision Rough Set Model", *Journal of Computer and System Sciences*, 40, 39–59.
42. A.K. Jain, *Fundamentals of Digital Image Processing,* Prentice Hall, Inc., 1989 .
43. Z.Q. Hong, "Algebraic Feature Extraction of Image for Recognition," Pattern Recognition, Vol. 24, No. 3, pp. 211-219, 1991 .
44. P. Brodatz, *Textures: A Photographic Album for Artists and Designers,* Dover, NY, 1965 .

Medical information systems – problems with analysis and way of solutions

Krzysztof Słowiński[1] *and Jerzy Stefanowski*[2]

[1] Clinic of Traumatology, K.Marcinkowski University of Medical Sciences in Poznań,
1/2 Długa Street, 61-848 Poznań, Poland,
E-mail: slowik@rose.man.poznan.pl

[2] Institute of Computing Science, Poznań University of Technology,
3A Piotrowo Street, 60-965 Poznań, Poland,
E-mail: Jerzy.Stefanowski@cs.put.poznan.pl

Abstract: This paper discusses some difficulties connected with using the rough set theory to analyse medical information systems. These difficulties mainly refer to ambiguity in identifying the most significant attributes for the patients' classification and discovery of too weak decision rules representing the dependencies between attributes describing the patient's conditions and results of the treatment. As the directed use of basic rough set approach may give a high number of reducts, two heuristic approaches are proposed. Both aim at selecting significant attributes that have also good clinical interpretation. Particular attention is paid to the problem of discovery of various types of decision rules, testing their classification ability and looking for their accepted clinical interpretation. The above considerations are illustrated on a medical example concerning patients with urolithiasis treated by extracorporeal shock wave lithotripsy.

Keywords: Medical Information Systems, Rough Sets, Knowledge Discovery, Decision Rules.

1 Introduction

Physicians examining patients from the point of view of final diagnosis and choice of the proper treatment have to take into account several factors: they collect data from interviewing the patients or clinical investigations, they analyse results of several laboratory tests and image techniques (as e.g. ultrasonography, computer topography and X-ray contrast tests). The patient's status resulting from the analysis of these data and the clinical experience of the physician leads to determination of indications and contraindications for the particular mode of treatment. The physician's clinical experience is mainly based on his current knowledge and experience with a given disease as well as the therapeutic efficiency of the chosen mode of its treatment.

The progress in medical sciences in last years has significantly increased the amount of available measurement data. These data may have, however, different practical importance. So, they are analysed in order to find and select the most important and valuable data elements for the medical interpretation. Typical tasks in the analysis of the medical data, in particular concerning the problems of diagnosing and/or treatment of a given disease, are the following:

- to identify the most significant attributes for the patients' classification (resulting from the diagnostic and therapeutic point of view),
- to discover the dependencies between values of the significant attributes and the patients' classification.

One of the possible data analysis methods which are used to solve the above tasks is the *rough set theory* introduced by Z.Pawlak [12]. In last years, the authors successfully applied it to several medical problems (see e.g. [5, 8, 16, 15, 17, 18, 19, 24, 25]). Such elements of the rough set theory as the *approximations of objects' classification*, the *quality of these approximations* and notions of *reducts* could help in evaluating the attributes. Moreover, combination of the rough set theory with *rule induction techniques* gives the representation of the important dependencies in the form of *decision rules*. They represent the description of the real clinical cases, summarizing the medical experience but are often very difficult for interpretation from the point of view of the cause-and-effect relationships.

It must be noticed that the medical data systems are characterized by many properties that make these data specific to analyse, e.g.

- the number of attributes is too large comparing to the number of patients,
- there are too many independent and 'noisy' attributes,
- most of the attributes have a qualitative character,
- classes of the patients' classification are non-balanced taking into account the number of individuals; one class is often a strong majority class,
- observations describing patients may have individual character and be difficult to generalize.

The application of the rough set theory to such data sets usually leads to finding a high number of possible reducts, makes it very difficult to evaluate the significance of attributes, produces too many decision rules which are too specific and refer to single cases.

As practitioners usually want to avoid the above ambiguity in interpreting results, it seems to be necessary to consider some extensions of rough set theory approach devoted to the analysis of such data sets. In the following paper, we discuss such an attempt.

First, we will present how to evaluate the significance of the attributes by using two approaches: the technique of adding to the core the attributes with best discriminatory properties and another technique based on dividing the set of attributes into disjoint subsets. Then, we will look for the readable decision rules referring to the acceptable clinical experience. The above problems will be illustrated on a practical medical example concerning clinical experience with *urolithiasis* treated by *extracorporeal shock wave lithotripsy* (**ESWL**).

The paper is organised as follows. Section 2 gives the brief description of the ESWL data set. Then, basic information about the chosen methodology are summarized in section 3. Section 4 describes the performed analysis of the significance of attributes. Specifity of rule discovery is considered in Section 5. Discussion of obtained results and final remarks are presented in Section 6.

2 The ESWL information systems - description of the data sets

Data concerning patients with urolithiasis treated by the extracorporeal shock wave lithotripsy (ESWL) [3, 29] were collected at the Urology Clinic of University of Medical Sciences in Poznań [9]. In order to qualify patients for the ESWL treatment different data are taken into account, i.e. anamnesis (i.e. information coming from investigating patients by the physician), laboratory and imaging tests. Although the current experience of the Urology Clinic includes over 1000 patients per year, we could analyse part of it only, i.e. data about patients with completely defined pre-operation attributes and with known and verified long term results of the treatment.

The patients are described by 33 pre-operation attributes currently considered in urological practice. These are the following attributes: 1 - age, 2 - sex, 3 - duration of disease, 4 - type of urolithiasis, 5 - lithuresis, 6 - operations in the past, 7 - nephrectomy, 8 - PCNL, 9 - number of the ESWL treatment previously done, 10 - evacuation of calculi by zeiss cathether, 11 - lumbar region pains, 12 - dyspeptic symptoms, 13 - basic dysuric symptoms, 14 - other dysuric symptoms, 15 - body temperature, 16 - general uriscopy, 17 - urine reaction, 18 - erythrocyturia, 19 - leucocyturia, 20 - bacteriuria, 21 - crystaluria, 22 - proteinuria, 23 - urea, 24 - creatininc, 25 - bacteriological test, 26 - kidney location, 27 - kidney size, 28 - kidney defect, 29 - status of urinary system, 30 - secretion of urinary contrast, 31 - location of the concrement, 32 - calixcalculus, 33 - stone size. Let us notice that nearly all of these attributes have a qualitative character. Their domains usually consist of a limited number of values which are qualitative and linguistic terms. In addition, the domains of many attributes cannot be ordered.

The post-operation conditions of the patients were described by two attributes having the following clinical meaning:

1. A patient's physical condition after the lithotripsy, i.e.:
   - without complications,
   - with complications.
2. Long term results of the treatment:
   - recovery (good results),
   - no recovery,
   - lack of effects.

The both post-operation attributes define two classifications of patients, denoted as $\mathcal{Y}_1$ and $\mathcal{Y}_2$ respectively. Values of these classifications will be further called *decision classes*. These classifications are typical standards used to evaluate the medical treatment.

The representation of the ESWL experience for a set of 343 patients and 43 attributes has been preliminary examined by the authors in the study [18]. This study resulted in discovery of very high number of reducts, difficulties in their interpreting, and serious problems with identifying groups of the most significant attributes. To avoid this ambiguity in interpreting results we have decided to:

- extend the number of considered patients up to 500 ones,
- redefine the original set of attributes to obtain the smaller number of attributes with better discriminatory properties,
- use additional heuristic strategies to identify the most significant attributes.

Some of these problems have been partially resolved by the authors in [19, 24]. In the current study, additionally, we focused on inducing decision rules from experience represented in the redefined information system.

3 Chosen methodology

The ESWL information system is analysed using the rough set theory [12]. From the rough set theory point of view, the analysis is connected with examining *dependencies between attributes* in the defined data set (called further an *ESWL information system*). More precisely, similarly to previous medical applications [5, 15, 16, 18], the following elements of rough set theory are used:

- creating classes of *indiscernibility relations* (atoms) and building *approximations* of the objects' classification,
- evaluating the ability of attributes to approximate the objects' classification; the measure of the *quality of approximation of the classification*, defined as the ratio of the number of objects in the lower approximations to the total number of objects, is used to for this aim,
- discovering *cores* and *reducts* of attributes (a reduct is the minimal subset of attributes ensuring the same quality of the classification as the entire set of attributes; a core is an intersection of all reducts in the information system),
- examining the *significance* of attributes by observing changes in the quality of approximation of the classification caused by removing or adding given attributes.

All necessary definitions could be found in [12, 20, 31, 14].

In a case of finding too many possible reducts and difficulties in interpreting of them, we propose to use the following heuristic strategies:

- The strategy based on adding to the core, the attributes of the highest increase of discriminatory power,
- The strategy based on dividing the set of attributes into disjoint subsets and analysing the significance of attributes inside subsets,

In the first strategy, the core of attributes is chosen as a starting reduced subset of attributes. It usually ensures lower quality of approximation of the objects' classification than all attributes. A single remaining attribute is temporarily added to the core and the influence of this adding on the change of the quality is examined. Such an examination is repeated for all remaining attributes. The attribute with the highest increase of the quality of classification is chosen to be added to the reduced subset of attributes. Then, the procedure is repeated for remaining attributes. It is finished when an acceptable quality of the classification is obtained. If there are ties in choosing attributes, several possible ways of adding are checked. This strategy has been introduced in [15] and successfully used to analyse the several medical problems.

The aim of the second strategy is to reduce the number of interchangeable and independent attributes in the considered information system. If the system contains too many such attributes, one usually gets as a result an empty core, high number of equivalent reducts. We propose to divide the set of all attributes into disjoint subsets. Each subset should contain attributes which are dependent each other in a certain degree and have a common characteristic for a domain expert. Such a division could be done either nearly automatically as in [32] or depending on the background domain knowledge [24].

In the current study, *discovery of decision rules* is also considered. Decision rules are logical statements expressed in the following form:

$$IF\ (a_1,v_1)\&(a_2,v_2)\&\ldots\&(a_n,v_n)\ THEN\ class_j$$

where a_i is the ith attribute, v_i is its value and $class_j$ is one of the decision classes in the objects' classification.

To discover the decision rules in the ESWL data set we use our implementation [11] of LEM2 algorithm introduced by Grzymala [6]. This algorithm induces from lower approximations of decision classes, so called, *discriminating rules* (also called certain rules). These rules distinguish *positive examples*, i.e. objects belonging to the lower approximation of the decision class, from other objects.

To interpret the discovered rules we use the measure of their *strength*. It is the number of objects in the information system whose description satisfy the condition part of the rule. Generally, one is interested in discovering the strongest rules [26, 10]. Discovery of such rules may be impossible for data sets like ESWL one where rules are weak and too specific.

We think that one of the possible solution to avoid these limitations is to induce *partly discriminating decision rules* instead of discriminating only. These are rules that besides positive examples belonging to the given decision class could cover a limited number of objects not belonging to it. The partly discriminating rules are characterized by a coefficient called *level of discrimination* defined as:

$$d = \frac{p}{n}$$

where p is a number of positive examples and n is a total number of examples covered by the rule.

We argue that establishing a proper threshold for the minimum value of the level of discrimination will result in discovering stronger rules having *good interpretation characteristics*. In this study we induce partly discriminating rules using a modified version of LEM2 algorithm.

The above motivation is somehow similar to the concept of so called *Variable Precision Rough Set Model* introduced by [30].

4 Selection of attributes in the ESWL information system

First, let us summarize the results already obtained in [19, 24] for the analysis of the first redefined ESWL information system (i.e. extended to 435 patients described by 33 attributes).

For both classifications $\mathcal{Y}_1$ and $\mathcal{Y}_2$ lower and upper approximations of decision classes were calculated. The qualities of the approximations of classification by the set of all 33 attributes in both cases were equal to 1.0. The number of atoms was equal to 434 for both classifications. So, they could not be treated as a good basis for expressing strong classification patterns.

Then, we looked for cores and reducts of attributes. Using the microcomputer program RoughDAS [21] we were able to conclude, that the core of the first classification $\mathcal{Y}_1$ was empty and the core of the second one $\mathcal{Y}_2$ consisted of two attributes only (i.e. 20 and 21). For both classifications, we found out that the number of the reducts was very high. We could suspect that the attributes used to construct the data are interchangeable. One can remove few of them and others will take their role and still give the highest classification ability. As a result, the number of reducts is very high and even finding all of them would not lead to any reasonably solution from the medical point of view.

Proceeding in the way described in section 3, for both classifications we obtained the most acceptable reduced subset of attributes by adding the most discriminating ones to the core. The obtained subsets are presented in Table 1. Here, due to the limited size of the paper, we give summarized results only. Details are presented in [24].

Table 1. Acceptable subsets of attributes for both classifications obtained as a result of adding the most discriminatory attributes to a core

Classification	Selected attributes
$\mathcal{Y}_1$	1 3 6 11 14 21 22 25 28 29 30 31 33
$\mathcal{Y}_2$	1 2 6 11 14 16 20 21 31 33

Then, we used the second strategy. According to it, we divided the set of all attributes into disjoint subsets. Each subset should contain attributes which are dependent each other in certain degree and have common characteristics for a domain expert. Thus, we chose two subsets which have a different medical source and interpretation:

- attributes coming from the physician's investigation of the patien; g i.e. these are attributes 1 - 14 and they create *information system A*,
- attributes obtained as results of laboratory tests and examinations; i.e. these are attributes 15 - 33 and they create *information system B*.

Then, for both information systems A and B we examined the significance of attributes by removing temporarily single attributes and observing the decrease of quality of classification. Results are presented in Table 2.

Table 2. Subsets of attributes resulting from examining the significance of attributes in subdivided systems A and B

Classification	Significance of attributes	Selected attributes
y_1	the most sign.	1 3 6 11 21 25 29 30 32
y_1	the less sign.	7 8 10 13 17 23 26 27 28
y_2	the most sign.	1 6 11 21 25 30 32 33
y_2	the less sign.	7 10 13 15 16 26 27 28

One can notice that results obtained using these two strategies, i.e. chosen subsets of attributes, are quite similar. This led us to the final redefinition of the ESWL information system which was also consulted with medical experts. As a result we decided to:

- remove from the information system the attributes: 7,10,17,20,26,27 ,
- create a new attribute referring to dysuric symptoms on the basis of previous attributes 13 and 14 (both also referring to different dysuric symptoms).

The ESWL information system was finally reduced to 26 attributes. In the following parts of the paper, we stay with the same number codes of attributes for the reduced system as for the original one to be more consistent. Moreover, in the same time the number of analysed patients was extended to 500 ones.

The rough set approach was applied to the reduced ESWL information system. The results are presented in Table 3. Let us notice that the quality of the classification slightly decreased as a result of redefining attributes. However, now we can obtain cores of attributes characterizing by more elements and the higher value of quality.

As the number of possible reducts was still high we repeated the strategy of adding the most discriminating attributes to the cores. It was noticed that now the choice of added attributes was more supported by higher difference in the increase of the quality of approximation of the classification than for 33 attributes case. This procedure led us to first reduced subsets of attributes characterized by the fastest increase of the quality of the classification - these are subsets 1 and 3 presented in Table 4. This procedure indicates that sometimes more than one attribute could be chosen (the difference between increase of the quality for possible attributes was smaller than 5%). So, few other reducts could be found. We summed up attributes occurring in the additional reducts and get subsets 2 and 4, showed in Table 4.

Table 3. Results of approximations of patients' classification for the reduced ESWL information system - 26 attributes

Classi-fication	Quality of approx. of classification	Core of attributes	Quality of classification for the core
$\mathcal{Y}_1$	0.996	11 29 31	0.174
$\mathcal{Y}_2$	0.992	1 11 21 29 31	0.448

Table 4. The most significant attributes for the reduced ESWL information system

Classi-fication	Number of subset	Selected attributes
$\mathcal{Y}_1$	1	1 2 6 11 13 22 29 31 33
$\mathcal{Y}_1$	2	1 2 3 6 11 13 21 22 25 29 30 31 32 33
$\mathcal{Y}_2$	3	1 2 11 13 21 29 31
$\mathcal{Y}_2$	4	1 2 3 6 11 21 25 29 30 31 32 33

It should be noticed that the selected attributes are quite similar to oness obtained for 33 attribute version of the ESWL system (compare Tables 2 vs 4).

5 Induction of decision rules

The implementation of LEM2 algorithm [6] was used to induce discriminating (certain) decision rules. The rules were induced both from the ESWL information system describing 500 patients by the selected 26 attributes and the information system built using reduced subset of attributes (subsets 2 and 4 from Table 4). The results are presented in Table 5. In this table numbers in brackets refer to decision classes in both classifications. The first classification consists of two decision classes and the second classification consists of 3 classes.

One can notice that the number of discriminating rules is quite large. Most of them refer to a small number of patients or even single cases. As we wanted to look for stronger decision rules we decided to induce partly discriminating ones (in the way described in section 3). Let us notice that the majority decision class, coded by 1, is equal to 68.8% and 60.4% patients for classifications $\mathcal{Y}_1$ and $\mathcal{Y}_2$, respectively. Therefore, we systematically tested the following values of the level of discrimination: 0.95, 0.90, 0.85, 0.80, 0.75. As values 0.95 and 0.90 do not influence the set of rules too much, we restricted out attention to values equal to 0.75, 0.8, 0.85. Similarly as before we considered the sets of all (26) and reduced attributes. Information about obtained sets of rules are presented in Table 6.

One can notice that for the level of discrimination equal to 0.8 it was possible

Table 5. The discriminating decision rules induced from the ESWL information system for 26 attributes and reduced subsets of attributes.

Classification	Subset of attributes	No. of rules: total and in classes	Average strength of the rule in classes [objects]	the strongest rule [objects]
$\mathcal{Y}_1$	26 attr.	119 (65/64)	(9.42/3.37)	23
$\mathcal{Y}_1$	subset 2	151 (79/72)	(6.81/2.5)	22
$\mathcal{Y}_2$	26 attr.	123 (55/51/17)	(9.15/4.24/1.71)	24
$\mathcal{Y}_2$	subset 4	160 (74/68/18)	(7.55/3.25/1.56)	18

Table 6. The partly discriminating decision rules induced from the ESWL information system for 26 attributes and reduced subsets of attributes.

Classification	level of discrimination	Subset of attributes	No. of rules: total and in classes	Average strength of the rule [objects]	the strongest rule [objects]
$\mathcal{Y}_1$	0.85	26 attr.	97 (44/53)	(17.93/3.43)	63
$\mathcal{Y}_1$	0.80	26 attr.	82 (33/49)	(27.18/4.22)	105
$\mathcal{Y}_1$	0.75	26 attr.	59 (20/39)	(39.4/5.46)	144
$\mathcal{Y}_1$	0.85	subset 2	126 (54/72)	(14.59/2.68)	31
$\mathcal{Y}_1$	0.80	subset 2	111 (44/67)	(20.1/2.91)	79
$\mathcal{Y}_1$	0.75	subset 2	87 (29/58)	(34.1/3.48)	140
$\mathcal{Y}_2$	0.85	26 attr.	114 (46/51/17)	(12.83/4.96/1.71)	72
$\mathcal{Y}_2$	0.80	26 attr.	95 (34/44/17)	(21.35/5.18/1.88)	72
$\mathcal{Y}_2$	0.75	26 attr.	80 (23/43/13)	(26.91/5.41/2.23)	126
$\mathcal{Y}_2$	0.85	subset 4	140 (53/69/18)	(13.21/3.33/1.56)	54
$\mathcal{Y}_2$	0.80	subset 4	118 (44/56/18)	(18.25/4.57/1.56)	70
$\mathcal{Y}_2$	0.75	subset 4	99 (30/54/16)	(32.97/4.8/1.69)	124

to discover smaller number of rules which are stronger at least three times than strictly discriminating rules. Moving the level of discrimination down to 0.75 resulted in obtaining even stronger rules. Moreover, among them there are rules which have very simple and general condition parts and refer to the high number of patients (see Table 6). For example, for class 1 in classification $\mathcal{Y}_1$ there are rules covering 144 or 102 patients (41.87% and 20% of all patients from this class).

On the other hand the medical experts met serious difficulties and troubles while interpreting the clinical meaning of condition parts of these strongest decision rules. For instance, the analysis of the strongest partly discriminating decision rules (in particular for the level 0.75) has led to discovery very trivial

Table 7. Classification accuracies of decision rules induced from the ESWL information system for 26 attributes and reduced subsets of attributes.

Classification	level of discrimination	Subset of attributes	Overall classification accuracy
y_1	1.00	26 attr.	62.80%
y_1	0.85	26 attr.	65.60%
y_1	0.80	26 attr.	65.40%
y_1	0.75	26 attr.	65.40%
y_1	1.00	subset 2	64.00%
y_1	0.85	subset 2	64.40%
y_1	0.80	subset 2	64.40%
y_1	0.75	subset 2	65.40%
y_2	1.00	26 attr.	62.60%
y_2	0.85	26 attr.	60.80%
y_2	0.80	26 attr.	62.40%
y_2	0.75	26 attr.	62.60%
y_1	1.00	subset 4	59.70%
y_2	0.85	subset 4	59.70%
y_2	0.80	subset 4	60.10%
y_2	0.75	subset 4	60.40%

and self-evident attribute dependencies that do not take into account accepted clinical knowledge in the urology.

Weaker decision rules (in particular strictly discriminating ones) are usually more interesting from the medical point of view and are closer to the current clinical practice.

For example one could analyse the strongest partly discriminating decision rule discovered in the ESWL information system:
if *(dyspeptic symptoms = no) and (urine reaction = acid) and (urea = normal)* **then** *(patient's condition after lithotripsy = no complications)* which covers 144 cases.

The above conditions are absolutely too obvious and in clinical practice do not have any real influence on the prognosis of the results of the treatment.

On the other hand, if one analyses the strongest strictly discriminating decision rules the interpretation is much better, e.g. look at the following decision rule:
if *(urolithiasis = primarily) and (nephrectomy in the past = no) and (lithuresis = spontaneously) and (dyspeptic symptoms = no) and (disuric symptoms = no) and (bacteriuria = no)* **then** *(patient's condition after lithotripsy = no complications)* which covers 23 cases.

The use of reduced subsets of attributes usually did not give rules with better properties (see Table 5 and 6).

Taking into account all above remarks it is also necessary to ask a question: what is the prognostic ability of sets of induced rules. It other words – how good are the decision rules for classifying new coming patients. To answer these questions we performed computational tests based on the standard *10-fold cross validation techniques* [28]. While performing the classification, possible ambiguity in matching the testing example to decision rules is solved by the Stefanowski's nearest rule approach where the decision is chosen on the basis of multiple or partly matched rules (see for more details [22]). The calculated classification accuracies are presented in Table 7.

Table 8. Classification accuracies for various induction system. Results for the first classification

Induction system	classification accuracy
LEM2 algorithm strict discrimination	62.80%
LEM2 algorithm reduct	64.40%
LEM2 algorithm partial discrimination	65.60%
IBL 1	56.80%
IBL 1 with reduction of attributes by wrapper approach	59.60%
IBL 3 with reduction of noisy cases	51.80%
C4.5 unpruned tree	59.60 %
C4.5 pruned tree	66.80%
RISE	59.20%

We should noticed that for both classifications the obtained accuracies are lower than the default accuracy (i.e. estimated frequency of majority decision class - equal to 68.8% and 60.4% for classifications $\mathcal{Y}_1$ and $\mathcal{Y}_2$, respectively). It means that the induced decision rules have unsatisfactory prognostic properties. For comparison purposes we also tried to use other well known machine learning classification algorithms, like Quinlan's C4.5 system for induction of decision trees [13], Aha's instance base learning approach IBL [1] or other rule inducer - system RISE [4]. It has be done to check whether these unsatisfactory properties are connected with rough set approach only or have more general character. We chosen C4.5 (in both versions - trees and rules) as it is one of the most efficient classification system. IBL and RISE approaches were considered because they pay a particular attention to single cases, not only general rules while making final classification (see the discussion in [4, 27]). Generally, we did not get much better classification performance. The experiments shown that the C4.5 system was the best – achieving accuracy 66.80% for the first classification and accuracy 63.80% for the second classification. Results for first classification are summarized in Table 8.

To sum up the experiments we have to conclude that the construction of decision support system based on the decision rules framework is still unjustified.

6 Conclusion and final remarks

Let us summarize conclusions and final remarks in the following points:

1. The aim of this paper is to discuss the difficulties in the analysis of the medical information systems by means of the rough set approach. The above problems have been illustrated on the real example concerning patients with urolithiasis treated by extracorporeal shock wave lithotripsy (ESWL). The ESWL information system is the typical example of a medical data set often met in clinical practice.
2. The use of the 'classical' rough set approach for such a data set has led to obtaining a high number of reducts which are very difficult for analysis and clinical interpretation. Simple increasing the number of patients and even attempts of reducing the number of considered attributes has not helped in the analysis.
3. We have shown that it is possible to avoid this difficulties by using two heuristic strategies: first based on adding the most discriminating attributes to the initial set and the other one based on dividing the set of attributes into disjoint subsets. Both strategies have given as a result accepted subsets of attributes establishing the hierarchy of significance between these attributes. In a case of the ESWL information system the proposed strategies provided better clinical interpretation. However the first startegy is more objective as it is based on the measure of quality of classification, while the second strategy is more subjective as it depends on the expert's decision concerning the initial division of attributes into two sets.
4. One could notice that the selection of relevant attributes in the case of the ESWL data is quite a hard problem. Thus, one has to be careful with analysing and interpreting these results. It could be interesting to extend this study by performing more exhaustive discussion and comparison with other methods of feature selection. For instance in a case of the rough set theory we can think about other approaches, e.g. looking for dynamic reducts [2] or hybrid approaches using statistical techniques [7, 8] and constructive induction tools - see, e.g. [25].
5. Analysis of the dependencies between data describing the patient's status and results of the treatment can be done on the basis of the decision rules. In our ESWL case we have obtained a high number of the decision rules which have been very weak and too specific. Discovery of partly discriminating rules have allowed to extract stronger ones. However, their interpretation

from the clinical point of view is also inconsistent with the current clinical practice.
6. The prognostic ability of the induced decision rules is unsatisfactory - as it has been checked in the classification tests.
7. Taking into account the above difficulties with induced decision rules one should remember that in medical practice it may be more interesting and even more instructive to analyse the weaker rules (referring to the single cases) reflecting anomalies and particular clinical cases. These cases often enrich the medical experience in the diagnosis and treatment of many diseases.

In this paper we wanted to show that besides simple and successful applications of the rough set theory to analyse medical data, one can meet some difficulties in the case of non-trivial data. The way of solution was presented and discussed but some next research on relevant feature extraction and rule discovery are still necessary as it is one of possibile way searching for succesful analysis of hard medical data.

Acknowledgments
Research of the second author of this paper was partly supported from the KBN grant no. 8-T11C 013 13 and CRIT 2 Esprit research project no. 20288.

References

1. Aha D.W., Kibler D., Albert M.K., Instance based learning algorithms. *Machine Learning*, 6, 1991, 37 – 66.
2. Bazan J., Skowron A., Synak P., Dynamic reducts as a tool for extracting laws from decision tables. In *Proc. of the Symp. on Methodologies for Intelligent Systems*, Charlotte, NC. October 16-19, 1994, Lecture Notes in Artificial Intelligence, no. 869, Springer-Verlag, Berlin 1994, 346-355.
3. Chaussy C., Schmiedt E., Jocham D., Walter V., Brendel W., *Extracorporeal Shock Wave Lithotripsy. New Aspects in the Treatment of Kidney Stones*. New York: S. Karger, 1982.
4. Domingos P., Unifying instance based and rule based induction. *Machine Learning*, 24, 1994, 141 – 168.
5. Fibak J., Pawlak Z., Slowinski K., Slowinski R., Rough sets based decision algorithm for treatment of duodenal ulcer by HSV. *Bulletin of the Polish Academy of Sciences. Biological Sciences* vol. 34, 10/12, 1986, 227-246.
6. Grzymala-Busse J.W., LERS - a system for learning from examples based on rough sets. In: Slowinski R. (ed.) *Intelligent Decision Support. Handbook of Applications and Advances of the Rough Sets Theory*. Kluwer, 1992, 3 – 18.
7. Krusinska E., Slowinski R., Stefanowski J., Discriminant versus rough sets approach to vague data analysis. *Applied Stochastic Models and Data Analysis*, 8, 1992, 43–56.
8. Krusinska E., Stefanowski J., Stromberg J.E., Comparibility and usefulness of newer and classical data analysis techniques. Application in medical domain classification. In Didey E. et al. (eds.) *New Approaches in Classification and Data*

Analysis, Springer Verlag, Studies in Classification, Data Analysis and Knowledge Organization, 1993, 644-652.

9. Kwias Z., et al. , Clinical experiences in treatment of urinary tract's stones using Dornier MPL 9000 lithotriptor, *Nowiny Lekarskie*: 2, 1992, 15-24 (in Polish).
10. Mienko R., Stefanowski J., Toumi K., Vanderpooten D., Discovery-Oriented Induction of Decision Rules. *Cahier du Lamsade* no. 141, Paris, Universite de Paris Dauphine, septembre 1996.
11. Mienko R., Slowinski R., Stefanowski J., Susmaga R., RoughFamily - software implementation of rough set based data analysis and rule discovery techniques. In Tsumoto S. (ed.) *Proceedings of the Fourth International Workshop on Rough Sets, Fuzzy Sets and Machine Discovery*, Tokyo Nov. 6-8 1996, 437-440.
12. Pawlak Z., *Rough sets. Theoretical aspects of reasoning about data.* Kluwer Academic Publishers, Dordrecht, (1991).
13. Quinlan J. R., *C4.5: Programs for Machine Learning.* Morgan Kaufmann, San Mateo CA, 1993.
14. Skowron A., Extracting Laws from Decision Tables. *Computational Intelligence: An International Journal* 11(2), 1995, 371-388.
15. Slowinski K., Slowinski R., Stefanowski J., Rough sets approach to analysis of data from peritoneal lavage in acute pancreatitis. *Medical Informatics*, 13, 1988, 143-159.
16. Slowinski K., Rough classification of HSV patients, In: Slowinski R. (ed.), *Intelligent Decision Support. Handbook of Applications and Advances of the Rough Sets Theory.* Kluwer Academic Publishers, Dordrecht, 1992, 363-372.
17. Slowinski K., El. Sanossy Sharif, Rough sets spproach to analysis of data of diagnostic peritoneal lavage applied for multiple injuries patients. In: Ziarko W. (ed.), *Rough Sets, Fuzzy Sets and Knowledge Discovery*, Springer-Verlag, London, 1994, 420-425.
18. Slowinski K., Stefanowski J., Antczak A., Kwias Z., Rough set approach to the verification of indications for treatment of urinary stones by extracorporeal shock wave lithotripsy (ESWL). In: Lin, T.Y, Wildberg A.M. (eds.), *Soft Computing*, Simulation Council Inc., San Diego, 1995, 93-96.
19. Slowinski K., Stefanowski J., On limitations of using rough set approach to analyse non-trivial medical information systems. In: Tsumoto (ed.) *Proceedings of the Fourth International Workshop on Rough Sets, Fuzzy Sets and Machine Discovery*, Tokyo Nov. 6-8 1996, The University of Tokyo Press, 176–184.
20. Slowinski R. (ed.), *Intelligent Decision Support - Handbook of Applications and Advances of the Rough Sets Theory*, Kluwer Academic Publishers, Dordrecht, 1992.
21. Slowinski R., Stefanowski J., RoughDas and RoughClass software implementation of the rough sets approach. In: Slowinski, R. (ed.), *Intelligent Decision Support - Handbook of Applications and Advances of the Rough Sets Theory*, Kluwer Academic Publishers, 1992, 445-456.
22. Stefanowski J., Classification support based on the rough sets. *Foundations of Computing and Decision Sciences*, vol. 18, no. 3-4, 1993, 371-380.
23. Stefanowski J., On rough set based approaches to induction of decision rules. In: Polkowski L., Skowron A. (eds.) *Rough Sets in Data Mining and Knowledge Discovery*, Series Soft Computing, Springer-Verlag, 1997 (to appear).
24. Stefanowski J., Slowinski K., Rough sets s a tool for studying attribute dependencies in the urinary stones treatment data set. In: T.Y. Lin, N. Cecrone (ed.), *Rough Sets and Data Mining*, Kluwer Academic Publishers, 1997, 177-198.

25. Stefanowski J., Slowinski K., Rough set theory and rule induction techniques for discovery of attributes dependencies in medical information systems. In: Komorowski J., Zytkow J. (eds.) *Principles of Knowledge Discovery.* Proceedings of the First European Symposium: PKDD'97, Trondhiem, Norway, Springer LNAI no. 1263, Springer Verlag, 1997, 36–46.
26. Stefanowski J., Vanderpooten D., A general two stage approach to rule induction from examples. In: Ziarko W. (ed.) *Rough Sets, Fuzzy Sets and Knowledge Discovery,* Springer-Verlag, 1994, 317-325.
27. Stefanowski J., Urbaniak S., Using case-based learning in decision support systems. *Proceedings of the VI Int. Symp. on Intelligent Information Systems,* Zakopane 9-13 June 1997. IPI PAN Press, 159-168.
28. Weiss S.M., Kulikowski C.A., *Computer Systems that Learn,* Morgan Kaufmann, San Mateo CA, 1991.
29. Wilbert D.M., Reinchreberger H., Noske E., Riedmiller, H., Alken, P., Hohenfellner R., New generation shock wave lithotripsy. *J.of Urology,* vol. 137, 1987, 563.
30. Ziarko W., Variable precision rough set model. *Journal of Computer and System Sciences,* 40, 1993, 39-59.
31. Ziarko W. (ed.), *Rough Sets, Fuzzy Sets and Knowledge Discovery,* Springer-Verlag, London, 1994.
32. Ziarko W., Shan N, On Discovery of Attributes Interactions and Domain Classifications, *Int. J.: Intelligent Automation and Soft Computing* 1996 (to appear).

Induction of Expert Decision Rules using Rough Sets and Set-Inclusion

Shusaku Tsumoto

Department of Information Medicine, Medical Research Institute,
Tokyo Medical and Dental University,
1-5-45 Yushima, Bunkyo-city Tokyo 113 Japan.
E-mail: tsumoto@computer.org

Abstract. *One of the most important problems on rule induction methods is that they cannot extract rules, which plausibly represent experts' decision processes. On one hand, rule induction methods induce probabilistic rules, the description length of which is too short, compared with the experts' rules. On the other hand, construction of Bayesian networks generates too lengthy rules. In this paper, the characteristics of experts' rules are closely examined and a new approach to extract plausible rules is introduced, which consists of the following three procedures. First, the characterization of decision attributes (given classes) is extracted from databases and the classes are classified into several groups with respect to the characterization. Then, two kinds of sub-rules, characterization rules for each group and discrimination rules for each class in the group are induced. Finally, those two parts are integrated into one rule for each decision attribute. The proposed method was evaluated on medical databases, the experimental results of which show that induced rules correctly represent experts' decision processes.*

Keywords: Rough Sets, Data Mining, Knowledge Discovery in Databases, Medical Knowledge Acquisition

1 Introduction

One of the most important problems in developing expert systems is knowledge acquisition from experts [3]. In order to automate this problem, many inductive learning methods, such as induction of decision trees [2, 12], rule induction methods [6, 7, 9, 12, 13] and rough set theory [10, 15, 18], are introduced and applied to extract knowledge from databases, and the results show that these methods are appropriate.

However, it has been pointed out that conventional rule induction methods cannot extract rules, which plausibly represent experts' decision processes [15, 16]: the description length of induced rules is too short, compared with the experts' rules (Those results are shown in Appendix B). For example, rule induction methods, including AQ15 [9] and PRIMEROSE [15], induce the following common rule for muscle contraction headache from databases on differential diagnosis of headache [16]:

[location = whole] ∧[Jolt Headache = no] ∧[Tenderness of M1 = yes]
→ muscle contraction headache.

This rule is shorter than the following rule given by medical experts.

[Jolt Headache = no] ∧[Tenderness of M1 = yes]
∧[Tenderness of B1 = no] ∧ [Tenderness of C1 = no]
→ muscle contraction headache,

where [Tenderness of B1 = no] and [Tenderness of C1 = no] are added.

These results suggest that conventional rule induction methods do not reflect a mechanism of knowledge acquisition of medical experts.

In this paper, the characteristics of experts' rules are closely examined and a new approach to extract plausible rules is introduced, which consists of the following three procedures. First, the characterization of each decision attribute (a given class), a list of attribute-value pairs the supporting set of which covers all the samples of the class, is extracted from databases and the classes are classified into several groups with respect to the characterization. Then, two kinds of subrules, rules discriminating between each group and rules classifying each class in the group are induced. Finally, those two parts are integrated into one rule for each decision attribute. The proposed method is evaluated on medical databases, the experimental results of which show that induced rules correctly represent experts' decision processes. The paper is organized as follows: in Section 2, we make a brief description about rough set theory and the definition of probabilistic rules based on this theory. Section 3 discusses interpretation of medical experts' rules. Then, Section 4 presents an induction algorithm for incremental learning. Section 5 gives experimental results. Section 6 discusses the problems of our work and related work, and finally, Section 7 concludes our paper.

2 Rough Set Theory and Probabilistic Rules

2.1 Rough Set Theory

Rough set theory clarifies set-theoretic characteristics of the classes over combinatorial patterns of the attributes, which are precisely discussed by Pawlak [10, 17]. This theory can be used to acquire some sets of attributes for classification and can also evaluate how precisely the attributes of database are able to classify data.

Let us illustrate the main concepts of rough sets which are needed for our formulation. Table 1 is a small example of database which collects the patients who complained of headache. First, let us consider how an attribute "loc" classifies the headache patients' set of the table. The set whose value of the attribute "loc" is equal to "who" is $\{2,4,5,6\}$, which shows that the second, fourth, fifth and sixth case are indiscernible (In the following, the numbers in a set are used to represent each record number). This set means that we cannot classify $\{2,4,5,6\}$ further solely by using the constraint $R = [loc = who]$. This set is defined as the indiscernible set over the relation R and described as follows: $[x]_R = \{2,4,5,6\}$. In this set, $\{2,5\}$ suffer from muscle contraction headache ("m.c.h."), $\{4\}$ from classical migraine ("migraine"), and $\{6\}$ from psycho ("psycho"). Hence we need

Table 1. An Example of Database

	age	loc	nat	prod	nau	M1	class
1	50-59	occ	per	no	no	yes	m.c.h.
2	40-49	who	per	no	no	yes	m.c.h.
3	40-49	lat	thr	yes	yes	no	migra
4	40-49	who	thr	yes	yes	no	migra
5	40-49	who	rad	no	no	yes	m.c.h.
6	50-59	who	per	no	yes	yes	psycho

DEFINITIONS: loc: location, nat: nature, prod: prodrome, nau: nausea, M1: tenderness of M1, who: whole, occ: occular, lat: lateral, per: persistent, thr: throbbing, rad: radiating, m.c.h.: muscle contraction headache, migra: migraine, psycho: psychological pain,

other additional attributes to discriminate between "m.c.h.", "migraine", and "psycho". Using this concept, we can evaluate the classification power of each attribute. For example, "nat=thr" is specific to the case of classic migraine ("migraine"). We can also extend this indiscernible relation to multivariate cases, such as $[x]_{[loc=who] \wedge [nau=no]} = \{2,5\}$ and $[x]_{[loc=who] \vee [nau=no]} = \{1,2,4,5,6\}$, where $\wedge$ and $\vee$ denote "and" and "or" respectively. In the framework of rough set theory, the set $\{2,5\}$ is called *strictly definable* by the former conjunction, and also called *roughly definable* by the latter disjunctive formula. Therefore, the classification of training samples D can be viewed as a search for the best set $[x]_R$ which is supported by the relation R. In this way, we can define the characteristics of classification in the set-theoretic framework. For example, accuracy and coverage, or true positive rate is defined as follows.

Definition 1 Classification Accuracy and Coverage.
Classification accuracy and coverage(true positive rate) is defined as:

$$\alpha_R(D) = \frac{|[x]_R \cap D|}{|[x]_R|} (= P(D|R)), \text{ and}$$

$$\kappa_R(D) = \frac{|[x]_R \cap D|}{|D|} (= P(R|D)),$$

where $|A|$ denotes the cardinality of a set A, $\alpha_R(D)$ denotes a classification accuracy of R as to classification of D, and $\kappa_R(D)$ denotes a coverage, or a true positive rate of R to D, respectively. It is notable that these two measures are equal to conditional probabilities: accuarcy is a proability of D under the condition of R, coverage is one of R under the condition of D.

For example, when R and D are set to $[nau = yes]$ and $[class = migraine]$, $\alpha_R(D) = 2/3 = 0.67$ and $\kappa_R(D) = 2/2 = 1.0$.

It is notable that $\alpha_R(D)$ measures the degree of the sufficiency of a proposition, $R \to D$, and that $\kappa_R(D)$ measures the degree of its necessity.[1] For example, if $\alpha_R(D)$ is equal to 1.0, then $R \to D$ is true. On the other hand, if $\kappa_R(D)$ is equal to 1.0, then $D \to R$ is true. Thus, if both measures are 1.0, then $R \leftrightarrow D$.

For further information on rough set theory, readers could refer to [10, 17, 18].

2.2 Probabilistic Rules

In this section, a probabilistic rule is defined by the use of the following three notations of rough set theory [10]. The main ideas of these rules are illustrated by a small database shown in Table 1.

First, a combination of attribute-value pairs, corresponding to a complex in AQ terminology [8], is denoted by a formula R. For example, $[age = 50 - 59] \wedge [loc = occular]$ will be one formula, denoted by $R = [age = 50 - 59] \wedge [loc = occular]$.

Secondly, a set of samples which satisfy R is denoted by $[x]_R$, corresponding to a star in AQ terminology. For example, when $\{2,3,4,5\}$ is a set of samples which satisfy $[age = 40 - 49]$, $[x]_{[age=40-49]}$ is equal to $\{2,3,4,5\}$.[2]

Finally, U, which stands for "Universe", denotes all training samples.

According to these notations, a rule is defined as follows:

Definition 2 Representation of Rule. Let R be a formula (conjunction of attribute-value pairs), D denote a set whose elements belong to a class d, or positive examples in all training samples (the universe), U. Finally, let $|D|$ denote the cardinality of D. A rule of D is defined as a tripule, $< R \xrightarrow{\alpha,\kappa} d, \alpha_R(D), \kappa_R(D) >$, where $R \xrightarrow{\alpha,\kappa} d$ satisfies the following conditions:

$$(1) \quad [x]_R \cap D \neq \phi,$$

$$(2) \quad \alpha_R(D) = \frac{|[x]_R \cap D|}{|[x]_R|},$$

$$(3) \quad \kappa_R(D) = \frac{|[x]_R \cap D|}{|D|}.$$

In the above definition, α corresponds to the accuracy measure: if α of a rule is equal to 0.9, then the accuracy is also equal to 0.9. On the other hand, κ is a statistical measure of what proportion of D is covered by this rule, that is, a coverage or a true positive rate: when κ is equal to 0.5, half of the members of a class belong to the set whose members satisfy that formula.

For example, let us consider a case of a rule $[age = 40 - 49] \to m.c.h.$ Since $[x]_{[age=40-49]} = \{2,3,4,5\}$ and $D = \{1,2,5\}$, accuracy and coverage are obtained as: $\alpha_{[age=40-49]}(D) = |\{2,5\}|/|\{2,3,4,5\}| = 0.5$ and $\kappa_{[age=40-49]}(D) = |\{2,5\}|/|\{1,2,5\}| = 0.67$. Thus, if a patient, who complains of a headache, is 40 to 49 years old, then m.c.h. is suspected, whose accuracy and coverage are equal to 0.5 and 0.67, respectively.

---

[1] These characteristics are from formal definiton of accuracy and coverage. In this paper, these measures are important not only from the viewpoint of propositional logic, but also from that of modelling medical experts' reasoning, as shown later.

[2] In this notation, "n" denotes the nth sample in a dataset (Table 1).

Probabilistic Rules The simplest probabilistic model is that which only uses classification rules which have high accuracy and high coverage.

This model is applicable when rules of high accuracy can be derived. Such rules can be defined as:

$$R \stackrel{\alpha,\kappa}{\rightarrow} d \quad \text{s.t.} \quad R = \vee_i R_i = \vee \wedge_j [a_j = v_k],$$
$$\alpha_{R_i}(D) \geq \delta_\alpha \text{ and } \kappa_{R_i}(D) \geq \delta_\kappa,$$

where δ_α and δ_κ denote given thresholds for accuracy and coverage, respectively. For the above example shown in Table 1, probabilistic rules for m.c.h. are given as follows:

$$[M1 = yes] \rightarrow m.c.h. \; \alpha = 3/4 = 0.75, \kappa = 1.0,$$
$$[nau = no] \rightarrow m.c.h. \; \alpha = 3/3 = 1.0, \kappa = 1.0,$$

where δ_α and δ_κ are set to 0.75 and 0.5, respectively.

It is notable that this rule is a kind of probabilistic proposition with two statistical measures, which is an extension of Ziarko's variable precision model (VPRS) [18]. [3]

3 Interpretation of Medical Experts' Rules

As shown in Section 1, rules acquired from medical experts are much longer than those induced from databases the decision attributes of which are given by the same experts.[4]

Those characteristics of medical experts' rules are fully examined not by comparing between those rules for the same class, but by comparing experts' rules with those for another class. For example, a classification rule for muscle contraction headache is given by:

[Jolt Headache = no] ∧([Tenderness of M0 = yes]
   ∨[Tenderness of M1 = yes] ∨ [Tenderness of M2 = yes])
   ∧[Tenderness of B1 = no] ∧ [Tenderness of B2 = no]
   ∧[Tenderness of B3 = no]
   ∧[Tenderness of C1 = no] ∧ [Tenderness of C2 = no]
   ∧[Tenderness of C3 = no] ∧ [Tenderness of C4 = no]
   → muscle contraction headache

---

[3] In VPRS model, the two kinds of precision of accuracy is given, and the probabilistic proposition with accuracy and two precision conserves the characteristics of the ordinary proposition. Thus, our model is to introduce the probabilistic proposition not only with accuracy, but also with coverage.

[4] This is because rule induction methods generally search for shorter rules, compared with decision tree induction. In the latter cases, the induced trees are sometimes too deep and in order for the trees to be learningful, pruning and examination by experts are required. One of the main reasons why rules are short and decision trees are sometimes long is that these patterns are generated only by one criteria, such as high accuracy or high information gain. The comparative study in this section suggests that experts should acquire rules not only by one criteria but by the usage of several measures.

This rule is very similar to the following classification rule for disease of cervical spine:

[Jolt Headache = no] ∧([Tenderness of M0 = yes]
 ∨[Tenderness of M1 = yes] ∨ [Tenderness of M2 = yes])
 ∧([Tenderness of B1 = yes] ∨ [Tenderness of B2 = yes]
 ∨[Tenderness of B3 = yes]
 ∨[Tenderness of C1 = yes] ∨ [Tenderness of C2 = yes]
 ∨[Tenderness of C3 = yes] ∨ [Tenderness of C4 = yes])
 → disease of cervical spine

The differences between these two rules are attribute-value pairs, from tenderness of B1 to C4. Thus, these two rules can be simplified into the following form:

$$a_1 \wedge A_2 \wedge \neg A_3 \rightarrow \text{muscle contraction headache}$$
$$a_1 \wedge A_2 \wedge A_3 \rightarrow \text{disease of cervical spine}$$

The first two terms and the third one represent different reasoning. The first and second term a_1 and A_2 are used to differentiate muscle contraction headache and disease of cervical spine from other diseases. The third term A_3 is used to make a differential diagnosis between these two diseases. Thus, medical experts firstly selects several diagnostic candidates, which are very similar to each other, from many diseases and then make a final diagnosis from those candidates.

In the next section, a new approach for inducing the above rules is introduced.

4 Rule Induction

Rule induction consists of the following three procedures. First, the characterization of each decision attribute (a given class), a list of attribute-value pairs the supporting set of which covers all the samples of the class, is extracted from databases and the classes are classified into several groups with respect to the characterization. Then, two kinds of sub-rules, rules discriminating between each group and rules classifying each class in the group are induced. Finally, those two parts are integrated into one rule for each decision attribute.

4.1 An Algorithm for Rule Induction

An algorithm for rule induction is given as follows.

1. Calculate $\alpha_R(D)$ and $\kappa_R(D)$ for each elementary relation R and each class D.
2. Make a list of R $L(D)$ the coverage of which is equal to 1.0 ($L(D) = \{R|\ \kappa_R(D) = 1.0\}$) for each class D.
3. For each class D, make a list $L_2(D)$, each element $L(D_j)$ of which is a subset of $L(D)$.

4. Make a new decision attribute D' for each $L_2(D)$ and search for a partition P of all the classes D such that $L_2(D_i) \cap L_2(D_j) \neq \phi$.
5. Construct a new table $(T(P))$ for P. Also construct a new table $(T(D'))$ for each decision attribute D'.
6. Induce classification rules R_p for each P in $T(P)$.
7. Induce classification rules R_d for each D in $T(D')$.
8. Integrate R_p and R_d into a rule $R(D)$.

Induction of Classification Rules For induction of rules, the algorithm introduced in PRIMEROSE [15] is applied, which is shown in Fig. 1.

```
procedure Induction of Classification Rules;
  var
    i : integer;    M, L_i : List;
  begin
    L_1 := L_er;  /* L_er: List of Elementary Relations */
    i := 1;   M := {};
    for i := 1 to n do      /* n: Total number of attributes */
      begin
        while ( L_i ≠ {} ) do
          begin
            Select one pair R = ∧[a_i = v_j] from L_i;
            L_i := L_i - {R};
            if   (α_R(D) ≥ δ_α)   and   (κ_R(D) ≥ δ_κ)
              then do S_ir := S_ir + {R}; /* Include R as Inclusive Rule */
              else M := M + {R};
          end
        L_{i+1} := (A list of the whole combination of the conjunction formulae in M);
      end
  end {Induction of Classification Rules };
```

Fig. 1. An Algorithm for Classification Rules

Integration of Rules An algorithm for integration is given as follows.

1. For each D_i, repeat the following step.
2. Select one rule $R \rightarrow D_i$.
3. Search for a rule in D_i, $R' \rightarrow d_i$, the supporting set of which is a subset of that of $R \rightarrow D_i$.
4. Integrate these two rules into one: $R \wedge R' \rightarrow d_i$.

4.2 Example

Let us illustrate how the introduced algorithm works by using a small database in Table 1. For simplicity, the threshold δ_α is set to 1.0, which means that only deterministic rules should be induced.

After the first and second step, the following three $L(D_i)$ will be obtained: $L(m.c.h.) = \{[prod = no], [M1 = yes]\}$, $L(migra) = \{[age = 40 - 49], [nat = thr], [prod = yes], [nau = yes], [M1 = no]\}$, and $L(psycho) = \{[age = 50 - 59], [loc = who], [nat = per], [prod = no], [nau = no], [M1 = yes]\}$.

Thus, since a relation $L(m.c.h.) \subset L(psycho)$ holds, a new decision attribute is $D_1 = \{m.c.h., psycho\}$ and $D_2 = \{migra\}$, and a partition $P = \{D_1, D_2\}$ is obtained. From this partition, two decision tables will be generated, as shown in Table 2 and Table 3 in the fifth step.

Table 2. A Table for a New Partition P

	age	loc	nat	prod	nau	M1	class
1	50-59	occ	per	0	0	1	D_1
2	40-49	who	per	0	0	1	D_1
3	40-49	lat	thr	1	1	0	D_2
4	40-49	who	thr	1	1	0	D_2
5	40-49	who	rad	0	0	1	D_1
6	50-59	who	per	0	1	1	D_1

Table 3. A Table for D_1

	age	loc	nat	prod	nau	M1	class
1	50-59	occ	per	0	0	1	m.c.h.
2	40-49	who	per	0	0	1	m.c.h.
5	40-49	who	rad	0	0	1	m.c.h.
6	50-59	who	per	0	1	1	psycho

In the sixth step, classification rules for D_1 and D_2 are induced from Table 2. For example, the following rules are obtained for D_1.

$[M1 = yes] \rightarrow D_1$ $\alpha = 1.0$, $\kappa = 1.0$, supported by $\{1,2,5,6\}$
$[prod = no] \rightarrow D_1$ $\alpha = 1.0$, $\kappa = 1.0$, supported by $\{1,2,5,6\}$
$[nau = no] \rightarrow D_1$ $\alpha = 1.0$, $\kappa = 0.75$, supported by $\{1,2,5\}$
$[nat = per] \rightarrow D_1$ $\alpha = 1.0$, $\kappa = 0.75$, supported by $\{1,2,6\}$
$[loc = who] \rightarrow D_1$ $\alpha = 1.0$, $\kappa = 0.75$, supported by $\{2,5,6\}$
$[age = 50 - 59] \rightarrow D_1$ $\alpha = 1.0$, $\kappa = 0.5$, supported by $\{2,6\}$

In the seventh step, classification rules for m.c.h. and psycho are induced from Table 3. For example, the following rules are obtained from m.c.h..

$[nau = no] \rightarrow m.c.h.$ $\alpha = 1.0$, $\kappa = 1.0$, supported by $\{1,2,5\}$
$[age = 40 - 49] \rightarrow m.c.h.$ $\alpha = 1.0$, $\kappa = 0.67$, supported by $\{2,5\}$

In the eighth step, these two kinds of rules are integrated in the following way. For a rule $[M1 = yes] \rightarrow D_1$, $[nau = no] \rightarrow m.c.h.$ and $[age = 40 - 49] \rightarrow m.c.h.$ have a supporting set which is a subset of $\{1,2,5,6\}$. Thus, the following rules are obtained:

$[M1 = yes]$ & $[nau=no] \rightarrow m.c.h.$ $\alpha = 1.0$, $\kappa = 1.0$, supported by $\{1,2,5\}$
$[M1 = yes]$ & $[age=40-49] \rightarrow m.c.h.$ $\alpha = 1.0$, $\kappa = 0.67$, supported by $\{2,5\}$

5 Experimental Results

The above rule induction algorithm is implemented in PRIMEROSE4 (Probabilistic Rule Induction Method based on Rough Sets Ver 4.0), [5] and was applied to databases on differential diagnosis of headache, meningitis and cerebrovascular diseases (CVD), whose precise information is given in Table 4. In these experiments, δ_α and δ_κ were set to 0.75 and 0.5, respectively. [6]

Table 4. Information about Databases

Domain	Samples	Classes	Attributes
headache	1477	20	20
meningitis	198	3	25
CVD	261	6	27

This system was compared with PRIMEROSE [15], C4.5 [12], CN2 [4], AQ15 and k-NN [7] with respect to the following points: length of rules, similarities between induced rules and expert's rules and performance of rules.

In this experiment, length was measured by the number of attribute-value pairs used in an induced rule and Jaccard's coefficient was adopted as a similarity measure, the definition of which is shown in the Appendix. Concerning the performance of rules, ten-fold cross-validation was applied to estimate classification accuracy.

Table 5 shows the experimental results, which suggest that PRIMEROSE4 outperforms the other four rule induction methods and induces rules very similar to medical experts' ones.

---

[5] The program is implemented by using SWI-prolog [14] on Sparc Station 20.
[6] These values are given by medical experts as good thresholds for rules in these three domains.
[7] The most optimal k for each domain is attached to Table 5.

Table 5. Experimental Results

Method	Length	Similarity	Accuracy
		Headache	
PRIMEROSE4	8.6 ± 0.27	0.93 ± 0.08	93.3 ± 2.7%
Experts	9.1 ± 0.33	1.00 ± 0.00	98.0 ± 1.9%
PRIMEROSE	5.3 ± 0.35	0.54 ± 0.05	88.3 ± 3.6%
C4.5	4.9 ± 0.39	0.53 ± 0.10	85.8 ± 1.9%
CN2	4.8 ± 0.34	0.51 ± 0.08	87.0 ± 3.1%
AQ15	4.7 ± 0.35	0.51 ± 0.09	86.2 ± 2.9%
k-NN (7)	6.7 ± 0.25	0.61 ± 0.09	88.2 ± 1.5%
		Meningitis	
PRIMEROSE4	2.6 ± 0.19	0.91 ± 0.08	92.0 ± 3.7%
Experts	3.1 ± 0.32	1.00 ± 0.00	98.0 ± 1.9%
PRIMEROSE	1.8 ± 0.45	0.64 ± 0.25	82.1 ± 2.5%
C4.5	1.9 ± 0.47	0.63 ± 0.20	83.8 ± 2.3%
CN2	1.8 ± 0.54	0.62 ± 0.36	85.0 ± 3.5%
AQ15	1.7 ± 0.44	0.65 ± 0.19	84.7 ± 3.3%
k-NN (5)	2.3 ± 0.41	0.71 ± 0.33	83.5 ± 2.3%
		CVD	
PRIMEROSE4	7.6 ± 0.37	0.89 ± 0.05	91.3 ± 3.2%
Experts	8.5 ± 0.43	1.00 ± 0.00	92.9 ± 2.8%
PRIMEROSE	4.3 ± 0.35	0.69 ± 0.05	84.3 ± 3.1%
C4.5	4.0 ± 0.49	0.65 ± 0.09	79.7 ± 2.9%
CN2	4.1 ± 0.44	0.64 ± 0.10	78.7 ± 3.4%
AQ15	4.2 ± 0.47	0.68 ± 0.08	78.9 ± 2.3%
k-NN (6)	6.2 ± 0.37	0.78 ± 0.18	83.9 ± 2.1%

k-NN (i) shows the value of i which gives the highest performance in k ($1 \leq k \leq 20$).

6 Discussion

6.1 Focusing Mechanism

One of the most interesting features in medical reasoning is that medical experts make a differential diagnosis based on focusing mechanisms: with several inputs, they eluminate some candidates and proceed into further steps. In this elimination, our empirical results suggest that grouping of diseases are very important to realize automated acquisition of medical knowledge from clinical databases. Readers may say that conceptual clustering or nearest neighborhood methods(k-NN) [1, 13] will be useful for grouping. However, those two methods are based on classification accuracy, that is, they induce grouping of diseases, whose rules are of high accuracy. Their weak point is that they do not reflect medical reasoning: focusing mechanisms of medical experts are chiefly based not on classification accuracy, but on coverage.

Thus, we focus on the role of coverage in focusing mechanisms and propose an algorithm on grouping of diseases by using this measure. The above experiments show that rule induction with this grouping generates rules, which are similar to medical experts' rules and they suggest that our proposed method should capture medical experts' reasoning.

6.2 Precision for Probabilistic Rules

In the above experiments, the thresholds δ_α and δ_κ for selection of inclusive rules were set to 0.75 and 0.5, respectively. Although this precision contributes to the reduction of computational complexity, this methodology, which gives a threshold in a static way, causes a serious problem. For example, there exists a case when the accuracy for the first, the second, and the third candidate is 0.5, 0.49, and 0.01, whereas accuracy for other classes is almost equal to 0. Formally, provided an attribute-value pair, R, the following equations hold: $\alpha_R(D_1) = 0.5, \alpha_R(D_2) = 0.49, \alpha_R(D_3) = 0.01$, and $\alpha_R(D_i) \approx 0 (i = 4, \cdots, 10)$. Then, both of the first and the second candidates should be suspected because those accuracies are very close, compared with the accuracy for the third and other classes. However, if a threshold is statically set to 0.5, then this pair is not included in positive rules for D_2. In this way, a threshold should be determined dynamically for each attribute-value pair. In the above example, an attribute-value pair should be included in positive rules of D_1 and D_2.

From discussion with domain experts, it is found that this type of reasoning is very natural, which may contribute to the differences between induced rules and the ones acquired from medical experts. Thus, even in a learning algorithm, comparison between the whole given classes should be included in order to realize more plausible reasoning strategy.

Unfortunately, since the proposed algorithm runs for each disease independently, the above type of reasoning cannot be incorporated in a natural manner, which causes computational complexity to be higher. It is our future work to develop such interacting process in the learning algorithm.

6.3 Rough Mereology

This research is closely related with rough mereology, which introduces rough inclusion [11]. Whereas rough mereology firstly applies to distributed information systems, its essential idea is rough inclusion: Rough inclusion focuses on set-inclusion to characterize a hierachical structure based on a relation between a subset and superset. On the other hand, in this paper, we focus on a coverage, a probability of a relation R under the condition of D ($P(R|D)$) and examine the set-inclusion relationship in terms of coverage, which is deterministic.

Although the relations between this research and rough mereology have not been examined yet, the experimental results suggest that medical experts also incorporate a decision process, which is similar to rough mereology. It will be our future work to formulate this empirical study from the viewpoint of rough mereology and extend our approach from deterministic inclusion to indeterministic one.

7 Conclusion

In this paper, the characteristics of experts' rules are closely examined, whose empirical results suggest that grouping of diseases are very important to realize automated acquisition of medical knowledge from clinical databases. Thus, we focus on the role of coverage in focusing mechanisms and propose an algorithm on grouping of diseases by using this measure. The above experiments show that rule induction with this grouping generates rules, which are similar to medical experts' rules and they suggest that our proposed method should capture medical experts' reasoning. Interestingly, the idea of this proposed procedure is very similar to rough mereology. The proposed method was evaluated on three medical databases, the experimental results of which show that induced rules correctly represent experts' decision processes and also suggests that rough mereology may be useful to capture medical experts' decision process.

References

1. Aha, D. W., Kibler, D., and Albert, M. K., Instance-based learning algorithm. *Machine Learning*, **6** (1991) 37–66.
2. Breiman, L., Freidman, J., Olshen, R., and Stone, C., *Classification And Regression Trees*. Wadsworth International Group, Belmont, 1984.
3. Buchnan, B. G. and Shortliffe, E. H., *Rule-Based Expert Systems*. Addison-Wesley, New York, 1984.
4. Clark, P. and Niblett, T., The CN2 Induction Algorithm. *Machine Learning* **3** (1989) 261–283.
5. Everitt, B. S., *Cluster Analysis*. 3rd Edition, John Wiley & Son, London, 1996.
6. Langley, P. *Elements of Machine Learning*. Morgan Kaufmann, CA, 1996.
7. Mannila, H., Toivonen, H., Verkamo, A.I., Efficient Algorithms for Discovering Association Rules. In: *Proceedings of the AAAI Workshop on Knowledge Discovery in Databases (KDD-94)*, AAAI press, Menlo Park (1994) 181–192.
8. Michalski, R. S., A Theory and Methodology of Machine Learning. *Machine Learning - An Artificial Intelligence Approach*. In: Michalski, R.S., Carbonell, J.G. and Mitchell, T.M., (eds.) Morgan Kaufmann, Palo Alto, 1983.
9. Michalski, R. S., Mozetic, I., Hong, J., and Lavrac, N., The Multi-Purpose Incremental Learning System AQ15 and its Testing Application to Three Medical Domains. In *Proceedings of the fifth National Conference on Artificial Intelligence*, AAAI Press, Menlo Park (1986) 1041–1045.
10. Pawlak, Z., *Rough Sets – Theoretical Aspects of Reasoning about Data*. Kluwer Academic Publishers, Dordrecht, 1991.
11. Polkowski, L. and Skowron, A., Rough mereology: a new paradigm for approximate reasoning. Intern. J. Approx. Reasoning **15** (1996) 333–365.
12. Quinlan, J.R., *C4.5 - Programs for Machine Learning*, Morgan Kaufmann, Palo Alto, 1993.
13. Shavlik, J. W. and Dietterich, T.G. (eds.), *Readings in Machine Learning*. Morgan Kaufmann, Palo Alto, 1990.
14. SWI-Prolog Version 2.0.9 Manual, University of Amsterdam, 1995.

15. Tsumoto, S. and Tanaka, H., PRIMEROSE: Probabilistic Rule Induction Method based on Rough Sets and Resampling Methods. *Computational Intelligence* **11** (1995) 389-405.
16. Tsumoto, S., Empirical Induction of Medical Expert System Rules based on Rough Set Model. PhD dissertation, 1997 (in Japanese).
17. Ziarko, W., The Discovery, Analysis, and Representation of Data Dependencies in Databases. In: Shapiro, G. P. and Frawley, W. J., (eds.), *Knowledge Discovery in Databases*, AAAI Press, Menlo Park (1991) 195-209.
18. Ziarko, W., Variable Precision Rough Set Model. *Journal of Computer and System Sciences* **46** (1993) 39-59.

A Comparison of Rule Length

Table 6 shows comparison between induced rules and medical experts' rules with respect to the number of attribute-value pairs used to describe. The most important difference is that medical experts' rules are longer than induced rules for diseases of high prevalence.

Table 6. Comparision of Rule Length between Induced Rules and Medical Experts' Rules

Disease	Samples	PRIMEROSE	RHINOS
Muscle Contraction Headache	923	3.00	9.00
Disease of Cervical Spine	163	5.50	3.50
Common Migraine	112	4.00	7.50
Psychological Headache	79	6.67	3.67
Tension Vascular Headache	79	11.00	10.50
Classical Migraine	49	4.50	9.00
Teeth Disease	21	3.25	6.00
Costen Syndrome	19	4.00	3.00
Sinusitus	11	4.50	5.00
Neuritis of Occipital Nerves	5	10.00	14.00
Ear Disease	5	8.50	7.00
Intracranial Mass Lesion	2	2.75	3.75
Intracranial Aneurysm	2	4.00	2.00
Autonomic Disturbance	1	5.25	3.50
Trigeminus Neuralgia	1	5.25	3.50
Inflammation of Eyes	1	6.00	8.00
Arteriosclerotic Headache	1	9.50	11.00
Herpes Zoster	1	3.00	1.00
Tolosa-Hunt syndrome	1	6.00	4.00
Ramsey-Hunt syndrome	1	3.00	7.00
Total	1477		

B Similarity Measure

PRIMEROSE4 calculates the following similarity measure from all the inputs. Although there are many kinds of similarities [5], a family of similarity measures based on a contingency table is adopted. Let us consider a contingency table for a rule of a certain disease (Table 6). The first and second column denote the positive and negative information of an experts' rule. The first and second row denote the positive and negative information of an induced rule. Then, for example, a denotes the number of attributes in an induced rule which matches an experts' rule. From this table, several kinds of similarity measures can be de-

Table 7. Contigency Table for Similarity

		Rule	
	1	0	Total
Sample 1	a	b	a+b
Sample 0	c	d	c+d

fined. The best similarity measures in the statistical literature are four measures shown in Table 7. In PRIMEROSE4, users can choose a similarity measure from

Table 8. Definition of Similarity Measures

(1) Matching Number	a
(2) Jaccard's coefficient	$a/(a+b+c)$
(3) χ^2-statistics	$N(ad-bc)^2/M$
(4) point correlation coefficient	$(ad-bc)/\sqrt{M}$
$N = a+b+c+d$, $M = (a+b)(b+c)(c+d)(d+a)$	

these four. As a default, Jaccard's coefficient, is used for defining similaritites, because it satisfies not only the low computational complexity, but also a good performance.

Modelling Customer Retention with Statistical Techniques, Rough Data Models, and Genetic Programming

A. E. Eiben[1] T. J. Euverman[2] W. Kowalczyk[3]* F. Slisser[4]

[1] Department of Computer Science, Leiden University
e-mail: gusz@wi.leidenuniv.nl

[2] Faculty of Economics and Econometry, University of Amsterdam
e-mail: teije@fee.uva.nl

[3] Faculty of Mathematics and Computer Science, Vrije Universiteit Amsterdam
De Boelelaan 1081A, 1081 HV Amsterdam, The Netherlands
e-mail: wojtek@cs.vu.nl

[4] VISOR, B.V., The Netherlands
e-mail: slisser@visor.nl

Abstract

This paper contains results of a research project aiming at modelling the phenomenon of customer retention. Historical data from a database of a big mutual fund investment company have been analyzed with three techniques: logistic regression, rough data models, and genetic programming. Models created by these techniques were used to gain insights into factors influencing customer behaviour and to make predictions on ending the relationship with the company in question. Because the techniques were applied independently of each other, it was possible to make a comparison of their basic features in the context of data mining.

1 Introduction

Banks, as many other companies, try to develop a long-term relationship with their clients. When a client decides to move to another bank it usually implies some financial loses. A climbing defection rate is namely a sure predictor of a diminishing flow of cash from customers to the company–even if the company replaces the lost customers–because older customers tend to produce greater cash flow and profits. They are less sensitive to price, they bring along new customers, and they do not require any acquisition or start-up costs. In some industries, reducing customer defections by as little as five percents can double profits, Reichheld (1996). Customer retention is therefore an important issue.

---
* The corresponding author

To be able to increase customer retention the company has to be able to predict which clients have a higher probability of defecting. Moreover, it is important to know what distinguishes a stopper from a non-stopper, especially with respect to characteristics which can be influenced by the company. Given this knowledge the company may focus their actions on the clients which are the most likely to defect, for example, by providing them extra advice and assistance. One way of obtaining such knowledge is analysis of historical data that describe customer behaviour in the past.

In our research, which was carried out in a cooperation with a big mutual fund investment company[1], we have analyzed a fragment of a database containing information about more than 500.000 clients. In our analysis we have used three different techniques: logistic regression, e.g., Hair et al. (1995), rough data modelling, Kowalczyk (1998b), and genetic programming, Koza (1992).

Logistic regression is a well-known, "classical" method of analyzing data and requires no further explanations.

Rough data models have been introduced recently by Kowalczyk (1998b, 1996). They consist of a simple partitioning of the whole data set, an ordering of elements of this partition and some cumulative performance measures. In a sense, rough data models can be viewed as an extension of the concept of rough classifiers, Lenarcik and Piasta (1994), Kowalczyk and Piasta (1998).

Genetic Programming, introduced by Koza (1992), is also a relatively new technique which is based on evolutionary principles. It aims at finding complex expressions which describe a given data set as good as possible (with respect to a predefined objective criterion).

All the techniques were applied to the same data set independently, providing us, in addition to the main objective of the project (analysis of retention), a unique opportunity of comparing their various features (accuracy, comprehensibility of results, speed, etc.).

Our research was carried out in different phases, similarly to an earlier project, Eiben et al. (1996):

1. defining the problem,
2. designing conceptual models,
3. acquiring and arranging data,
4. exploratory data analysis,
5. building models by three techniques,
6. analysis and interpretation of the obtained models.

The organization of this paper reflects the order of these steps. In the next section we describe the problem and the available data. Sections 3-5 describe results obtained by the three techniques. In section 6 we discuss all the results and compare the techniques.

---

[1] For confidentiality reasons we are not allowed to disclose the name of this company. Also other details like the exact meaning of some variables and their actual values are not given.

2 Problem and Data Description

The company collects various data about their clients since many years. On the basis of these historical data we were supposed to investigate the following issue:

> *What are the distinguishing (behavioural) variables between investors that ended their relationship with the company (stoppers) from investors that continued (non-stoppers) and how well can different techniques predict that investors will stop the relation within the next month.*

The company offers at this moment about 60 different investment forms which attract customers with different profiles. Due to this diversity of clients and investment forms we had to restrict our research to a homogeneous group of clients that invest money in a specific form. In particular, we have focused on clients which were "real investors" (i.e., clients which had only a simple savings account or a mortgage were not considered). Further, we restricted our attention to clients that stopped their relation between January 1994 and February 1995 (14 possible "stop months"). These restrictions led to a data set with about 7.000 cases (all stoppers). As we were interested in discriminating stoppers from non-stoppers, the data set has been extended by about 8000 "non-stopper" cases. Each record in the dataset contained the history of a single client over a period of 24 months before the moment of stopping (for non-stoppers dummy "stop-moments" were generated at random). By a "history" we mean here sequences (of length 24) of values which represent various measurements like the number of transactions, monthly profit, degree of risk, etc. In addition to these "dynamic" attributes also some "static" attributes were stored, e.g., client's age, starting capital, etc. The dataset we finally extracted from databases consisted of 15.000 records, each record having 213 fields. Some of the most relevant variables are listed in Table 1.

3 Statistical analysis of data

When statistical methods are applied on very large data sets, the emphasis is on the explanatory significance rather than statistical significance. For example, a correlation of .001 can be statistically significant without having any explanatory significance in a sufficiently large data set. Statistical estimation becomes computation of meaningful statistics. In this section we shall describe the sequence of actions which we undertook in order to arrive at an intelligible model.

3.1 Data reduction

We separately analysed each set of dynamic attributes trying to reduce the number of variables in each set. Using growth curve analysis, see, e.g., Timm (1975), we tried to discover different average polynomial trends for the two groups of stoppers and non-stoppers in order to retain only those that were discriminating between groups. Necessary for the existence of such salient components is a sufficiently large multivariate difference between the group averages. A meaningful statistic is Wilks' $1 - \Lambda$, which ranges from zero to one. It may be conceived as a multivariate generalisation of $1 - R^2$. A value near zero means that the difference between groups is negligible compared to the differences within groups. In that case there is no gain in information when two groups are distinguished instead of envisaging just one group of clients with stoppers and non stoppers mixed together. The values of $1 - \Lambda$ of the sets of dynamic variables ranged from .003 to .06. These meaningless magnitudes led us to consider only aggregate values over time of each set.

3.2 Univariate exploration

Inspection of the outcomes of standard data exploration techniques has led us to categorise some of the aggregated variables in order to enhance the interpretabily. For this categorisation we took into consideration the distributional characteristics as well as the domain of content. For example, the following five variables were categorised as follows:

investments
One category was formed for clients for which this variable did not change over time. The remaining five categories were based on quintiles.

risk
One category was formed for clients for which this variable did not change over time. Quintiles were used for the remaining clients.

number of transactions A
A two-category variable was constructed based on the number of transactions of type A.

number of transactions B
A two-category variable similar to the preceding one.

funds
A variable with 6 categories was formed on the basis of the number of funds in which the client invested his/her money.

We verified the meaningfulness of the recodings in two ways. First, by a visual inspection of the corresponding relative frequency bar charts. Second, by performing a two-dimensional correspondence analysis, Krzanowski (1993), in order to study the placing of the categories in one-dimensional space in relation to stoppers and non-stoppers categories.

3.3 Simple logistic regressions

We carried out twelve logistic regressions, one for each variable. We used three indices for judging the results:

1. $R^2_{logistic\ regression}$, which can be interpreted as the proportional reduction of the lack-of-fit by incorporating the variable of interest above a model based only on the intercept parameter, Agresti (1990). It is defined as:

$$R^2_{logistic\ regression} = 1 - \frac{log(likelihood_{intercept+variable})}{log(likelihood_{intercept})}.$$

2. $\lambda_{logistic\ regression}$, which can be interpreted as the proportional reduction of errors in classification by incorporating the variable of interest above a model with only an intercept. A model with only an intercept classifies all clients in the group with the largest observed frequency, being the non-stoppers in the present case. This means that all stoppers are misclassified and regarded as errors (see, e.g., Menard (1995)). So it is defined as:

$$\lambda_{logistic\ regression} = \frac{\#errors_{intercept} - \#errors_{intercept+variable}}{\#errors_{intercept}}.$$

3. $\gamma_{logistic\ regression}$, which can be interpreted as a measure of ordinal association between the predicted probabilities of being a stopper and actually being a stopper. The measure is widely used for cross-tabulations. It was proposed by Goodman and Kruskal (1954). It measures a weak monotonicity and ignores ties. It is easily interpretable as it ranges from -1 to +1. It is based on the number of concordant pairs C and discordant pairs D. A pair (non-stopper, stopper) is concordant when the predicted probability for a non-stopper is lower than for a stopper and disconcordant in the opposite case (see, e.g., Coxon (1982)), for a discussion on measures for association). It is defined as:

$$\gamma_{logistic\ regression} = \frac{C - D}{C + D}.$$

An overview of the results is given in Table 1.

Except of the categorised aggregated dynamic variables the table displays bad results for R^2 and λ. These figures gave us reasons to try to improve the results. We decided to do a quintile-categorisation for all variables for which R^2 and λ were zero in two decimals. A quintile-categorisation of the duration of relation B did not produce interpretative results in the sense that the differences between the values were too small to justify an interpretation in a scale running, for example, from "extremely short" to "extremely long". The results of the categorisations are displayed in Table 2. As we can see some slight improvements were achieved.

3.4 Multiple logistic regression

In the final model we used only one profit and one emotional variable in order to prevent redundancy in the model and for reasons of interpretation. We used *profit B* and *emotion index A* because they had the best performance in the univariate regressions. Consequently, ten variables were put into the logistic

Table 1. Some variables and their significance.

	R^2	λ	γ
investments	.07	.10	.42
risk	.05	.11	.35
transactions A	.02	.00	.32
transactions B	.08	.18	.59
funds	.04	.05	.37
profit A	.00	.00	.18
profit B	.00	.00	.11
emotion index A	.00	.00	.31
emotion index B	.00	.00	.34
duration of relation A	.02	.00	.15
duration of relation B	.00	.00	.06
starting capital	.00	.00	.31

Table 2. Results of quintile categorization.

	R^2	λ	γ
profit A	.02	.00	.24
profit B	.01	.00	.14
emotion index A	.05	.14	.31
emotion index B	.06	.17	.37
duration of relation A	.01	.00	.13
starting capital	.04	.07	.31

regression model. The use of categories for most of the variables means that choices for interesting contrasts were made possible. We do not go into details here, because this article is mainly meant to compare different methods of modelling on the same data set. It is possible to improve (slightly) the predictive power of logistic regression, but the main advantage of classical techniques in the present context is to build an interpretative model with sufficient predictive power. Note that individual statistical significance of each variable or their categories is of less importance, although one can state that non-significant results for such large data sets may cast doubts on the importance about the inclusion of the variable. The following values of the indices were obtained for the training set: $R^2 = .15, \lambda = .25, \gamma = .52$. Another useful statistic is the Spearman's correlation coefficient ρ between the predicted probabilities and the observed relative frequencies. We categorised the predicted probabilities using percentiles. Within each category the observed relative frequency was computed and based on these 100 values the value of ρ was calculated. For the training set $\rho = .974$. For the validation set we obtained the following values for the various statistics: $\lambda = .24, \gamma = .53, \rho = .923$.

4 Analysis of data with Rough Data Models

This section contains a brief presentation of the concept of Rough Data Models, Kowalczyk (1998b, 1996), and results our experiments. A more detailed description of these experiments can be found in Kowalczyk and Slisser (1997).

4.1 Rough Data Models

Informally, a Rough Data Model consists of a collection of clusters that form a partition of the data set, some statistics calculated for every cluster (e.g., cluster size, number of elements of specific type), and a linear ordering on clusters. This ordering is supposed to reflect cluster importance and is used for calculating various cumulative performance measures. To define the concept of RDM more formally we need some notation and terminology used in the theory of rough sets, Pawlak (1991). Let us consider a decision table

$$\mathbf{T} = (\mathbf{U}, \mathbf{A}, \mathbf{d}),$$

where U is a finite collection of objects (the universe), $A = \{a_1, \ldots, a_k\}$ is a set of attributes on U, i.e., every a_i is a function from U into a corresponding set of attribute values V_i, $a_i : U \to V_i$, for $i = 1, \ldots, k$, and d is a decision function which takes values in a finite set of decisions $D = \{d_1, \ldots, d_n\}$, $d : U \to D$. Elements of U are often called *patterns* and associated decision values *types*, thus if $d(u) = d_1$ then u is called a pattern of type d_1. Let R denote the indiscernibility relation which is defined by the set of attributes A, i.e., for any $u_1, u_2 \in U$, $R(u_1, u_2)$ iff $a_i(u_1) = a_i(u_2)$, for $i = 1, \ldots, k$. The relation R determines a partition of U into a number of (pairwise disjoint) equivalence classes $C_1, \ldots, C_m$, which will further be called *clusters*. Every cluster may contain elements of different types. However, elements that belong to the same cluster are, by definition, not distinguishable, so they will be classified (by any classifier) as elements of the same type. Therefore, any classifier is determined by assigning to every cluster C its type, $class(C)$, which is an element of D. Given a partitioning of the universe and a classification function $class$, a number of useful parameters which characterise clusters can be introduced:

- cluster size, $size(C_i)$, which is just the number of elements of C_i,
- number of elements of a given type, $size(C_i, d_j)$, which is the number of elements of type d_j that are members of C_i,
- number of correctly classified elements, $corr(C_i)$, which is the number of elements of C_i which are of type $class(C_i)$,
- cluster accuracy, $accuracy(C_i)$, defined as the ratio $corr(C_i)/size(C_i)$.

These parameters can be used for ranking clusters according to some, user specified, criteria. For example, clusters might be ordered according to their size (the bigger the better), according to their accuracy or according to the percentage of elements of specific type.

Now we can formally define a *rough data model* of a decision table $\mathbf{T} = (U, A, d)$ as a triple:

$$\mathbf{T} =< C, class, \leq >,$$

where

- C is a set of clusters,
- $class : C \to D$ is a function that assigns to every cluster its type,
- $\leq$ is a linear ordering on C.

Performance of rough data models can be measured in many different ways, Kowalczyk (1996). In addition to some problem independent measures like cumulative accuracy, gain curves, response curves, etc., one can introduce problem specific measures, for example, the percentage of elements of specific type in "best" (in sense of the $\leq$ relation) clusters which cover 10% of all cases.

There are two important features of RDMs:

1. there are almost no restrictions on the form of performance measure which is used for evaluating model quality; this measure is defined by the user and is problem dependent,
2. computational complexity of generating RDMs is very low (linear in the size of the data set); this feature allows for exploring huge number of alternative RDMs and focusing on these models that optimise the given performance criterion.

In practice, the process of generating high quality models consists of three major steps:

1. formulation of a performance measure that should be optimised (e.g., classification rate, percentage of correctly classified cases of the given type in specific fragment of the model, total misclassification cost, etc.).
2. determination of a search space, i.e., a collection of models which should be searched to find an optimal one (for example, a collection of models which are based on k attributes which are taken from a set of n attributes, or a collection of models determined by various discretization procedures, etc.)
3. determination of a search procedure (for example, exhaustive search, local search, branch & bound, etc.)

Usually, rough data models are used as an efficient tool which helps to get an insight into data sets. The user first specifies some objective function, then proposes a number of data transformations, formulates some restrictions on model complexity (e.g., "the model should be based on at most four attributes") and then models which satisfy all these criteria are automatically generated and evaluated. In spite of its simplicity, this approach often provides models which have relatively high accuracy, Kowalczyk (1998a).

4.2 Retention and Rough Data Models

To get some idea about the importance and relationships between various attributes a number of standard tests were carried out. First of all, we have generated numerous plots which are routinely used in statistical data analysis: frequency histograms, means, density estimates, etc., see Hair et al. (1995). Visual inspection of these plots led to the discovery of a large group of clients (4809) which behaved differently from the rest. Therefore, we decided to split the whole data set into two subsets and analyse them independently. We will refer to both groups as to A-clients and B-clients. In order to identify most important attributes we have calculated, for every attribute, values of three importance measures: correlation coefficients, coefficients of concordance and information gain. Correlation coefficients measure linear dependency between attributes, are widely used and require no further explanations. Coefficient of concordance (sometimes called the CoC index or just the c index) measures the degree of similarity of an ordering (of all cases) which is induced by values of the measured attribute and the ordering induced by the decision attribute. Information gain measures the amount of information provided by a (discrete-valued) attribute and is explained in Quinlan (1986). As a result of this analysis we have identified 8 attributes which were used as the basis for construction of RDMs. To guide the search process we had to specify some performance measure that should be optimized. After some discussions with bank experts we took as our objective function the percentage of stoppers that can be found in the top 10% of cases.

We have restricted our attention to models that were based on all combinations of 2, 3 or 4 attributes taken from the set of 8 important attributes mentioned above. Each attribute has been discretized into 5 intervals, according to the "equal frequency" principle. Unfortunately, a model which is based on 4 variables which are discretized into 5 intervals may have 5*5*5*5=625 clusters–too many to expect good generalisation. Therefore, we allowed each attribute to be split into 3 intervals only; ends of these intervals were taken from the 6 points which were determined by the discretization into 5 intervals. Thus every attribute could be partitioned into 15 ways, which leads to 15*15*15*15=50.625 various models which are based on 4 variables. Moreover, there are 70 ways of selecting 4 attributes out of 8, so the total number of models based on 4 variables is about 3.5 million; adding models which are based on 2 or 3 variables does not increase this figure too much. Due to computational simplicity of RDMs we could systematically generate all these models, evaluate them and select the best one. It turned out that the performance of best models which were based on 4 attributes was almost the same as of models based on 3 attributes. Moreover, in both groups of models there were several models which were very close to the optimal ones. All these models have been carefully analysed on basis of their performance curves and the structure of clusters.

Clusters, together with their definitions (formulated in terms of values of attributes which determine them) can be used for formulating some rules about the data. For example, the best cluster from the model of B-clients captured clients who were investors for a long time, invested money in funds with very small risk, and got small profits-all of them have stopped their relation with the company. Clearly, a detailed analysis of all clusters provided a good insight into customer behaviour.

Additionally, the models have been tested on an independent validation set in order to evaluate their generalisation capabilities. Not surprisingly (models based on 3 attributes had only 27 clusters), they generalised very well (performance dropped less than 1%).

4.3 Rule extraction

As mentioned above, clusters which are determined by best models can be directly translated into decision rules. However, such rules do not cover large fragments of the model. In order to identify some general rules we have run a systematic search algorithm which generated rules in the form

if $(a < X_1 < A) \& (b < X_2 < B) \& (c < X_3 < C)$ **then** *decision*

(where X_1, X_2 and X_3 are attribute names and $a, A, \ldots, c, C$ are some numbers), and tested them in terms of the number of covered cases and accuracy. The search process was restricted to rules such that:

1. attributes X_1, X_2 and X_3 were arbitrary combinations of attributes taken from the set of 8 most important attributes,
2. splitting points $a, A, \ldots, c, C$ were determined by an "equal frequency" discretization of the corresponding attributes into 7 intervals: they could be chosen from the set of ends of these intervals,
3. rules were allowed to involve only 2, 3, 4 or 5 "splitting points".

Out of several million rules generated in this way (only for the group of B-clients) we have focused on rules which were "interesting" in the following sense: they had to cover at least 10% of all cases and had accuracy at least 80% (i.e., at least 80% of all cases which were covered by the rule had to be "stopper" cases). The resulting collections of rules were relatively small (1, 24, 98 and 132 rules which involved 2, 3, 4 and 5 splitting points, resp.). A similar collection of rules has been found for A-clients. All rules have been carefully analysed by experts and their analysis led to the discovery of some interesting patterns in customer behaviour.

5 Genetic Programming

Genetic Programming is a new search paradigm which is based on evolutionary principles, Koza (1992). Potential solutions (individuals) are represented by (usually complex) expressions which are interpreted as definitions of functions (models). The quality of individuals (fitness) is measured by evaluating performance of the corresponding models (e.g., classification rate). The search process mimics the evolution: a collection of individuals (population) "evolves" over time and is subjected to various genetic operators. In our research we used the system OMEGA, Walker at al. (1995).

OMEGA is a genetic programming system that builds prediction models in two phases. In the first phase a data analysis is performed, during which several statistical methods are carried out to find the variables with the highest individual predictive power. In the second phase the genetic modelling engine initializes with a targeted first generation of models making use of the information obtained in the first phase. After initialization, further optimization then takes place with a fitness definition stated in terms of practical objectives.

In modelling applications the most frequently applied measure for evaluating the quality of models is 'accuracy', which is the percentage of cases where the model correctly fits. When building models for binary classification the so-called CoC measure is a better option for measuring model quality. The CoC (Coefficient of Concordance) actually measures the distinctive power of the model, i.e., its ability to separate the two classes of cases, see Walker at al. (1994) for the definition. Using the CoC prevents failures caused by accepting opportunistic models. For instance, if 90% of the cases to be classified belongs to class A and 10% to class B, a model simply classifying each case as A would score 90% accuracy. When the CoC is used, this cannot occur. Therefore, we used the the CoC value of the models as fitness function to create good models. Let us note that selection of the best model happened by measuring accuracy (on the test set), and that the final comparison between the four different techniques was also based on the accuracy of the models. Yet, we decided to use the CoC in order to prevent overfitting and some control experiments (not documented here) confirmed that evolving models with a fitness based on accuracy results in inferior performance.

As stated previously, OMEGA builds prediction models in two phases. The first phase, a data analysis, selects the variables that are of interest out of all supplied variables. Of these selected variables a summarisation of their performance measured in CoC is shown in Table 3.

Table 3. Values of CoC for 6 variables.

duration relation A	.60
duration relation B	.62
start capital	.65
funds	.63
investments	.71
risk	.71

By the data analysis carried out in the first phase OMEGA is able to create a good initial population by biasing the chances of variables to be included in a tree. The initial models were generated into 2 populations of 20 expressions each. During this initialization, the 40 models varied in performance from .72 till

.80 measured in CoC. Computing the accuracy of the best model on the training set gave the value of .73. The relatively good CoC values after initialization show an increased joint performance compared to the best individual variable performance (which was .71 for *investments*). This is due to the use of special operators that act on the variables and the optimised interactions between the selected variables in the models. Till sofar, no genetic optimization has taken place and several satisfactory models have been found.

During the genetic search we were using 0.5 crossover rate, 0.9 mutation rate and 0 migration rate between the two sub-populations, that is no migration took place. The two populations were only used to maintain a higher level of diversity. The maximum number of generations was set to 2000. After the genetic search process the best performing tree had a CoC value of .81 and a corresponding accuracy of .75 on the training set. For this particular optimization problem the genetic optimization phase does not show a dramatic improvement in performance as the initialization did.

6 Analysis of the results

In this section we evaluate the outcomes from two perspectives. Firstly, we will concentrate on the original problem of modelling customer behavior. Secondly, we compare the predictive power of our genetically created model to the best models the other techniques obtained. In this analysis other models are competitors of the genetic model.

6.1 Interpretation of the models

When evaluating the results it is important to keep in mind that the company providing the data is not only interested in good predictive models, but also in conclusions about the most influential variables. Namely, these variables belong to customer features that have the most impact on customer behaviour. In case of models built with logistic regression the interpretation of results was straightforward: we simply had to look into coefficients involved in these models. Rough data models were also easy to interpret: the meaning of cluster characteristics and extracted rules was obvious. The situation was a bit more complicated with genetic programming. The resulting models (complex expressions) were very difficult to interpret. Therefore we performed a sensitivity analysis on the variables involved in the generated models by fixing all but one variables of the best model and varying the value of the free variable through its domain. The changes in the performance of the model are big for very influential variables, and small for variables with less impact. The results of the sensitivity analysis are given in Table 4, where a high value indicates a high influence, while lower values show a lower impact.

Comparing the interpretations of the different techniques, i.e., (dis)agreement on the importance of variables we observed a high level of agreement. Based on this observation it was possible to bring out a well-founded advice for the

Table 4. Results of sensitivity analysis.

duration relation B	4.32%
duration relation A	14.37%
start capital	13.50%
funds	2.88%
investments	18.56%
risk	82.35%

company that specific risk values in the portfolio substantially raises the chance of ending the business relationship. This conclusion allows the company to adapt its policy and perform directed actions in the form of advising customers to change the risk value of their portfolio.

6.2 Comparison of applied techniques

The three techniques can be compared to each other with respect to various criteria. In our study we have focussed on the following aspects: accuracy of the generated models, their interpretability, time needed for their construction and expertise required by each technique.

Accuracy. Model accuracy was measured by creating cumulative response rate tables for each technique. The results are presented in Figure 1. It can be noticed that the model provided by rough data modelling outperforms the other two models on the first 6 percentiles. On the other hand, on the remaining percentiles the model constructed with genetic programming outperforms the other two. Logistic regression provided a model which is slightly worse than the two others. It should be noticed that the best rough data model was based on 3 variables only, in contrast to 9 variables used in the model generated by logistic regression and 24 variables used by the "genetic" model. Finally, let us note that the applied techniques were focused on optimizing 3 different objective criteria: response rate in the first 10 percentiles (rough data modelling), the CoC factor (genetic programming), and the mean squared error (logistic regression).

Interpretability. In all cases it was possible to provide an interpretation of the generated model. Logistic regression provided a list of most influential variables (or their combinations) together with their weights. The OMEGA system generated a complex formula that was too difficult to interpret. Instead, a sensitivity analysis provided a list of most significant variables. Rough Data Models provided the best insight into the analyzed data. They provided a lot of information about meaningful combinations of variables, lists of significant clusters and explicit rules.

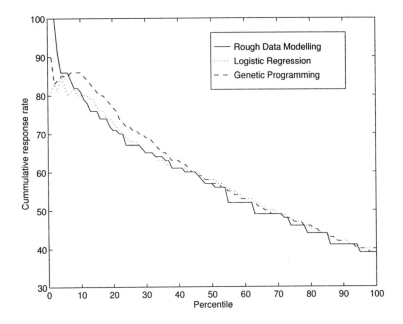

Fig. 1. Predictive power of the best models created by 3 techniques.

Time. The computer time needed for generating models was different for different techniques. Calculations necessary for logistic regression took about 30 minutes, but a lot of conceptual work (about one week) was needed first. The time required by the other two techniques was significantly longer (a few days). Precise comparison is not possible because both systems (OMEGA and TRANCE) were running on different computers (PC and UltraSparc).

Necessary expertise. In case of logistic regression an extensive statistical knowledge and some acquaintance with a statistical package (in our case: SPSS) was indispensable. Almost no knowledge was necessary for building rough data models. However, due to the current status of the TRANCE system (a prototype implemented in MATLAB), a lot of programming work (scripting) and knowledge of the MATLAB system were necessary. In contrast, the use of the OMEGA system is very simple. This commercial tool has a friendly user interface and experiments can be run by a user with no extensive knowledge of genetic programming.

7 Concluding Remarks

In this paper we described an application oriented research project on applying different modelling techniques in the field of marketing. Our conclusions and recommendations can be summarized as follows.

- Cross-validation on the most influential variables based on models developed with other techniques raises the level of confidence. In our project we observed good agreement between conclusions of the three approaches.
- Non-linear techniques such as genetic programming and rough data modelling proved to perform better than linear ones with respect to the predictive power of their models on this problem. This observation is in full agreement with the outcomes of an earlier comparative research on a different problem, see Eiben et al. (1996).
- It is advisable to use CoC as fitness measure in the GP. Control runs with accuracy as fitness led to models that had a worse accuracy than those evolved with CoC.
- Simple statistical data analysis can distinguish powerful variables. Using this information during the initialization of the GP works as an accelerator, by creating a relatively good initial population.
- It is somewhat surprising that even a long run of the GP could only raise the performance of the initial population by approximately 2% in terms of accuracy on the training set. Note, however, that in financial applications one percent gain in predictive performance can mean millions of dollars in cost reduction or profit increase.

The global evaluation of the project is very positive. We have gain a good insight into the phenomenon of retention and constructed models with satisfactory accuracy. The company in question highly appreciated obtained results and decided on further research which should lead to an implementation of a system that could be used in the main business process.

Ongoing and future research concerns adding new types of variables, validation of the models for other time periods as well as developing models for a longer time horizon.

References

Agresti, A. *Categorical Data Analysis*, John Wiley & Sons, New York, 1990.

Coxon, A.P.M. *The User's Guide to Multidimensional Scaling*, Heinemann, London, 1982.

Goodman, L.A. and Kruskal, W.H. Measures of Association for cross-classifications, Journal of the American Statistical Association, **49**, pp. 732–764, **54**, pp. 123–163, and **58**, pp. 310–364.

Eiben, A.E., T.J. Euverman, W. Kowalczyk, E. Peelen, F. Slisser, and J.A.M. Wesseling. Comparing Adaptive and Traditional Techniques for Direct Marketing, in H.-J. Zimmermann (ed.), In *Proceedings of the 4th European Congress on Intelligent Techniques and Soft Computing*, Verlag Mainz, Aachen, pp. 434-437, 1996.

Haughton, D. and S. Oulida, Direct marketing modeling with CART and CHAID, In Journal of direct marketing, volume 7, number 3, 1993.

Hair, J.F., Jr., Anderson, R. E., Anderson, T.R.L., Black, W. *Multivariate Data Analysis* (fourth edition), Prentice Hall, Englewood Cliffs, New Jersey, 1995.

Hosmer, D.W., and L. Lemeshow. *Applied logistic regression*, John Wiley & Sons, New York, 1989.

Kowalczyk, W. and Piasta, Z. Rough set-inspired approach to knowledge discovery in business databases. In X. Wu, R. Kotagiri, and K. B. Korb (eds.), *Research and Development in Knowledge Discovery and Data Mining. Proceedings of the 2nd Pacific-Asia Conference on Knowledge Discovery and Data Mining*, PAKDD-98, Melbourne, Australia, Lecture Notes in AI 1394 (1998), Springer, pp. 186-197.

Kowalczyk, W. and Slisser, F. Analyzing customer retention with rough data models. In J. Komorowski and J. Zytkow (eds.), *Proceedings of the 1st European Symposium on Principles of Data Mining and Knowledge Discovery*, PKDD'97, Trondheim, Norway, Lecture Notes in AI 1263 (1997), Springer, pp. 4–13.

Kowalczyk, W. TRANCE: A tool for rough data analysis, classification and clustering. In S. Tsumoto, S. Kobayashi, T. Yokomori, H. Tanaka and A. Nakamura (eds.), *Proceedings of the Fourth International Workshop on Rough Sets, Fuzzy Sets, and Machine Discovery, RSFD'96*, Tokyo University, (1996), pp. 269–275.

Kowalczyk, W. An Empirical Evaluation of the Accuracy of Rough Data Models. In *Proceedings of the 7th International Conference on Information Processing and Management of Uncertainty in Knowledge-based Systems*, IPMU'98, Paris, La Sorbonne, (1998), pp. 1534–1538.

Kowalczyk, W. Rough Data Modeling: a new technique for analyzing data. In: L. Polkowski and A. Skowron (eds.) *Rough Sets in Knowledge Discovery*, Physica–Verlag, 1998.

Koza, J. *Genetic Programming*, MIT Press, 1992.

Krzanowski, W.J. *Principles of Multivariate Analysis: A User's Perspective*, Clarendon Press, Oxford, 1993.

Lenarcik, A. and Piasta, Z. Rough classifiers. In: Ziarko, W. (ed.), *Rough Sets, Fuzzy Sets and Knowledge Discovery*, Springer-Verlag, London, (1994), pp. 298–316.

Magidson, M. Improved statistical techniques for response modeling. In *Journal of direct marketing*, volume 2, number 4 (1988).

Menard, S. *Applied Logistic Regression Analysis*, Sage, London, 1995.

Pawlak, Z. *Rough Sets: Theoretical Aspects of Reasoning About Data*, Kluwer Academic Publishers, Dordrecht, 1991.

Quinlan, R. Induction of decision trees, *Machine Learning* 1, 81-106 (1986).

Reichheld, F.F. Learning from Customer Defections, in *Harvard Business Review*, March-April (1996).

Timm, N.H. *Multivariate Analysis with Applications in education and psychology*, Brooks/Cole, Monterey, 1975.

Walker, R., Barrow, D., Gerrets, M., and Haasdijk, E. Genetic Algorithms in Business. In J. Stender, E. Hillebrand and J. Kingdon (eds.), *Genetic Algorithms in Optimisation, Simulation and Modelling*, IOS Press, 1994.

Walker, R., Haasdijk, E., and Gerrets, M. Credit Evaluation Using a Genetic Algorithm. In S. Goonatilake and P. Treleaven (eds.), *Intelligent Systems for Finance and Business*, John Wiley & Sons, 1995.

Rough-Fuzzy Application

A Rough Sets Approach to Assessing Software Quality: Concepts and Rough Petri Net Models

James F. Peters III[1] and Sheela Ramanna[2]

[1] Computational Intelligence Laboratory, Department of Electrical and Computer Engineering, University of Manitoba, Winnipeg, R3T 2N2, Canada,
[2] Department of Business Computing, University of Winnipeg, Winnipeg, R3B 2E9, Canada and Department of Electrical and Computer Engineering, University of Manitoba, Winnipeg, R3T 3E2, Canada

Abstract

This paper introduces an approximate reasoning system for assessing software quality. Based on observations concerning software quality and the granulations of measurements in an extended form of McCall software quality measurement framework, an approach to deriving rules about software quality is given. Quality decision rules express relationships between evaluations of software quality criteria measurements. Software quality assessment is carried out in the context of a decision system of the form (U, A ∪ {d}), where U is a universe of objects, A is a set of attributes which map objects to values, and d is a special form of attribute called a decision. The approach is illustrated relative to software usability assessments for a collection of software projects. A rough Petri net model of the basic structure of a software quality decision system is also given. The contribution of this paper is the introduction of an approach to designing software quality decision systems.
Keywords: Decision system, discernability, fuzzy sets, information granule, metric, Petri net, rough sets, software quality.

1 Introduction

Software quality enhancement is a continuous process, which is usually guided by some form of decision process. Decisions concerning aspects of software requiring improvement are based on assessments of the degree to which software possesses particular quality attributes. It should be noted that software quality improvements commonly result from an evolution of software through gradual rather than radical change [1]-[2]. Quality itself is identified with the degree to which a system, component or process satisfies specified requirements [3]. Software quality is defined in terms of high-level attributes called factors and lower-level attributes called criteria (see Table 1). A software attribute is a characteristic of the software such as usability or maintainability. Typically, factors are assessed in terms of criteria measurements [4]-[10]. Metrics are used to assess quality attributes. A software metric maps software data to a value representing the degree to which software possesses a particular quality attribute. It has been observed that software quality is a user-oriented rather than a developer-oriented property of software, and that there is an absence of an adequate measure of software quality [11]. Software quality criteria and factors tend to have a cause-effect relationship. That is, a decision about the degree to which a software product possesses a particular factor results from the aggregation of software criteria measurements. Operational factors (reliability, efficiency, correctness, integrity and usability), revision factors (maintainability, flexibility and testability), and transition factors (portability, interoperability and reusability) are determined by lower level criteria. For example, the correctness factor can be measured relative to measures of traceability, completeness, consistency criteria.

Frameworks for software quality measurement have reached a high degree of

Table 1 Quality Factors and Criteria

Quality factors (effect)	Criteria (cause)
correctness: extent that software satisfies its requirements.	traceability, completeness, consistency
reliability: frequency of occurrence of software problems [11].	error tolerance, consistency, accuracy, simplicity, % of cleanroom engineering
maintainability: effort required to locate and fix a software error.	consistency, conciseness, simplicity, modularity, number of comments, low cyclomatic complexity, understandability, modularity, scalability
testability: effort required to ensure that software performs intended its functions	simplicity, modularity, instrumentation, self-descriptiveness
efficiency: amount of resources that software requires.	execution efficiency, storage efficiency
integrity: extent of control of unauthorized access.	access control, access audit
usability: effort required to learn, operate, prepare input, and interpret output of a software system.	operability, training, friendliness, efficiency, understandability, durability, human engineering, productivity, system maintenance, syntax flexibility, adaptability, ease-of-learning, familiarity, error recovery, and helpfulness, communicativeness
portability: effort required to transfer software between platforms.	software system independence, machine independence, self-descriptiveness, modularity
interoperability: effort required to couple software systems.	communications commonality, data commonality
reusability: extent to which software modules can be used in different applications.	self-descriptiveness, generality, modularity, software system independence, machine independence

maturity and detail [4]-[10], [12]-[18]. This study is based on an extension of the McCall software quality measurement framework, which forms a hierarchy: software product → factors → criteria → metrics [1]-[5]. The McCall framework points to relationships between factors, criteria and metrics and serves as a valuable aid in organizing software quality assessment. However, there is no formal model for the McCall framework, and there is no provision for investigating granulations of measurements or dependencies between quality assessments. The weighting formulas for each factor in the McCall framework have the form $w_1 c_1 + w_2 c_2 + ... + w_n c_n$, where $w_1, ..., w_n$ are weights and $c_1, ..., c_n$ are criteria. A weighting formula measures the aggregative effect of weighted criteria. In the McCall framework, relationships between a factor such as usability and criteria such as operability, training, communicativeness, input/output volume, input/output rate are identified. An assigned weight reflects the contribution of a criterion to a factor goal. A weakness in the approach to software quality evaluation described in volume II in [1] is that coefficients in weighting formulas for factors such as usability are derived intuitively or experimentally and the intertwining of dependencies among criteria in weighting formulas are not considered. In addition, the approximate character of software quality measurements is not taken into account. Finally, the McCall framework does not solve the problem that many individual quality criteria are in conflict with each other [19]. For example, added efficiency sometimes diminishes understandability. Again, for instance, conciseness can conflict with understandability and human engineering. The need to unravel the dependencies among criteria measurements in arriving at an assessment of the quality of software provides a rather natural basis for a software quality decision system in the context of rough sets [20]-[30]. The contribution of this paper is the introduction of an approach to designing software quality decision systems.

This paper is structured as follows. In Section 2, granulation of software criteria measurements and the distribution of quality scores are considered. The methodology for an approximate reasoning system for software quality are presented in Section 3. The basic concepts of fuzzy sets and rough sets needed to establish an approximate reasoning system for software quality are also presented in Section 3. An illustration of the construction of a sample decision system relative to measurements of software usability criteria is given in Section 4. Rough Petri nets and an extension of rough Petri nets called rough fuzzy Petri nets are given in Section 5, which also presents a high-level model of a software quality rule generation system.

2 Imprecision in Measuring Quality

Two common approaches to assessing ratings of software quality measurements are either to display the criteria scores graphically in a Kiviat diagram (see Fig. 1) or to compute a weighted average [15]. To begin the evaluation process with a Kiviat diagram, each of the criteria are rated on a scale from 0 to 100. Quality is then assessed in terms of the uniformity (or lack of uniformity) of the stellated areas of the diagram, and observed vertex values relative to minimum values (e.g., score = 30) and estimated maximum values (e.g., score = 90) specified by the lengths of the radii of the inner and outer circles. A sample Kiviat diagram is shown in Fig 1. In Fig. 1, system maintenance has a "low" score because it lies inside the inner circle (i.e., it falls below the minimum value), whereas scores for efficiency, understandability, durability and human engineering have "medium" scores because they lie within the band bounded by minimum and (estimated) maximum scoring ranges.

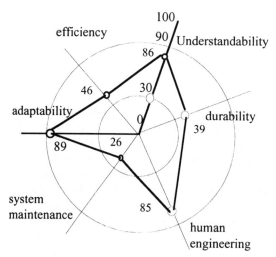

Fig. 1 Sample Kiviat Diagram

Not surprisingly, efficiency has a considerably lower score than understandability and human engineering, which tend to work against efficiency by consuming both time and space. In the quantitative approach, weighted averages provide a quantitative scoring of the usability factor of a software project. Let w1, ..., w6 be weights for system maintenance (c1), adaptability (c2), efficiency (c3), understandability (c4), durability (c5), human engineering (c6) usability criteria, respectively. In Table 2, a sample evaluation of the usability factor is given for a particular software module computed as a weighted average of criteria scores shown in Fig. 1.

Table 2 Usability Factor Measurement

w1 c1	w2 c2	w3 c3	w4 c4	w5 c5	w6 c6	usability $\frac{\sum_{i=1}^{n} w_i c_i}{\sum_{i=1}^{n} w_i}$
0.7 26	.85 90	0.5 46	0.9 86	0.8 39	0.9 85	302.8/4.65 = 65.1187

Using the same scoring scheme shown in the Kiviat diagram in Fig. 1, the usability quality factor of 65.1187 in Table 2 has a "medium" value, since it lies between the estimated minimum and maximum values. However, in both the Kiviat diagram and weighted average interpretation of quality, no consideration is given to interactions between and possible dependencies among quality criteria.

It should also be observed that there is evidence that indicates that one cannot expect to measure software quality exactly [8, 19]. With either a Kiviat diagram or a pure numerical assessment, there is a certain level of imprecision. In other words, there is a certain amount of vagueness, non-crispness, uncertainty, approximation concerning our knowledge of software quality in assessing usability. As Cavano and McCall observe, every software quality measure is partially imperfect [8]. Software quality measurements tend to be inexact, approximate. The imprecise character of software quality measurements leaves the way open to an approximate reasoning approach to assessing software quality. In this paper, the approach to assessing software quality is limited to development of decision systems based on observations concerning efficiency, understandability, and human engineering criteria.

2.1 Granulating Quality Assessments

There is some justification for evaluating quality assessments in the context of information granules, a clustering of measurements with vaguely defined boundaries (see Table 3 and Fig. 2). An information granule consists of a "clump" of similar values [31, 32].

Table 3. Usability Criteria Raw Scores
Legend: e= efficiency, u =understandability, h=human engineering

The point-plots (i.e., scattergrams) of the raw scores in Table 3 are given in Figures 2(a), 2(b) and 2(c). It is possible to granulate quality criteria measurements, where similar quality measurements are "clumped together" in granules to provide a form of computing with words [17]. For example, in considering software understandability, a quantitative assessment is made relative to information granules with labels low, medium, and high. Sample raw scores for three usability criteria (efficiency, understandability, and human engineering) are given in Table 3. The next step is to granulate each collection of measurements for a particular quality factor criterion. We illustrate this in terms of the granulation of efficiency scores in Fig. 2(a). A graphical representation of efficiency granules is shown in Fig. 3. Each of the efficiency raw scores in a particular granule are evaluated in terms of its degree-of-membership in the granule relative some distribution. For example, a sketch of three overlapping Gaussian distributions are shown in the left-hand scale of Fig. 3. Then an efficiency score of 23 belongs to the granule named "low" in Fig. 3, and this score has a degree of membership of 0.95, approximately, in this granule.

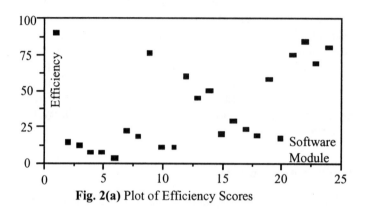

Fig. 2(a) Plot of Efficiency Scores

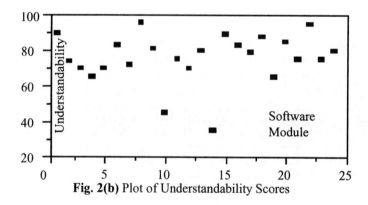

Fig. 2(b) Plot of Understandability Scores

2.2 Distribution of Quality Scores

It is assumed that quality criteria raw scores for evolving software modules are approximately normally distributed. Hence, the degree-of-membership of a quality criterion score in a granule is computed relative to a particular Gaussian

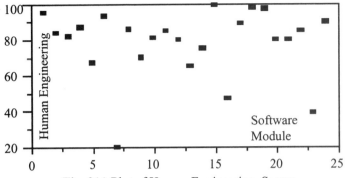

Fig. 2(c) Plot of Human Engineering Scores

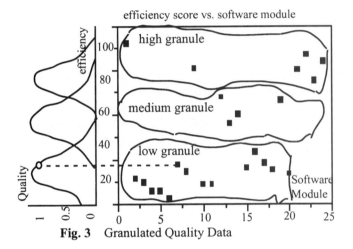

Fig. 3 Granulated Quality Data

distribution of the form given for the granule labeled low (1)

$$low(x) = \exp\left[\frac{-(x-m)^2}{s^2}\right] \qquad (1)$$

The mean m and variance s^2 in (1) determine the modal point and spread in the graph of degrees of membership of quality data. The modal point m and spread s of the Gaussian distribution in (1) will vary depending on the distribution of the data in a granule. For example, let $m_{low} = 20$, $m_{med} = 45$, $m_{high} = 85$ be the modal points for three quality granules (each with s = 15) named low, med, and high, respectively. The membership functions of low, medium, and high granules are defined as follows:

$$low(x) = e^{\left[\frac{-(x-20)^2}{15^2}\right]}, med(x) = e^{\left[\frac{-(x-45)^2}{15^2}\right]}, high(x) = e^{\left[\frac{-(x-85)^2}{15^2}\right]}$$

Quality assessments are also aggregated together relative to an information granule labeled acceptable (not shown in Fig. 4). This approach to software quality measurements provides the basis for designing an information system in

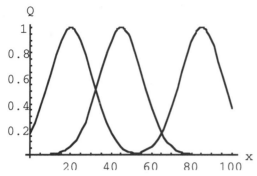

Fig. 4 Quality Measurement Distributions

the context of rough sets [9]-[11] and the derivation of rules representing dependencies among evaluation values in an information systems table [12]-[19]. Based on observations concerning software quality factors during stages in the life cycle and computation of degrees of acceptability derived from quality metrics, a framework for a software quality decision system is constructed. A certain amount of preprocessing of software criteria measurements is performed in the context of fuzzy sets and logic processing in constructing a quality decision system. This preprocessing is done to provide a mechanism for calibrating the strengths-of-connections of criteria to reflect the intentions of system designers.

3 Approximate Reasoning

The basic concepts of fuzzy sets and rough sets underlying the creation of an approximate reasoning system for making decisions about software quality are presented in this section. In the construction of quality decision systems, there are a variety of sensors which are sources of system inputs. Sensors aggregate information about software quality, and provide a means of drawing conclusions about quality. Sensors are modeled with operators from fuzzy set theory. A complete framework for approximate reasoning about software quality is provided by rough sets theory. Dependencies among sensor computations relative to decisions about software quality are expressed as rules. Rough sets theory provides a means of deriving rules from software quality decision system tables. Derived rules provide a basis for planning changes to software as part of a continuous effort to improve software quality. A high-level flowchart of the methodology underlying the approximate reasoning system for software quality is given in Fig. 5a. Recall that a flowchart specifies operations, data (input and output), and control flow needed to compute a result. The basic flowchart symbols are given in Fig. 5b. In the flowchart in Fig. 5b, software quality criteria measurements are accumulated and granulated. Next, models of sensors needed to aggregate information about quality criteria measurements are selected. This modeling is performed concurrently relative to selected granules.

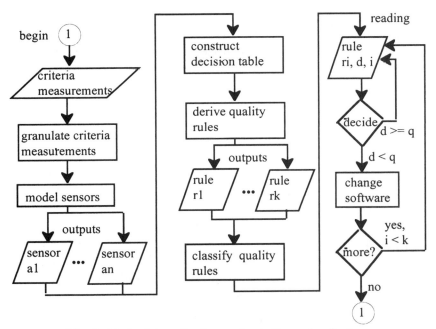

Fig. 5a Methodology for Approximate Reasoning System

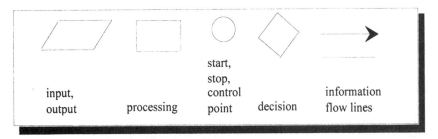

Fig. 5b Basic Flowcharting Symbols

Once the sensor modeling has been completed, it is possible to construct a quality decision system table for a particular set of criteria measurements. Rough sets theory then makes it possible to derive quality decision rules. Concurrent derivation of rules is performed relative to what are known as reducts. After classifying derived rules, decisions about changes to software are made relative to each class of rules. This process is continuous. That is, the completion of one set of software changes begins a new cycle in the pursuit of software quality.

The task named "change software" in Fig. 5a requires a methodology which is tailored to the needs, requirements engineering, design and quality engineering of particular software systems. Although this task is not considered in this paper, it should be mentioned that the problem of rule-driven changes to software can be solved stepwise by an agent which coordinates the concurrent actions of other agents. An agent is modeled as a discrete event system capable of communicating with other agents and its environment as in [59]. Agents communicate with each other over hidden channels. The architecture of a coordinator agent would include a blackboard architecture designed to identify and schedule software changes. A blackboard architecture is a knowledge-based form of repository appropriate in applications requiring cooperative problem-solving either by virtual minds or human minds or both [60]. A typical blackboard system has a structure like the one shown in Fig. 6. A blackboard has three basic components

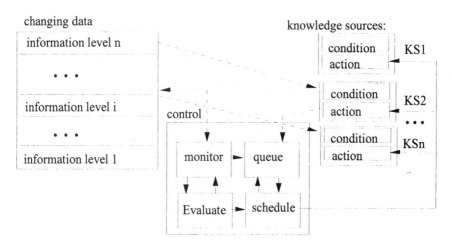

Fig. 6 Basic Blackboard Architecture

· Knowledge sources: independent processes whose actions are triggered by satisfaction of a particular conditions.
· Blackboard: repository of problem-solving state data (organized in an application-dependent hierarchy: level n (highest), ..., level 1 (lowest)).
· Control: monitors information in the blackboard, maintains permissible combinations of knowledge source activations, schedules pending Knowledge-Source Activations (KSAs), evaluates local, problem-solving goals, and executes scheduled KSAs to solve some problem specific to the blackboard.
The activity of a knowledge source is event-driven. Each change in the blackboard is an event which can satisfy the condition of one or more knowledge sources. Whenever the condition in a knowledge source is satisfied, this adds a KSA record to a queue, and eventually triggers an action by the knowledge source. Briefly, we describe how a blackboard would function in scheduling software changes. The information levels in Fig. 7 are designed relative to sensor activation levels extracted by a coordinator agent from conditions of rules in a software quality decision system. The derivation of software quality rules is described in this article. The KSAs in a software quality in the blackboard of software quality coordinator agent are tuples of the form

(degreeCriterionQ $<$ limit $\wedge$ req$_1 \wedge$ ... $\wedge$ req$_n$,
initiate_change(criterion_i, (limit - degreeCriterionQ))

The blackboard schedules initiate_change( ) actions whenever degreeCriterionQ < limit and each of the requirements in req$_1$ ∧ ... ∧ req$_n$ are satisfied. The assumption here is that each knowledge source has been designed to resolve conflicts between criteria (conflict resolution is expressed by rcq$_1$ ∧ ... ∧ req$_n$). In a largescale software development project, each agent (other than the coordinator) would maintain a dynamically-changing decision system for some part of the system being developed. Using the methods described in this article, each agent would derive its own set of decision rules, classify and analyze derived rules, and communicate the results to the coordinator agent. A formal model for a coordinator agent which uses a blackboard to coordinate tasks in a multiagent system is given in [61].

3.1 Fuzzy Sets: Basic Concepts

Fuzzy sets are distinguished from the classical notion of a set (also called a crisp set) by the fact that the boundary of a fuzzy set is not precise [33]. The characteristic function for a set X returns a value indicating the degree of membership of an element x in X. For a crisp set, the characteristic function returns a value in $\{0, 1\}$. A fuzzy set is non-crisp, and was introduced by Zadeh [34]. By contrast with a crisp set, the characteristic function for a fuzzy set returns a value in $[0, 1]$. Let U, X, Ã, x be a universe of objects, subset of U, fuzzy set in U, and an individual object x in X, respectively. For a set X, $\mu_{\tilde{A}} : X \to [0,1]$ is a function which determines the degree of membership of an object x in X. A fuzzy set Ã is then defined to be a set of ordered pairs as in (2).

$$\tilde{A} = \{(x, \mu_{\tilde{A}}(x)) \mid x \in X\} \tag{2}$$

The set X is called the reference set. The counterpart of intersection and union (crisp sets) are the t-norm and s-norm operators in fuzzy set theory. For the intersection of fuzzy sets, the min operator was suggested by Zadeh [34], and belongs to a class of intersection operators (min, product, bold intersection) known as triangular or t-norms. A t-norm is a mapping $t : [0,1]^2 \to [0, 1]$. The algebraic sum (also called probabilistic sum) is commonly used for the union of fuzzy sets [34] as in (3). The probabilistic sum belongs to a class of union operators called triangular co-norms (or s-norms). An s-norm is a mapping $s : [0,1]^2 \to [0, 1]$. For example, let x, y belong to a fuzzy set Ã, and compute the s-norm relative to x and y as in (3)

$$\mu_{\tilde{A}}(x) \, s \, \mu_{\tilde{A}}(y) = \mu_{\tilde{A}}(x) + \mu_{\tilde{A}}(y) - \mu_{\tilde{A}}(x)\mu_{\tilde{A}}(y) \tag{3}$$

In fuzzy set theory, the symbol "→" denotes a multivalued implication operation. Many forms of implication are possible [35]-[37]. For simplicity, the Lukasiewicz and Gaines forms of implication are compared in (4) and (5).

$$(r \to x)_{Lukasiewicz} = \begin{cases} 1 - r + x & \text{if } r > x \\ 1 & \text{if } r \leq x \end{cases} \tag{4}$$

$$(r \to x)_{Gaines} = \begin{cases} \frac{x}{r} & \text{if } x < r \\ min(1, \frac{x}{r}) & \text{otherwise} \end{cases} \qquad (5)$$

In the case of Gaines, $r_i \to x_i = x/r$ for values of $x < r$ (see Fig. 7). Otherwise, $r_i \to x_i = min(1, \frac{x_i}{r_i})$, $r \leq x$, and the $\to$ is induced by the product operation. Hence, values of $r_i \to x_i$ rise smoothly along the r-axis to the 45° line. Notice, for example, that for $r = 0.5$ and $x = 0$, we have $r_i > x_i = 0$. By contrast, the Lukasiewicz form of implication $r_i \to x_i = 1 - r_i + x_i$ for $r_i \to x_i$ "ramps up" more rapidly than the Gaines form of implication (see Fig. 8). Otherwise for Lukasiewicz, $r_i \to x_i = 1$, $r \leq x$, the $\to$ has a constant value of 1. As a result, for $r = 0.5$ and $x = 0$, we have $r_i \to x_i = 0.5$.

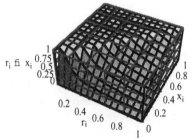

Fig. 7 $r_i \dashrightarrow x_i$ [Gaines]

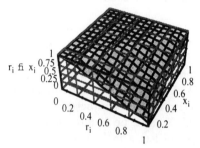

Fig. 8 $r_i \dashrightarrow x_i$ [Lukasiewicz]

3.2 Rough Sets: Basic Concepts

To begin, let $S = (U, A)$ be an information system with set U (universe of objects) and set A (attributes). Then let R be a relation defined on U. For x, y $\in$ U, let xRy indicate that x has relation R to y. R is a tolerance relation, if xRx (reflexivity) for x $\in$ U and for all x, y $\in$ U, if xRy, then yRx (symmetry). In the case where transitivity also holds, R is an equivalence relation over U. The notation U/R (known as the quotient set) denotes the family of equivalence classes of R. For x $\in$ U, the notation $[x]_R$ identifies an equivalence class in U/R. The equivalence class $[x]_R$ is called an elementary category or concept of R [20]. A subset X in U is called a reference set, which can be approximated with two other sets in (6) and (7).

$$\underline{R}X = \{x \in U \mid [x]_R \subseteq X\}, \textit{ lower approximation} \qquad (6)$$

$$\overline{R}X = \{x \in U \mid [x]_R \cap X \neq \phi\}, \textit{ upper approximation} \qquad (7)$$

The pair $(\underline{R}X, \overline{R}X)$ is a rough set with reference set X. The vagueness of a set stems from its borderline region. A measure of the accuracy of a set $X \subseteq U$ is computed using $\alpha_R(X)$ (see (8)).

$$\alpha_R(X) = \frac{|\underline{R}X|}{|\overline{R}X|}, \ accuracy\ measure \tag{8}$$

The accuracy measure $\alpha_R(X)$ captures the degree of completeness of our knowledge represented by X [26]. The degree of incompleteness of our knowledge represented by a set X (its roughness) is computed using $\rho_R(X)$ given in (9).

$$\rho_R(X) = 1 - \alpha_R(X), \ R - roughness\ of\ X \tag{9}$$

Similarity among members of an equivalence class provides the basis for what is known as the indiscernability relation. Let B be a subset of the set of attributes A, and let Ind(B) be the set of all elements of X that match each other relative to B (see (10)).

$$Ind(B) = \{(x,y) \mid \forall a \in B, a(x) = a(y)\} \tag{10}$$

The Ind(B) relation simplifies the investigation of a particular information system, where the representatives of U/Ind(B) are studied. Knowledge reduction is possible using the method shown in [26]. A minimal subset $B \subseteq A$ such that Ind(B) = Ind(A) is called a reduct of A. Any set of attributes has one or more reducts [38]. Let a $\in$ P in A. The attribute a is indispensable in P if Ind(P) $\neq$ Ind(P - {a}). The set of all indispensable attributes in P is called the core of P (denoted CORE(P)), which can be considered the most important part of knowledge [20]. For an information system S, the set of all reducts in S is denoted RED(S) [29]. In deriving decision system rules, the discernability matrix and discernabiliy function are essential [25]. Given an information system S = (U, A), the nxn matrix (cij) is called the discernability matrix of S (denoted M(S)) defined in (11).

$$c_{ij} = \{a \in A : a(x_i) \neq a(x_j)\}, for\ i,j = 1,...,n. \tag{11}$$

A discernability function fM(S) for information S is a boolean function of m boolean variables a_1^*, ..., a_m^* corresponding to attributes $a_1,...,a_m$ respectively, and defined in (12).

$$f_{M(S)}(a_1^*,...,a_m^*) =_{df} \wedge\{\vee c_{ij}^* \mid 1 \leq j < i \leq n, c_{ij} \neq \phi, c_{ij}^* = \{a^* \mid a \in c_{ij}\} \tag{12}$$

Precise conditions for decision rules can be extracted from a discernability matrix as in [23]. For the information system S = (U, A), let $B \subseteq A$ and let P (V_a) denote the powerset of V_a, where V_a is the value set of a. For every d $\in$ A - B, a decision function d_d^b : U $\to$ P (V_a) is defined in (13).

$$d_d^b(u) = \{v \in V_d \mid \exists u' \in U, (u,u') \in Ind_B \bullet d(u') = v\} \tag{13}$$

In other words, $d_d^b(u)$ is the set of all elements of the decision column of S such that the corresponding object is a member of the same equivalence class as argument u. The next step is to determine a decision rule with a minimal number of descriptors on the left-hand side. Pairs (a, v), where a $\in$ A, v $\in$ V are called descriptors. A decision rule over the set of attributes A and values V is an expression of the form given in (14).

$$a_{i_1}(u_i) = v_{i_1} \wedge ... \wedge a_{i_1}(u_i) = v_{i_1} \wedge ... \wedge a_{i_r}(u_i) = v_{i_r} \underset{S}{\Rightarrow} d(u_i) = v \qquad (14)$$

where $u_i \in U$, $v_{i_j} \in V_{a_{i_j}}$, $v \in V_d$, $j = 1,...,r$ and $r \leq |A|$. Let $\| \tau \|_S$ denote the meaning of term τ. A rule is true in system S if (15) holds.

$$\|(a_{i_1} = v_{i_1}) \wedge ... \wedge (a_{i_r} = v_{i_r})\| \subseteq \|(a_p = v_p)\| \qquad (15)$$

The fact that a rule is true is indicated by writing it in the form given in (16).

$$(a_{i_1} = v_{i_1}) \wedge ... \wedge (a_{i_r} = v_{i_r}) \underset{S}{\Rightarrow} (a_p = v_p) \qquad (16)$$

Let $R \in RED(S)$ be a reduct in the set of all reducts in an information system S. For an information system S, the set of decision rules constructed with respect to a reduct R is denoted OPT(S, R) [27]-[28]. Then the set of all decision rules derivable from reducts in RED(S) is the set in (17).

$$OPT(S) = \cup \{OPT(S, R) | R \in RED(S)\} \qquad (17)$$

4 Software Usability Measurements

By way of illustration of the derivation of rules useful in software quality assessments, a sample evaluation of a subset of possible usability criteria: efficiency (e), understandability (u), and human engineering (h). The evaluation is carried out relative to a collection of 24 Java programs. These criteria are represented in a decision system shown in Table 4.

Each object x_i in Table 3 is a 3-tuple (e_i, u_i, h_i) representing particular measurements of efficiency, understandability, and human engineering, respectively. The quality criteria scores for each tuple (e_i, u_i, h_i) provide input to sensors to compute the values in Table 4. Columns labeled a, b, c represent computations performed by sensors low_e, med_e, high_e which capture quality estimates in granules labeled low, med, and high. Let r_i, w_i, $g(x)$ be a modulator, strength-of-connection, and degree-of-membership function $g(x)$ for a quality estimate x in a particular granule g, respectively. Modulators and strengths-of-connections have the effect of hiding information which is not of interest to a designer and to highlight information considered crucial in deciding which features of a system need to be changed. Designer preferences affect decisions about quality.

Each sensor in Table 4 is modeled relative to the s-norm of a fuzzy implication $r_i \to g(x)$ and a strength-of-connection w_i. For example, the sensor named low_e ("low efficiency") in column a of Table 4 is modeled with the formula given in (18).

$$low\_e(low(x)) = ((r_i \to low(x)) \; s \; w_i) \qquad (18)$$

Table 4. Software Quality Information System

	a	b	c	d	e	f	g	h	i		k	l
	low_e	med_e	high_e	low_u	med_u	high_u	low_h	med_h	high_h			
m=	20.0	45.0	85.0	20.0	45.0	85.0	20.0	45.0	85.0			
s=	15	20	15	15	20	15	15	20	15			
r=	0.5	0.8	0.5	0.9	0.8	0.5	0.9	0.8	0.5			
w=	0.4	0.75	0.85	0	0.75	0.35	0	0	0.35		deg. of acceptance	Q
	0.400	0.750	0.970	0.000	0.750	0.870	0.000	0.000	0.617		0.442	0.442
	0.836	0.762	0.850	0.000	0.762	0.572	0.000	0.009	0.994		0.326	0.326
	0.740	0.758	0.850	0.000	0.758	0.438	0.000	0.014	0.950		0.155	0.155
	0.534	0.753	0.850	0.000	0.753	0.369	0.000	0.004	0.977		0.542	0.369
	0.534	0.753	0.850	0.000	0.753	0.438	0.000	0.220	0.386		0.429	0.386
	0.445	0.751	0.850	0.000	0.751	0.977	0.000	0.001	0.718		0.647	0.647
	0.979	0.798	0.850	0.000	0.798	0.495	1.000	0.142	0.350		0.671	0.350
	0.979	0.776	0.850	0.000	0.750	0.572	0.000	0.005	0.994		0.326	0.326
	0.400	0.762	0.923	0.000	0.754	0.914	0.000	0.142	0.438		0.155	0.155
	0.692	0.757	0.850	0.046	1.000	0.350	0.000	0.017	0.914		0.671	0.350
	0.692	0.757	0.850	0.000	0.765	0.617	0.000	0.007	1.000		0.442	0.442
	0.400	0.874	0.851	0.000	0.785	0.438	0.000	0.022	0.870		0.155	0.155
	0.402	1.000	0.850	0.000	0.795	0.870	0.000	0.287	0.369		0.542	0.369
	0.400	0.961	0.850	0.329	0.933	0.350	0.000	0.060	0.617		0.671	0.350
	1.000	0.785	0.850	0.000	0.751	0.914	0.000	0.000	0.464		0.040	0.040
	0.692	0.862	0.850	0.000	0.753	0.977	0.027	0.968	0.350		0.671	0.350
	0.954	0.805	0.850	0.000	0.757	0.822	0.000	0.002	0.914		0.805	0.805
	0.995	0.780	0.850	0.000	0.751	0.950	0.000	0.000	0.495		0.081	0.081
	0.400	0.897	0.850	0.000	0.822	0.369	0.000	0.000	0.531		0.542	0.369
	0.954	0.772	0.850	0.000	0.752	1.000	0.000	0.022	0.870		0.841	0.841
	0.400	0.765	0.912	0.000	0.765	0.617	0.000	0.022	0.870		0.442	0.442
	0.400	0.752	0.999	0.000	0.750	0.617	0.000	0.007	1.000		0.442	0.442
	0.400	0.791	0.865	0.000	0.765	0.617	0.168	0.894	0.350		0.671	0.350
	0.400	0.755	0.970	0.000	0.755	0.870	0.000	0.002	0.870		0.866	0.866

The implication $r_i \to low(x)$ is modeled with a Gaines form of fuzzy implication shown in (19).

$$(r_i \to low(x)) = min\left(1, \frac{low(x)}{r_i}\right) \qquad (19)$$

A graph of a sample computation for the Gaines form of implication is given in Fig. 9. For values of $x \in [0, 45]$, low(x) computed with m = 20, s = 15, and r = 0.5 from Table 4, the Gaines form of implication focuses on values of the ratio low(x)/0.5 such that low(x) is less than 0.5. In this sample quality assessment scheme, values of low(x) $\geq$ 0.5 are hidden (and ignored). That is, 0.5 $\to$ low(x) returns 1 whenever low(x) $\geq$ 0.5. The choice of an r-value serves as a threshhold,

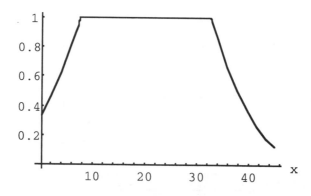

Fig. 9. Sample Gaines Graph

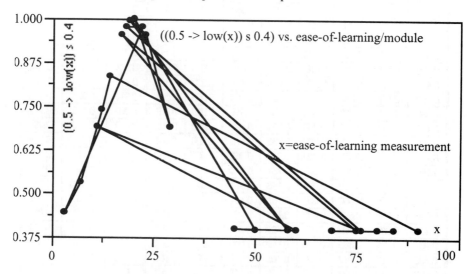

Fig. 10 Gaines Computation

and reflects the preferences of a software system designer. Notice that in the case where $r = 1$, then $1 \rightarrow \text{low}(x) = \text{low}(x)$. In Table 4, $r_i = 0.5$ and $w_i = 0.4$, and the s-norm operation is computed as a probabilistic sum for a specific Gaines computation (see Fig. 10).

The selection of values of modulator r and strength-of-connection w is vital in assessing the impact of software quality criteria measurements on the software quality enhancement process. The values of r and w in Table 4 would normally be derived during off-line training relative to selected training sets and corresponding target values in software quality assessments. Values in the degree-of-acceptance column (i.e., column k) in Table 4 are computed with a sensor modeled with the formula in (20).

$$deg-of-acceptance = |0.9 - MRE_{high}| \qquad (20)$$

where MRE_{high} stands for Magnitude of Relative Error of the degree of membership of the quality criteria scores in the granule named high, and provides an approach to error estimation. MRE_{high} is computed using the formula in (21).

$$MRE_{high} = \frac{|0.9 - \min(high\_e, high\_u, high-h)|}{\min(high\_e, high\_u, high-h)} \qquad (21)$$

The final column labeled l in Table 4 is a software quality decision column. Next, the reducts and core are derived from Table 4. Using Rosetta, the reducts and CORE(S) derived from Table 4 are given in Table 5.

Table 5. RED(S) and CORE(S)

reducts containing f (high_u)	reducts containing h (med_h)	reducts containing i (high_h)	reducts containing k (deg of accept.)	CORE(S)
{b, f} {c, e, f} {a, e, f, g} {f, h} {f, i}	{a, c, h} {a, e, h} {b, h}	{c, h, i} {a, c, i} {b, i} {c, d, i} {e, i}	{k}	{ }

4.1 Analysis of Rules

A total of 318 rules were derived for the software quality decision system represented by Table 4. Among these, the rule in (22)

$$a(0.995) \text{ } AND \text{ } e(0.751) \text{ } AND \text{ } f(0.950) \text{ } AND \text{ } g(0.000) \Longrightarrow l(0.081) \qquad (22)$$

is a deduction made in terms of row 18 of Table 3 (see Fig. 11). Analysis of attributes for rules relative to an assessment of low quality leads to a continuation of the software enhancement process and evolution of software reflected in Fig. 12 exhibiting the results software enhancement. Measurements from the software quality enhancement process lead to a new set of quality decision rules. The rule in (22) represented in line 18 of Table 4asserts that a software module with a combination of a very low efficiency score and high understandability score leads to a very low quality assessment. Remarkably, a software module with lower scoring human engineering and essentially the same evaluations as in module represented by line 18 of Table 4, receives a high software quality evaluation.

				attributes								
			a	b	c	d	e	f	g	h	i	k	l
			low_e	med_e	high_e	low_u	med_u	high_u	low_h	med_h	high_h		
	m=		20.0	45.0	85.0	20.0	45.0	85.0	20.0	45.0	85.0		
	s=		15	20	15	15	20	15	15	20	15		
	r=		0.5	0.8	0.5	0.9	0.8	0.5	0.9	0.8	0.5		
	w=		0.4	0.75	0.85	0	0.75	0.35	0	0	0.35		
Object													
	e	u	h										deg	Q
18	19	88	98	0.995	0.780	0.850	0.000	0.751	0.950	0.000	0.000	0.495	0.081	0.081
19	58	65	97	0.400	0.897	0.850	0.000	0.822	0.369	0.000	0.000	0.531	0.542	0.369
20	17	85	80	0.954	0.772	0.850	0.000	0.752	1.000	0.000	0.022	0.870	0.841	0.841

Fig. 11 Rows 18-20 of Table 4

This phenomenon is reflected in the rule in (23) with quality decision $l = 0.841$ in line 20 of Table 4.

$$a(0.954) \; AND \; e(0.752) \; AND \; f(1.000) \; AND \; g(0.000) \Longrightarrow l(0.841) \quad (23)$$

The raw scores (19, 88, 98) and (17, 85, 80) in lines 18 and 20 of Table 4 underly rules in (12) and (13), respectively. The raw scores for efficiency and understandability in lines 18 and 20 are almost the same.
row 1 2 3 4 5 6 7 8 9 10 11 12 13 14 15 16 17 18 19 20 21 22 23 24

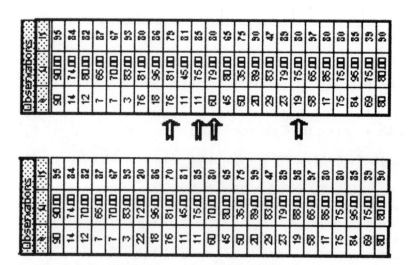

Fig. 12.. Evolving Raw Quality Criteria Scores

The apparent mystery in rules (22) and (23) is dispelled, if one takes into account the intentions of a software developer, namely, that nearly perfect human engineering in a software product is discouraged. By lowering the human engineering raw score from 98 to 80 in line 18, the quality assessment increases to $l = 0.442$. In other words, the quality assessment framework has been designed to reward less effort (and cost) being directed toward human engineering.

Rules showing dependencies among attributes leading to low software quality assessments provide indicators of weaknesses in a software module relative to the intentions expressed in a software quality assessment framework. Rightly or wrongly, the assessment of the quality of a software module sample framework represented by Table 4 pivots on the understandability criterion score. This can be seen in the following two rules in lines (24) and (25).

$$a(0.740) \; AND \; e(0.758) \; AND \; f(0.438) \; AND \; g(0.000) \Longrightarrow l(0.155) \qquad (24)$$

$$a(0.400) \; AND \; e(0.785) \; AND \; f(0.438) \; AND \; g(0.000) \Longrightarrow l(0.155) \qquad (25)$$

Notice that even though the efficiency and human engineering scores differ (namely, 76, 70 in line 9 are higher than scores 12, 82 in line 3 of Table 4), the modules represented by lines 3 and 9 receive the same quality assessment (see Fig. 13).

						attributes						
		a	b	c	d	e	f	g	h	i	k	l
		low_e	med_e	high_e	low_u	med_u	high_u	low_h	med_h	high_h		
m=		20.0	45.0	85.0	20.0	45.0	85.0	20.0	45.0	85.0		
s=		15	20	15	15	20	15	15	20	15		
r=		0.5	0.8	0.5	0.9	0.8	0.5	0.9	0.8	0.5		
w=		0.4	0.75	0.85	0	0.75	0.35	0	0	0.35		
Objects												
e	u	h									deg	Q
3	12	70	82	0.740	0.758	0.850	0.000	0.758	0.438	0.000	0.014	0.950	0.155	0.155
9	76	81	70	0.400	0.762	0.923	0.000	0.754	0.914	0.000	0.142	0.438	0.155	0.155

Fig 13 Sample Quality Decisions

4.2 Software Evolution

Based on what we learned about the intentions of the designer of the framework represented by Table 4, and by making appropriate changes in the software module evaluated in line 3 of Fig. 13, we can continue the software enhancement process. The "evolution" of the raw quality criteria scores are shown in Fig. 14. In fact, this is what will happen if changes are made in the software so that the understandability criterion score is increased by 10 points to achieve a new score of 80, and the scores of the other two criteria are left unchanged. It is easily verified that this leads to a software quality evaluation score of 0.841 (see Table 6). The lesson learned from rules (22) through (24) do not appear to carry over

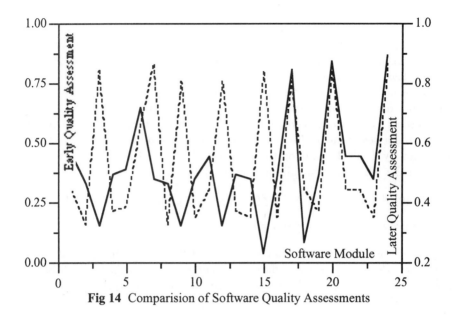

Fig 14 Comparision of Software Quality Assessments

in responding to rule (25). Notice that the first conjunct in rules (24) and (25) are quite different in value, where the raw score of 76 for the module in line 9 leads to the conjunct a = 0.40. By contrast, the raw score of 12 for the module in line 3 corresponds to a = 0.740 in rule (24). In other words, the software module represented by line 9 is more understandable than the module represented by line 3. The raw scores for a second generation decision system is given in Table 4.

Remarkably, substantial increases in the raw score for understandability in Table 4 do not change the quality decision. For example, changing the raw score for understandability in line 9 of Fig. 13 from 81 to 98 leaves the quality decision the same, namely, $l = 0.155$. However, by making appropriate changes in the software module represented by line 9 so that the human engineering raw score increases from 70 to 79, it can be verified that the quality decision changes to $l = 0.805$. This is a quite remarkable result. By continuing to make adjustments in software modules as a result of the analysis of the rules derived from the decision system in Table 4, a new decision system table is constructed. The plot in Fig. 14 shows sample quality assessments in 24 software modules resulting from selected changes triggered by derived quality rules. In Fig. 14, the solid line connects software quality assessments from Table 4 while the dotted line connects quality assessments found in Table 6. This approach to software quality assessment makes it possible for system designers to identify and evaluate dependencies between granulated quality measurements.

Table 6. Evolved Decision System Table

					attributes					
	a	b	c	d	e	f	g	h	i	k	l
	low e	med e	high e	low u	med u	high u	low h	med h	high h		
m=	20.0	45.0	85.0	20.0	45.0	85.0	20.0	45.0	85.0		
s=	15	20	15	15	20	15	15	20	15		
r=	0.5	0.8	0.5	0.5	0.8	0.5	0.5	0.8	0.5		
w=	0.4	0.75	0.85	0	0.75	0.35	0	0	0.35	deg of accept- ance	Q
	0.400	0.750	0.970	0.000	0.750	0.870	0.000	0.000	0.617	0.442	0.442
	0.836	0.762	0.850	0.000	0.762	0.572	0.000	0.009	0.994	0.326	0.326
	0.740	0.758	0.850	0.000	0.758	0.870	0.000	0.014	0.950	0.841	0.841
	0.534	0.753	0.850	0.000	0.753	0.369	0.000	0.004	0.977	0.542	0.369
	0.534	0.753	0.850	0.000	0.753	0.438	0.000	0.220	0.386	0.429	0.386
	0.446	0.751	0.850	0.000	0.751	0.977	0.000	0.001	0.718	0.647	0.647
	0.400	0.762	0.923	0.000	0.762	0.914	0.000	0.022	0.870	0.866	0.866
	0.979	0.776	0.850	0.000	0.750	0.572	0.000	0.005	0.994	0.326	0.326
	0.400	0.762	0.923	0.000	0.754	0.914	0.000	0.027	0.822	0.805	0.805
	0.692	0.757	0.850	0.045	1.000	0.350	0.000	0.017	0.914	0.671	0.350
	0.692	0.757	0.850	0.000	0.765	0.617	0.000	0.007	1.000	0.442	0.442
	0.400	0.874	0.851	0.000	0.757	0.822	0.000	0.022	0.870	0.805	0.805
	0.402	1.000	0.850	0.000	0.755	0.870	0.000	0.287	0.369	0.542	0.369
	0.400	0.981	0.850	0.329	0.933	0.350	0.000	0.060	0.617	0.671	0.350
	1.000	0.785	0.850	0.000	0.751	0.914	0.000	0.002	0.870	0.841	0.841
	0.692	0.862	0.850	0.000	0.753	0.977	0.027	0.988	0.350	0.671	0.350
	0.954	0.805	0.850	0.000	0.757	0.822	0.000	0.002	0.914	0.805	0.805
	0.995	0.780	0.850	0.000	0.765	0.617	0.000	0.022	0.870	0.442	0.442
	0.400	0.897	0.850	0.000	0.822	0.369	0.000	0.000	0.531	0.542	0.369
	0.954	0.772	0.850	0.000	0.752	1.000	0.000	0.022	0.870	0.841	0.841
	0.400	0.765	0.912	0.000	0.765	0.617	0.000	0.022	0.870	0.442	0.442
	0.400	0.752	0.999	0.000	0.750	0.617	0.000	0.007	1.000	0.442	0.442
	0.400	0.791	0.865	0.000	0.765	0.617	0.168	0.894	0.350	0.671	0.350
	0.400	0.755	0.970	0.000	0.755	0.870	0.000	0.002	0.870	0.866	0.866

5 Process Modeling with Petri Nets

Considerable work has already been carried out in modeling decision system rules with Petri nets [23, 29, 30]. This aim of the earlier as well as the current research has been to simplify the analysis of large information systems, and the transformation of such systems as well as derived rules into corresponding concurrent models. The motivation for introducing rough Petri nets stems from an effort to capture the understandings and operations from rough set theory, which have been used to construct both general-purpose as well as highly-specialized decision systems. To facilitate the design and analysis of approximate reasoning systems, and the process of deriving decision rules, this section introduces rough Petri nets and an extension of such nets called rough fuzzy Petri nets. A brief introduction to classical as well as coloured Petri nets is also given in this section. An overview of the types of Petri nets and related set theories leading to a Petri net of model of the software quality assessment process, is given in Fig. 15.

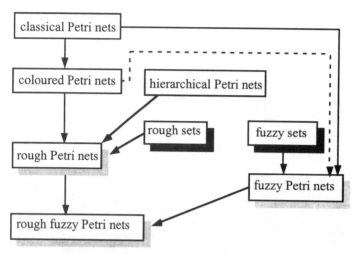

Fig. 15 Lineage of Rough Petri Nets

5.1 Classical and Coloured Petri nets

Petri nets were introduced in 1962 by Petri to describe and analyze the structure and information flow in systems containing concurrent processes [40, 41]. A Petri Net is a structure (P, T, A, I, O, W, M_o) where P is a finite set of places; T, a finite set of transitions; mapping I: T→ P to a collection of input places; mapping O: T → P to a collection of output places; arcs A ⊆ (P x T)∪ (T x P); weight function W: A → {1, 2, ...}; and initial marking M_o:P→ {0, 1, 2, ...}. A marking is an assignment of tokens to places of a net. A token represents a typeless fragment of information. A black dot •symbolizes a single token. Tokens are used to define the execution of a Petri net. Places represent storage for input or for output. Transitions represent activities (transformations) which transform input into output. A sample Petri net is given in Fig. 16.

Fig. 16. Sample Petri Net

There are three transition firing rules for Petri nets. Petri nets are governed by transition firing rules:

- A transition is enabled if each of its input places is marked with at least $W(p, t)$ tokens, where the weight function $W(p, t)$ specifies the weight of the arc from input place p to transition t.
- A transition can only fire if it is enabled.
- Whenever a transition t fires, $W(p, t)$ tokens are removed from each input place p, and transition t adds $W(t, p)$ to each output place p, where $W(t, p)$ specifies the weight of the arc from transition t to output place p.

The significance of the first firing rule is that more than one transition can be enabled at the same time (concurrent processing is possible). Whenever transition t fires, it removes the token in place p1, and adds a token to place p2. In addition, $P \cap T = \phi$ and $P \cup T \neq \phi$. To facilitate formal specification and analysis of the structure, information flow, control and computation in systems, coloured Petri nets were introduced in 1986 by Jensen [42]-[43]. Formally, a coloured Petri net is a structure $(\Sigma, P, T, A, N, C, G, E, I)$ where P, T, A are the same as in a Petri Net (PN) and

- Σ is a finite set of non-empty data types called color sets.
- N is a node function where N: $A \to (P \times T) \cup (T \times P)$.
- C is a color function where C: $P \to \Sigma$.
- G is a guard function where G: $T \to$ Boolean expressions.
- E is an arc expression function where E: $A \to$ expression $E(a)$ of type $C(p(a))$.
- I is an initialization function where I: $P \to$ closed expressions $p(a)$ of type $C(p)$.

A guard is a Boolean expression on a transition t which must be satisfied before t can fire. A CPN provides data typing (colour sets) and sets of values of a specified type for each place. The expression $E(p, t)$ is the name of a variable associated with the arc from input place p to transition t, and the expression $E(t, p)$ is associated with the transformation (activity) performed by transition t on its inputs to produce an output for place p. A guard $G(t)$ is an enabling condition associated with transition t. Each place in a CPN is associated with a data type. A sample CPN is given in Fig. 17, where place p_1 supplies a value of x of type item to transition t, which outputs complement$(x) = 1 - x$ to place p_2 whenever the token x in place p_1 satisfies the guard $[x >= 0.45]$. The notation $1`0.4 + 4`0.3$ specifies that a multiset contains 1 element with the value 0.4 and four elements with value 0.3. A multiset is a set which can have multiple appearances of the same element. The prefix 1 indicates the number of tokens in the multiset (in this case, one token in place p_1), and the suffix indicates that x has been assigned an x-value. Whenever transition t fires, it augments the multiset associated with place p_2.

To simplify coloured Petri net models of complex, largescale systems, hierarchical Petri Nets (hPNs) with transitions representing subnets are introduced [44].

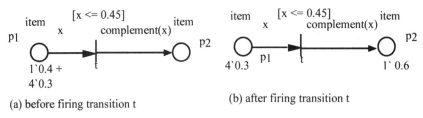

Fig. 17 Sample Coloured Petri Net

5.2 Rough Petri Nets

A rough Petri net (rPn) is a structure $(\Sigma, P, T, A, N, C, E, I, W, R, \rho)$ where S (data types), P (places), T (transitions), A (arcs), N (maps A to (PxT) ∪ (TxP)), C (maps P to Σ), E (maps A to expression E(a) of type C(p(a))), I (maps P to closed expressions p(a) of type C(p)) are as in a coloured Petri net (CPN). Strengths-of- connections (chosen from a finite set of weights W) are assigned to arcs with ρ: A → W. A strength of connection $w_i \in W$ specifies the relative importance of input to a sensor (a form of attribute in rough sets), and guarantees a certain magnitude of input to a transition. Weights are restricted to values in the interval [0, 1]. Let X, S,ξ be a set of inputs, information system S, and reduct ξ belonging to set of all reducts RED(S) of S, respectively. Let S = (U, A) be an information system and let R be an equivalence relation which forms the quotient set X/R, where R ⊆ A and X ⊆ U. Also, recall that the notation X/R (known as the quotient set) denotes the family of equivalence classes of R. The set $POS_R(X) = \underline{R}X$ is the set of all elements of U which can classified as elements of X. Similarly, the set $NEG_R(X) = U\text{-}\overline{R}\,X$ is the set of those elements of U which can be classified as elements of U - X [21]. Further, let the set R consist of $\sigma_{X/R}, \rho_{POS_R(X)}, \rho_{NEG_R(X)}, \rho_{\inf_R(X)}, \rho_{dec_A(X)}, \rho_{M(S)}, \rho_{RED(S)}, \rho_{fM(RED(S))}, \rho_{OPT(S)}, \rho_{OPT(R,S)}$ which identify distinguished procedures used to construct the quotient set X/R, set $POS_R(X)$ and set $NEG_R(X)$, as well as procedures to construct an information system table, decision system table, discernability matrix, set all rules for a decision system, and set of all rules relative to a reduct R, respectively. The prescription of the elements of R is non-exhaustive. The arc expression function E has been specialized relative to a finite set of rough set operations R such that E: A → R. The operations in R are used to describe processes which are the key components of an approximate reasoning system, namely, decision tables, discernability matrices and functions, reducts, and rules. Assume that each input x has a strength-of-connection w. Further, let S = ({x}, {a}∪ {d}) be a decision system with a single input, attribute a, and decision d. An example of a rough Petri net with a single transition is given in Fig. 18. The output of the net in Fig. 18 is a tuple (w □ x, a(w □ x), d) representing a decision system table with a single row. The computation w □ x aggregates w and x with the anonymous operation □: $[0,1]^2 \to [0,1]$. In the absence of a specific strength-of-connection, the default value of w is 0 and w □ x equals x. This is the case in Fig. 19. To construct a multi-transition rough Petri net, let $a_1, a_2,...,a_i,...a_n$, represent a collection of sensors A in a decision

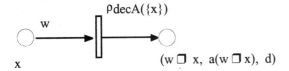

Fig. 18. Single Transition Rough Petri Net

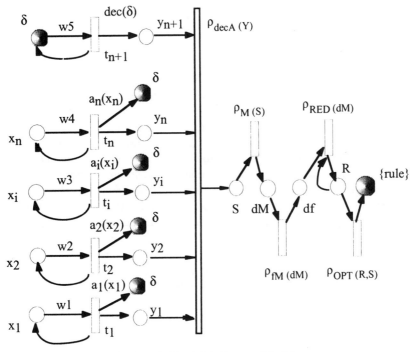

Fig. 19 Sample Decision System Model

system $S = (U, A \cup \{dec\})$. Also, let dM, df, ξ, {rule} represent discernability matrix, discernability function, reduct ξ in RED(S) for a decision system S, and set of rules derived from ξ. The rough Petri net in Fig. 9 describes the process of constructing a set of rules derived from S.

To simplify the rough Petri net model in Fig. 9, aliasing of input place δ of transition t_{n+1} has been used. Then the output place labeled δ of transitions t_1, t_2, ..., t_i, ..., t_n provides input to transition t_{n+1}. By allowing transitions in a rough Petri net to represent subnets, it is possible to model complex information systems concisely. This is in fact what has been done in Fig. 19, where each of the transitions represents a subnet modeling a process needed to carry out necessary computations. For example, the transition labeled $\rho_M(S)$ decomposes into a subnet designed to model the process which constructs a discernability matrix.

5.3 Fuzzy Petri Nets

Research concerning fuzzy Petri nets is quite extensive [45]-[54]. Fuzzy Petri nets offer a concise means of modeling the interpretation of data points which have been granulated. In the case where the objects of a universe are granulated, sensors in a decision system aggregate weighted degree-of-membership computations. Further, a fuzzy Petri net makes it possible to model a form of neural processing where modulators and strengths-of-connections are calibrated [49-51]. Such calibrations make it possible to express the intentions of system designers in assessing software quality, and to construct a variety of highly-specialized software quality measurement frameworks.

A Fuzzy Petri Net (FPN) is a structure (Σ, P, T, A, N, C, E, I, M, W, Z, ζ, ρ) where Σ, P, T, A, N, C, E, I are as in a CPN [42]. Annotations of arcs with strengths of connections chosen from a finite set of weights W are determined by ρ: A $\to$ W, and modulators (also called reference points) chosen from the finite set M are determined by ζ: A $\to$ M. A strength of connection $w_i \in W$ specifies the relative importance of input, and guarantees a certain magnitude of input to a transition. A modulator $r_i \in M$ prescribes a certain magnitude of the level of marking of a place which must be maintained. Weights and modulators are restricted to values in the interval [0, 1]. The arc expression function E has been specialized relative to a finite set Z such that E:A $\to$ Z. The expressions in Z make it possible to compute degrees of membership of values in a universe of discourse in fuzzy sets, to perform aggregations, and any other necessary operations for the functioning of a particular system. Minimally, Z has four operations consisting of what are known as a dominance AND {OR} as well as conjunctive {disjunctive} ways of aggregating weighted inputs to a transition (see 26).

$$\begin{aligned} Z = \{ &T_{i=1}^n((r_i \to x_i)\ s\ w_i), &&\text{dominance AND operation} \\ &S_{i=1}^n((r_i \to x_i)\ t\ w_i), &&\text{dominance OR operation} \\ &S_{i=1}^n(x_i\ t\ w_i), &&\text{OR operation} \\ &T_{i=1}^n(x_i\ s\ w_i), &&\text{AND operation} \} \end{aligned} \qquad (26)$$

The operations in Z employ triangular norms s,t as well as the implication operator $\to$ where r_i specifies a threshold level which modulates the strength of firing coming from the ith input place. Depending on the marking of the input places, a transition can fire. In contrast to two-values Petri Nets, the generalized version studied includesa gradual firing (strength of firing) of transitions together with the level of marking of places. First, let us discuss a generic model of a transition represented in Fig. 20. An elementary FPN has a single multivalued(fuzzy) transition z_i with inputs x_i(input signal), r_i (reference point), w_i (weight), and single output place out_i. Each input x_i is a fuzzy number (i.e., result of applying a membership function to an element of a universe of discourse which consists of real numbers). The results out1, ..., outn of elementary FPNs are aggregated. The level of firing of transition in a fuzzy Petri net is determined by (27).

$$Z = T_{i-1}^n[(r_i \to x_i)\ s\ w_i] \qquad (27)$$

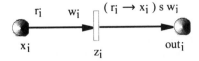

Fig. 20. Elementary Fuzzy Petri Net

For the computation (17) associated with transition Z, the limit "n" denotes the number of input places; x_i, a level of marking at the i-th place; r_i, a level of modulation of the input; and w_i, an associated degree of contribution of the x_i to the overall firing of the transition. Here "s" and "T" (or t) denote s- and t-norms. Similarly, "$\rightarrow$" denotes a multivalued implication operation. Many forms of implication are possible. For simplicity, the Gaines form of $r_i \rightarrow x_i = \min(1, \frac{x_i}{r_i})$ has been used in aggregating granulated software quality measurements.

5.4 Roughly Fuzzy Petri Nets

Roughly fuzzy Petri nets were introduced as an extension of fuzzy Petri Nets in [55]. In this section, rough fuzzy Petri nets are presented as a straightforward

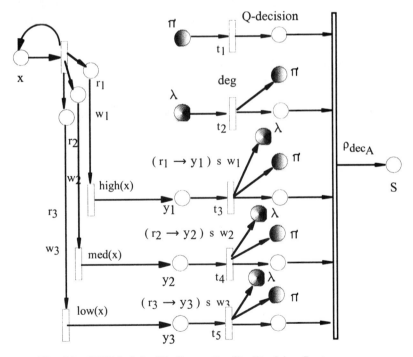

Fig. 21 rfPN Model of Software Quality Decision System

extension of rough Petri nets. The introduction of rough fuzzy Petri Nets (rfPNs) is motivated by the need to develop mathematical models of decision-making system S relative to aggregations of granulated inputs such that the models are capable of learning, and are designed to react to dynamic changes in the reduct set for different samples of decision tables. A roughl fuzzy Petri net (rfPn) is a structure given in (28)

$$(\Sigma, P, T, A, N, C, E, I, M, W, R \cup Z, \zeta, \rho) \tag{28}$$

where Σ, P, T, A, N, C, E, I, W, R, ρ are as in a rough Petri net. The set R is augmented with operations in Z from fuzzy Petri nets to handle aggregations of granulated inputs. The modulators (also called reference points) of operations in Z are in the set M. The operation ζ also comes from fuzzy Petri nets such that modulators chosen from the finite set M are assigned to arcs by $\zeta: A \rightarrow M$. In effect, a rough fuzzy Petri net is an extension of the rough Petri net model, which provides a concise means of modeling the process of deriving rules for an approximate reasoning system for software quality (see Fig. 21). To simplify the rough fuzzy Petri net model in Fig. 21, aliasing of input places λ and π of transitions t1 and t2, respectively, has been used. For example, the output place labeled λ of transitions t3, t4, t5 provides input to transition t2. The rfPn in Fig. 20 models the operations of the sensors for a decision system S. The place labeled S in Fig. 20 becomes input to the back end of the rough Petri net in Fig. 18. Observe that raw software criteria measurements are viewed in terms of degree of each measurement being a part of a granule, and the estimate of degree of membership initializes a rough fuzzy Petri net. In the case where modulators and strengths of connections are incorporated into the rough fuzzy Petri net for processing quality measurements, then learning is possible. Second, the outputs of a calibrated rough fuzzy Petri net provide inputs to a decision system table. In this case, granulated quality measurements are computed off-line as a result of supervised learning relative to target values as in [49]-[51].

6 Concluding Remarks

A software quality decision rule makes it possible for a designer to evaluate the consistency and correctness of assessments of software characteristics for a particular system. In addition, decision rules reveal dependencies among software evaluations across granulated measurements. Derived rules provide pointers to particular aspects of software requiring further tuning. That is, it is the evaluation of the degree that a raw software quality attribute measurement is a part of a particular granule (especially granules representing lower- and medium-valued measurements) that signal not only the need for further tuning but also the extent of change required to induce higher software quality. This approach to software quality assessment has been illustrated in terms of a small subset of measurements relative to software usability, and is easily extended to a more comprehensive evaluation of software criteria measurements.

A variety of issues concerning the software quality decision system framework have been left for future research. Dominant among the outstanding issues is a formal treatment of a software criteria measurements being a part of a granule to a degree examined in the context of rough inclusions and rough mereology. It is also possible to envision the new approach to assessing software quality extended to a system of intelligent cooperating agents such as the one described in [57], and in [61]. The aim of this extension would be to develop an approach to approximate reasoning by a system of agents designed to assist in assessing approximate software quality. The agents in this system would continuously carry out all of the computations needed by humans. A system of cooperating "software quality" agents is symmetric with a system of collaborating humans engaged in continuous software enhancement. Another issue which is part of the current research on developing a software quality decision system is the quantization of the real values of software quality criteria sensors so that sensor values are partitioned into intervals. Let R be the set of real numbers, and let R^k be a k-dimensional affine space. An object in a decision table is treated as a point in R^k where k is the number of conditional sensors a $\in$ A such a $\neq$ d in S = (U, A $\cup$ {d}). Then the objects in are partitioned into r decision classes using the method described by N.H. Son in [58]. The important consequence of this classification scheme is that all objects belonging to the same class have the same decision. This simplifies the problem of determining what aspects of software represented by decision system objects need to be changed to improve software quality. Another outstanding issue currently being considered is the modeling of software quality decision rules with rough Petri nets. This problem entails not only a consideration of the analysis of rough Petri net models of rules but also the subtler problem of taking into account rough mereology in the conceptualization of rough Petri nets themselves.

Acknowledgement. We want to thank the anonymous reviewers for their helpful comments and corrections for this article. We also thank Prof. Skowron for the invitation to write this paper and for his insights, suggestions and comments related to this research. We also wish to thank Prof. Zbigniew Suraj, Institute of Mathematics, Rzeszow and members of the Institute of Mathematics at Warsaw University for discussions related to research on Petri nets. Finally, we gratefully acknowledge the funding for this research provided by the Natural Sciences and Engineering Research Council of Canada (NSERC) operating funds, the University of Manitoba research grants committee, and the University of Winnipeg research grants committee.

References

[1] Bandinelli, S., Fuggetta, A., Lavazza, L., Loi, M., Picco, G.P.: Modeling and improving an industrial software process. IEEE Trans. on Software Engineering 21/ 5 (1995) 440-453

[2] Cusumano, M.A.: Japan's Software Factories. Oxford, Oxford University Press (1991)

[3] IEEE,: Standard glossary of software engineering terminology. In: IEEE Standards Collection Software Engineering, New York, IEEE (1997)
[4] McCall, J. A., Richards, P. K., Walters, G. F.: Factors in Software Quality, vols. I, II, III. U.S. Rome Air Development Center Reports, NTIS AD/A- 049 014, 015, 055 (1977)
[5] McCall, J. A., et al.: Software Quality Measurement Manual. RADC Technical Report TR-80-109, U.S. Rome Air Development Center (1980)
[6] McCall, J. A., et al.: Methodology for Software Reliability Prediction. RADC-TR-87-171, 2 volumes, U.S. Rome Air Development Center (1987)
[7] McCall, J. A.: Quality factors. In: Encyclopedia of Software Engineering, Marciniak, J.J. (Ed.), New York, John Wiley & Sons, 2 (1994) 959-969
[8] Cavano, J., McCall, J. A.: A framework for the measurement of software quality. Proc. Software Quality Assurance Workshop (1978)
[9] Bowen, T.P.: Software Quality Measurement for Distributed Systems. RADC-TR-83-175, U.S. Rome Air Development Center (1983)
[10] Bowen, T.P., Wigle, G.B., Tsai, J.T.: Specification of Software Quality Attributes. RADC-TR-85-37, 3 volumes, U.S. Rome Air Development Center, February 1985.
[11] Musa, J.D., Iannino, A., Okumoto, K.: Software Reliability: Measurement, Prediction, Application. New York, McGraw-Hill Publishing Co. (1990)
[12] Hiering, V., Bennett, D.: A developer's perspective on software quality metrics. IEEE ComImunications Magazine 24/9 (1986) 66-71
[13] Pedrycz, W., Peters, J.F.: Computational intelligence in software engineering. Proc. Canadian Conf. on Electrical & Computer Engineering (1997) 253-256
[14] IEEE,: Standards for a software quality metrics methodology, P-1061/D20. In: IEEE Standards Collection Software Engineering, New York, IEEE (1997)
[15] Yoshida, T.: Attaining higher quality in software development–evaluation in practice. Fijitsu Scientific Technical Journal 21/3 (1985) 305-316
[16] Pedrycz, W., Peters, J.F., Ramanna, S.: Neuro-fuzzy approach to software quality. Proc. Annual Oregon Workshop on Software Metrics (AOWSM), Coeur d'Alene, Idaho (1997) 1-9
[17] Peters, J.F., Ramanna, S.: Software Deployability Decision Framework: A Rough Sets Approach. Proc. IPMU'98 Paris, France (1998)
[18] Pedrycz, W., Peters, J.F., Ramanna, S.: Design of a software quality decision system: A computational intelligence approach. Proc. CCECE'98 Waterloo, Ontario (1998)
[19] Boehm, B.W., Brown, J.R., Kaspar, H., Lipow, M., Macleod, G.J., Merrit, M.J.: Characteristics of Software Quality. Amsterdam, North-Holland (1985)
[20] Pawlak, Z.: Rough Sets: Theoretical Aspects of Reasoning About Data. Boston, MA, Kluwer Academic Publishers (1991)
[21] Z. Pawlak.: Rough sets: present state and future prospects. ICS Research Report 32/95, Institute of Computer Science, Warsaw Institute of Technology, (1995)
[22] Pawlak, Z., Grzymala-Busse, J.W., Slowinski, R., Ziarko, W.: Rough Sets. Communications of the ACM 38 (1995) 88-95
[23] Skowron, A.: Extracting laws from decision tables: a rough set approach. Computational Intelligence 11/2 (1995) 371-388
[24] Skowron, A., Polkowski, L.: Rough mereology: A new paradigm for approximate reasoning. Journ. of Approximate Reasoning 15/4 (1996) 333-365
[25] Skowron, A., Rauszer, C.: The discernability matrices and functions in information systems. In: Intelligent Decision Support, Handbook of Applications and

Advances of the Rough Sets Theory, Slowinski, R. (Ed.), Dordrecht, Kluwer Academic Publishers (1992) 331-362
[26] Skowron, A., Suraj, Z.: A rough set approach to real-time state identification. Bulletin EATCS 50 (1993) 264-275
[27] Skowron, A., Suraj, Z.: Synthesis of concurrent systems specified by information systems. ICS Research Report 39/94, Institute of Computer Science, Warsaw Institute of Technology (1994)
[28] Skowron, A., Suraj, Z.: Discovery of concurrent data models from experimental data tables: a rough set approach. Institute of Computer Science Research Report 15/95, Warsaw Institute of Technology (1995)
[29] Skowron, A., Suraj, Z.: A parallel algorithm for real-time decision making: a rough set approach. Journal of Intelligent Information Systems 7 (1996) 5-28
[30] Skowron, A., Suraj, Z.: A rough set approach to real-time state identification for decision making. Institute of Computer Science Research Report 18/93, Warsaw University of Technology (1993)
[31] Zadeh, L.A.: Fuzzy logic = computing with words. IEEE Trans. on Fuzzy Systems 4/2 (1996) 103-111
[32] Zadeh, L.A.: Toward a theory of fuzzy information granulation and its certainty in human reasoning and fuzzy logic. Fuzzy Sets and Systems 90/ 2 (1997) 111-128
[33] Klir, G.J., Wierman, M.J.: Uncertainty-Based Information: Elements of Generalized Information Theory. Report, Center for Research in Fuzzy Mathematics and Computer Science, Creighton University, Omaha, Nebraska 63178, U.S.A. (1997)
[34] Zadeh, L.: Fuzzy sets, Information and Control 8 (1965) 338-353
[35] Gaines, B.R.: Multivalued logics and fuzzy reasoning. BCS AISB Summer School, Cambridge (1975).
[36] Lukasiewicz, J.: Logic and the problem of the foundations of mathematics. In: Jan Lukasiewicz, Borkowski, L. (Ed.), Amsterdam, North-Holland Pub. Co. (1970) 278-294
[37] Ruan, D.: A critical study of widely used fuzzy implication operators and their influence on the inference rules in fuzzy expert systems. Ph.D. thesis, Gent (1990).
[38] Sienkiewicz, J.: Rough sets for boolean functions minimization. Research Report, Warsaw Institute of Technology (1995)
[40] Murata, T.: Petri nets: properties, analysis and applications. Proceedings of the IEEE 77/4 (1989) 541-580
[41] Petri, C.A.: Kommunikation mit Automaten. Schriften des IIM Nr. 3, Institut fŸr Instrumentelle Mathematik, Bonn, West Germany. See, also, Communication with Automata (in English). Griffiss Air Force Base, New York Technical Report RADC-Tr-65-377, 1, Suppl. 1 (1962)
[42] Jensen, K.: Coloured Petri nets. In: Advances in Petri Nets 254 (1986) 288-299
[43] Jensen, K.: Coloured Petri Nets–Basic Concepts, Analysis Methods and Practical Use 1. Berlin, Springer-Verlag (1992)
[44] Huber, P., Jensen, K., Shapiro, R.M.: Hierarchies in coloured Petri nets. Proc. Int. Conf. Science on Application and Theory of Petri Nets. In: Rozenberg, G. (Ed.), Lecture Notes in Computer Science 483 (1986) 261-292
[45] Scrinivan, P., Gracarin, D.: Approximate reasoning with fuzzy Petri nets. Proc. IEEE Int. Conf. on Fuzzy Systems, San Francisco, CA (1993) 396-401
[46] Scarpelli, H., Gomide, F.: Relational calculus in designing fuzzy Petri nets. In: W. Pedrycz (Ed.), Fuzzy Modelling: Paradigms and Practice. Boston, MA, Kluwer Academic Publishers (1996) 70-89

[47] Scarpelli, H., Gomide, F.: Fuzzy reasoning and high level fuzzy Petri nets. In: Proc. First European Congress on Fuzzy and Intelligent Technologies, Aachen, Germany (1993) 600-605
[48] Scarpelli, H., Gomide, F. Yager, R.: A reasoning algorithm for high-level fuzzy Petri nets. IEEE Trans. on Fuzzy Systems 4/3 (1996) 282-295
[49] Pedrycz, W., Peters, J.F., Ramanna, S., Furuhashi, T.: From data to fuzzy Petri nets: generalized model and calibration abilities. Proc. of Seventh Int. Fuzzy Systems Association World Congress (IFSA'97) III (1997) 294-299
[50] Pedrycz, W., Gomide, F.: A generalized fuzzy Petri net model. IEEE Trans. on Fuzzy Systems 2/4 (1994) 295-301
[51] Pedrycz, W., Peters, J.F.: Learning in fuzzy Petri nets. In: Fuzzy Petri Nets, Cardoso, J., Sandri, S. (Eds.). Berlin, Physica Verlag [in press]
[52] Pedrycz, W., Peters, J.F.: Information Granularity Uncertainty Principle: Contingency Tables and Petri Net Representations. Proc. Proc. North American Fuzzy Information Processing Society NAFIPS'97, Syracuse, NY, (1997) 222-226
[53] Garg, M.L., Ahson, S.I., Gupta, P.V.: A fuzzy Petri net for knowledge representation and reasoning. Information Processing Letters 39 (1991) 165-171
[54] Son, H.S., Seong, P.H.: A safety analysis method using fuzzy Petri nets. Proc. North American Fuzzy Information Processing Society (NAFIPS'97), Syracuse, NY (1997) 412-417.
[55] J.F. Peters.: Time and clock information systems: Concepts and roughly fuzzy Petri net models. In: Rough Sets and Knowledge Discovery, Kacprzyk, J., Berlin, Physica Verlag, a division of Springer Verlag [in press].
[56] Bazan, J.G., Skowron, A., Synak, P.: Discovery of decision rules from experimental data. Institute of Mathematics Report, Warsaw University (1994)
[57] Polkowski, L., Skowron, A.: Approximate reasoning about complex objects in distributed systems: Rough mereological formalization. Research Report, Institute of Mathematics, Warsaw University (1997)
[58] Son, N.H.: Discretization of real-valued attributes: Boolean reasoning approach. Doctoral Thesis, Faculty of Mathematics, Computer Science and Mechanics, Warsaw University (1997)
[59] Milner, R.: Communication and Concurrency. NJ, Prentice-Hall (1989)
[60] Hayes-Roth, B., Pfleger, K., Lalanda, P., Morignot, P., Balabanovic, M.: A domain-specific software architecture for adaptive intelligent systems. IEEE Transactions on Software Engineering 21/4 (1995) 288-301
[61] Peters, J.F., Sohi, N.: Coordination of multiagent systems with fuzzy clocks, Concurrent Engineering: Research and Applications 4/1 (1996) 73-88

Assessment of Concert Hall Acoustics Using Rough Set and Fuzzy Set Approach

Bozena Kostek

Technical University of Gdansk, Faculty of Electronics, Telecommunications and Informatics, Sound Engineering Dept., 80-952 Gdansk, Poland

1 Introduction

Modern acoustics encompass computer studies as applied to this domain. Only recently, however, have a variety of computer programs become available to assist the designer concerned with architectural acoustics. It should be remembered that the calculations performed by such programs are only approximate, and thus they may not suit a particular application and may even result in substantial error. Another difficulty in acoustical design is that it is rare for a hall to be used for a single purpose. Thus, the so-called "optimum" requirements vary for each type of usage, and as a result the acoustical solutions are very often compromises. Since the primary aim in the design of an acoustical space is sound quality, it is therefore necessary to correlate objective measurements to subjective impressions of an interior space. Correlating an objective measurement to an expert's subjective assessment, however, is not an easy process. Many literature references already exist as to how this process may be carried out, but there is not yet any consensus on this still unresolved acoustical subject [1][2][3]. It seems that one of possible solutions is to build a knowledge base directly upon interviewing experts. As a result one can model the practices of experts, which is more important in quality evaluation than discovering exact relationship between physical measures.

However, relationships between objectively measured parameters of acoustical objects (concert halls, sound processing programs, loudspeakers, etc.) and their subjective quality as assessed by listeners (preferably experts) cannot in most cases be crisply defined, leaving a wide margin of uncertainty which depends on individual subjects' preferences and the unknown influences of individual parameter values on the overall acoustic quality of the tested object. Consequently, results of subjective tests have to be processed statistically (hitherto used approach) in order to find links between preference results and concrete values of parameters representing the objective features of tested objects. The statistical approach, which is commonly used in these studies does not allow one to predict the result of preference when parameters are varied (for example in designing concert halls or changing sound processing algorithms settings during computer simulations). Consequently, a decision system is needed that would be able to provide an automatic assessment of sound quality basing on both: the objective (measurable) values of acoustical parameters and expert opinions concerning the resulting sound quality.

Recently, a novel approach to computer assessment of acoustical quality has been made using the soft computing approach [4]. Rough set and fuzzy set theories were used for the purpose of processing subjective evaluation results. A prototype system based on the rough set theory was used to induce generalized rules that describe the relationship between acoustical parameters of concert halls and sound processing algorithms. The rough set approach produces also reducts and a set of rules allowing one to study principles underlying experts' decisions. On the other hand, a fuzzy-based system proved to be applicable to revealing which of the parameters mostly contributed to the overall sound quality [4]. However, the introduced methods have not been directed towards making automatic decisions of acoustical quality. Their role was limited to the replacement of statistical processing of subjective testing results by the soft computing (non-statistical) approach. Therefore, in this paper a method is introduced by which an influence of chosen parameter values on the quality ratings of an acoustical object can be automatically determined.

First section of this paper provides notions of the subjective evaluation of sound quality, the acoustical background and statistical processing of results. In the next section the organization of the analysis procedure of subjective testing results is presented. Having collected the assessment results of the overall acoustical quality of the tested objects from all of the experts, it is possible to create a decision table and then to process this table using the rough set method. In this way a set of rules may be created, that may be subsequently verified by experts. The next step is to analyze objective parameters step-by-step, trying to obtain subjective ratings for each of them when assessed separately from other ones. The mapping of objective parameter values to their subjective assessments by many experts creates some fuzzy dependencies which can be represented by the fuzzy membership functions. Then, having rules determined from the rough set decision table and membership functions determined empirically for studied parameters one can create an expert system providing automatic decision on acoustical quality each time when the concrete set of parameters is presented to its inputs. This system uses fuzzy logic principles for automatic determination of the acoustical quality.

A concise description of the engineered expert system is given below.

Knowledge acquisition phase:

Selection of acoustical objects to be tested;

Choice of subjective parameters describing acoustical quality of these objects;

Subjective listening tests carried out with regard to the object quality to be assessed (various acoustic interiors, either existing or simulated). The tests should use subjectively defined parameters that can be expressed in terms of objective measures. Parameter values should be expressed in ranges labeled descriptively as *low*, *medium*, and *high*;

Collecting all experts' answers related to the overall quality in tables together with the descriptively labeled values of parameters;

Creating a rough set decision table from the collected data;

Rough set processing of the above decision table (derivation of reducts and rules);

Measuring objective characteristics of investigated acoustical objects;

Calculating histograms from the experts' vote results for separated parameters;

Defining universe and domain of objectively measured parameters, labeling membership functions representing subdomains (ranges, scopes) of objectively measured parameters (the most typical labels are: *low*, *medium*, and *high*);

Defining fuzzy sets on the basis of subjective voting in such a way that each assessed parameter value is mapped to the number of votes assigned to it by experts;

Estimation of membership functions shapes based on the probability density approach;

Statistical validation of obtained membership functions by means of the Pearson's χ^2 test (especially important in the case of statistically small number of tested objects).

Automatic quality assessment phase:

Collecting new set of parameter values related to an object that was not measured previously;

Calculating the degree of membership for each parameter and for each predefined membership function;

Applying the rules stored in the knowledge base (derived using the rough set method and validated by experts during the learning phase);

Calculating the value assigned to each rule (using the fuzzy-set defined AND function in the conditional part of the rules);

Finding a rule which was assigned the maximum value (the winning rule);

Applying the λ-cut to the output membership function associated with the winning rule (one of the membership functions describing the overall preference);

Calculating the centroid value on the basis of the λ-cut as above;

Mapping the centroid onto the 100 point subjective grade scale. The obtained crisp value provides a measure of the automatically assessed acoustical quality of the tested object.

The detailed description of all procedures is beyond this presentation. Hence, it will be limited to the most important issues only, especially these ones which are related to the system implementation.

2 Statistical Analysis of Test Results

2.1 Subjective Evaluation of Sound Quality

Subjective listening tests are a part of assessment of sound production in acoustic environments. They are designed to reveal the presence of differences between objects under tests or are intended to yield subjectively scaled ratings according to some chosen criteria. In overall, they consist of a test procedure, data acquisition and analysis, and interpretations of results. Analysis and result interpretation parts are still only recommendations and not standards in the acoustical practice. The experimental procedure aims to identify certain physical and psycho-physiological variables, then to isolate them and in the last, to control them. The objective is to minimize the biases and variations in listeners' judgments that contribute to factors other than those under test. In the acoustical practice a standard statistical processing of subjective evaluation results is usually performed, however it will be shown that methods based on soft computing can be also valuable in this domain or even they can outperform traditional methods in some issues.

There are several problems related to the object evaluation process. First of all, experts should be trained and experienced in critical listening. An additional procedure based on a blind listening test provides an evaluation of the listener's self-consistency. If a listener always ranks the same system in the same way, his/her reliability is thus undoubted. It is desirable for experts to be of the same background, i.e. acousticians or musicians, etc., but in practice it is not easy to find such a group that is willing to participate in a series of experiments. It should be remembered that listening sessions are quite tiring and time consuming.

In the experiment one should make some assumptions, such as: number of experts taking part in the subjective evaluation, number of parameters to be tested, number of music types, number of points to be tested within the parameter range. Additionally, in the case of multidimensional subjective tests, when subject's task is to vote for several parameters at the same time, a technique called multidimensional scaling (MDS) is highly recommended [5]. This technique aims at specific arrangement of tested objects in multidimensional Euclidean space basing on the fact that distances between objects correspond monotonically to the perceived degree of dissimilarity between them. The space derived from listener's ratings is interpreted as reflecting the listener's actual perceptual space.

2.2 Measurable Data Analysis

When considering the evaluation of an acoustical hall, both measurement procedures and listening tests are carried out, resulting in a set of data. Usually, some statistical tests are employed in order to check the reliability of the obtained results. In the first step of statistical analysis of such elementary measures as mean and variance of distributions are calculated. Next, in order to check whether two selected parameters are dependent one on another, the degree of correlation is to be calculated for pairs of quantities (x_i, y_i), $i=1,...,n$. Most widely used is the linear correlation coefficient r (Pearson's) calculated according to the formula:

$$r = \frac{\sum_{i=1}^{n}(x_i - \bar{x}) \cdot (y_i - \bar{y})}{\sqrt{\sum_{i=1}^{n}(x_i - \bar{x})^2 \cdot \sum_{i=1}^{n}(y_i - \bar{y})^2}} \quad (1)$$

where:

$$\bar{x} = \frac{1}{n} \cdot \sum_{i=1}^{n} x_i, \quad \bar{y} = \frac{1}{n} \cdot \sum_{i=1}^{n} y_i \tag{2}$$

In the case of binomial or two-dimensional Gaussian distributions some additional statistical tests can be proceeded. In order to verify whether the assumed null hypothesis H_0 is valid, another expression is calculated, namely Student's test t with a number of degrees of freedom equal to $n-2$:

$$t = r \cdot \frac{\sqrt{(n-2)}}{\sqrt{1-r^2}} \tag{3}$$

In Table I a set of acoustical data is presented. The data were collected on the basis of acoustical references and measurements made by the author [4]. In both cases this set of acoustical data represents parameter values measured in various acoustical objects, the number of which was equal to 24. Therefore, the total number of parameter values equals 168. Parameters from Tab. I are described below.

The reverberation time, originally introduced by Sabine, is given by the relation:

$$RT = 0.161 \cdot \frac{V}{A} \tag{4}$$

where: A - total area of absorption.

This factor is defined as time needed to decrease energy by 60dB from its original level after an instantaneous termination of the excitation signal. Equation (4) assumes that the sound energy is equally diffused throughout the room (i.e. homogenous and isotropic). Actually, this condition is rarely fulfilled because of large absorption areas existing in a hall. Therefore, the decay of sound may be described by another parameter, namely EDT (Early Decay Time) computed within the range of (0dB, -5dB). On the other hand, parameter called *Loudness* is both measurable and subjective criterion. In some sources on acoustics, it is represented as the distance between the source and the listener's place [1].

The proposed quantity: $Definition$ - C_{def} introduced by Thiele [6] is expressed as follows:

$$C_{def} = 10 \log \frac{\int_{0ms}^{50ms} p^2(t) dt}{\int_{50ms}^{\infty} p^2(t) dt} \tag{5}$$

The *Spatial Impression* parameter - C_{SI} introduced by Ando is often taken into account when analyzing room quality. Adequate C_{SI} implies sufficient early lateral sound component energy.

$$C_{SI} = 10 \log \frac{\int_{0ms}^{80ms} p^2(t) \cos \Theta dt}{\int_{0ms}^{80ms} p^2(t)(1 - \cos \Theta) dt} \tag{6}$$

where: $p(t)$ is the instantaneous pressure value produced by an impulsive sound source at the listener location, and Θ is the angle between an incoming acoustic sound wave and an axis parallel to the listener's ears.

One of the highest importance is the delay of the first reflection expressed in miliseconds, called Initial-Time-Delay Gap ($ITDG$).

A parameter called $Diffusion$ was defined by Kutruff. This criterion is calculated on the basis of the autocorrelation function ($\Psi(t)$) of the impulse response $h(t)$. It is expressed as the ratio of the autocorrelation function value at $\tau = 0$ to its maximum value. The diffusion is small in the case when the flutter echo appears in the hall (nonhomogenous distribution of sound), the value of the autocorrelation function being at the same time large one.

$$\Delta = \frac{\Psi(0)}{\Psi_{max}(\tau \neq 0)}, \quad \Delta \geq 1 \quad when \quad \Psi(\tau) = \int_{-\infty}^{+\infty} h(\tau) \cdot h(t+\tau) dt \qquad (7)$$

For this exemplary set of data some basic statistical measures were calculated (mean values and dispersion). They are presented in Table II. Additionally, values of correlation coefficients and corresponding values of Student's t are shown respectively in Tab. III and IV. As is seen from Tab. III some values of r are quite large. Therefore, their significance was checked according to expression (3) and with regard to inequality $|t_0| > t_\alpha$. It was found that correlation coefficients between pairs of parameters: 1st-2nd, 1st-3rd, 1st-7th, 2nd-4th, 2nd-7th, 4th-5th, and 6th-7th, are significant. A significance test has shown that these parameters are strongly correlated (at the significance level equals 0.01), while other significant correlations were not found.

Tab. I Exemplary acoustical data

Hall	C_{def}	$Diff.$	$ITDG$	EDT	RT	$Loudn.$	C_{SI}
1	0.5076	0.4321	7.1	1.950	2.14	0.2837	0.3583
2	0.5256	0.2868	14.9	1.83	2.25	0.3429	0.3565
i	....	....	....	....	....	....	....
n	0.6695	0.2719	27.7	1.413	1.617	0.2058	0.1981

Tab. II Mean values and dispersions of the data set presented in Tab. I

Par.	C_{def}	$Diff.$	$ITDG$	EDT	RT	$Loudn.$	C_{SI}
Mean	0.628	0.305	19.061	1.643	1.888	0.278	0.290
Disper.	0.105	0.106	8.630	0.369	0.374	0.053	0.091

Tab. III Correlation coefficients r of the data set presented in Tab. I

r	C_{def}	$Diff.$	$ITDG$	EDT	RT	$Loudn.$	C_{SI}
C_{def}	1						
$Diff.$	-0.817	1					
$ITDG$	0.368	-0.295	1				
EDT	-0.748	0.746	0.099	1			
RT	-0.413	0.493	0.131	0.828	1		
$Loudn.$	-0.581	0.322	-0.251	0.298	0.051	1	
C_{SI}	-0.769	0.638	-0.544	0.418	0.126	0.744	1

Tab. IV Student's statistic t of the data set presented in Tab. I

t	C_{def}	Diff.	ITDG	EDT	RT	Loudn.	C_{SI}
C_{def}	—						
Diff.	-5.294	—					
ITDG	1.479	-1.154	—				
EDT	-4.217	4.191	0.372	—			
RT	-1.695	2.122	0.496	5.534	—		
Loudn.	-2.673	1.274	-0.968	1.167	0.192	—	
C_{SI}	-4.508	3.098	-2.427	1.724	0.476	4.163	—

Distributions of values of parameters respectively: $Definition$ (C_{def}) versus $Diffusion$ and EDT versus RT are shown in Fig. 1. As is seen from Fig. 1 in both cases data are strongly correlated, a. negative correlation, b. positive correlation.

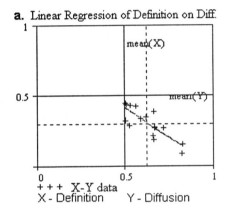
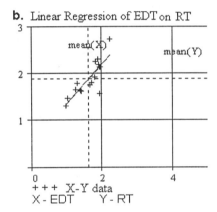

Fig. 1 Scattering of data from Tab. I (a. C_{def} versus $Diff.$, b. EDT versus RT)

According to the above statistical considerations it may be assumed that the number of measured parameters may be reduced to only 2 or 3 parameters for a given hall. It should be however emphasized that performed calculations were limited to some available data, therefore the definite conclusions on the number of parameters needed to describe a hall cannot be drawn, yet.

3 Rough Set Processing of Data

In the next steps of analysis, the results of the listening test sessions should be collected into tables, separately for each expert and for each of the various music excerpts. Then these tables should be transformed into the format of decision tables used in the rough set decision systems (Tab. V) [7]. There exists a practical way to carry evaluation procedure in laboratory conditions. Such an experiment can be based on computer simulations of a hall acoustics. In this case, the sound excerpts recorded in an anechoic chamber, thus without any reverberation, are used. Therefore, objects t_1 to t_n represent various simulated acoustical interiors. Attributes A_1 to A_m are to be denoted as tested parameters, introduced previously. They are used as conditional attributes. The expert's scoring is defined by a_{11} to a_{nm} grades (quantized values - labeled descriptively as *low*, *medium*, and *high*). The decision D is understood as a value assigned to the overall quality of sound (*Quality*). The decision system based on the rough set theory engineered at the Technical University of Gdansk will be used for further investigations [8]. It consists of learning and testing algorithms. During the first phase rules are derived that form a basis for the second phase performance. The generation of decision rules starts from rules of the length equals 1, then the system generates rules of the length equals 2, etc. The maximum rule length may be defined by the operator. The system induces both certain and possible rules. It is assumed that the rough set measure (μ_{RS}) for possible rules should exceed the value 0.5.

Tab. V Decision table for testing sound quality

object/ attr.	A_1	A_2	.....	A_m	D
t_1	a_{11}	a_{12}	.....	a_{1m}	d_1
t_2	a_{21}	a_{22}	.....	a_{2m}	d_2
.....	.....	.....	.....	.....	.....
t_n	a_{n1}	a_{n2}	.....	a_{nm}	d_n

Moreover, only such rules are taken into account, that were preceded by any shorter rule operating on the same parameters. The system produces rules of the following form:

$$(attr\_A_1) = (grade\_a_{11}) \& ... \& (attr\_A_m) = (grade\_a_{nm}) => (Qual\_d_i) \quad (8)$$

The next step is the testing phase in which the leave-one-out procedure is usually performed. During j-th experiment, the j-th object is removed from every class contained in the database, then the learning procedure is performed for remaining objects, and the result of classification of the previously omitted objects by produced rules is recorded. The result of such a processing is the set of rules that will be later used to assess the quality of an object unseen by the system.

The questionnaire form used in listening tests was as presented in Table VI. The same descriptors (attributes) as previously shown in Tab. I were used in subjective assessments. Subjects were asked to fill in the questionnaire. The expert decision set was limited to 3 grades. Having results of several simulated halls and at the same time collected from several subjects, these data are then processed by the rough set algorithm.

Tab. VI Listening test results for an exemplary hall

Attr.	C_{def}	$Diff.$	$ITDG$	EDT	RT	$Loudn.$	C_{SI}	$Qual.$
1	low	high	low	med.	high	med.	med.	med.
i	....	....	....	....	....	....	....	....
n	med.	low	high	high	med.	high	med.	high

The first step is the elimination of rows in decision tables that are duplicated (superfluous data elimination). The second step of data processing is the calculation of rules. In the discussed example the following strongest rules were obtained:

if (C_{SI} med) then (Quality good), $\mu_{RS} = 1$

if (EDT low) & (C_{SI} low) then (Quality fair), $\mu_{RS} = 0.9$

if (C_{def} med) & (C_{SI} high) & (Loudn. high) then (Quality good), $\mu_{RS} = 0.8$

if (Loudn. high) & (RT med) then (Quality very good), $\mu_{RS} = 0.8$

if (Loudn. med) & (EDT med) then (Quality good), $\mu_{RS} = 0.7$

if (C_{def} med) & (C_{SI} high) then (Quality very good), $\mu_{RS} = 0.7$

4 Mapping Test Results to Fuzzy Membership Functions

In this step of experiment, acoustical simulations were used instead of real hall measurements in order to minimize the costs of this experiment. The sound samples recorded in an anechoic chamber are then processed by adding some portions of artificially generated reverberation. Experts, while listening, are instructed to rate their judgements of the performances using descriptive notions such as low, medium, high. They vote separately for each attribute to be assessed. This results in the relation of semantic descriptors to the particular parameter quantities. Moreover, this process transforms real value parameter domain into perceptions.

Some exemplary data are graphed in Fig. 2. As is seen from figure, the distribution of observed instances may suggest the trapezoidal shape of a membership functions. In the next step of analysis such membership functions will be defined by the use of some statistical methods.

In any case, it is necessary to define also the set of output membership functions representing the grades of overall subjective preference. It was assumed that this preference is expressed in 100 point linear scale subdivided to 3 ranges mapped non-exclusively to 3 typical membership functions.

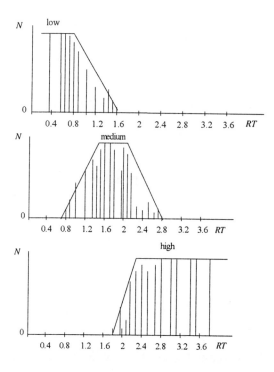

Fig. 2 Experts' votes for the parameter RT; N - number of experts voting for individual values of RT

5 Analysis Procedure

One of the main tasks of subjective tests result analysis is to approximate the tested parameter distribution. It can be done by several techniques. First of all, the linear approximation is the most common technique, that transforms the original data range to the interval $[0, 1]$, thus the triangular or trapezoidal membership functions may be used in this case. In the linear regression method one should assign the minimum and maximum attribute values. Assuming that distribution of parameters provides the triangular membership function of the estimated parameter, thus in this case the maximum value may be assigned as the average value of obtained results. This may cause, however, a loose of information and a bad convergence. The second technique uses the bell shaped functions. The initial values of parameters can be derived from the statistics of the input data. Further, the polynomial approximation of data, both ordinary or Chebyshev may be used. The polynomial approximation of degree k approximates a given set of parameter values by $k+1$ coefficients assuming the least-square error. This technique is justified by a sufficiently big number of results or by increasing the order of polynomials, however the latter direction may lead to a weak generalization of results. Coefficients of a linear combination of Chebyshev polynomials of the degree 0,1,...,k may be also used for data representation purposes. As is seen from the presented considerations, there are some advantages and disadvantages concerning the mentioned methodologies. Another approach to define the shape of the membership function consists in the use of the probability density function. The last mentioned technique will be discussed more thoroughly.

Approximation of obtained results calculated on the basis of the least-square criterion is shown in Fig. 3 (data from Fig. 2 - "*medium*" membership function). Intuitively, it seems appropriate to built the initial membership function using the probability density function and assuming that the parameter distribution is trapezoidal or triangular. The result of observed assumption is given by the function shown in Fig. 4.

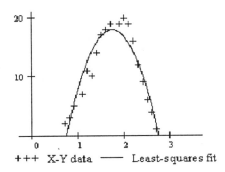

+++ X-Y data ——— Least-squares fit

Fig. 3 Least-square approximation of obtained results (quadratic fit)

The subsequent $f_1...f_3$ membership functions from Fig. 4 are defined by a set of parameters, A, b, c, d and e, and are determined as follows:

$$f_1(x, A, b, c) = \begin{cases} A & if \ x < b \\ \frac{A \cdot (c-x)}{(c-b)} & if \ b \leq x \leq c \\ 0 & if \ x > c \end{cases} \quad (9)$$

$$f_2(x, A, b, c, d, e) = \begin{cases} 0 & if \ x < b \ or \ x > e \\ \frac{A \cdot (x-b)}{(c-b)} & if \ b \leq x \leq c \\ A & if \ c < x < d \\ \frac{-A \cdot (x-e)}{(e-d)} & if \ d \leq x \leq e \end{cases} \quad (10)$$

$$f_3(x, A, d, e) = \begin{cases} 0 & if \ x < d \\ \frac{A \cdot (x-d)}{(e-d)} & if \ d \leq x \leq e \\ A & if \ x > e \end{cases} \quad (11)$$

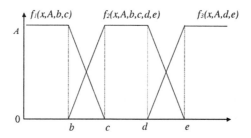

Fig. 4 Trapezoidal membership functions

The trapezoidal function $f_2(x, A, b, c, d, e)$, represents the estimation of the "*medium*" membership function. The m-th moment of the probability density for the function $f_2(x, A, b, c, d, e)$ is calculated as follows:

$$m_n = \int_{-\infty}^{+\infty} x^n f_2(x) dx \qquad (12)$$

The estimate of the m-th moment of the probability density function from the test (assuming that all observation instances fall to the interval j, $j=1,2,...,k$) is calculated according to the formula:

$$\hat{m}_n = \sum_{j=1}^{k} x^n (P(x_j)) \qquad (13)$$

where: $P(x = x_j)$ represents the probability that the attribute value of instance x falls to the interval j.

The subsequent moments of order from 0 to 4 for this function are given as a set of equations:

$$m_{0...4} = \left\{ \begin{array}{c} \frac{A \cdot (d-c+e-b)}{2} \\ \frac{A \cdot (d^2+e^2-de-b^2-c^2-bc)}{6} \\ \frac{A \cdot (d^3+e^3+de^2+d^2e-b^3-c^3-bc^2-b^2c)}{12} \\ \frac{A \cdot (d^4+e^4+de^3+d^2e^2+d^3e-b^4-c^4-bc^3-b^2c^2-b^3c)}{20} \\ \frac{A \cdot (d^5+e^5+de^4+d^2e^3+d^3e^2+d^4e-b^5-c^5-bc^4-b^2c^3-b^3c^2-b^4c)}{30} \end{array} \right\} \qquad (14)$$

Next, using values from observations in Eq. (14), the consecutive values of m_n are calculated. From this the set of 5 linear equation with 5 unknown variables A, b, c, d, e is to be determined. After solving this set of equation numerically, the second task of analysis is to validate the observed results using the Pearson's χ^2 test of k-1 degrees of freedom:

$$\chi^2 = \sum_{j=1}^{k} \frac{(n_j - np_j)^2}{np_j} \tag{15}$$

where: n_j - number of observed instances within the interval j, p_j - probability of the instance falling within the interval j (estimated by the probability density function), n - number of intervals j ($j=1,2,...,k$). Additionally, the n_{pj} should have the same value for all observation intervals.

Furthermore, it is assumed that the significance level should be set at 5%. After calculating the value of χ^2 from Eq. (15) it is to compare this value with the critical one given in the statistical tables. If the computed value is smaller than the one from the statistical tables, then the null hypothesis is valid. In the contrary case the hypothesis should be rejected and the assumed trapezoidal membership function is not a valid model of the measured phenomenon.

Using the above statistical method, the approximation of fuzzy membership function for the studied parameter can be done. The membership functions reflect the number of subjective votings given by experts to individual values of the assessed parameters (RT as discussed in the example in Section 4).

6 Automatic Quality Assessment Phase

In order to enable the automatic quality assessment procedure, new data representing parameter values of a given concert hall is fed to system inputs (Tab. VII). The first step is the fuzzification process in which degrees of membership are assigned for each crisp input value (as in Fig. 5). As was mentioned previously, the number of membership functions was limited to three. Therefore, for the data presented in Tab. VII, the degree of membership for each input value (for a given label) has to be determined.

Tab. VII Set of parameter values presented to system inputs

Hall	C_{def}	$Diff.$	$ITDG$	EDT	RT	$Loudn.$	C_{SI}
i	0.562	0.422	2.7	1.66	1.75	0.282	0.407

Pointers in Fig. 5 refer to the degree of memberships for the crisp value (RT=1.75). Thus, the value of RT equals 1.75 belongs respectively to the low fuzzy set with the degree of 0, to the medium fuzzy set with the degree of 0.65 and to the high fuzzy set with the degree of 0.25. The same procedure is applied to other parameters of Tab. VII.

It is to remember that after the rough set processing only the strongest rules were considered, thus these ones with the rough set measure exceeding value of 0.5. This rough measure will be treated further as the weight applied to this rule when it is used for the fuzzy processing. Additionally, the strength of the rule in the fuzzy processing is determined, according to fuzzy logic principles, on the basis of the smallest value of the degree of memberships found in the rule premise. Therefore, the overall strength of a rule is equal to the product of the rough set measure and the strength derived from the fuzzy processing (as shown below):

if ($C_{SI} = 0.35$) then (*Quality good*) => rule strength = $0.35 \cdot \mu_{RS} = 0.35 \cdot 1 = 0.35$,

if ($EDT = 0$) & ($C_{SI} = 0$) then (*Quality fair*) => rule strength = $0 \cdot \mu_{RS} = 0$

if ($C_{def} = 0.65$) & ($C_{SI} = 0.65$) & (*Loudn.* = 0) then (*Quality good*) => rule strength = $0 \cdot \mu_{RS} = 0$

if (*Loudn.* = 0.55) & ($RT = 0.65$) then (*Quality very good*) => rule strength = $0.55 \cdot \mu_{RS} = 0.55 \cdot 0.8 = 0.44$

if (*Loudn.* = 0.45) & ($EDT = 0.45$) then (*Quality good*) => rule strength = $0.45 \cdot \mu_{RS} = 0.45 \cdot 0.7 = 0.32$

if ($C_{def} = 0.65$) & ($C_{SI} = 0.65$) then (*Quality very good*) => rule strength = $0.65 \cdot \mu_{RS} = 0.65 \cdot 0.7 = 0.46$

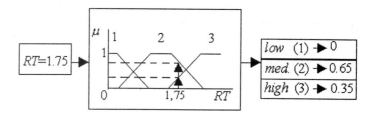

Fig. 5 Fuzzification process of the parameter RT

The next step is to determine the maximum rule strength of all rules involved in the same output action. The winning rule is found on the basis of the maximum value. Therefore, for the data set given in the example, the fuzzy output is defined in Tab. VIII.

Tab. VIII Fuzzy output of data from Tab. VII

Membership function	Fair	Good	Very good
Fuzzy output	0	0.35	0.46

The defuzzification method used in the system is based on the calculation of the centroid value, given by formula:

$$G = \frac{\int_a^b \mu(x) \cdot x \, dx}{\int_a^b \mu(x) \, dx} \tag{16}$$

However, in most cases it is sufficient to use the estimate of centroid, according to expression:

$$\hat{G} = \frac{\sum_a^b \mu(x) \cdot x}{\sum_a^b \mu(x)} \tag{17}$$

In this case, only a finite number of points from the output domain is taking into account.

A graphic illustration of the defuzzification process is shown in Fig. 6. In this figure the λ-cut resulting from the analysis is applied to each output membership function. The estimated value of centroid calculated for the data from Tab. VII is equal to 85. Thus, the obtained crisp value provides a measure of the automatically assessed acoustical quality of the tested object. As may be seen, the overall quality of the object from Tab. VII belongs to the scope of the fuzzy set labeled as *"very good"*.

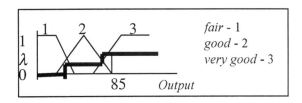

Fig. 6 Deffuzification process

7 Conclusions

The proposed method of automatic acoustical quality assessment provides a combination of the rough set decision system and the fuzzy logic inference. The rough set algorithm is applied to the decision table containing subjectively quantized parameters and the results of overall subjective preference of acoustical objects described by these parameters. The fuzzy membership functions are determined on the basis of the separate subjective testing of individual parameters underlying overall preference. In such a way the knowledge base is built containing both objective and subjective values and their hidden relations. Then, in the testing of the created expert system fuzzy logic is used which automatically provides the quality assessment using the 100 preference point scale. The fuzzy system uses the membership functions determined empirically for tested parameters and the rules generated in the training phase by the rough set algorithm.

As is seen from the discussion presented in this paper soft computing methods can find their way to acoustical quality assessments. There are many potential applications of intelligent algorithms applied to the sound quality evaluation, such as testing electroacoustic devices and newly created bit-rate compression algorithms or standards, etc. In the future, the engineered system will be further developed and tested by performing systematic experiments in the domain of automatic acoustical quality assessment.

References

1. Beranek L. , "Music, acoustics and architecture", J. Wiley & Sons, New York, 1962.
2. Ando Y., "Calculation of subjective preference at each seat in the concert hall", J. Acoust. Soc. Amer., Vol. 74, No. 3, 1983.
3. Schroeder M.R., Gottlob D., and Siebrasse K.F., "Comparative study of European concert halls, correlation of subjective preference with geometric and acoustic parameters", J. Acoust. Soc. Amer., Vol. 56, No. 4, 1974.
4. Kostek B., "Rough Set and Fuzzy Set methods Applied to Acoustical Analyses", J. Intell. Automation and Soft Computing - Autosoft, Vol. 2, No. 2, pp. 147-158, 1996.
5. Grey J.M., Multidimensional perceptual scaling of musical timbres, J. Acoust. Soc. Amer., Vol. 61, No. 5, 1977.
6. Thiele R., "Richtungsverteilung und Zeitvolge der Schallruckwurfe in Raumen", Acustica, Vol. 3, 1975.
7. Pawlak Z., "Rough sets", J. of Computer and Information Science, Vol. 11, No. 5, 1982.
8. Czyzewski A., Kaczmarek A. , "Speech Recognition Systems Based on Rough Sets and Neural Networks, 3rd Intern. Workshop on Rough Sets and Soft Computing, San Jose, California, USA, 1994.

APPLICATION OF FUZZY LOGIC AND ROUGH SETS TO AUDIO SIGNAL ENHANCEMENT

Andrzej CZYZEWSKI, Rafal KROLIKOWSKI

Technical University of Gdansk,
Faculty of Electronics, Telecommunications and Informatics,
Sound Engineering Dept.,
Narutowicza 11/12; 80-952 Gdansk, Poland.

1 INTRODUCTION

The noise and distortion reduction is important among others in wireless telephony, long distance wire transmission, in signal recording and in hearing prostheses. The majority of methods applied to this task are based on the estimation of signal representation in the spectral domain, adaptive filtration or on the predictive coding of audio. However, there are not many approaches which directly employ soft computing systems to this task. In the paper, two approaches to noise reduction are presented which are based on reasoning systems employing intelligent inference engines. A number of experiments related to the cancellation of stationary and non-stationary noise have been carried out using fuzzy logic and rough set systems. The paper presents some details of the engineered systems for noise and distortion reduction in audio signals.

2 NOISE AND DISTORTION REDUCTION METHOD

2.1 General Description of the Method

It is assumed that the useful signal s (not distorted) is affected by an additive noise n. Thus, the resultant signal y is represented in the time domain by the following formula:

$$y(k) = s(k) + n(k) \tag{1}$$

The spectrum of the signal y is expressed as follows:

$$Y(\omega) = S(\omega) + N(\omega) \tag{2}$$

In order to obtain a formula for the spectral representation of the non-distorted signal portion $S(\omega)$, the relationship (2) may be transformed as follows:

$$S(\omega) = Y(\omega) - N(\omega) = \left(1 - \frac{N(\omega)}{Y(\omega)}\right) \cdot Y(\omega) \tag{3}$$

However, as the explicit form of the disturbing noise $N(\omega)$ is difficult to obtain in practice, thus there is a need to introduce an estimator $\tilde{N}(\omega)$ of the noise $N(\omega)$. In such a case, the restored signal is only the estimate $\tilde{S}(\omega)$ of the original (non-distorted) signal $S(\omega)$:

$$N(\omega) \to \tilde{N}(\omega) \Longrightarrow S(\omega) \to \tilde{S}(\omega) \qquad (4)$$

Thus, the estimated non-distorted signal $\tilde{S}(\omega)$ may be expressed as follows:

$$\tilde{S}(\omega) = \left(1 - \frac{\tilde{N}(\omega)}{Y(\omega)}\right) \cdot Y(\omega) = H(\omega) \cdot Y(\omega), \qquad (5)$$

where: $H(\omega)$ - represents transfer function
$Y(\omega)$ represents the spectrum of the signal s affected by the noise n.

The filtration is performed with regard to the possibility distribution p. A spectral component can be removed (when it is qualified as noise) or can pass the filtration procedure unaffected depending on this distribution. If the estimated signal and noise components are similar, then the possibility p_o that the output $Y(\omega)$ represents noise in a given moment of time is: $p_o = p\left(\tilde{N}(\omega), Y(\omega)\right)$. Hence, the transfer function of the noise cancellation filter should be defined as follows:

$$H(\omega) = f\left(\tilde{N}(\omega), Y(\omega)\right) = \begin{cases} 0 & \text{with } p_o = p\left(\tilde{N}(\omega), Y(\omega)\right) \\ 1 & \text{with } p_o = 1 - p\left(\tilde{N}(\omega), Y(\omega)\right) \end{cases}, \qquad (6)$$

Since the signal $\tilde{N}(\omega)$ provides only an estimate of distortions, thus it is difficult to calculate the function $p\left(\tilde{N}(\omega), Y(\omega)\right)$ on the basis of hard-defined rules especially when $N(\omega)$ is non-stationary. This is the reason why intelligent inference engine with soft-defined rules is applied to determine $p\left(\tilde{N}(\omega), Y(\omega)\right)$.

2.2 Scheme of the Noise Reduction Method

As is seen in Fig. 1, noise patterns $\tilde{n}$ and a signal s affected by noise n are delivered to the system inputs. The noise patterns $\tilde{n}$ are correlated with the noise n and may be fed simultaneously with the signal $s + n$ or can be derived from the "silence" passages of a transmission. The patterns $\tilde{n}$ are analyzed in subbands of the frequency (corresponding to the so called critical bands of hearing). The distribution of the noise density $\rho(A, t)$ is evaluated which depends on distribution of magnitudes of spectral components A in time t in consecutive subbands. In turn, the signal affected by noise $s + n$ is analyzed in these subbands (critical bands) and in result, the distribution of the signal density $current(A, t)$ in each subband is evaluated. Next, these density distributions of the noise and the signal are compared inside the inference engine module. The first task of the inference engine is to discover the implicit relationship between them, i.e. to assess how similar each to other they are. Then, on the basis of the reasoning about this assessment, the estimated distribution $p(\rho, current)$ used for the removal of spectral components is determined. Thus, some components of the corrupted signal $s+n$ are removed by the filter depending on the distribution of $p(\rho, current)$. It means that a single spectral component of the signal $s+n$ is considered noise with the confidence level $p_o = p(\rho, current)$ or passes to the resynthesis with the confidence level $1-p_o$. In result, a denoised signal $\tilde{s}$ is obtained at the output of the system.

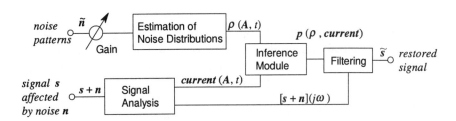

Fig. 1. General data flow in the intelligent spectral substraction noise reduction algorithm. Denotations:
$\rho(A,t)$ - estimated noise density distribution,
$current(A,t)$ - computed distribution of current signal density,
$p(\rho, current)$ - soft computed distribution of signals similarity of $(s+n)$ to $\tilde{n}$

2.3 Estimation of Signal and Noise Distribution

Let the critical band cb^{th} be concerned. The question is how many percent ρ of spectral components of a given magnitude A in this subband represent noise. In a given moment of time t_j, this ratio may be expressed as follows:

$$\rho(A,t) = \frac{k(A,t)}{X}|_{t=t_j}, \tag{7}$$

where: $k(A,t)$ - number of noise components of the magnitude value equals to A,
X - number of spectral components in a given critical band,
$j = 1, 2, ...$ - consecutive indexes of time.

Generally, the distribution introduced by the formula (7) satisfies the following relationship:

$$\int_0^{+\infty} \rho(A,t)\, dA\big|_{t=t_j} \leq 1, \tag{8}$$

However, in practice of digital signal processing, the values of magnitudes are quantized and discretized in time. Thus, the formula (7) can be transformed to:

$$\rho(\Delta_i, t_j) = \frac{k(\Delta_i, t_j)}{X}\big|_{t=t_j}; \qquad i = 1, ..., I, \tag{9}$$

where: Δ_i - interval of values from the range: $[(i-1)\cdot\Delta, (i+1)\cdot\Delta]$ of spectral magnitudes and Δ is the length of this constant interval,
$k(\Delta_i, t_j)$ - number of noise components which magnitude values belong to the ith interval Δ_i,
I - number of the intervals Δ_i the whole range of spectral magnitudes is divided into. For the maximum value MAX of the magnitudes in the cb^{th} critical band, it may be written that:

$$\lceil MAX \rceil = \Delta \cdot I, \tag{10}$$

where $\lceil\ \rceil$ denotes the entier operator. The procedure defined by the relationship (10) is illustrated in Fig. 2.

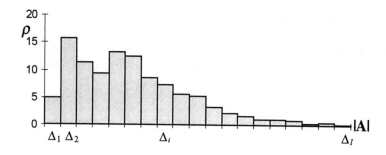

Fig. 2a. Exemplary distribution of noise density ρ in a single critical band within a given time-interval

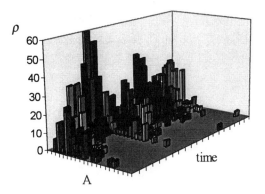

Fig. 2b. Exemplary distribution of noise density ρ in a single critical band plotted versus time

Consequently, the relationships in the formula (8) can be transformed to the following form:

$$\sum_{i=1}^{I} \rho(\Delta_i, t_j)|_{j=k} \leq 1; \qquad j, k = 1, 2, ..., \qquad (11)$$

where descriptions are the same as in relation (9).

The estimation of current signal distribution is performed in a similar way as the noise distribution was estimated. For the cb^{th} critical band, in a given moment of time t_j and by the analogy to the formula (7), the spectral magnitude density *current* may be expressed as follows:

$$current(A, t) = \frac{k(A, t)}{X}|_{t=t_j}, \qquad (12)$$

where: $k(A, t)$ - number of signal components which magnitude values are equal to A.

In practice, the expression of the formula (12) can be transformed to the following form:

$$current(\Delta_i, t_j) = \frac{k(\Delta_i, t_j)}{X}|_{j=k}; \qquad i = 1, 2, ..., I \wedge j, k = 1, 2, ..., \qquad (13)$$

where: $k(\Delta_i, t_j)$ - number of signal components which magnitude values belong to the ith interval Δ_i.

2.4 Implementations of the Inference Engine

Let the cb^{th} critical band and the ith interval Δ_i in this subband be concerned. Next, assume the following denotations:

- ρ is density of spectral magnitudes of components related to noise patterns, given by the formula (9),
- *current* is density of spectral magnitudes of components related to signal affected by noise (13),
- p is possibility that the output $Y(\omega)$ represents noise in a given moment of time

The variables: ρ, *current* are inputs and p is an output of the inference module as in Fig. 1.Practically, the noise density distribution $\rho(t)$ should not be calculated on the basis of a single noise pattern, but it should use a set of noise samples edited from different intervals of the received audio. Since the noise can be non-stationary, thus there is a need to take some exemplary portions of noise in such a way that the distribution ρ would be constantly correlated with the distibution *current*. Keeping this correlation high ensures that some characteristic combinations of signal and noise would be represented in the knowledge base of the system. This feature will be supported, provided the following terms are fulfilled:

$$\rho = MAX_{I} \left(\rho_k(j) \right)$$
$$\rho_k(j) = \frac{1}{I} \sum_{i=1}^{I} \rho_k(\Delta_i) \cdot current(\Delta_{i+j}); \qquad j = 1, ..., I/4 \text{'} \qquad (14)$$

where: ρ_k – represents distribution of noise in the pattern k
K- is the number of used noise samples,
other denotations are the same as in formulas (9) and (13).

2.5 Application of Fuzzy Reasoning

The first tested inference engine used for the assessment of similarity of ρ to *current* was based on fuzzy logic. Since rules reflecting the dependency p on ρ and *current* are easy to define for the expert, the application of fuzzy logic seems the most natural way to obtain values of p. The set of fuzzy rules was used as follows:

- if ρ is Low and *current* is Low then p is High
- if ρ is Low and *current* is Medium then p is Medium
- if ρ is Low and *current* is High then p is Low
- if ρ is Medium and *current* is Low then p is Medium
- if ρ is Medium and *current* is Medium then p is High
- if ρ is Medium and *current* is High then p is High
- if ρ is High and *current* is Low then p is Low
- if ρ is High and *current* is Medium then p is Medium
- if ρ is High and *current* is High then p is High

Membership functions defined experimentally for fuzzy variables are presented in Fig. 3. Some noisy speech patterns were analyzed in order to optimize shapes of the membership functions. The speech signal patterns mixed with stationary noise of different levels were used to find the widths and shapes of membership functions. The expert listened to the results of the speech enhancement when changing these parameters. An example of rules firing and evaluation of the value p by the fuzzy inference module for $\rho = 0.3$ and $current = 0.2$, for a given interval Δ and the selected critical band is presented in Fig. 4. Those rules which are not fired are omitted in Fig. 4. The example shows that since the distribution of the current signal portion $current$ is similar to the noise distribution ρ, it may be expected that majority of spectral components which magnitudes are located in the interval Δ are noisy ones. This is why the value of p related to this case is high (equal to 0.851). Consequently, spectral magnitudes from the interval Δ are removed with the confidence level of 0.851.

As the most flexible defuzzification method is the COG (*Center of Gravity Method*) [4] the concrete, crisp values p_o to be used in the formula (6) are calculated as follows:

$$p_o\left(\tilde{N}(\omega), Y(\omega)\right) = \frac{\int_P \mu_P\left(p, \tilde{N}(\omega), Y(\omega)\right) \cdot p\, dp}{\int_P \mu_P\left(p, \tilde{N}(\omega), Y(\omega)\right) dp} = \frac{\int_P \mu_P\left(p, \rho, current\right) \cdot p\, dp}{\int_P \mu_P\left(p, \rho, current\right) dp}, \quad (15)$$

where: p_o - crisp value of the output variable p,
$\mu_P(\cdot)$ - membership function for the variable p,
P - fuzzy domain of the fuzzy variable p,
other denotations mean the same as in the formula (6).

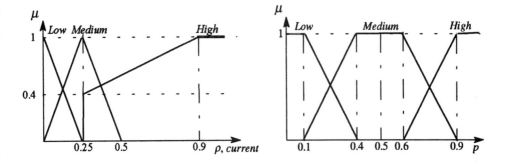

Fig. 3 Membership functions for variables: ρ, $current$ and for the output variable p

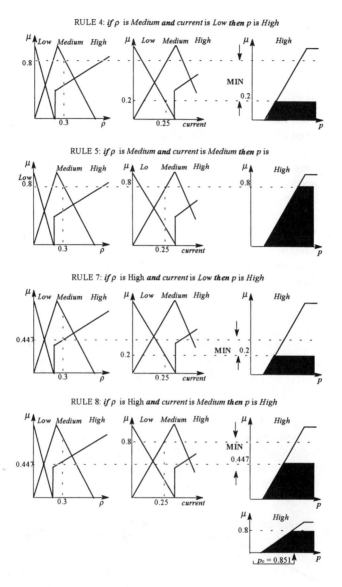

Fig. 4 Exemplary rules firing and evaluation of the confidence level p for $\rho = 0.3$, $current = 0.2$ by the fuzzy inference module

2.6 Application of Rough Set Based Inference System

The second tested inference module was based on rough sets. The rule induction algorithm was engineered which is presented in details further in this paragraph.

For practical reasons crisp values of the variables: ρ, $current$ and p should be discretized according to a selected discretization scheme $\Im$ [5]. In such a case, the output value $p_o(\rho, current)$ is defined by the formula:

$$p_o\left(\tilde{N}(\omega), Y(\omega)\right) = p_o(\rho, current) = f\left(\Im(\rho, current), \mathbf{DecisionTable}\right), \quad (16)$$

where: $\Im$ - discretization scheme,

DecisionTable – Pawlak's information table.

Subsequently rules are induced from the above decision table using the following rough set algorithm prepared for this task.

Rough Set-Based Rule Induction Algorithm:

The elaborated rule induction algorithm is based on the rough set methodology. Since the basic rough operators (the partition of a universe into classes of equivalence, C-lower approximation of X and calculation of a positive region) can be performed more efficiently when objects are ordered, the algorithm often executes sorting of all objects with respect to a set of attributes. Reducing of values of attributes requires that all combinations of the conditional attributes must be analysed. In general case, the decision table should be sorted as many times as is the number of all these combinations. Therefore, in every set of attributes A ($A \subseteq C$), a subset of the conditional attributes C should be exploited as many times as possible. The scheme of the optimal sequence of the analysis of the conditional attributes is illustrated in Fig. 5. Numbers in brackets indicate the order in which procedures *left*() and *_P*() are executed in *P*().

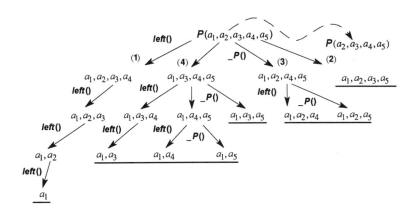

Fig. 5. Simplified scheme of analysis of decision table computationally optimised with regard to a given subset of attributes

The algorithm splits the decision table T into two tables: consisting of only certain rules (T_{CR}) and of only uncertain rules (T_{UR}). For them both, there is additional information associated with every object in them. The information concerns the minimal set of indispensable attributes and the rough measure μ_{RS}. The latter case is applied only for uncertain rules.

The elaborated algorithm consists of the following main group of modules:

a) Master procedure of all procedures related to the rough set-based induction algorithm

– **procedure** RS_algorithm

It is assumed that a decision table is fed to the procedure either at a call or supplied during its execution. By analogy, the procedure returns a table with generated rules.

b) Initial procedures

– **procedure** preprocessing

The procedure prepares 2 tables of certain T_{CR} and uncertain rules T_{UR}. In addition, it computes the set of concepts V with respect to the set of attributes D.

– **procedure** generate_rules(C : set_of_attributes)

Input: C - the set of the conditional attributes

The procedure is a master procedure of all those procedures and functions which task is to generate rules by removing superfluous values of attributes. At the end, these values are replaced by 'do not care' value. However, the procedure affects the tables T_{CR} and T_{UR} in an implicit way.

– **procedure** postprocessing

The procedure prepares the output table of generated rules T in such a way that a rule could be accessed in at most $|C| \cdot \log_2 N$ comparisons, where N is the number of objects in T.

c) Procedures preserving the proper depth of the analysis of the conditional attributes

– **procedure** P (C, A : set_of_attributes)

Input: C - the set of the conditional attributes,
         A - an arbitrary set of attributes, where $A \subseteq C$
Output: potentially modified auxiliary data associated with
         the tables T_{CR}, T_{UR}

The procedure provides the proper depth of the analysis of all combinations of the conditional attribute set C and is executed recursively.

- **procedure** _P (A : set_of_attributes; *depth* : integer)

    Input: A - an arbitrary set of conditional attributes,
               depth - depth of recursions,
    Output: potentially modified auxiliary data associated with
                 the tables T_{CR}, T_{UR}

Similarly to the procedure P(), this procedure provides the proper order of the analysis of all combinations of the conditional attribute set C. It contributes to a further recurrent processing of the conditional attributes and is similar to P().

- **procedure** left (A : set_of_attributes; *depth* : integer)

    Input: A - an arbitrary set of conditional attributes,
               depth - depth of recursions,
    Output: potentially modified auxiliary data associated with
                 the tables T_{CR}, T_{UR}

The procedure is strictly related to the proper depth of the analysis of the conditions and concerns recurrent processing of the left branch of the tree from Fig. 5. The execution of the procedure enables the analysis of some subsets of the attributes without sorting the table .

d) Procedures eliminating superfluous values of attributes in certain and uncertain rules

- **function** process_certain_rules (A : set_of_attributes) : set_of_sets_of_objects

    Input: A - an arbitrary set of conditional attributes,
    Output: Z - set of sets of objects belonging to the same equivalence class,
                potentially modified auxiliary data associated with the table T_{CR}

The function performs all necessary operations in order to reduce values of attributes. In consequence, it may cause a removal of an attribute. The procedure affects only certain rules. The output value $Z = U_{CR}/IND(A)$ is returned so that the procedure **process_uncertain_rules** doesn't need to compute the partition $U_{CR}/IND(A)$.

- **procedure** process_uncertain_rules
    (Z : set_of_sets_of_objects; A : set_of_attributes)

    Input: Z - set equivalence classes for the table T_{CR} ($U_{CR}/IND(A)$) ,
               A - an arbitrary set of conditional attributes
    Output: potentially modified auxiliary data associated with the table T_{UR}

The procedure calculates the rough measure for each object of the decision table T_{UR} (uncertain rules) and for each combination of attributes. The largest values of the measure are stored.

e) Functions which are related to the basic rough set operators

- **function** U_IND (T : table; A : set_of_attributes) : set_of_sets_of_objects

    Input: T - a decision table,
    A - an arbitrary set of attributes,
    Output: a set of objects belonging to the same equivalence class

    The function parts the universe U (table T) into classes of equivalence $U/IND(A)$ according to indiscernibility relation with respect to a set of attributes A.

- **function** _CX (T : table; C, X : set_of_attributes) : set_of_objects

    Input: T - a decision table,
    C, X - arbitrary sets of attributes,
    Output: a set of objects in T which belong to the C-lower approximation of the set X

    The function computes the C-lower approximation of X.

- **function** POS_REG (T : table; A1, A2 : set_of_attributes) : set_of_objects

    Input: T - a decision table,
    $A1, A2$ - arbitrary sets of attributes,
    Output: a set of objects in T which constitute the positive region

    The function calculates the positive region of classification $U/IND(A2)$ for the set of attributes $A1$.

f) Auxiliary procedure and function

- **procedure** checkU_IND (Y : set_of_sets_of_objects; A : set_of_attributes)

    Input: Y - a set of equivalence classes after the partitioning $U_{CR}/IND(A)$,
    A - an arbitrary set of conditional attributes
    Output: potentially modified auxiliary data associated with the table T_{CR}

    The procedure checks whether all objects belonging to the partition $U_{CR}/IND(A)$
    have the same decision values. If so and if the number of current dispensable attributes is less than the number of such attributes for these objects, stored so far, it stores the information of the dispensable attributes.

- **function** intersection (X, Y : set_of_objects) : set_of_objects

    Input: X, Y - ascending sorted sets of objects
    Output: Z - the intersection of X and Y

    The function calculates the intersection of the input sets X and Y. The product is returned by the output set $Z = X \cap Y$. The assumption is that the input sets X and Y are ascending sorted sets and therefore in the worst case, there are $|X|+|Y|$ comparisons necessary. However this case is unlikely to occur.

3 RESULTS AND CONCLUSIONS

The noise reduction system was implemented in the C++ computer language. In order to test the engineered methods, a number of experiments was carried out. At the beginning, the experiments concerned signals affected by stationary or almost stationary noise. Some of them were presented in the previous papers [1][2][3]. Recently, the elaborated noise cancellation systems were optimized to restore audio signals affected by non-stationary noise. These corrupted signals were obtained from really existing telecommunication channels.

The experiments confirmed that the reduction of noise which is augmented by intelligent inference engines can work quite efficiently. Despite the simplicity of audio signal and noise representations, the elaborated system is able to restore audio signals affected by both stationary and non-stationary noise. Both presented soft computing based inference engines brought subjectively similar results, although there are some differences between them. The fuzzy logic implementation is simpler, however the COG deffuzzyfying procedure makes difficult real-time applications of this method. It does not concern the rough set system which can work extremely fast in the rule matching phase. Due to the use of rough set system the process of building the knowledge base of signal and of distortions may be augmented by including into decision table opinions provided by different experts. Currently, this subject is in the scope of research conducted by the authors.

References

1. Czyzewski A., Krolikowski R., " Simultaneous Noise Reduction and Data Compression in Old Audio Recordings", 101st AES Conv., Los Angeles 1996, Preprint 4338.
2. Czyzewski A., Krolikowski R., " New Methods of Intelligent Filtration and Coding of Audio", 102nd AES Conv., Munich 1997, Preprint 4482.
3. Czyzewski A., Krolikowski R., " Application of Intelligent Decision Systems to the Perceptual Noise Reduction of Audio Signals", Proceedings of Fifth European Congress on Intelligent Techniques and Soft Computing, Aachen, Germany, vol. 1, pp. 188-192.
4. Kosko B., " Neural networks and fuzzy systems. A dynamical system approach to machine intelligence.", Prentice-Hall Int., New York, 1992.
5. Nguyen H.S., Nguyen S.H.: Discretization of Real Value Attributes for Control Problems, Proc. of 4th European Congress on Intelligent Techniques and Soft Computing, Aachen, Germany, Sept., (1996).

4 ACKNOWLEDGMENTS

Research was sponsored by the Committee for Scientific Research, Warsaw, Poland. Grant No. 8 T11D 021 12

Application of Fuzzy-Rough Sets in Modular Neural Networks

Manish Sarkar and B. Yegnanarayana

Neural Networks Laboratory
Department of Computer Science & Engineering
Indian Institute of Technology, Madras
Chennai - 600 036, India
{manish@bronto, yegna}.iitm.ernet.in

Abstract

In a modular neural network, the conflicting information supplied by different information sources, i.e., the outputs of the subnetworks, can be combined by applying Sugeno's fuzzy integral. Here the fuzzy integral works as a nonlinear aggregation operator. To compute the fuzzy integral, it is essential to know the importance of each subset of the information sources in a quantified form. In the fuzzy integral approach the importance of a particular information source is considered to be independent of the other information sources. Therefore, determination of the importance of each information source should be based on the incomplete knowledge supplied by the source itself. Using this fact, this article proposes a fuzzy-rough entropic measure to find the importance of each subset of the information sources from this incomplete knowledge.

1 Introduction

Generally it is difficult to train a monolithic feedforward neural network, i.e., *All-Class-One-Network* (ACON) [10], for a complex classification task involving large number of similar classes. Such cases arise in script recognition [3], speech recognition [20], game playing [26]. If the number of classes is large, in many cases, the ACON classifier either does not converge or takes lot of time to converge. It is possible to develop a classifier based on the concept of *One-Class-One-Network* (OCON) [10] architecture, where a separate network is trained for each class. But, this approach requires a large number of subnetworks, and in addition, the discrimininatory capability of the OCON classifiers is poor [20]. Modular approach is one viable alternative. Modularity can be viewed as a *divide and conquer* approach, which permits one to solve a complex classification task by dividing it into simple subtasks. Specifically, in the modular approach of classification the large number of classes are grouped into smaller subgroups, and a separate neural network is trained for each subgroup [20]. The principle of modular networks can be thought of as the working principle of *Some-Class-One-Network* (SCON) architecture. The outputs of the modules are mediated by an integrating unit, which is not permitted to feed the information back to the modules. The modular networks have the advantages of both ACON and

OCON approaches, like quick convergence, parallel training, better generalization. One way to decompose a monolithic network is to create modules that serve very different functions. The top-down structure of a large software project is an example, where each procedure has its own function. This is called *functional* modularisation [4]. Another way is to decompose the problem such that the modules perform different versions of the same job. It is called *categorical* modularisation [4]. It can be thought of as a set of experts giving their individual opinions on the same subject.

This article proposes a technique to fuse the information supplied by the subnetworks of a modular network with functional modularisation. The proposed method interprets each subnetwork as a nonlinear filter tailored to the subgroup. The outputs of all the filters are viewed as a feature vector representing the input. Each module classifies the input pattern from different angles. In other words, each feature, i.e., the output of each module, can be considered as an evidence in classifying the input. Each of these evidences may support or contradict one another. The conflicts arise primarily due to the following two reasons:

- Ideally, when an input pattern does not belong to any class of a module, all the outputs of the module are supposed to be close to zero. However, if the output of the module represents the *a posteriori* probability of belongingness of the input pattern into a class, then the sum of the outputs of the module will be equal to one. Thus, some or all the outputs of the modules will be nonzero, which indicates that the input belongs to some classes of the module. Consequently, each module claims that the input is from the module.
- The modules are trained locally, and hence, the modules cannot resolve the global uncertainties. For example, if the classes of any two modules are close or overlapping, then for an input, outputs of both the modules may be high. In other words, each of these two modules claims that the input belongs to some class of the module. Moreover, due to *roughness* [17], an input may completely belong to two different classes. If these classes are from two different modules, then for a similar test input, both the modules will produce high outputs. It indicates that the input belongs to both the modules.

Due to the presence of conflicts and cooperations, each feature would have a different degree of importance in classifying the input to a particular class. The classification capability of a feature for a particular class is known as *partial evaluation*. In this article, Sugeno's fuzzy integral [8] [6] is used to combine the partial evaluations of all the features with the importance of the subsets of the features to yield the final classification result. Henceforth, we use the term fuzzy integral to mean Sugeno's fuzzy integral.

The behavior of the fuzzy integral in an application depends critically on the importance of the subsets of the features, which further depends on the importance of each feature. Therefore, determination of the worth of each feature is very important. In some applications of the fuzzy integral, the importances is supplied subjectively by an expert or it is estimated directly from the data [24] [21]. These methods require some kind of prior knowledge about the behavior of the outputs generated by the modules. In many applications, it may be difficult to obtain the prior knowledge. However, it is interesting to note that in the fuzzy integral approach, influence of the other features on a given feature is not considered. Hence, determination of the importance of a particular feature is based on the partial information supplied by the feature itself. A feature is important for a particular class when all the input patterns can be classified correctly to that class based only on this feature value. This is possible when all the input patterns are clustered based only on this feature value and all the input patterns from each cluster have the same class label. When this does not happen, the relationship between the clusters and the output class labels becomes one-to-many. It results in *rough ambiguity* [15]. In this case, the notion of *rough sets* [15] can be effectively exploited to determine the importance of each feature for a particular output class. Moreover, the information supplied by each feature is inherently fuzzy. In other words, the clusters formed in the input space based on each feature value is fuzzy. Therefore, in this article, an attempt is made to determine the importance of each feature using *fuzzy-rough set* [5] theoretic technique. The performance of the proposed scheme is studied on a Contract Bridge Opening Bid problem.

The organization of the article is as follows: Section 2 discusses the background materials for fuzzy measure, fuzzy integral, rough sets, rough-fuzzy sets and fuzzy-rough sets. Section 3 describes the architecture, training strategy and testing of the proposed modular network. Section 4 illustrates the experimental results.

2 Background

2.1 Fuzzy Measure

Let Ξ be a finite set of elements. A set function $g : 2^\Xi \to [0, 1]$ with the following properties is called a fuzzy measure [23]:

P1: $g(\phi) = 0$
P2: $g(\Xi) = 1$
P3: If $U \subseteq V$, then $g(U) \subseteq g(V)$, where $U, V \subseteq \Xi$

The fuzzy measure generalizes the classical measure which plays a crucial role in probability and integration theory. A probability measure P is characterized by the property of additivity: For all sets U and V, if $U \cap V = \phi$, then $P(U \cup V) = P(U) + P(V)$. In the fuzzy measure, this property of additivity is weakened by the more general property of monotonicity (property *P3*). Sugeno's g_λ measure is a special type of fuzzy measure [23] which satisfies all the properties of the fuzzy measure, in addition to the following:

$$g(U \cup V) = g(U) + g(V) + \lambda g(U)g(V) \tag{1}$$

where $\lambda > -1$, $U, V \subseteq \Xi$ and $U \cap V = \phi$. By varying the values of λ, one can obtain different types of fuzzy measure. For example, $\lambda = 0$ produces the probability measure.

2.2 Fuzzy Integral

Let $\Xi = \{\xi_1, \xi_2, \ldots, \xi_S\}$ be a finite set of elements, $h : \Xi \to [0, 1]$ be a mapping and g be a fuzzy measure on Ξ. Then the fuzzy integral (over Ξ) of the function h with respect to the fuzzy measure g is defined as

$$\mathcal{F} = h(\Xi) \circ g() \qquad (2)$$

$$= \max_{\Omega \subseteq \Xi} \left[\min \left(\min_{\xi_s \in \Omega} (h(\xi_s)), g(\Omega) \right) \right] \qquad (3)$$

where $1 \leq s \leq S$. Since both h and g map to $[0, 1]$, $\mathcal{F}$ also lies in $[0, 1]$. The above integral can be seen as an extension of Lebesgue integral if the product and summation operators are substituted for min and max, respectively. Intuitively the interpretation of the above relation is as follows: Let us suppose, an object is evaluated from the point of view of a set of information sources Ξ. Let $h(\xi_s) \in [0, 1]$ denote the decision for the object when a single information source $\xi_s \in \Xi$ is considered. Moreover, suppose $g(\{\xi_s\})$, known as *fuzzy density*, denotes the importance of the source ξ_s. Instead of a single information source, if a set of sources, namely $\Omega \subseteq \Xi$, is taken to evaluate the object, then it is reasonable to consider $\min_{\xi_s \in \Omega} h(\xi_s)$ as the largest security decision. Evidently, $g(\Omega)$ expresses the degree of importance or the expected worth of the set Ω. Therefore, $\min \left(\min_{\xi_s \in \Omega} (h(\xi_s)), g(\Omega) \right)$ denotes the grade of agreement between the real possibility h and the expectation g. Thus, the fuzzy integral can be interpreted as a search for the maximal grade of agreement between the objective evidence and the expectation. However, the definition can further be simplified if $h(\xi_s)$, $s = 1, 2, \ldots, S$ are ordered in a decreasing manner. Let $h(\xi_1) \geq h(\xi_2) \geq \ldots \geq h(\xi_S)$ (if not, Ξ is rearranged so that this relation holds). Then the relation (3) is simplified to

$$\mathcal{F} = h(\Xi) \circ g() = \max_s \left[\min \left(h(\xi_s), g(\Omega_s) \right) \right] \qquad (4)$$

where $\Omega_s = \{\xi_1, \xi_2, \ldots, \xi_s\}$.

In order to evaluate the fuzzy integral, i.e., $\mathcal{F}$, we should have some way to determine $g(\Omega_s)$ from $g(\{\xi_s\})$. For that, we need to use the concept of the fuzzy measure. In the next section we will show how to determine the individual fuzzy densities $g(\{\xi_s\})$, $s = 1, 2, \ldots, S$ for each information source from the given data. For the time being, let us suppose that we know the fuzzy densities of the individual sources. But, $g(\Omega_s)$ is not necessarily equal to $g(\{\xi_1\}) + g(\{\xi_2\}) + \ldots + g(\{\xi_s\})$. The simple additive property may not hold because there may be some interactions among ξ_s. If the interactions are cooperative, then $g(\Omega_s) \geq g(\{\xi_1\}) + g(\{\xi_2\}) + \ldots + g(\{\xi_s\})$. On the contrary, if the interactions are noncooperative, then $g(\Omega_s) \leq g(\{\xi_1\}) + g(\{\xi_2\}) + \ldots + g(\{\xi_s\})$ [11]. From this discussion, note that probability theory cannot be used to determine the value of $g(\Omega_s)$. However, the concept of Sugeno's g_λ fuzzy measure can be exploited here to find the value of $g(\Omega_s)$. The procedure is as follows:

$$g(\Omega_1) = g(\{\xi_1\})$$
$$g(\Omega_2) = g(\{\xi_2\}) + g(\{\xi_1\}) + \lambda g(\{\xi_2\})g(\{\xi_1\})$$
$$= g(\{\xi_2\}) + g(\Omega_1) + \lambda g(\{\xi_2\})g(\Omega_1)$$
$$\cdots \quad \cdots \quad \cdots \quad \cdots$$
$$g(\Omega_s) = g(\{\xi_s\}) + g(\Omega_{s-1}) + \lambda g(\{\xi_s\})g(\Omega_{s-1}) \quad \text{for } 1 < s \leq S \qquad (5)$$

One problem remains still unresolved; that is, how to determine λ, which is the key term to decide the amount of interactions among the information sources. In order to find λ, we use the equation (5), and we express $g(\Xi)$ in terms of the individual fuzzy densities as follows:

$$g(\Xi) = g(\{\xi_S\}) + g(\{\xi_1, \xi_2, \ldots, \xi_{S-1}\})$$
$$+ g(\{\xi_S\})\lambda g(\{\xi_1, \xi_2, \ldots, \xi_{S-1}\}) \qquad (6)$$
$$= \sum_{s=1}^{S} g(\{\xi_s\}) + \lambda \sum_{s=1}^{S-1} \sum_{k=s+1}^{S} g(\{\xi_s\})g(\{\xi_k\}) + \ldots$$
$$+ \lambda^{S-1} g(\{\xi_1\})g(\{\xi_2\}) \cdots g(\{\xi_S\}) \qquad (7)$$
$$= \left[\prod_{s=1}^{S} (1 + \lambda g(\{\xi_s\})) - 1 \right] \Big/ \lambda \qquad \text{where } \lambda \neq 0 \qquad (8)$$

From $(P2)$, we know that the value of g over the whole set Ξ must be one as no uncertainty is involved. Hence, using $g(\Xi) = 1$ and the equation (6), we obtain

$$\prod_{s=1}^{S} (1 + \lambda g(\{\xi_s\})) = \lambda + 1 \qquad (9)$$

It is possible to find the value of λ after solving the above $(S-1)$th degree equation. In [24], it has been shown that λ has a unique value in $(-1, 0) \cup (0, +\infty)$ when $0 < g(\{\xi_s\}) < 1$, $\forall s = 1, 2, \ldots, S$.

2.3 Rough, Rough-Fuzzy and Fuzzy-Rough Sets

In any classification task the aim is to form various classes where each class contains objects that are not significantly different. These *indiscernible* or indistinguishable objects can be viewed as basic building blocks (concepts) used to build a knowledge base about the real world. For example, if the objects are classified according to color (red, black) and shape (triangle, square and circle), then the classes are: red triangles, black squares, red circles, etc. Thus, these two attributes make a *partition* in the set of objects and the universe becomes coarse. Now, if two red triangles with different areas belong to different classes, it is impossible for anyone to correctly classify these two red triangles based on the given two attributes. This kind of uncertainty is referred to as *rough uncertainty* [15]. The rough uncertainty is formulated in terms of *rough sets* [16]. Obviously, the rough uncertainty can be completely avoided if we can successfully extract the essential features so that distinct feature vectors are used to represent different objects. But, it may not be possible to guarantee as our knowledge about the system generating the data is limited [22]. Therefore, rough sets are essential to deal with a classification system, where we do not have complete knowledge about the system.

Let R be an equivalence relation on a universal set X. Moreover, let X/R denote the family of all equivalence classes induced on X by R. One such equivalence class in X/R, that contains $\mathbf{x} \in X$, is designated by $[\mathbf{x}]_R$. For any output class $C_c \subseteq X$, we can define the lower $\overline{R}(C_c)$ and upper $\underline{R}(C_c)$ approximations, which approach C_c as closely as possible from inside and outside, respectively [9]. Here,

$$\underline{R}(C_c) = \cup\{[\mathbf{x}]_R \mid [\mathbf{x}]_R \subseteq C_c, \mathbf{x} \in X\} \qquad (10)$$

is the union of all equivalence classes in X/R that are contained in C_c, and

$$\overline{R}(C_c) = \cup\{[\mathbf{x}]_R \mid [\mathbf{x}]_R \cap C_c \neq \phi, \mathbf{x} \in X\} \qquad (11)$$

is the union of all equivalence classes in X/R that overlap with C_c. A rough set $R(C_c) = \langle \overline{R}(C_c), \underline{R}(C_c) \rangle$ is a representation of the given set C_c by $\underline{R}(C_c)$ and $\overline{R}(C_c)$ [16]. The set $BN(C_c) = \overline{R}(C_c) - \underline{R}(C_c)$ is a rough description of the boundary of C_c by the equivalence classes of X/R. The approximation is rough uncertainty free if $\overline{R}(C_c) = \underline{R}(C_c)$. Thus, when all the patterns from an equivalence class do not carry the same output class label, rough ambiguity is generated as a manifestation of the one-to-many relationship between that equivalence class and the output class labels.

One possible way to measure the rough uncertainty associated with a pattern is through *rough membership function* [25]. The rough membership function $r_{C_c}(\mathbf{x}) : C_c \to [0, 1]$ of a pattern $\mathbf{x} \in X$ for the output class C_c is defined by [25]

$$r_{C_c}(\mathbf{x}) = \frac{\|[\mathbf{x}]_R \cap C_c\|}{\|[\mathbf{x}]_R\|} \qquad (12)$$

where $\|C_c\|$ denotes the cardinality of the set C_c.

When the output class $C_c \subseteq X$ is a fuzzy set, the formulation of rough set can be generalized to *rough-fuzzy* set. A rough-fuzzy set is a tuple $\langle \overline{R}(C_c), \underline{R}(C_c) \rangle$, where the lower approximation $\underline{R}(C_c)$ and the upper approximation $\overline{R}(C_c)$ of C_c are fuzzy sets of X/R, with membership functions defined by [5]

$$\mu_{\underline{R}(C_c)}([\mathbf{x}]_R) = \inf\{\mu_{C_c}(\mathbf{x}) \mid \mathbf{x} \in [\mathbf{x}]_R\} \qquad (13\text{-a})$$

$$\mu_{\overline{R}(C_c)}([\mathbf{x}]_R) = \sup\{\mu_{C_c}(\mathbf{x}) \mid \mathbf{x} \in [\mathbf{x}]_R\} \qquad (13\text{-b})$$

Here, $\mu_{\underline{R}(C_c)}([x]_R)$ and $\mu_{\overline{R}(C_c)}([x]_R)$ are the membership values of $[\mathbf{x}]_R$ in $\underline{R}(C_c)$ and $\overline{R}(C_c)$, respectively.

Similar to rough membership function, it is possible to quantise the rough-fuzzy uncertainty associated with a pattern in form of *rough-fuzzy membership function* [19]. The rough-fuzzy membership value of a pattern $\mathbf{x} \in X$ for the fuzzy output class $C_c \subseteq X$ is defined by [19]

$$\iota_{C_c}(\mathbf{x}) = \frac{\|F \cap C_c\|}{\|F\|} \tag{14}$$

where $F = [\mathbf{x}]_R$ and $\|C_c\|$ is equal to the cardinality of the fuzzy set C_c. One possible way to determine the cardinality is to use [27] $\|C_c\| \stackrel{def}{=} \sum_{\mathbf{x} \in X} \mu_{C_c}(\mathbf{x})$. For the '∩' (intersection) operation, we can use [27] $\mu_{C_c \cap B}(\mathbf{x}) \stackrel{def}{=} \min\{\mu_{C_c}(\mathbf{x}), \mu_B(\mathbf{x})\}$ $\forall \mathbf{x} \in X$. When the output class is crisp, the definition (14) reduces to the definition (12).

When the equivalence classes are not crisp, i.e., they are in form of fuzzy clusters, the concept of rough-fuzzy set can be further generalized to *fuzzy-rough set* [5]. Let the fuzzy clusters $\{F_1, F_2, \ldots, F_H\}$ be generated by a *fuzzy weak partition* [5] of the input set X. The term fuzzy weak partition means that each F_j is a normal fuzzy set, i.e., $\max_\mathbf{x} \mu_{F_j}(\mathbf{x}) = 1$ and $\inf_x \max_j \mu_{F_j}(\mathbf{x}) > 0$ while

$$\sup_\mathbf{x} \min\{\mu_{F_i}(\mathbf{x}), \mu_{F_j}(\mathbf{x})\} < 1 \quad \forall\, i,\, j \in \{1, 2, \ldots, H\} \tag{15}$$

Here $\mu_{F_j}(\mathbf{x})$ is the fuzzy membership function of the pattern $\mathbf{x}$ in the cluster F_j. In addition, the output classes C_c, $c = \{1, 2, \ldots, C\}$ may be fuzzy too. Then the fuzzy set C_c can be described by means of the fuzzy partitions under the form of an upper and a lower approximation $\overline{C_c}$ and $\underline{C_c}$ is as follows:

$$\mu_{\underline{C_c}}(F_j) = \inf_\mathbf{x} \max\{1 - \mu_{F_j}(\mathbf{x}), \mu_{C_c}(\mathbf{x})\} \; \forall\, j \tag{16-a}$$

$$\mu_{\overline{C_c}}(F_j) = \sup_\mathbf{x} \min\{\mu_{F_j}(\mathbf{x}), \mu_{C_c}(\mathbf{x})\} \; \forall\, j \tag{16-b}$$

The tuple $\langle \underline{C_c}, \overline{C_c} \rangle$ is called a fuzzy-rough set. Here, $\mu_{C_c}(\mathbf{x}) = \{0,\, 1\}$ is the fuzzy membership of the input $\mathbf{x}$ to the class C_c. Fuzzy-roughness appears when a fuzzy cluster contains patterns that belong to different classes.

Now, the definition (14) can be generalized to the following definition of the fuzzy-rough membership function [18]:

$$\tau_{C_c}(\mathbf{x}) = \begin{cases} \frac{1}{\hat{H}} \sum_{j=1}^{\hat{H}} \mu_{F_j}(\mathbf{x}) \iota_{C_c}^j(\mathbf{x}) & \text{if } \exists j \text{ with } \mu_{F_j}(\mathbf{x}) > 0 \\ 0 & \text{otherwise} \end{cases} \tag{17}$$

where $\hat{H}$ ($\leq H$) is the number of clusters in which $\mathbf{x}$ has nonzero memberships and $\iota_{C_c}^j(x) = \frac{\|F_j \cap C_c\|}{\|F_j\|}$. Here, $\tau_{C_c}(\mathbf{x})$ represents the fuzzy-rough uncertainity of $\mathbf{x}$ in the class C_c.

Some important properties of the fuzzy-rough membership functions are [18]

1. $0 \leq \tau_{C_c}(\mathbf{x}) \leq 1$
   Proof: Since $\phi \subseteq F_j \cap C_c \subseteq F_j$, $0 \leq \iota_{C_c}^j(\mathbf{x}) \leq 1$. Moreover, $0 \leq \mu_{F_j}(\mathbf{x}) \leq 1$. Hence, the proof follows.

2. $\tau_{C_c}(\mathbf{x}) = 1$ or 0 if and only if no fuzzy-rough uncertainty is associated with the pattern $\mathbf{x}$.
   Proof:
   If part: If no fuzzy-rough uncertainty is involved, then $\mathbf{x}$ must belong completely to all the clusters in which it has non-zero belongingness. It implies $\mu_{F_j}(\mathbf{x}) = 1$ for which $\mu_{F_j}(\mathbf{x}) > 0$. Moreover, all the clusters in which $\mathbf{x}$ has non-zero belongingness either (a) must be the subsets of the class C_c, or (b) must not share any pattern with the class C_c. In other words, the condition (a) implies that $F_j \subseteq C_c \ \forall \ j$ for which $\mu_{F_j}(\mathbf{x}) > 0$. Hence, $\tau_{C_c}(\mathbf{x}) = \frac{1}{\hat{H}}\sum_{j=1}^{\hat{H}} 1.1 = 1$. Similarly the condition (b) expresses that $F_j \cap C_c = \phi \ \forall \ j$ for which $\mu_{F_j}(\mathbf{x}) > 0$. Hence, $\tau_{C_c}(\mathbf{x}) = \frac{1}{\hat{H}}\sum_{j=1}^{\hat{H}} \mu_{C_c}(\mathbf{x}).0 = 0$
   Only if part: If $\tau_{C_c}(\mathbf{x}) = 0$, then $\mathbf{x}$ may not belong to any cluster. Therefore, in this case there is no fuzzy-roughness associated with $\mathbf{x}$. Otherwise, each term under the summation symbol, i.e., $\mu_{F_j}(\mathbf{x})\iota_{C_c}^j(\mathbf{x})$ is separately zero. It implies that either $\mu_{F_j}(\mathbf{x})$ or $\iota_{C_c}^j(\mathbf{x})$, or both $\mu_{F_j}(\mathbf{x})$ and $\iota_{C_c}^j(\mathbf{x})$ are zero. If $\mu_{F_j}(\mathbf{x}) = 0$, then the pattern $\mathbf{x}$ does not belong to the cluster F_j, and hence, no fuzzy-rough uncertainty is involved with $\mathbf{x}$. If $\iota_{C_c}^j(\mathbf{x}) = 0$, then F_j and C_c do not have any pattern common, and therefore, no fuzzy-rough uncertainty exists with $\mathbf{x}$. Thus, $\tau_{C_c}(\mathbf{x}) = 0$ implies that fuzzy-roughness is not associated with the pattern $\mathbf{x}$. If $\tau_{C_c}(\mathbf{x}) = 1$, then $\mu_{F_j}(\mathbf{x}) = 1$ and $\iota_{C_c}^j(\mathbf{x}) = 1$, $\forall j = 1, 2, \ldots, \hat{H}$. It also indicates the absence of fuzzy-roughness.

3. *If no fuzzy linguistic uncertainty is associated with the pattern* $\mathbf{x}$, *then* $\tau_{C_c}(\mathbf{x}) = \iota_{C_c}^j(\mathbf{x})$ *for some* $j \in \{1, 2, \ldots, H\}$.
   Proof: If no fuzzy linguistic uncertainty is involved, then $\mu_{F_j}(\mathbf{x}) = 1$ for some $j \in \{1, 2, \ldots H\}$, and $\mu_{F_k}(\mathbf{x}) = 0$, for $k \in \{1, 2, \ldots H\}, j \neq k$. Hence, $\tau_{C_c}(\mathbf{x}) = \iota_{C_c}^j$, $j \in \{1, 2, \ldots, H\}$.

4. *If no fuzzy linguistic and fuzzy classification uncertainties are associated with the pattern* $\mathbf{x}$, *then* $\tau_{C_c}(\mathbf{x}) = r_{C_c}(\mathbf{x})$.
   Proof: If no fuzzy linguistic uncertainty is involved, then each cluster is crisp. Consequently, the input pattern belongs to only one cluster. Let it be the jth cluster. So, $\mu_{F_j}(\mathbf{x}) = 1$ and $\mu_{F_k}(\mathbf{x}) = 0 \ \forall k \neq j$. Since the classification is crisp, $\tau_{C_c}(\mathbf{x}) = \frac{\|F_j \cap C_c\|}{\|F_j\|} = r_{C_c}(\mathbf{x})$ (see the definition (12)).

5. *When each cluster is crisp and fine, that is, each cluster consists of a single pattern and the associated cluster memberships are crisp,* $\tau_{C_c}(\mathbf{x})$ *is equivalent to the fuzzy membership of* $\mathbf{x}$ *in the class* C_c.
   Proof: Since each cluster is crisp and fine, $\tau_{C_c}(\mathbf{x}) = 1.\frac{\mu_{C_c}(\mathbf{x}).}{1} = \mu_{C_c}(\mathbf{x})$.

These properties illustrate that the fuzzy-rough membership function characterises the fuzzy-rough uncertainty associated with an input pattern and it works even if only rough, fuzzy or rough-fuzzy uncertainties are present.

It is to be noted here that if $\hat{H} > 1$, $\tau_{C_c}(\mathbf{x}) \neq 0$ and fuzzy-rough uncertainty is absent, then $\tau_{C_c}(\mathbf{x})$ never becomes one, rather it approaches towards one. It is because, the condition expressed in (15) does not allow $\mu_{F_j}(\mathbf{x}) = 1$ to be true for more than one cluster. However, it hardly happens in practice as it needs two cluster centers to be same.

3 Modular Neural Networks

3.1 Architecture

The given pattern classification task is initially subdivided into several subtasks, and one subnetwork is assigned for each subtask. Let, the original problem has M output classes $\{C_1, C_2, \ldots, C_M\}$, and these classes are divided into S subnetworks (Fig. 1). The sth subnetwork is assigned to classify a group of classes, represented by $\{C_{c_{s-1}+1}, \ldots, C_{c_s}\}$ with $c_0 = 0$ and $c_S = M$. The output of the sth subnetwork is $\{y_{c_{s-1}+1}, \ldots, y_{c_s}\}$, which is expressed in a vector notation as $\xi_s = [y_{c_{s-1}+1}, \ldots, y_{c_s}]$. The proposed method interprets each subnetwork as a nonlinear filter tailored to the subgroup. Thus, the outputs of all the filters corresponding to an input $\mathbf{x}$ is viewed as an S dimensional feature vector This feature vector is presented as an input to a fuzzy integrator, which computes the value of the fuzzy integral with the help of fuzzy densities. The class label of the input $\mathbf{x}$ is the class index that yields the maximum value of the fuzzy integral corresponding to ψ.

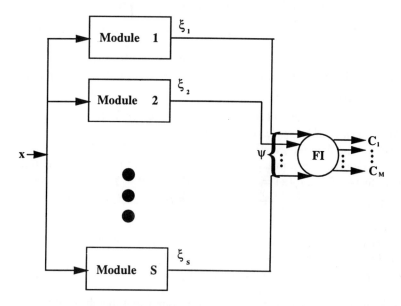

Fig. 1. A typical modular neural network with S modules. The output of the sth module is represented by ξ_s. All the outputs are combined in a nonlinear manner through a fuzzy integrator (FI). ψ denotes $[\xi'_1, \xi'_2, \ldots, \xi'_S]$.

3.2 Training

When a modular network is used for classification, a given training pattern is input to all the subnetworks and the outputs of the subnetworks are processed to determine the class. We can decide the class label of the input based on *winner-take-all* policy. It means that the class label of the input pattern is assigned as j, $1 \leq j \leq M$, if $y_j = \max_{k=1,2,\ldots,M}\{y_k\}$. However, this kind of assignment is not proper as all the subnetworks are independently trained on different sets of data. A better approach is to declare the jth class winner, if the jth class correspondences to $\max_{k=1,2,\ldots,M}\{g_k y_k\}$, where g_k is the importance associated with the class C_k. One possible choice for g_k is the *a priori* probability of the class C_k. However, the constraint $\sum_{k=1,2,\ldots,M} g_k = 1$ used in the probability theory cannot distinguish between lack of evidence and ignorance. Therefore, the concept of the fuzzy integral is appealing here. In the fuzzy integral approach, the outputs of the modules are processed further so that the interactions among the outputs are also exploited for the final classification result. Hence, the term g_k is replaced by a more specific term $g_k(\{\xi_s\})$, where $g_k(\{\xi_s\})$ denotes the importance of ξ_s in characterizing the class C_k. With the help of $g_k(\{\xi_s\})$, $s = 1, 2, \ldots, S$, the fuzzy integral $\mathcal{F}_k$ for the class C_k combines the outputs of all the modules, i.e., ξ_s, $s = 1, 2, \ldots, S$, in a nonlinear fashion. The final class label corresponding to the input is j, if $\mathcal{F}_j = \max_{k=1,2,\ldots,M}\{\mathcal{F}_k\}$. Specifically, the training of the whole modular network is comprised of the following two stages:

Training of each subnetwork: For this stage, separate data sets are prepared to train the subnetworks independently. The training data set for a subnetwork generally consists of patterns belonging to the classes in its subgroup only. Then, each subnetwork is trained by the conventional backpropagation algorithm [7] to form the decision surfaces for the classes in its subgroup. It may be necessary to train each subnetwork with a few patterns belonging to the output classes of the other subnetworks. These patterns may be considered as negative examples.

Pattern matching: This stage of training is needed to compare the kth class prototype and the feature vector ψ. Here, the partial evaluation $h_k(\xi_s)$ implies how good the feature ξ_s alone is to classify the patterns from the class C_k, and the individual fuzzy density $g_k(\{\xi_s\})$ signifies the importance of the feature ξ_s for the class C_k. Hence, the comparison between the prototypes of C_k and ψ can be accomplished in terms of closeness. Roughly speaking, the closeness can be expressed as $h_k(\xi_1)g_k(\{\xi_1\}) + h_k(\xi_2)g_k(\{\xi_2\}) + \ldots + h_k(\xi_S)g_k(\{\xi_S\})$. When the domain is continuous and the continuity of the function $h_k(\xi_s)$ is not guaranteed, the closeness can be represented more comfortably by the fuzzy integral $\mathcal{F}_k$. Therefore, in this stage it is essential to know the values of: (a) class prototypes, from which the partial evaluation $h_k(\xi_s)$ can be obtained, and (b) the individual fuzzy density $g_k(\{\boldsymbol{\xi}_s\})$.

Class prototype selection: The set $\{\mathbf{x}\}$, that contains training inputs from all the classes, are passed through all the subnetworks to generate a set of feature vectors $\{\boldsymbol{\psi}\}$. All the feature vectors corresponding to each class (say C_k) is collected separately, and the mean of the vectors (say $\mathbf{m}_k$) is calculated. This mean represents the class prototype. In the testing phase, these class prototypes will help us to compute the partial evaluation $h_k(\boldsymbol{\xi}_s)$.

Evaluation of fuzzy density $g_k(\{\boldsymbol{\xi}_s\})$: The individual fuzzy densities are calculated based on how well the outputs generated by the subnetworks separate all the classes for the training data. Since we have already mentioned that each $\boldsymbol{\xi}_s$ can be considered as a feature, determining individual fuzzy densities are equivalent to the determination of the importance of each feature. We propose a fuzzy-rough set theoretic approach to determine the individual fuzzy densities, i.e., the importance of the features for a particular class. This approach is described below for the feature $\boldsymbol{\xi}_s$ and the output class C_k.

A set of features $\{\boldsymbol{\xi}_s\}$ is collected by passing a set of training inputs $\{\mathbf{x}\}$ through the sth subnetwork. Fuzzy K-means algorithm [1] is applied on this feature set. Since the number of clusters is not known, we assume K is equal to the number of classes M. While applying the fuzzy K-means clustering on the set $\{\boldsymbol{\xi}_s\}$, we can observe the following two points:

1. Some $\boldsymbol{\xi}_s$ belong to more than one cluster partially as the clusters are overlapping.
2. All $\boldsymbol{\xi}_s$ from the same cluster may not belong to the same class.

The first type of uncertainty, known as fuzzy uncertainty, is generated because the outputs of the subnetworks are not from $\{0, 1\}$. It may be due to the fuzziness, if the outputs of the subnetworks are considered to be fuzzy. Or it may be due to the lack of confidence, if the outputs of the subnetworks are viewed as *a posteriori* probabilities. The second type of uncertainty is known as rough uncertainty, which is generated as the feature $\boldsymbol{\xi}_s$ is not sufficient to classify all the input patterns $\{\mathbf{x}\}$. Hence, two different $\boldsymbol{\xi}_s$ belonging to the same cluster may represent two different classes. Thus, the relationship between the sth feature and the class labels may be a one-to-many mapping. In other words, the classes are *indescernible* or not distinguishable with respect to the sth feature. The sth feature $\boldsymbol{\xi}_s$ is an important feature if

1. The clusters are compact and wide apart. The less is the fuzzy uncertainty, the more important the feature is [13] [14].
2. All the elements from a particular cluster belong to the same class. The less is the rough uncertainty, the more important the feature is.

That is, the feature $\boldsymbol{\xi}_s$ is important if each cluster, generated by the feature, is compact and isolated, and if all the patterns from each cluster represent the same class. Therefore, the more fuzzy and rough the uncertainty is, the less is the importance. Note that the presence of any one or both of these uncertainties change the importance of the feature for a particular class. We seek to measure the amount of fuzzy and rough uncertainties involved by using fuzzy-rough sets. Later we will use the quantified value to determine the importance of the sth feature for the kth class.

Based on the feature ξ_s, the approximation of C_k by the set of feature vectors $\{\psi\}$ is expressed here as a fuzzy-rough set. The lack of discriminatory power of the feature ξ_s is due to the fact that we are not considering the other features ξ_j, $j \neq s$, $j = 1, 2, \ldots, S$ into account. Here we do not have complete information to classify a particular pattern in the class C_k based on the information supplied by ξ_s. To determine the importance of the feature ξ_s for the class C_k with such incomplete knowledge, the concept of rough sets can be used. In the terminology of rough set, two patterns $\psi_p \in \{\psi\}$ and $\psi_q \in \{\psi\}$ are called *indiscernible* with respect to the sth feature when the sth component of these two patterns have the same value. Mathematically, it can be stated as

$$\psi_p R^s \psi_q \quad \text{iff} \quad \xi_{sp} = \xi_{sq} \tag{18}$$

where R^s is a binary relation over $\{\psi\} \times \{\psi\}$. Obviously, R^s is an equivalent relation. Therefore, R^s partitions $\{\psi\}$ into a set of equivalent classes, namely $\{F_1^s, F_2^s, \ldots, F_K^s\}$, where K is greater than one but less than the cardinality of $\{\psi\}$. For continuous features, it is better to consider that ψ_p and ψ_q are related if the sth component of the two features are similar (not necessarily strictly equal as in (18)). Two patterns from the same cluster can be considered similar as they have spatial similarity. The resultant equivalence classes become fuzzy clusters. It can be proved [5] that the fuzzy clusters $F_1^s, F_2^s, \ldots, F_M^s$ will be present if and only if there exists some similarity relation like (18). Moreover, it can be shown that [5] the generated clusters will follow weak fuzzy partitioning [5]. This situation can be formulated in terms of fuzzy-rough sets. One obvious problem is to decide the number of clusters needed for the task. We are assuming that the number of clusters is equal to the number of classes, i.e., $K = M$.

After showing that the fuzzy-rough uncertainty is associated with each ξ_s, we have represented the approximation of C_k by $\{\psi\}$ in terms of fuzzy-rough sets. Now we are ready to quantify the fuzzy-rough uncertainty associated with each ξ_s. In the section 2.3, we have observed that using the fuzzy-rough membership values, we can measure the fuzzy-rough uncertainty associated with each input pattern. It works even when only fuzzy or only rough uncertainty is present. The fuzzy-rough uncertainty associated with a pattern is the highest when the fuzzy-rough membership value is 0.5, and it decreases as the fuzzy-rough membership values approach towards 0 or 1. Now, a measure of fuzzy-roughness is needed to estimate the average ambiguity in the output class C_k for the input feature ξ_s. As a measure we use the concept of *fuzzy-rough entropy* for the sth feature and the kth class as

$$\mathcal{H}_k^s = -\frac{1}{\tilde{n}\log 2} \sum_{\boldsymbol{\xi}_s} \left[\tau_{C_k}(\boldsymbol{\xi}_s)\log(\tau_{C_k}(\boldsymbol{\xi}_s)) + (1-\tau_{C_k}(\boldsymbol{\xi}_s))\log(1-\tau_{C_k}(\boldsymbol{\xi}_s))\right] \quad (19)$$

where $\tau_{C_k}(\boldsymbol{\xi}_s)$ is the fuzzy-rough membership value of the feature $\boldsymbol{\xi}_s$ in the class C_k and $\tilde{n}$ is the number of feature vectors used to determine the importance of the feature. It can be noticed that $\mathcal{H}_k^s$ increases monotonically in [0, 0.5] and decreases monotonically in [0.5, 1]. It reaches the maximum value when $\tau_{C_k}(\boldsymbol{\xi}_s) = 0.5 \; \forall \boldsymbol{\xi}_s$, and minimum value when $\tau_{C_k}(\boldsymbol{\xi}_s) = 0$ or $1 \; \forall \boldsymbol{\xi}_s$ [12]. The lower the value of $\mathcal{H}_k^s$ is, the greater is the number of $\boldsymbol{\xi}_s$ having $\tau_{C_k}(\boldsymbol{\xi}_s) \approx 1$ or $\tau_{C_k}(\boldsymbol{\xi}_s) \approx 0$, i.e., less is the difficulty in deciding whether $\boldsymbol{\xi}_s$ can be considered a member of C_k or not. In particular, when $\tau_{C_k}(\boldsymbol{\xi}_s) \approx 1$, greater is the tendency of $\boldsymbol{\xi}_s$ to form a compact class C_k in the sth subspace, resulting in less internal scatter in the sth subspace. Moreover, when $\tau_{C_k}(\boldsymbol{\xi}_s) \approx 0$, $\boldsymbol{\xi}_s$ is far away from the kth class, and hence, the interclass distance increases in the sth subspace. On the otherhand, when $\tau_{C_k}(\boldsymbol{\xi}_s) \approx 0.5$, $\boldsymbol{\xi}_s$ lies in between C_k and the other classes in the sth subspace; hence, compactness and interclass distance both decrease in the sth subspace. Therefore, the reliability of $\boldsymbol{\xi}_s$, in characterizing the class C_k, increases as the corresponding $\mathcal{H}_k^s$ value decreases. Thus, $\mathcal{H}_k^s$ quantifies the importance of $\boldsymbol{\xi}_s$ in characterizing the kth class. One way to determine the importance of the sth feature in the kth class is by the term $(1 - \mathcal{H}_k^s)$. Hence, the fuzzy densities can be determined as

$$g_k(\{\boldsymbol{\xi}_s\}) = 1 - \mathcal{H}_k^s \quad \forall \; s, \; k \quad (20)$$

The procedure to find the fuzzy density can be summarised as follows: We interpret the fuzzy density of a module with respect to an output class as the importance of the module for that class. It is equivalent to the importance of the feature generated by the module (since the module is treated as a feature extractor). The importance of the feature for an output class depends on the fuzzy-roughness associated with the output class for the given feature. We have demonstrated that a set of input patterns can be clustered based on the feature value, and as a consequence, the approximation of the output class by these clusters can be expressed in terms of a fuzzy-rough set. The fuzzy-roughness associated with each input pattern for the output class can be quantified in terms of fuzzy-rough membership functions. The fuzzy-rough ambiguity associated with the output class for the given set of input patterns is measured using the fuzzy-rough entropy. The fuzzy density for the output class is found from the fuzzy-rough entropy.

The complete training procedure, consisting of training the subnetworks and matching the patterns, is shown in Fig. 2.

3.3 Testing

A separate set of test patterns is used as inputs to all the subnetworks. The outputs of all the subnetworks corresponding to the input test pattern **x** form the

```
Use different training sets T_s, s = 1,2,...,S to train all the
subnetworks using the backpropagation algorithm. The training
set T_s contains the training input-output pairs only for the
sth subnetwork.

Prepare another training set {x_1,x_2,...,x_ñ}, that contains the
training input-output pairs for all the subnetworks. Pass this
training set through all the
subnetworks to collect the feature vectors ψ_p, p = 1,2,...,ñ as
the outputs.

DO for each k = 1,2,...,M
      Record the class prototype m_k.
END DO

DO for each s
      Apply the fuzzy K-means clustering algorithm with M
      clusters on {ξ_sp | p = 1,2,...,ñ}.
      DO for each class C_k, k = 1,2,...,M
            Use (20) to compute the fuzzy density g_k({ξ_s})
            from {ξ_sp | p = 1,2,...,ñ}.
      END DO
END DO
```

Fig. 2. Training of the proposed modular neural network.

feature vector $\psi = [\acute{\xi}_1, \acute{\xi}_2, \ldots, \acute{\xi}_S]$. To determine the partial evaluation $h_k(\xi_s)$ from the already recorded class prototypes, we use the following relation [1]:

$$h_k(\xi_s) = 1 \bigg/ \sum_{h=1}^{M} (d_k/d_h)^{2/(q-1)} \quad (21)$$

where d_k is the distance between the feature ξ_s and the prototype of the kth class, i.e., $d_k = ||\xi_s - \mathbf{m}_k^s||^2 = (\xi_s - \mathbf{m}_k^s)'\Sigma(\xi_s - \mathbf{m}_k^s)$, with $\mathbf{m}_k^s = [m_{c_{s-1}+1,k}, \ldots, m_{c_s,k}]'$. Here, Σ is a positive definite matrix and $q \in (1, \infty)$ is an index. Generally, Σ is taken as the covariance matrix of ξ_s and $\mathbf{m}_k^s$, and q is taken as 2. The value of $h_k(\xi_s)$ is an indication of how certain we are in the classification of the input $\mathbf{x}$ into the class C_k using the feature ξ_s. Here, 1 indicates with absolute certainty that the input $\mathbf{x}$ is from the class C_k, and 0 means that the input certainly does not belong to the class C_k. Moreover, from the training the fuzzy densities $g_k(\{\xi_s\})$, $\forall s, k$, are known. Hence, using the equation (4), the fuzzy integral value of $\mathbf{x}$ corresponding to each output class can be computed. The class label corresponding to the test input is the class index which yields the maximum fuzzy integral value. The fuzzy integral value corresponding to a particular class can also be used as the confidence level in classifying the input to that class. The testing procedure is given in form of an algorithm in Fig. 3.

For the test input pattern x, find the outputs ξ_s, $s = 1, 2, \ldots, S$ for all the subnetworks.

DO for each output class C_k, $k = 1, 2, \ldots, M$

    DO for each ξ_s, $s = 1, 2, \ldots, S$

        Compute $h_k(\xi_s)$ from (21).

    END DO

    Calculate λ from (9).

    Calculate $\mathcal{F}_k$ from (4).

END DO

The class label of x is j if $\mathcal{F}_j = \max\limits_{k=1}^{M}\{\mathcal{F}_k\}$.

Fig. 3. Testing of the proposed modular neural network.

4 Results and Discussion

The proposed algorithm is tested on a Contract Bridge Bidding example. In Contract Bridge, a player makes a bid to convey information about the pattern of the thirteen cards in his hand. If he is the first player to make a bid, it is called "opening bid", which he makes based only on the pattern of the cards he is holding. He has no *a priori* knowledge of the rest of the cards in the other players' hands at this stage. The objective of this experiment is to explore the possibility of capturing the reasoning process of a player in making an opening bid based on the pattern of the cards presented to him. This can be interpreted as a task of building a classifier that can classify an input hand pattern into the output classes, where each output class corresponds to one bid. This classification process becomes complicated because for a given hand, the same player may make different bids at different times.

It has been observed [26] that, compared to a single monolithic network, a modular neural network is more appropriate for the bidding task. Here, we study the performance of the proposed fuzzy-rough set based method and one existing method on a two-module network. We have taken four classes corresponding to three Club (3C), three Diamond (3D), three Heart (3H) and three Spade (3S) card hands. One module is supposed to classify 3C and 3D and the other one 3H and 3S. In this study the Standard American System [26] is used as a bidding convention as it is less artificial than the other systems. Input of the network is represented as a series of fifty two 1s and 0s, where the presence or absence of a card is denoted by 1 or 0 [26]. Only one hidden layer with 50 hidden nodes is used in each module. Number of output nodes for both the subnetworks are 2. Both the modules are trained with backpropagation learning law on a training set of 232 card hands. Some negative examples are used in both the modules to fine tune the decision boundaries. The outputs of these two modules are fused by the fuzzy integral approach. To demonstrate the relative performance of the proposed fuzzy-rough method, the densities of the fuzzy integrator are calculated by the frequency-based method (used in [24] [3]) and the proposed method. The fuzzy densities are calculated in [24] [3] using $g_k(\{\xi_s\}) = p_{s,k} \bigg/ \sum_{j=1}^{S} p_{j,k} \cdot d_s$, where $p_{s,k}$ is the classification performance of ξ_s for the class C_k on a validation data set and d_s is the desired sum of the fuzzy densities. The classification results on a test set of 60 card hands are given in Table 1. In the Table-1, we can observe that the proposed method is performing better than the frequency-based method.

Table 1. Results of Card Data Classification Based on the Frequency Method and the Proposed Method

Class	Frequency-based Method	Proposed Method
'3C'	80.39%	83.87%
'3D'	74.41%	79.74%
'3H'	82.01%	82.92%
'3S'	80.32%	85.87%
Overall	79.28%	83.10%

5 Summary and Conclusion

This article applies a fuzzy integral-based technique to combine the outputs of the modules in a modular neural networks. The modules are viewed as nonlinear feature extractors. Hence, for each input the modules generate a feature vector. The fuzzy integral acts here as a weighted closeness measure between the feature vector and the class prototypes. The weights are determined based on how important the features are for a particular class. The importance of a feature for a particular class is measured in terms of fuzzy-rough ambiguity associated with the concerned output class for the given input feature. The class prototype, that is nearest to the feature vector is designated as the class label of the input pattern corresponding to the feature vector.

While determining the fuzzy density, we are assuming that the number of clusters is equal to the number of classes. This assumption may not be valid in some cases. Hence, cluster validity measures are needed to be applied to determine the number of clusters accurately. In the definition of fuzzy integral, we are using max and min operators which are noninteractive. It makes the fuzzy integral less sensitive towards the training data. A better approach may be to use the fuzzy integral with OWA operators [2].

Acknowledgment

This work was carried out while Manish Sarkar held a Dr. K. S. Krishnan Research Fellowship from the Department of Atomic Energy, Government of India.

References

1. J. C. Bezdek. *Pattern Recognition with Fuzzy Objective Function Algorithms.* Plenum Press, New York, 1981.
2. S. B. Cho. Fuzzy aggregation of modular neural networks with ordered weighted averaging operators. *Approximate Reasoning*, 13:359–375, 1995.
3. S. B. Cho and J. H. Kim. Multiple network fusion using fuzzy logic. *IEEE Transactions on Neural Networks*, 6(2):497–501, March 1995.
4. P. J. Darwen and X. Yao. Speciation as automatic categorical modularisation. *IEEE Transactions on Evolutionary Computation*, 1(2):101–108, July 1997.
5. D. Dubois and H. Prade. Putting rough sets and fuzzy sets together. In R. Slowinski, editor, *Intelligent Decision Support. Handbook of Applications and Advances of the Rough Set Theory.* Kluwer Academic Publishers, Dordrecht, 1992.
6. M. Grabisch and J. M. Nicolas. Classification by fuzzy integral: Performance and test. *Fuzzy Sets and Systems*, (65):255–271, 1994.
7. S. Haykin. *Neural Networks - A Comprehensive Foundation.* Macmillan College Publishing Company, New York, 1994.
8. J. M. Keller, P. Gader, H. Tahani, J. H. Chiang, and M. Mohamed. Advances in fuzzy integration for pattern recognition. *Fuzzy Sets and Systems*, (65):273–283, 1994.
9. G. S. Klir and B. Yuan. *Fuzzy Sets and Fuzzy Logic - Theory and Applications.* Prentice-Hall, Englewood Cliffs, NJ, 1995.
10. S. Y. Kung. *Digital Neural Networks.* Prentice Hall, Englewood Cliffs, New Jersey, 1993.
11. T. Murofushi and M. Sugeno. An interpretation of fuzzy measure and the Choquet integral as an integral with respect to a fuzzy measure. *Fuzzy Sets and Systems*, 29:201–227, June 1989.
12. N. R. Pal and J. C. Bezdek. On cluster validity for the fuzzy C-means model. *IEEE Transactions on Fuzzy Systems*, 3(3):330–379, August 1995.
13. S. K. Pal and B. Chakraborty. Fuzzy set theoretic measure for automatic feature evaluation. *IEEE Transactions on System, Man and Cybernetics*, 16(5):754–760, September/October 1986.
14. S. K. Pal and D. Dutta Majumder. *Fuzzy Mathematical Approach to Pattern Recognition.* Wiley (Halsted Press), New York, 1986.
15. Z. Pawlak. Rough sets. *International Journal of Computer and Information Science*, 11:341–356, 1982.
16. Z. Pawlak. *Rough Sets: Theoretical Aspects of Reasoning About Data.* Kluwer, Dordrecht, 1991.
17. Z. Pawlak, J.G. Grzymała-Busse, R. Słowiński, and W. Ziarko. Rough sets. *Communications of the ACM*, 38(11):89–95, November 1995.

18. M. Sarkar and B. Yegnanarayana. Fuzzy-rough membership functions. *Accepted in IEEE International Conference on Systems, Man and Cybernetics, San Diego, California, USA*, October 11-14 1998.
19. M. Sarkar and B. Yegnanarayana. Rough-fuzzy membership functions. In *Proceedings of IEEE International Conference on Fuzzy Systems (Anchorage, Alaska, USA)*, May 4-9 1998.
20. C. C. Sekhar and B. Yegnanarayana. Recognition of stop-consonant-vowel (svc) segments in continuous speech using neural network models. *Journal of Institution of Electronics and Telecommunication Engineers (IETE)*, 42(4 & 5):269–280, July-October 1996.
21. C. C. Sekhar and B. Yegnanarayana. Modular networks and constraint satisfaction model for recognition of stop consonant-vowel (scv) utterances. In *Proceedings of IEEE International Conference on Neural Networks (Anchorage, Alaska, USA)*, May 4-9 1998.
22. R. Słowiński and J. Stefanowski. *Foundations of Computing and Decision Sciences (eds.)*, 18(3-4), Fall 1993.
23. Michio Sugeno. *Theory of fuzzy integrals and its applications*. PhD thesis, Tokyo Institute of Technology, 1974.
24. H. Tahani and J. K. Keller. Information fusion in computer vision using fuzzy integral. *IEEE Transactions on System, Man and Cybernetics*, 20(3):733–741, May/June 1990.
25. S. K. M. Wong and W. Ziarko. Comparison of the probabilistic approximate classification and fuzzy set model. *Fuzzy Sets and Systems*, 21:357–362, 1987.
26. B. Yegnanarayana, D. Khemani, and M. Sarkar. Neural networks for contract bridge bidding. *Sadhana*, 21(3):395–413, June 1996.
27. L. A. Zadeh. Fuzzy sets as a basis for a theory of possibility. *Fuzzy Sets and Systems*, 1:3–28, 1978.

Rough Fuzzy Knowledge-based Network - A Soft Computing Approach

Sushmita Mitra,[1] *Sankar K. Pal*[1] *and Mohua Banerjee*[2]

[1] Machine Intelligence Unit
Indian Statistical Institute
Calcutta 700035, INDIA
Email : {sushmita,sankar}@isical.ac.in

[2] Department of Mathematics
Indian Institute of Technology
Kanpur 208016, INDIA
Email : mohua@iitk.ac.in

Abstract. A method of integrating rough sets with a fuzzy MLP for generating a knowledge-based network in soft computing paradigm is described. Fuzzy logic helps in handling uncertainty in the input description and output decision. The role of rough sets is in extracting crude domain knowledge from the data set in the form of rules. The syntax of these rules automatically determines the appropriate number of hidden nodes while the dependency factors are used in the initial weight encoding. Performance of the system is compared with that of the fuzzy and conventional versions of the MLP (involving no initial knowledge), Bayes' classifier, and k-nearest neighbors classifier, for speech and synthetic data.

Keywords: Knowledge-based networks, rough sets, fuzzy MLP, pattern recognition.

1 Introduction

There has recently been a spurt of activity to integrate different computing paradigms such as fuzzy set theory, neural networks, genetic algorithms and rough set theory, for generating more efficient hybrid systems that can be classified as *soft computing* methodologies [1, 2]. The purpose is to provide flexible information processing systems that can exploit the tolerance for imprecision, uncertainty, approximate reasoning and partial truth in order to achieve tractability, robustness and low cost solution in real life ambiguous situations [3].

Various hybridizations include *neuro-fuzzy* [4, 5], *genetic-neural* [6], *fuzzy-genetic* [7], *neuro-fuzzy-genetic* [8], *rough-fuzzy* [9] and *rough-neuro-fuzzy* [10] approaches. Here the individual tools act synergetically, not competitively, for enhancing the application domain of each other. One can thereby incorporate the generic advantages of artificial neural networks (ANNs) like massive parallelism, robustness and learning in data-rich environments into the system. The modeling of imprecise and qualitative knowledge, ambiguous decision, and the transmission of uncertainty are possible through the use of fuzzy logic. Moreover, certain application specific merits of artificial neural networks and fuzzy sets, can also be exploited. For example, in the case of pattern classification ANNs can help generate highly nonlinear decision boundaries and fuzzy sets can handle uncertainties in the input description and output decision. The primary role of rough sets is for management of uncertainty, knowledge reduction and knowledge extraction while that of genetic algorithms is to provide techniques for efficient searching and optimization. However, hybridizations involving these two are relatively new as compared to the well-known neuro-fuzzy approaches.

The theory of rough sets [11] has recently emerged as a mathematical approach for managing uncertainty that arises from inexact, noisy or incomplete information. It has been investigated in the context of expert systems, decision support systems, machine learning, inductive learning and various other areas of application. It is found to be particularly effective in the area of knowledge reduction. The focus of rough set theory is on the ambiguity caused by limited discernibility of objects in the domain of discourse. The intention is to approximate a *rough* (imprecise) concept in the domain of discourse by a pair of *exact* concepts, called the lower and upper approximations. These exact concepts are determined by an *indiscernibility* relation on the domain, which, in turn, may be induced by a given set of *attributes* ascribed to the objects of the domain. These approximations are used to define the notions of *discernibility matrices*, *discernibility functions* [12], *reducts* and *dependency factors* [11], all of which play a fundamental role in the reduction of knowledge.

Many have looked into the implementation of decision rules extracted from operation data using rough set formalism, especially in problems of machine learning from examples and control theory [13]. In the context of neural networks, an attempt of such implementation has been made by Yasdi [14]. The intention was to use rough sets as a tool for structuring the neural networks. The methodology consisted of generating rules from training examples by rough-set learning, and mapping the dependency factors of the rules into the connection weights of a four-layered neural network. Application of rough sets in neurocomputing has also been made in [15]. However, in this method, rough sets were used for knowledge discovery at the level of data acquisition, (*viz.*, in preprocessing of the feature vectors), and not for structuring the network.

Generally, ANNs consider a fixed topology of neurons connected by links in a pre-defined manner. These connection weights are usually initialized by small random values. Knowledge-based networks [16, 17] constitute a special class of ANNs that consider crude domain knowledge to generate the initial network architecture which is later refined in the presence of training data. This process helps in reducing the searching space and time while the network traces the optimal solution. Node growing and link pruning are also made in order to generate the optimal network architecture. When one encodes the connection weights of a fuzzy neural network with the domain knowledge, one generates a fuzzy knowledge-based network [18, 19]. In this article, we demonstrate how the theory of rough sets can be utilized for efficiently extracting domain knowledge for such a fuzzy knowledge-based network.

The fuzzy multilayer perceptron (MLP) [20, 21] is a fuzzy neural network that incorporates fuzzy set-theoretic concepts at the input and output levels and during learning. The input is modeled in terms of the $3n$-dimensional linguistic feature space while the output consists of class membership values. The feature space gives us the condition attributes and the output classes the decision attributes, so as to result in a decision table. This table, however, may be transformed, keeping the complexity of the network to be constructed in mind. Rules are then generated from the (transformed) table by computing relative reducts. The dependency factors of these rules are encoded as the initial connection weights of the fuzzy MLP. The network is next trained to refine its weight values.

Here the knowledge encoding procedure, unlike most other methods [16, 17], involves a non-binary weighting mechanism based on a detailed and systematic estimation of the available domain information. It may be noted that the appropriate number of hidden nodes is automatically determined. The model is capable of handling input in numerical, linguistic and set forms, and can tackle uncertainty due to overlapping classes. The classification performance is found to be better than the conventional and fuzzy versions of the MLP (involving no initial knowledge). It is also compared with the standard Bayes' classifier and the k-nearest neighbors (k-NN) classifier.

Section 2 provides a review of the current literature on fuzzy knowledge-based networks. This is followed in Section 3 by a brief description of the fuzzy MLP used here. The basics of rough set theory are presented in Section 4. In Section 5, we describe the knowledge encoding methodology. The model is implemented on real life speech data as well as synthetic data (in Section 6) for classification.

2 Fuzzy Knowledge-based Networks

Encoding the crude domain knowledge, extracted from the data, among the connection weights of an ANN, leads to the design of a knowledge-based network. Such a network is found to be more efficient than the conventional versions for the following reason. During learning an ANN (say, an MLP) searches for the set of connection weights that corresponds to some local minima. In other words, it searches for that set of weights that minimizes the difference between the target vector and the actual output (obtained by the MLP). Note that there may be a large number of such minimum values corresponding to various *good* solutions. If we initially set these weights so as to be near one such *good* solution, the searching space may be reduced and learning thereby becomes faster. The architecture of the network becomes simpler due to the inherent reduction of the redundancy among the connection weights.

Let us first of all mention a few of the conventional knowledge-based nets, before moving on to fuzzy knowledge-based networks. The models of [16, 17, 22, 23] involve *crisp* inputs and outputs. The initial domain knowledge, in the form of rules, is mapped into the multilayer feedforward network topology using binary link weights to maintain the semantics. Nodes and/or connections are introduced appropriately when the network performance stagnates during training using backpropagation. The knowledge of the trained neural model is then used to extract revised rules for the problem domain.

A brief survey on the knowledge-based networks involving fuzziness at different stages is provided here based on the studies of Masuoka *et al.* [24], Kasabov [25, 26], Kosko [27], Machado and Rocha [28, 19], Pedrycz and Rocha [29], and Hirota and Pedrycz [30].

Knowledge extracted from experts in the form of membership functions and fuzzy rules (in *And-Or* form) is used to build and preweight the neural net structure which is then tuned using training data. The model by Masuoka *et al.* [24] consists of the input variable membership net, the rule net, and the output variable net. Kasabov [25] uses three neural subnets, *viz.*, production memory, working memory and variable binding space to encode the production rules, that can be later updated. Fuzzy signed digraph with feedback, termed fuzzy cognitive map, has been used by Kosko [27] to represent knowledge. Additive combination of augmented connection matrices are employed to include the views of a number of experts for generating the knowledge network.

Machado and Rocha [28] have used a connectionist knowledge base involving fuzzy numbers at the input layer, fuzzy *And* at the hidden layers and fuzzy *Or* at the output layer. The hidden layers chunk input evidences into clusters of information for representing regular patterns of the environment. The output layer computes the degree of possibility of each hypothesis. The initial network architecture is generated using *knowledge graphs* elicited from experts by the application of the knowledge acquisition technique of [31]. The experts express their knowledge about each hypothesis of the problem domain by selecting an appropriate set of evidences and building an acyclic weighted *And-Or* graph (knowledge graph) to describe how these must be combined to support decision making.

Pedrycz and Rocha [29] have used basic aggregation neurons (And/Or) and referential processing units (matching, dominance and inclusion neurons) to design knowledge-based networks. The inhibitory and excitatory characteristics are captured by embodying direct and complemented input signals, and fully supervised learning is employed. Another related approach by Hirota and Pedrycz [30] has incorporated the use of fuzzy clustering for developing the geometric constructs leading to the design of knowledge-based networks.

Most of these models are mainly concerned with the encoding of initial knowledge by a fuzzy neural network followed by refinement during training. Extraction of fuzzy rules in this framework has been attempted in [24, 25, 28]. Connection weights above a pre-set threshold determine the *condition* or *action* elements in the extracted rules, along with the corresponding *degrees of importance* and *confidence factors* [26]. Inference, inquiry and explanation are possible during consultation in [28].

Machado and Rocha [19] have also used an interval-based representation for membership grades to allow reasoning with different types of uncertainty, *viz.*, vagueness, ignorance and relevance. The model reported in [28] is used as the building block for developing the facilities for incremental learning, inference, inquiry, censorship of input information, and explanation, as in expert systems. The utility-based inquiry process permits significant reduction of consultation cost/ risk and gives the system the common sense property possessed by experts when selecting tests to be performed. The ability to criticize input data when they disrupt a trend of acceptance/ rejection observed for a hypothesis mimics the behavior of experts, who are often able to detect suspicious input data and either reject them or ask for their confirmation.

Mitra *et al.* [18] have recently designed a knowledge-based neuro-fuzzy system for classification and rule generation. Here crude initial domain knowledge is encoded among the connection weights using the a priori class information (and their complements) and the distribution of pattern points in the feature space. An accurate estimation of the links connecting the output and hidden layers (in terms of the preceding layer link weights and node activations) is provided. Node growing and link pruning are incorporated to generate the optimal network architecture. Inferencing, querying and rule generation are demonstrated for recognizing vowels and diagnosing *hepatobiliary disorders*. Negative rules, indicative of cases where a pattern does not belong to a class, can also be generated. This is specially suitable in the ambiguous cases where positive rules (dealing with the belongingness of a pattern to a particular class) cannot be obtained. The performance of the knowledge-based net is seen to be superior as compared to the models incorporating no initial knowledge. ♣

Before describing the Rough-Fuzzy MLP in Section 5, let us briefly present the basics of fuzzy MLP and rough set theory in Sections 3 and 4 respectively. This will help one to understand the rule generation and knowledge encoding algorithm of Section 5.

3 Fuzzy MLP Model

In this section we describe, in brief, the fuzzy MLP [20] which is used for designing the knowledge-based network. Consider the layered network given in Fig. 1. The output of a neuron in any layer (h) other than the input layer $(h = 0)$ is given as

$$y_j^{(h)} = \frac{1}{1 + exp\left(-\sum_i y_i^{(h-1)} w_{ji}^{(h-1)}\right)}, \qquad (1)$$

where $y_i^{(h-1)}$ is the state of the ith neuron in the preceding $(h-1)$th layer and $w_{ji}^{(h-1)}$ is the weight of the connection from the ith neuron in layer $(h-1)$ to the jth neuron in layer (h). For nodes in the input layer, $y_j^{(0)}$ corresponds to the jth component of the input vector. Note that $x_j^{(h)} = \sum_i y_i^{(h-1)} w_{ji}^{(h-1)}$, as depicted in Fig. 1. The Mean Square Error in output vectors is minimized by the backpropagation algorithm using a gradient descent with a gradual decrease of the gain factor.

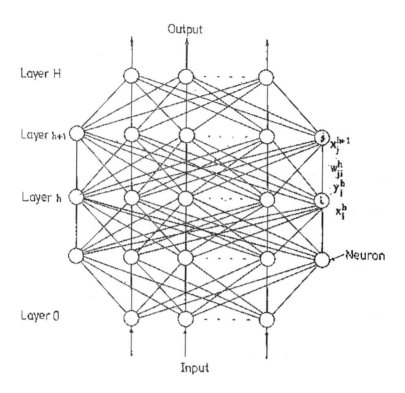

Fig. 1. The fuzzy MLP

3.1 Input vector

An n-dimensional pattern $\mathbf{F}_i = [F_{i1}, F_{i2}, \ldots, F_{in}]$ is represented as a $3n$-dimensional vector [32]

$$\mathbf{F}_i = \left[\mu_{\text{low}(F_{i1})}(\mathbf{F}_i), \ldots, \mu_{\text{high}(F_{in})}(\mathbf{F}_i)\right] = \left[y_1^{(0)}, y_2^{(0)}, \ldots, y_{3n}^{(0)}\right], \quad (2)$$

where the μ values indicate the membership functions of the corresponding linguistic π-sets *low*, *medium* and *high* along each feature axis and $y_1^{(0)}, \ldots, y_{3n}^{(0)}$ refer to the activations of the $3n$ neurons in the input layer. The three overlapping π-sets along a feature axis are depicted in Fig. 2.

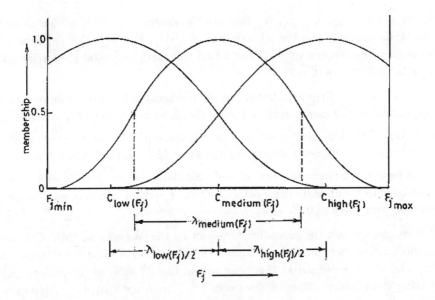

Fig. 2. Overlapping π-sets

When the input feature is numerical, we use the π-fuzzy sets (in the one dimensional form), with range [0,1], represented as

$$\pi(F_j; c, \lambda) = \begin{cases} 2\left(1 - \frac{\|F_j - c\|}{\lambda}\right)^2, & \text{for } \frac{\lambda}{2} \leq \|F_j - c\| \leq \lambda \\ 1 - 2\left(\frac{\|F_j - c\|}{\lambda}\right)^2, & \text{for } 0 \leq \|F_j - c\| \leq \frac{\lambda}{2} \\ 0, & \text{otherwise} \end{cases} \quad (3)$$

where $\lambda(> 0)$ is the radius of the π-function with c as the central point.

When the input feature F_j is linguistic, its membership values for the π-sets *low* (L), *medium* (M) and *high* (H) are quantified as

$$low \equiv \left\{ \frac{0.95}{L}, \frac{\pi\left(F_j\left(\frac{0.95}{L}\right); c_{j_m}, \lambda_{j_m}\right)}{M}, \frac{\pi\left(F_j\left(\frac{0.95}{L}\right); c_{j_h}, \lambda_{j_h}\right)}{H} \right\}$$

$$medium \equiv \left\{ \frac{\pi\left(F_j\left(\frac{0.95}{M}\right); c_{j_l}, \lambda_{j_l}\right)}{L}, \frac{0.95}{M}, \frac{\pi\left(F_j\left(\frac{0.95}{M}\right); c_{j_h}, \lambda_{j_h}\right)}{H} \right\} \quad (4)$$

$$high \equiv \left\{ \frac{\pi\left(F_j\left(\frac{0.95}{H}\right); c_{j_l}, \lambda_{j_l}\right)}{L}, \frac{\pi\left(F_j\left(\frac{0.95}{H}\right); c_{j_m}, \lambda_{j_m}\right)}{M}, \frac{0.95}{H} \right\}$$

where $c_{j_l}, \lambda_{j_l}, c_{j_m}, \lambda_{j_m}, c_{j_h}, \lambda_{j_h}$ indicate the centers and radii of the three linguistic properties along the jth axis, and $F_j\left(\frac{0.95}{L}\right)$, $F_j\left(\frac{0.95}{M}\right)$, $F_j\left(\frac{0.95}{H}\right)$ denote the corresponding feature values F_j at which the three linguistic properties attain membership values of 0.95.

For example, the linguistic feature *low* is represented by three components corresponding to the membership values of the three π-sets *low* (L), *medium* (M) and *high* (H) (Fig. 2). $\frac{0.95}{L}$ means a membership of 0.95 for L. $\frac{\pi\left(F_j\left(\frac{0.95}{L}\right);c_{j_m},\lambda_{j_m}\right)}{M}$ refers to the membership attained by the π-set M for that F_j which caused π-set L to have a membership value of 0.95. Similarly, $\frac{\pi\left(F_j\left(\frac{0.95}{L}\right);c_{j_h},\lambda_{j_h}\right)}{H}$ refers to the membership attained by the π-set H for that F_j which caused π-set L to have a membership value of 0.95.

Let us now explain the procedure for selecting the centres and radii of the overlapping π-sets. Let m_j be the mean of the pattern points along the jth axis. Then m_{j_l} and m_{j_h} are defined as the mean (along the jth axis) of the pattern points having co-ordinate values in the range $[F_{j_{min}}, m_j)$ and $(m_j, F_{j_{max}}]$ respectively, where $F_{j_{max}}$ and $F_{j_{min}}$ denote the upper and lower bounds of the dynamic range of feature F_j (for the training set) considering numerical values only. For the three linguistic property sets, the centres and the corresponding radii are defined as

$$c_{medium(F_j)} = m_j$$
$$c_{low(F_j)} = m_{j_l} \qquad (5)$$
$$c_{high(F_j)} = m_{j_h}$$

and

$$\lambda_{low(F_j)} = 2\left(c_{medium(F_j)} - c_{low(F_j)}\right)$$
$$\lambda_{high(F_j)} = 2\left(c_{high(F_j)} - c_{medium(F_j)}\right) \qquad (6)$$
$$\lambda_{medium(F_j)} = \frac{\lambda_{low(F_j)} * \left(F_{j_{max}} - c_{medium(F_j)}\right) + \lambda_{high(F_j)} * \left(c_{medium(F_j)} - F_{j_{min}}\right)}{F_{j_{max}} - F_{j_{min}}}$$

respectively. Here we take into account the distribution of the pattern points along each feature axis while choosing the corresponding centres and radii of the linguistic properties. Besides, the amount of overlap between the three linguistic properties can be different along the different axes, depending on the pattern set.

3.2 Output representation

Consider an l−class problem domain such that we have l nodes in the output layer. Let the n−dimensional vectors $\mathbf{o}_k = [o_{k1} \ldots o_{kn}]$ and $\mathbf{v}_k = [v_{k1} \ldots v_{kn}]$ denote the mean and standard deviation respectively of the numerical training data for the kth class c_k. The weighted distance of the training pattern $\mathbf{F_i}$ from the kth class c_k is defined as

$$z_{ik} = \sqrt{\sum_{j=1}^{n}\left[\frac{F_{ij}-o_{kj}}{v_{kj}}\right]^2} \quad \text{for } k=1,\ldots,l \qquad (7)$$

where F_{ij} is the value of the jth component of the ith pattern point.

The membership of the ith pattern in class k, lying in the range [0,1], is defined as [33]

$$\mu_k(\mathbf{F_i}) = \frac{1}{1+\left(\frac{z_{ik}}{f_d}\right)^{f_e}} \qquad (8)$$

where positive constants f_d and f_e are the denominational and exponential fuzzy generators controlling the amount of fuzziness in this class-membership set.

Then, for the ith input pattern, the desired output of the jth output node is defined as

$$d_j = \mu_j(\mathbf{F_i}) \qquad (9)$$

According to this definition a pattern can simultaneously belong to more than one class, and this is determined from the training set used during the learning phase.

4 Rough Set Preliminaries

Let us present some requisite preliminaries of rough set theory. For details one may refer [11] and [12].

An *information system* is a pair $S = <U, A>$, where U is a non-empty finite set called the *universe* and A a non-empty finite set of *attributes*. An attribute a can be regarded as a function from the domain U to some value set V_a.

An information system may be represented as an *attribute-value table*, in which rows are labelled by objects of the universe and columns by the attributes.

With every subset of attributes $B \subseteq A$, one can easily associate an equivalence relation I_B on U:
$$I_B = \{(x,y) \in U : \text{ for every } a \in B, a(x) = a(y)\}.$$
Then $I_B = \bigcap_{a \in B} I_a$.

If $X \subseteq U$, the sets $\{x \in U : [x]_B \subseteq X\}$ and $\{x \in U : [x]_B \cap X \neq \emptyset\}$, where $[x]_B$ denotes the equivalence class of the object $x \in U$ relative to I_B, are called the *B-lower* and *B-upper approximation* of X in S and denoted $\underline{B}X, \overline{B}X$ respectively.

$X(\subseteq U)$ is *B-exact* or *B-definable* in S if $\underline{B}X = \overline{B}X$. It may be observed that $\underline{B}X$ is the greatest B-definable set contained in X, and $\overline{B}X$ is the smallest B-definable set containing X. Let us consider the following simple example.

Consider an *information system* $<U,\{a\}>$, where the domain U consists of the students of a school, and there is a single attribute a – that of 'belonging to a class'.
Then U is partitioned by the classes of the school.

Now take the situation when an infectious disease has spread in the school, and the authorities take the two following steps.
(i) If at least one student of a class is infected, all the students of that class are vaccinated. Let $\overline{B}$ denote the union of such classes.
(ii) If every student of a class is infected, the class is temporarily suspended. Let $\underline{B}$ denote the union of such classes.

Then $\underline{B} \subseteq \overline{B}$. Given this information, let the following problem be posed:
Identify the collection of infected students.
Clearly, there cannot be a unique answer. But any set I that is given as an answer, must contain $\underline{B}$ *and* at least one student from each class comprising $\overline{B}$. In other words, it must have $\underline{B}$ as its *lower approximation* and $\overline{B}$ as its *upper approximation*.
I is then a *rough* concept/set in the information system $<U,\{a\}>$.
Further, it may be observed that any set I' given as another answer, is *roughly equal* to I, in the sense that both are represented (characterized) by $\overline{B}$ and $\underline{B}$.

We now define the notions relevant to knowledge reduction. Let $U = \{x_1, ..., x_n\}$ and $A = \{a_1, ..., a_m\}$ in the information system
$S = <U, A>$. By the discernibility matrix (denoted $\mathbf{M}(S)$) of S is meant an $n \times n$-matrix such that

$$c_{ij} = \{a \in A : a(x_i) \neq a(x_j)\}, \ i,j = 1, ..., n. \tag{10}$$

A discernibility function f_S is a boolean function of m boolean variables $\bar{a}_1, ..., \bar{a}_m$ corresponding to the attributes $a_1, ..., a_m$ respectively, and defined as follows :

$$f_S(\bar{a}_1, ..., \bar{a}_m) = \bigwedge \{\bigvee(c_{ij}) : 1 \leq j < i \leq n, \ c_{ij} \neq \emptyset\}, \tag{11}$$

where $\bigvee(c_{ij})$ is the disjunction of all variables $\bar{a}$ with $a \in c_{ij}$.
An attribute $b \in B(\subseteq A)$ is *dispensable* in B if $I_B = I_{B\setminus\{b\}}$, otherwise b is *indispensable* in B.
$B(\subseteq A)$ is *independent* in S if every attribute from B is indispensable in B. Otherwise B is *dependent* in S.
B is called a *reduct* in S if B is independent in S and $I_B = I_A$.

It is seen in [12] that $\{a_{i_1}, ..., a_{i_p}\}$ is a reduct in S if and only if $a_{i_1} \wedge ... \wedge a_{i_p}$ is a prime implicant (constituent of the disjunctive normal form) of f_S.

Let $B, C \subseteq A$. C *depends* on B if and only if $I_B \subseteq I_C$, i.e. information due to the attributes in C is derivable from that due to the attributes in B. This dependency can be partial, in which case one introduces a dependency factor $df, 0 \leq df \leq 1$. Formally,

$$df = \frac{card(POS_B(C))}{card(U)}, \qquad (12)$$

where $POS_B(C) = \bigcup_{X \in I_C} \underline{B}X$, and *card* denotes cardinality of the set.

We are concerned with a specific type of information system $\mathcal{S} = <U, A>$, called a *decision table*. The attributes in such a system are distinguished into two parts, viz. *condition* and *decision* attributes. Classification of the domain due to decision attributes could be thought of as that given by an expert.

Let $C, D \subseteq A$ be the sets of condition and decision attributes of $\mathcal{S}$ respectively. The *rank* of a decision attribute $d \in D$, $r(d)$, is the cardinality of the image $d(U)$ of the function d on the value set V_d. One can then assume that $V_d = \{1, ..., r(d)\}$. The *generalised decision* in $\mathcal{S}$ is a function $\partial_\mathcal{S} : U \to \mathcal{P}(\{1, ..., r(d)\})$ defined as

$$\partial_\mathcal{S}(x) = \{i : \exists x' \in [x]_C \text{ and } d(x) = i\}.$$

A decision table $\mathcal{S}$ with $D = \{d\}$ is called *consistent (deterministic)* if $card(\partial_\mathcal{S}(x)) = 1$ for any $x \in U$, or equivalently, if and only if $POS_C(d) = U$. Otherwise, $\mathcal{S}$ is *inconsistent (non-deterministic)*.

Knowledge reduction now consists of eliminating superfluous values of the condition attributes by computing their reducts, and we come to the notion of a *relative reduct*.

An attribute $b \in B(\subseteq C)$ is *D-dispensable* in B, if $POS_B(D) = POS_{B \setminus \{b\}}(D)$; otherwise b is *D-indispensable* in B.

If every attribute from B is D-indispensable in B, B is *D-independent* in $\mathcal{S}$. A subset B of C is a *D-reduct* in $\mathcal{S}$ if B is D-independent in $\mathcal{S}$ and $POS_C(D) = POS_B(D)$.

Relative reducts can be computed by using a *D-discernibility matrix*. If $U = \{x_1, ..., x_n\}$, it is an $n \times n$ matrix (denoted $\mathbf{M}_D(\mathcal{S})$), the ijth component of which has the form

$$c_{ij} = \{a \in C : a(x_i) \neq a(x_j) \text{ and } (x_i, x_j) \notin I_D\} \qquad (13)$$

for $i, j = 1, ..., n$.

The relative discernibility function f_D is constructed from the D-discernibility matrix in an analogous way as $f_\mathcal{S}$ from the discernibility matrix of $\mathcal{S}$ (cf. eqns. (10)-(11)). It is once more observed that [12] $\{a_{i_1}, ..., a_{i_p}\}$ is a D-reduct in $\mathcal{S}$ if and only if $a_{i_1} \wedge ... \wedge a_{i_p}$ is a prime implicant of f_D.

5 Network Configuration using Rough Sets

Here we formulate rule generation and knowledge encoding for configuring a network. The algorithm is able to deal with multiple objects corresponding to one decision attribute. From the perspective of pattern recognition, this implies using multiple prototypes to serve as representatives of any arbitrary decision region. A block diagram in Fig. 3 illustrates the entire procedure.

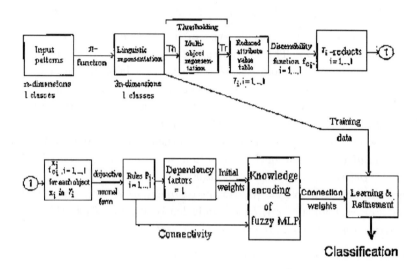

Fig. 3. Block diagram of the procedure

Let $\mathcal{S} = <U, A>$ be a decision table, with C and $D = \{d_1, ..., d_n\}$ its sets of condition and decision attributes respectively. We divide the decision table $\mathcal{S} = <U, A>$ into n tables $\mathcal{S}_i = <U_i, A_i>, i = 1, ..., n$, corresponding to the n decision attributes $d_1, ..., d_n$, where
$U = U_1 \cup ... \cup U_n$ and $A_i = C \cup \{d_i\}$.

5.1 Rule generation

The size of each $\mathcal{S}_i$ $(i = 1, ..., n)$ is first reduced with the help of a threshold on the number of occurrences of the same pattern of attribute values. This will be elicited in the sequel. Let the reduced decision table be denoted by $\mathcal{T}_i$, and $\{x_{i1}, ..., x_{ip}\}$ be the set of those objects of U_i that occur in $\mathcal{T}_i, i = 1, ..., n$.

For each d_i-reduct $B = \{b_1, ..., b_k\}$ (say), we define a discernibility matrix (denoted $\mathbf{M}_{d_i}(B)$) from eqn. (13) as follows:

$$c_{ij} = \{a \in B : a(x_i) \neq a(x_j)\}, \qquad (14)$$

for $i, j = 1, ..., p$.

Now for each object $x_j \in \{x_{i1}, ..., x_{ip}\}$, we consider the discernibility function $f_{d_i}^{x_j}$ which is defined as

$$f_{d_i}^{x_j} = \bigwedge\{\bigvee(c_{kj}) : 1 \leq j \leq p,\ j \neq k,\ c_{kj} \neq \emptyset\}, \qquad (15)$$

where $\bigvee(c_{kj})$ is the disjunction of all members of c_{kj}.

$f_{d_i}^{x_j}$ is brought to its conjunctive normal form (c.n.f.) P_i. One thus obtains a dependency rule r_i, viz. $P_i \to d_i$, where P_i is the disjunctive normal form (d.n.f.) of $f_{d_i}^{x_j}, j \in \{i_1, ..., i_p\}$.

It may be noticed that each component of P_i induces an equivalence relation on U_i as follows. If a component is a single attribute b, the relation I_b is taken. If a component of the c.n.f. is a disjunct of attributes, say $b_{i_1}, ..., b_{i_p} \in B$, we consider the transitive closure of the union of the relations $I_{b_{i_1}}, ..., I_{b_{i_p}}$. Let I_i denote the intersection of all these equivalence relations.

The dependency factor df_i for r_i is then given by

$$df_i = \frac{card(POS_i(d_i))}{card(U_i)}, \qquad (16)$$

where $POS_i(d_i) = \bigcup_{X \in I_{d_i}} l_i(X)$, and $l_i(X)$ is the lower approximation of X with respect to I_i. Note that in this case $df_i = 1$ for each r_i.

5.2 Knowledge encoding

Here, we formulate a methodology for encoding initial knowledge in the fuzzy MLP of [20], following the above algorithm.

Let us consider the case of feature F_j for class c_k in the l-class problem domain. The inputs for the ith representative sample $\mathbf{F}_i$ are mapped to the corresponding 3-dimensional feature space of $\mu_{low(F_{ij})}(\mathbf{F}_i)$, $\mu_{medium(F_{ij})}(\mathbf{F}_i)$ and $\mu_{high(F_{ij})}(\mathbf{F}_i)$, by eqn. (2). Let these be represented by L_j, M_j and H_j respectively. We consider only those attributes which have a numerical value greater than some threshold Th ($0.5 \leq Th < 1$). This implies clamping those features demonstrating high membership values with a 1, while the others are fixed at 0. We generate a separate $n_k \times 3n$-dimensional attribute value table for each class c_k (where n_k indicates the number of objects in c_k).

Let there be m sets $O_1, ..., O_m$ of objects in the table having identical attribute-values, and $card(O_i) = n_{k_i}, i = 1, ..., m$, such that $n_{k_1} \geq ... \geq n_{k_m}$ and $\sum_{i=1}^{m} n_{k_i} = n_k$. The attribute-value table can now be represented as an $m \times 3n$ array. Let $n_{k'_1}, n_{k'_2}, ..., n_{k'_m}$ denote the distinct elements among $n_{k_1}, ..., n_{k_m}$ such that $n_{k'_1} > n_{k'_2} > ... > n_{k'_m}$. We apply a heuristic threshold function defined by

$$Tr = \left\lceil \frac{\sum_{i=1}^{m} \frac{1}{n_{k'_i} - n_{k'_{i+1}}}}{Th} \right\rceil \tag{17}$$

All entries having frequency less than Tr are eliminated from the table, resulting in the reduced attribute-value table. Note that the main motive of introducing this threshold function lies in reducing the size of the resulting network. We attempt to eliminate noisy pattern representatives (having lower values of n_{k_i}) from the reduced attribute-value table. The whole approach is, therefore, data dependent. The dependency rule for each class is obtained by considering the corresponding reduced attribute-value table. A smaller table leads to a simpler rule in terms of conjunctions and disjunctions, which is then translated into a network having fewer hidden nodes. The objective is to strike a balance by reducing the network complexity and reaching a *good* solution, perhaps at the expense of not achieving the *best* performance.

As sketched in the previous section, one generates the dependency rules for each of the l classes, such that the antecedent part contains a subset of the $3n$ attributes, along with the corresponding dependency factors.

Let us now design the initial structure of the three-layered fuzzy MLP. The input layer consists of the $3n$ attribute values and the output layer is represented by the l classes. The hidden layer nodes model the first level (innermost) operator in the antecedent part of a rule, which can either be a conjunct or a disjunct. The output layer nodes model the outer level operator, which can again be either a conjunct or a disjunct. As mentioned earlier, the dependency factor of any rule is 1 in this method. Only those input attributes that appear in a conjunct/disjunct are connected to the appropriate hidden node, which in turn is connected to the corresponding output node. Note that a single attribute (involving no operator) is directly connected to the appropriate output node via a hidden node.

Next we proceed to the description of the initial weight encoding procedure. Let the dependency factor for a particular dependency rule for class c_k be α by eqn. (16). The weight $w_{ki}^{(1)}$ between a hidden node i and output node k is set at $\frac{\alpha}{fac} + \epsilon$, where fac refers to the number of operands and ϵ is a small random number taken to destroy any symmetry among the weights. Note that $fac \geq 1$ and each hidden node is connected to only one output node. Let the initial weight so clamped at a hidden node be denoted as β. The weight $w_{ia_j}^{(0)}$ between an attribute a_j (where a corresponds to low (L), medium (M) or high (H))

and hidden node i is set to $\frac{\beta}{facd} + \epsilon$, such that $facd$ is the number of attributes connected by the corresponding conjunct/ disjunct. Note that $facd \geq 1$. The sign of the weight is set randomly. Thus for an l-class problem domain we have at least l hidden nodes. All other possible connections in the resulting fuzzy MLP are set as small random numbers. It is to be mentioned that the number of hidden nodes is determined from the dependency rules.

The connection weights, so encoded, are then refined by training the network on the pattern set supplied as input.

6 Implementation and Experimental Results

Here we implement the algorithm on real life and artificial data. The initial weight encoding scheme is demonstrated and recognition scores are presented.

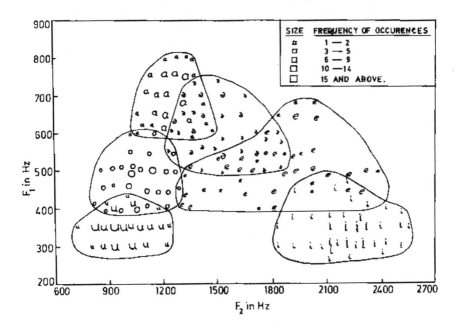

Fig. 4. Vowel data projection in F_1-F_2 plane

The speech data **Vowel** [34] deals with 871 Indian Telugu vowel sounds. These were uttered in a Consonant-Vowel-Consonant context by three male speakers in the age group of 30 to 35 years. The data set (depicted in $F_1 - F_2$ plane in Fig. 4 for ease of understanding) has three features corresponding to the first, second and third vowel formant frequencies obtained through spectrum analysis of the speech data. The overlapping vowel classes – ∂, a, i, u, e, o, shall be denoted by $c_1, \ldots, c_6$ respectively in the sequel.

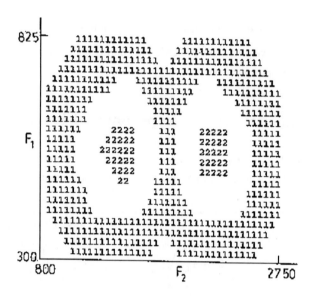

Fig. 5. Synthetic data (Pat)

The synthetic data **Pat** consists of 880 pattern points in the two-dimensional space F_1–F_2, as depicted in Fig. 5. There are three linearly nonseparable pattern classes. The figure is marked with classes 1 (c_1) and 2 (c_2), while class 3 (c_3) corresponds to the background region.

The training set considered 50% of the data selected randomly from each of the pattern classes. The remaining 50% data constituted the test set. It is found that the knowledge-based model converges to a good solution with a small number of training epochs (iterations) in both cases.

The data is first transformed into the $3n$-dimensional linguistic space of eqn. (2). A threshold of $Th = 0.8$ was used for the Vowel data, such that $y_i^{(0)} = 1$ if $y_i^{(0)} \geq 0.8$ and 0 otherwise. It can be observed from Fig. 5 that the synthetic data set is uniformly distributed over the entire feature space. Therefore, setting a threshold greater than 0.5 caused problems here, such that for certain objects all three input components corresponding to a feature became clamped at zero. To circumvent this, we set Th at 0.5 for Pat.

Table 1. Attribute-value table for class c_6 (Vowel)

	L_1	M_1	H_1	L_2	M_2	H_2	L_3	M_3	H_3
$x_1 - x_{20}$	1	0	0	1	0	0	1	0	0
$x_{21} - x_{29}$	1	1	0	1	0	0	1	0	0
$x_{30} - x_{36}$	1	1	0	1	0	0	0	0	1
$x_{37} - x_{41}$	1	1	0	1	0	0	0	1	0
$x_{42} - x_{45}$	1	1	0	1	0	0	0	1	1
$x_{46} - x_{49}$	1	0	0	1	0	0	0	1	1
$x_{50} - x_{51}$	0	1	0	1	0	0	0	1	1
$x_{52} - x_{53}$	0	1	0	1	0	0	1	0	0
x_{54}	1	0	0	1	0	0	1	1	0
x_{55}	0	1	0	1	0	0	0	0	1
x_{56}	1	0	0	1	0	0	0	0	1

6.1 Vowel data

Table 2. Reduced attribute-value table for class c_6 (Vowel)

	L_1	M_1	H_1	L_2	M_2	H_2	L_3	M_3	H_3
$x_1 - x_{20} : (y_1)$	1	0	0	1	0	0	1	0	0
$x_{21} - x_{29} : (y_2)$	1	1	0	1	0	0	1	0	0
$x_{30} - x_{36} : (y_3)$	1	1	0	1	0	0	0	0	1
$x_{37} - x_{41} : (y_4)$	1	1	0	1	0	0	0	1	0

Each class had a separate attribute value table consisting of multiple objects. Let us consider class c_6 as an example. The condition attributes are L_1, L_2, L_3, M_1, M_2, M_3, H_1, H_2, H_3. Note that these inputs are used only for the knowledge encoding procedure. During the refinement phase, the network learns from the original $3n$-dimensional training set with $0 \leq y_i^{(0)} \leq 1$ (eqn. (2)).

The first column of Table 1 corresponds to the objects which have the attribute-values indicated in the respective rows. We observe that the rows correspond to

20, 9, 7, 5, 4, 4, 2, 2, 1, 1, 1 objects respectively.

After applying the threshold Tr of eqn. (17), objects $x_{42} - x_{56}$ are eliminated from the table. Hence the reduced attribute-value table (Table 2) now consists of four rows only.

The discernibility matrix for class c_6 is

	y_1	y_2	y_3	y_4
y_1	ϕ			
y_2	$\{M_1\}$	ϕ		
y_3	$\{M_1, L_3, H_3\}$	$\{L_3, H_3\}$	ϕ	
y_4	$\{M_1, L_3, M_3\}$	$\{L_3, M_3\}$	$\{M_3, H_3\}$	ϕ

The discernibility function f for c_6 is

$$M_1 \wedge (M_1 \vee L_3 \vee H_3) \wedge (L_3 \vee H_3) \wedge (M_1 \vee L_3 \vee M_3) \wedge (L_3 \vee M_3) \wedge (M_3 \vee H_3)$$
$$= M_1 \wedge (L_3 \vee H_3) \wedge (L_3 \vee M_3) \wedge (M_3 \vee H_3).$$

The disjunctive normal form of f is

$$(M_1 \wedge L_3 \wedge M_3) \vee (M_1 \wedge L_3 \wedge H_3) \vee (M_1 \wedge M_3 \wedge H_3) \vee (M_1 \wedge L_3 \wedge M_3 \wedge H_3).$$

The resultant reducts are

$$M_1 \wedge L_3 \wedge M_3, \quad M_1 \wedge L_3 \wedge H_3, \quad M_1 \wedge M_3 \wedge H_3, \quad M_1 \wedge L_3 \wedge M_3 \wedge H_3.$$

The reduced attribute value table for reduct $M_1 \wedge L_3 \wedge M_3$ is

	M_1	L_3	M_3
y_1	0	1	0
y_2	1	1	0
y_3	1	0	0
y_4	1	0	1

The reduced discernibility matrix for $M_1 \wedge L_3 \wedge M_3$ is

	y_1	y_2	y_3	y_4
y_1	ϕ			
y_2	$\{M_1\}$	ϕ		
y_3	$\{M_1, L_3\}$	$\{L_3\}$	ϕ	
y_4	$\{M_1, L_3, M_3\}$	$\{L_3, M_3\}$	$\{M_3\}$	ϕ

The discernibility functions f_{y_i} for each object y_i, $i = 1, 2, 3, 4$ are

$$f_{y_1} = M_1 \wedge (M_1 \vee L_3) \wedge (M_1 \vee L_3 \vee M_3) = M_1$$
$$f_{y_2} = M_1 \wedge L_3 \wedge (L_3 \vee M_3) \qquad\qquad = M_1 \wedge L_3$$
$$f_{y_3} = (M_1 \vee L_3) \wedge L_3 \wedge M_3 \qquad\qquad = M_3 \wedge L_3$$
$$f_{y_4} = (M_1 \vee L_3 \vee M_3) \wedge (L_3 \vee M_3) \wedge M_3 = M_3.$$

A dependency rule thus generated for class c_6 is

$$M_1 \vee (M_1 \wedge L_3) \vee (M_3 \wedge L_3) \vee M_3 \rightarrow c_6,$$

i.e., $M_1 \vee M_3 \rightarrow c_6$.

The other rules for c_6 are

$$M_1 \vee H_3 \rightarrow c_6,$$

$$M_1 \vee M_3 \vee H_3 \rightarrow c_6.$$

Similarly, we obtain 1, 2, 1, 1, 2 dependency rules for classes c_1, c_2, c_3, c_4, c_5 respectively. The dependency factor of each rule is 1. So, considering all possible combinations we generate 12 sets of rules for the six classes. This leads to 12 possible network encodings.

A sample set of dependency rules generated for the six classes is
$H_1 \wedge L_2 \wedge L_3 \rightarrow c_1$, $M_1 \vee L_3 \rightarrow c_2$, $M_3 \vee H_3 \rightarrow c_3$, $M_3 \vee H_3 \rightarrow c_4$, $M_3 \rightarrow c_5$, $M_1 \vee M_3 \rightarrow c_6$.
This corresponds to the network represented in column 1 (of Rough-Fuzzy MLP) in Table 3.

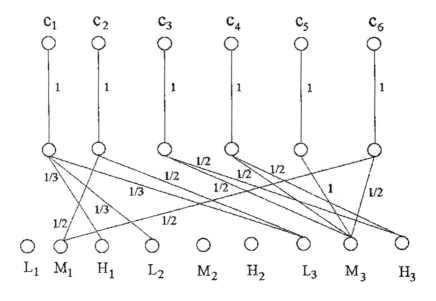

Fig. 6. Initial weight encoding for a sample network. Remaining weights are initialized to small random values

To encode the rule for class c_6 we require one hidden node for modeling the conjunct. The corresponding output node is connected to the hidden node with initial link weight of $df_6 = 1$. Then the input attribute pair (M_1, M_3) is connected to this hidden node with link weights $\frac{df_6}{2}(= 0.5)$. A sample network is illustrated in Fig. 6.

Table 3 demonstrates sample results obtained by the three-layered knowledge-based network, at the end of 150 sweeps. In all the cases the algorithm constructed a network with six hidden nodes and six input nodes. The performance (at the end of 150 sweeps) was compared with those of a conventional MLP and a fuzzy MLP [20], having the same number of hidden nodes but with no initial knowledge encoding. It was seen that the conventional MLP with six hidden nodes is unable to classify the data. Hence this is not included in the table. The performance of the Bayes' classifier for multivariate normal patterns, using different covariance matrices are for each pattern class, is demonstrated. The choice of normal densities for the vowel data has been found to be justified [35]. The performance of the Rough-Fuzzy MLP is found to improve on that of the fuzzy and conventional versions of the MLP and the Bayes' classifier.

Table 3. Recognition scores (%) for Vowel

Attribs. for c_1, c_3, c_4 c_2 c_5 c_6	Bayes' class-ifier	Fuzzy MLP	Rough-Fuzzy MLP						
			$H_1 \wedge L_2 \wedge L_3$; $M_1 \vee L_3$		$M_3 \vee H_3$; $M_1 \vee M_3$		$M_3 \vee H_3$		
			M_3				H_3		
			$M_1 \vee M_3$	$M_1 \vee M_3 \vee H_3$	$M_1 \vee M_3$	$M_1 \vee H_3$	$M_1 \vee M_3$	$M_1 \vee M_3 \vee H_3$	
# links	-	90	18	19	18	18	18	19	
Training	-	80.88	85.48	80.65	81.11	80.19	83.18	82.72	
T e s t s e t	∂	41.6	21.6	51.4	21.6	56.8	43.2	54.1	59.5
	a	91.1	82.2	84.4	88.9	82.2	88.9	82.2	75.6
	i	93.0	94.1	94.1	85.9	95.3	85.9	94.1	87.1
	u	94.7	87.8	90.2	86.6	87.8	87.8	90.2	87.8
	e	71.1	88.7	84.0	97.2	78.3	93.4	82.1	84.9
	o	71.1	95.1	93.9	93.9	95.1	93.9	93.9	93.9
	Net	79.2	84.44	86.27	85.13	85.13	86.27	85.81	84.44

6.2 Synthetic Data

The attribute-value table for class c_2 is depicted in Table 4. The rows correspond to 16, 12, 9, 8, 3, 2, 1 objects respectively. Application of eqn. (17) results in the elimination of objects $x_{46} - x_{51}$. The D-reducts generated are
$L_1 \wedge L_2, H_1 \wedge L_2, H_1 \wedge H_2, L_1 \wedge H_2$.
We obtain four D-reducts for each of the other two classes. Considering all possible combinations, we generate 64 sets of rules for the three classes. This results in 64 possible network encodings.

Table 4. Attribute-value table for class c_2 (Pat)

	L_1	M_1	H_1	L_2	M_2	H_2
$x_1 - x_{16}$	1	1	0	1	1	0
$x_{17} - x_{28}$	1	1	0	0	1	1
$x_{29} - x_{37}$	0	1	1	1	1	0
$x_{38} - x_{45}$	0	1	1	0	1	1
$x_{46} - x_{48}$	1	1	0	0	0	1
$x_{49} - x_{50}$	0	1	1	0	0	1
x_{51}	1	1	0	1	0	0

A sample set of dependency rules for the three classes is
$(L_1 \wedge M_2 \wedge H_2) \vee (L_1 \wedge M_1 \wedge H_2) \rightarrow c_1$, $H_1 \wedge H_2 \rightarrow c_2$, and
$(M_1 \wedge H_1) \vee (H_1 \wedge M_2 \wedge H_2) \vee (M_1 \wedge M_2 \wedge H_2) \rightarrow c_3$.
This corresponds to column 2 of Table 5.

Table 5. Recognition scores (%) for Pat

		k-NN classifier		FUZZY MLP	Rough-Fuzzy MLP			
c_1				$(L_1 \wedge M_2 \wedge H_2)$ $\vee(L_1 \wedge M_1 \wedge H_2)$	$(H_1 \wedge M_2 \wedge H_2) \vee (M_1 \wedge M_2 \wedge H_2)$		$(H_1 \wedge L_2 \wedge M_2)$ $\vee(M_1 \wedge H_1 \wedge L_2$
c_2				$H_1 \wedge H_2$		$L_1 \wedge H_2$	$H_1 \wedge H_2$
c_3				$(M_1 \wedge H_1)$ $\vee(H_1 \wedge M_2 \wedge H_2)$ $\vee(M_1 \wedge M_2 \wedge H_2)$	$(L_1 \wedge M_1)$ $\vee(L_1 \wedge M_2 \wedge H_2)$ $\vee(M_1 \wedge M_2 \wedge H_2)$	$(L_1 \wedge M_1)$ $\vee(L_1 \wedge L_2 \wedge M_2)$ $\vee(M_1 \wedge L_2 \wedge M_2)$	$(M_1 \wedge H_1)$ $\vee(H_1 \wedge L_2 \wedge M_2$ $\vee(M_1 \wedge L_2 \wedge M_2$
		$k=1$	3	MLP				
# links		-	-	54	22	22	22	22
# inpts		-	-	6	5	5	6	5
Train		-	-	82.0	84.05	77.68	80.87	83.83
T	c_1	89.6	94.2	88.08	93.08	81.92	84.23	91.54
e	c_2	88.5	88.5	69.23	69.23	61.54	65.38	57.69
s	c_3	81.9	87.7	68.39	68.39	80.65	81.94	67.1
t	Net	86.8	91.6	80.05	82.99	80.27	82.31	80.95

The sub-network for class c_3 consists of three hidden nodes, each with initial output link weight of $\frac{df_3}{3}(=0.33)$. The input attribute pair (M_1, H_1) is connected to the first of these hidden nodes with link weights $\frac{df_3}{6}(=0.17)$. The remaining attributes (H_1, M_2, H_2) and (M_1, M_2, H_2) are connected to the next two hidden nodes with link weights $\frac{df_3}{9}(=0.11)$.

Table 5 provides a sample set of results obtained by a three-layered knowledge-based network. Note that we have simulated all 64 networks. In all cases the algorithm generated six hidden nodes. The performance was compared with that of a conventional and fuzzy MLP (all at the end of 1000 sweeps), and the k-NN classifier. The conventional MLP failed to recognize class c_2 (e.g., the scores for classes c_1, c_2 and c_3 are 87.1, 0.0, 51.6 respectively, for the test set). The Rough-Fuzzy MLP generalizes better than the fuzzy MLP (with one hidden layer having six hidden nodes) for the test patterns considering the overall scores (Net). However, the k-NN classifier is found to provide better performance. Note that the k-NN classifier is reputed to be able to generate piecewise linear decision boundaries and is quite efficient in handling concave and linearly nonseparable pattern classes. It may also be mentioned that the fuzzy MLP with more hidden layers/ nodes provides results better than that of the k-NN [36]. However, we are restricted here in using six hidden nodes for maintaining parity with the Rough-Fuzzy MLP. It is revealed under investigation that the method of knowledge extraction using rough sets can lead to over-reduction for the data shown in Fig. 5.

Remarks :

1. We have transformed the decision table constructed from the initial data by dividing it into subtables, each corresponding to a decision attribute of the given system. The initial table gave rise to discernibility functions (computed by eqn. (15)) with too large a number of components and hence a network with a huge number of hidden nodes. The computational complexity of such a network was not considered to be feasible. On the contrary, the subtables resulted in the generation of discernibility functions with less components and thus finally, a less cumbersome (more efficient) network.

2. Each decision table considered so far, due to its binary nature, is trivially consistent.

3. Any comparative study of the performance of our model should consider the fact that here the appropriate number of hidden nodes is automatically generated by the rough-set theoretic knowledge encoding procedure. On the other hand, both the fuzzy and conventional versions of the MLP are required to empirically generate a suitable size of the hidden layer(s). Hence this can be considered to be an added advantage.

7 Conclusions

A methodology integrating rough sets with fuzzy MLP for designing a knowledge-based network is presented. The effectiveness of rough set theory is utilized for encoding the crude domain knowledge through concepts like discernibility matrix and function, reducts and dependency factors. This investigation not only demonstrates a way of integrating rough sets with neural networks and fuzzy sets, but also provides a method that is capable of generating the appropriate network architecture and improving the classification performance. The incorporation of fuzziness at various levels of fuzzy MLP also helps the resulting knowledge-based system to efficiently handle uncertain and ambiguous information both at the input and the output.

As was remarked earlier, a study of an integration, involving only neural nets and rough sets, was presented by Yasdi [14]. However, only one layer of adaptive weights was considered while the input and output layers involved fixed binary weights. *Max, Min* and *Or* operators were applied at the hidden nodes. Besides, the model was not tested on any real problem and no comparative study was provided to bring out the effectiveness of this hybrid approach. We, on the other hand, consider here an integration of the three paradigms, viz., neural nets, rough sets and fuzzy sets. The process of rule generation and mapping of the dependency factors to the connection weight values is novel to our approach. Moreover, the three-layered MLP used has adaptive weights at all layers. These are initially encoded with the knowledge extracted from the data domain in the form of dependency rules, and later refined by training. Effectiveness of the model is demonstrated on both real life and artificial data.

Acknowledgements

This work is partly supported by the CSIR grant no. 25(0093)/97/EMR-II. A part of this research was done when Dr. M. Banerjee held a Research Associateship of the CSIR, New Delhi.

References

1. *Proc. of Third Workshop on Rough Sets and Soft Computing (RSSC'94)*, (San José, U.S.A.), November 1994.
2. *Proc. of Fourth International Conference on Soft Computing (IIZUKA96)*, (Iizuka, Fukuoka, Japan), October 1996.
3. L. A. Zadeh, "Fuzzy logic, neural networks, and soft computing," *Communications of the ACM*, vol. 37, pp. 77–84, 1994.
4. J. C. Bezdek and S. K. Pal, eds., *Fuzzy Models for Pattern Recognition : Methods that Search for Structures in Data*. New York: IEEE Press, 1992.
5. *Proc. of IEEE International Conference on Fuzzy Systems (FUZZ-IEEE)*, (USA), September 1996.
6. S. K. Pal and D. Bhandari, "Selection of optimum set of weights in a layered network using genetic algorithms," *Information Sciences*, vol. 80, pp. 213–234, 1994.
7. A. Homaifar and E. McCormick, "Simultaneous design of membership functions and rule sets for fuzzy controllers using genetic algorithms," *IEEE Transactions on Fuzzy Systems*, vol. 3, pp. 129–139, 1995.
8. S. K. Pal and D. Bhandari, "Genetic algorithms with fuzzy fitness function for object extraction using cellular neural networks," *Fuzzy Sets and Systems*, vol. 65, pp. 129–139, 1994.
9. M. Banerjee and S. K. Pal, "Roughness of a fuzzy set," *Information Sciences (Informatics & Computer Science)*, vol. 93, pp. 235–246, 1996.
10. S. Mitra, M. Banerjee, and S. K. Pal, "Rough knowledge-based network, fuzziness and classification," *Neural Computing and Applications*.

11. Z. Pawlak, *Rough Sets, Theoretical Aspects of Reasoning about Data.* Dordrecht: Kluwer Academic, 1991.
12. A. Skowron and C. Rauszer, "The discernibility matrices and functions in information systems," in *Intelligent Decision Support, Handbook of Applications and Advances of the Rough Sets Theory* (R. Slowiński, ed.), pp. 331–362, Dordrecht: Kluwer Academic, 1992.
13. R. Slowiński, ed., *Intelligent Decision Support, Handbook of Applications and Advances of the Rough Sets Theory.* Dordrecht: Kluwer Academic, 1992.
14. R. Yasdi, "Combining rough sets learning and neural learning method to deal with uncertain and imprecise information," *Neurocomputing*, vol. 7, pp. 61–84, 1995.
15. A. Czyzewski and A. Kaczmarek, "Speech recognition systems based on rough sets and neural networks," in *Proc. of Third Workshop on Rough Sets and Soft Computing (RSSC'94)*, (San José, U.S.A.), pp. 97–100, 1994.
16. L. M. Fu, "Knowledge-based connectionism for revising domain theories," *IEEE Transactions on Systems, Man, and Cybernetics*, vol. 23, pp. 173–182, 1993.
17. G. G. Towell and J. W. Shavlik, "Knowledge-based artificial neural networks," *Artificial Intelligence*, vol. 70, pp. 119–165, 1994.
18. S. Mitra, R. K. De, and S. K. Pal, "Knowledge-based fuzzy MLP for classification and rule generation," *IEEE Transactions on Neural Networks*, vol. 8, pp. 1338–1350, 1997.
19. R. J. Machado and A. F. da Rocha, "Inference, inquiry, evidence censorship, and explanation in connectionist expert systems," *IEEE Transactions on Fuzzy Systems*, vol. 5, pp. 443–459, 1997.
20. S. K. Pal and S. Mitra, "Multi-layer perceptron, fuzzy sets and classification," *IEEE Transactions on Neural Networks*, vol. 3, pp. 683–697, 1992.
21. S. Mitra and S. K. Pal, "Fuzzy multi-layer perceptron, inferencing and rule generation," *IEEE Transactions on Neural Networks*, vol. 6, pp. 51–63, 1995.
22. S. I. Gallant, "Connectionist expert systems," *Communications of the Association for Computing Machinery*, vol. 31, pp. 152–169, 1988.
23. H. F. Yin and P. Liang, "A connectionist incremental expert system combining production systems and associative memory," *International Journal of Pattern Recognition and Artificial Intelligence*, vol. 5, pp. 523–544, 1991.
24. R. Masuoka, N. Watanabe, A. Kawamura, Y. Owada, and K. Asakawa, "Neurofuzzy system – fuzzy inference using a structured neural network," in *Proceedings of the 1990 International Conference on Fuzzy Logic and Neural Networks, Iizuka*, (Japan), pp. 173–177, 1990.
25. N. K. Kasabov, "Adaptable neuro production systems," *Neurocomputing*, vol. 13, pp. 95–117, 1996.
26. N. K. Kasabov, "Fuzzy rules extraction, reasoning and rules adaptation in fuzzy neural networks," in *Proceedings of IEEE International Conference on Neural Networks*, (Houston, USA), pp. 2380–2383, 1997.
27. B. Kosko, "Hidden patterns in combined and adaptive knowledge networks," *International Journal of Approximate Reasoning*, vol. 2, pp. 377–393, 1988.
28. R. J. Machado and A. F. Rocha, "A hybrid architecture for fuzzy connectionist expert systems," in *Intelligent Hybrid Systems* (A. Kandel and G. Langholz, eds.),

pp. 136–152, Boca Raton, FL: CRC Press, 1992.
29. W. Pedrycz and A. F. Rocha, "Fuzzy-set based models of neurons and knowledge-based networks," *IEEE Transactions on Fuzzy Systems*, vol. 1, pp. 254–266, 1993.
30. K. Hirota and W. Pedrycz, "Knowledge-based networks in classification problems," *Fuzzy Sets and Systems*, vol. 59, pp. 271–279, 1993.
31. B. F. Leao and A. F. Rocha, "Proposed methodology for knowledge acquisition : a study on congenital heart disease diagnosis," *Methods of Information in Medicine*, vol. 29, pp. 30–40, 1990.
32. S. K. Pal and D. P. Mandal, "Linguistic recognition system based on approximate reasoning," *Information Sciences*, vol. 61, pp. 135–161, 1992.
33. S. K. Pal and D. Dutta Majumder, *Fuzzy Mathematical Approach to Pattern Recognition*. New York: John Wiley (Halsted Press), 1986.
34. S. K. Pal and D. Dutta Majumder, "Fuzzy sets and decision making approaches in vowel and speaker recognition," *IEEE Transactions on Systems, Man, and Cybernetics*, vol. 7, pp. 625–629, 1977.
35. S. K. Pal, *Studies on the Application of Fuzzy Set-Theoretic Approach in Some Problems of Pattern Recognition and Man-Machine Communication by Voice*. PhD thesis, University of Calcutta, Calcutta, India, 1978.
36. S. K. Pal and S. Mitra, "Fuzzy versions of Kohonen's net and MLP-based classification : Performance evaluation for certain nonconvex decision regions," *Information Sciences*, vol. 76, pp. 297–337, 1994.